Glossary of Symbols

$\bar{A}$	Complement of set A
AD	Absolute deviation
ANOVA	Analysis of variance
α (alpha)	Probability of a type I error
β (beta)	Probability of a type II error
$1 - \beta$	Power of a statistical test
β_0	y-intercept of the true linear relationship
β_1	Slope of the true linear relationship
b_0	y-intercept for the line of best fit for the sample data
b_1	Slope for the line of best fit for the sample data
c	Column number or class width
d	Difference in value between two paired pieces of data
df or df( )	Number of degrees of freedom
d_i	Difference in the rankings of the ith element
E	Expected value or maximum error of estimate
$e = y - \hat{y}$	Error (observed)
ε (epsilon)	Experimental error
ε_{ij}	Amount of experimental error in the value of the jth piece of data in the ith row
F	F distribution statistic
$F(\text{df}_n, \text{df}_d, \alpha)$	Critical value for the F distribution
f	Frequency
H	Value of the largest-valued piece of data in a sample
H_a	Alternative hypothesis
H_0	Null hypothesis
i	Index number when used with Σ notation
i	Position number for a particular data
k	Identifier for the kth percentile
k	Number of cells or variables
L	Value of the smallest-valued piece of data in a sample
m	Number of classes
MAD	Mean absolute deviation
MS( )	Mean square
MSE	Mean square error
MPS	Marginal propensity to save
μ (mu)	Population mean
μ_d	Mean value of the paired differences

(continued on back cover)

Elementary Statistics for Business

Elementary Statistics for Business

Second Edition

Robert R. Johnson
Monroe Community College

Bernard R. Siskin
Temple University

Duxbury Press • Boston

PWS PUBLISHERS

Prindle, Weber & Schmidt • Duxbury Press • PWS Engineering • Breton Publishers
20 Park Plaza • Boston, Massachusetts 02116

This book is dedicated to
Marlene Savino, Barbara Siskin and friend Tom Cardea
and the Johnson and Siskin children

All rights reserved. No part of this book may be reproduced or transmitted in any form or by any means, electronic or mechanical, including photocopying, recording, or any information storage and retrieval system, without permission, in writing, from the publisher.

© Copyright 1985 PWS Publishers
© Copyright 1980 Wadsworth, Inc.

PWS is a division of Wadsworth, Inc.

86 87 88 89 — 10 9 8 7 6 5 4

Library of Congress Cataloging in Publication Data

Johnson, Robert Russell,
    Elementary Statistics for Business.

    Bibliography: p.
    Includex index.
    1. Commercial statistics. 2. Statistics.
 I. Siskin, Bernard R.,   II. Title.
HF1017.J63   1985   519.5   84-20671
ISBN 0-87150-851-6

ISBN 0-87150-851-6

Text design by Sara Waller. Cover design by Susan London. Production by Ian List. Composition by Composition House Limited. Artwork by Julie Gecha. Text printed and bound by R.R. Donnelley & Sons. Cover printed by Lehigh Press Lithographers.

The cover and title page photographs by Andrew Brilliant. Used with permission of the artist.

Special thanks to Nicosia Development for the use of their wall and the MBTA for the use of their garage and the kindness of their personnel.

Preface

We believe that we have written a truly readable introductory business statistics text that successfully relates statistics to the students' experiences and future career goals. By emphasizing managerial decision-making concepts and applications, we acknowledge today's dynamic business environment and stress the "how" and "when" of particular statistical concepts. We have found that students are more motivated to learn statistics when a realistic assessment of their priorities is complemented by an understanding of the business environment. To this end, we have integrated computer printouts with the text to present practical examples of the use of statistical concepts in business decision making.

This book was written for use in an introductory course for students who need a working knowledge of statistics, but do not have a strong mathematics background. Since statistics requires the use of many formulas, those students who have not had intermediate algebra should complete at least one semester of college mathematics before beginning this course.

The Changes in the Second Edition

The teaching objectives of this edition are the same as those of the first edition. The following significant changes made in this revision should be helpful in attaining these teaching objectives:

1. Chapter 1 has been reorganized to include a discussion of data collection.
2. The descriptive statistics for single-variable data have been reorganized and combined into one chapter with a "two-unit" approach. Unit 1 deals with graphical techniques, including the stem-and-leaf technique; unit 2 deals with calculated techniques.
3. Chapter 4, on Probability, now merges the General and Specific Addition (and Multiplication) Rules under one heading. The exposition has also been divided into additional sections to clarify the discussion and increase the number of drill exercises.
4. A section discussing the Poisson probability distribution has been added to chapter 5.
5. P-values have been added to hypothesis testing.

6. Chapter 12, on ANOVA, has been expanded to include unequal replicates and two-factor crossed experiments.
7. Computer output has been included for Descriptive Statistics, Hypothesis Testing, Chi-Square, ANOVA, and Regression Analysis.
8. Chapter 15, on Multiple Regression, has been revised and improved.
9. The exercise sets have been altered substantially. Many of the exercises are new, and others have been modified so that the solutions are different from the solutions in the first edition.
10. Challenging Problems, offering more complex exercises, are provided at the end of many of the chapters.

To the Instructor: The Text As a Teaching Tool

As stated earlier, one of our primary objectives in writing this book was to produce a truly readable presentation of elementary statistics for business. It is this specific intent that brought about the chapter format. Each chapter is designed to interest and involve students, and guide them step by step through the material in a logical manner. Each chapter includes:

- A **Chapter Outline** that shows students what to look for in the following material.
- A **News Article** that shows students how statistics are actually applied in the real world and demonstrates the type of statistic to be studied.
- **Chapter Objectives** that tell students the specific information to be learned upon completion of the chapter.
- **Worked-Out Examples with Solutions** to illustrate concepts as well as demonstrate the applications of statistics in real-world situations.
- **End-of-Section Exercises** to facilitate practice in the use of concepts as they are presented.
- **In Retrospect,** a section that provides a summary of the material just provided and relates the material to the chapter objectives.
- **Chapter Exercises** that give students further opportunity to master conceptual and computational skills.
- **Hands-On Problems** to provide a more personalized learning experience by directing students to collect their own data and apply the techniques they have been studying by use of those data. (Chapter 20 does not have a Hands-On Problem.)

The first three chapters are introductory in nature. Chapter 3 is a descriptive (first-look) presentation of bivariate data. We present this material at this point in the book for two reasons. First, students often ask about the relationship between two sets of data (such as heights and weights). Second, it affords us an opportunity to present a decision-making process (a hypothesis test with critical value) without confusion. This seems to reduce the resistance that always persists later when the formal hypothesis test procedure is introduced. (Of course, the instructor must make reference to this previous decision-making process.)

In the chapters on probability (4 and 5), we deliberately avoid the concepts of permutations and combinations. They are of no help in understanding statistics. Thus only the binomial coefficient is introduced in connection with the binomial probability distribution.

The instructor has several options in the selection of topics to be studied in a given course. We consider chapters 1 through 9 to be the core of a first course (some sections of chapters 2, 3, and 6 may be omitted without affecting continuity). After chapter 9 is completed, chapters 10 through 20 may be studied in almost any order. There are a few restrictions, however: chapter 3 must be studied prior to chapters 13, 14, and 15; chapter 14 must be studied before chapter 17; and chapter 10 must precede chapter 12.

The suggestions of instructors using the previous edition have been invaluable in helping us improve the text for the present revision. Should you, in teaching from this edition, have comments or suggestions, we would be most grateful to receive them. Please address such communications to Robert Johnson at Monroe Community College, Rochester, New York 14623.

To the Student: The Text As a Learning Tool

We believe that plain talk and a stress on common sense are this book's main merits as a learning tool. Such a treatment should allow you to work your way through the course with relative ease, provided that you have the necessary basic mathematics skills. Examples of this procedure are (1) illustration 1-4 (p. 9), which is used to reemphasize the meanings of the eight basic definitions presented in chapter 1, and (2) the use of previously completed homework exercises to introduce a new concept (see the introduction of hypothesis testing on page 268).

It is our aim to motivate and involve you in the statistics that you are learning. The chapter format reflects these aims and can best promote learning if each part of each chapter is used as indicated.

1. Read the annotated outline to gain an initial familiarity with several of the basic terms and concepts to be presented.
2. Read the news article, which puts to practical use some of the concepts to be learned in the chapter.
3. Use the chapter objectives as a guide to map out the direction and scope of the chapter.
4. Learn and practice using the concepts of each section by doing the exercise set at the end of the respective section. Answers and partial solutions are provided at the back of the book to complement the study illustrations and to enable you to work independently. While working within a chapter, it will be helpful to save the results of the exercises, since some results will be used again in later

exercise sets of the chapter. When this situation occurs, the later use has been cross-referenced.

5. Use the In Retrospect section to reflect on the concepts you have just learned and the relationship of the material in this chapter to the material of previous chapters. At this point it would be meaningful to reread the news article.

6. The Chapter Exercises at the end of the chapter offer additional learning experiences, since in these exercises you must now identify the technique to be used and must be able to apply it. The exercises are graded; everyone will be able to complete the first exercises with reasonable ease, but succeeding exercises become more challenging. Occasionally the results of an exercise should be saved for use in later exercises.

7. The Hands-On Problems direct you to collect a set of data, often of your own interest, and to apply the techniques you have studied. This opportunity should (i) reinforce the concepts studied and (ii) result in an interesting and informative statistical experience with real data.

8. After completing the Chapter Exercises, try the Challenging Problem, when provided, to expand your understanding of the chapter material.

The Instruction Package

To support our presentation of elementary business statistics, we have prepared the following supplements:

1. The **Study Guide with Self-Correcting Exercises** offers students an alternative approach to mastering difficult concepts. In addition, the study guide presents numerous worked-out, self-correcting exercises for each chapter. Also included in the study guide is a review of elementary algebra.

2. The **Solutions Manual** shows at least one complete solution to each exercise in the textbook. Occasionally we have added some parenthetical comments to aid the teacher in such areas as when to assign specific problems and how some problems can be of greatest use.

3. The **MINITAB Student Supplement** by Kenneth Bond and James Scott of Creighton University. For those interested in teaching or learning the course interactively with the computer, this supplement is a text-specific introduction to the Minitab Statistical Software Package and is keyed to text discussion and examples.

Acknowledgments

We owe a debt to many other books. Many of the ideas, principles, examples, and developments that appear in this text stem from thoughts provoked by these sources.

It is a pleasure to acknowledge the aid and encouragement we have received throughout the development of this text from our students and colleagues. Special thanks go also to those who read the previous edition and offered suggestions. We also want to acknowledge and thank the reviewers for this edition: Robert Kurys, Ryerson Polytechnical Institute; William Jedlicka, William Rainey Harper College; George Vlahos, University of Dayton; Fike Zahroon, Moorhead State University; Herbert Hooper, Jr., Chattanooga State Technical Community College; Alan L. Gordon, Bentley College; Ram Tripathi, University of Texas—San Antonio; Ira Perelle, Mercy College; Jack P. Suyderhoud, University of Hawaii—Manoa; Thomas L. Case, Georgia Southern University; Robert Pavur, North Texas State University; Harry Wilson, California Polytechnic University; Donald Tavares, Ryerson Polytechnical Institute; Paul Guy, California State University—Chico; and R. C. Patel, California State University—Chico.

We would also like to express our appreciation for the fine work that Sara Waller, Susan London, and Ian List have put into the editing, design, and production of the book. To the Duxbury Staff, we would like to say thanks for all your assistance and encouragement.

Thanks also to the many authors and publishers who so generously extended reproduction permission for the news articles and tables used in the text. These acknowledgments are specified individually throughout the text.

Robert R. Johnson

Bernard R. Siskin

Contents

Chapter 1   **Statistics**   2
       Chapter Objectives   4
   Unit 1   **Introduction to Statistics**   4
     1-1   What Is Statistics?   4
     1-2   Uses and Misuses of Statistics   6
     1-3   Introduction to Basic Terms   7
     1-4   Comparison of Probability and Statistics   13
   Unit 2   **Sampling**   13
     1-5   Population and Sampling Frame   13
     1-6   Errors in Sampling   16
     1-7   Judgment Versus Probability Samples   18
     1-8   Simple Random Sampling   19
     1-9   Systematic Sampling   21
     1-10   Stratified Sampling   22
     1-11   Cluster Sampling   23
       In Retrospect   26   Hands-On Problem   28
       Chapter Exercises   26

Chapter 2   **Descriptive Analysis and Presentation of Single-Variable Data**   30
       Chapter Objectives   32
   Unit 1   **Graphic Presentation of Data**   33
     2-1   Graphs and Stem-and-Leaf Displays   33
     2-2   Frequency Distributions, Histograms, Ogives   38
   Unit 2   **Calculated Descriptive Statistics**   50
     2-3   Measures of Central Tendency   50
     2-4   Measures of Dispersion   58
     2-5   Measures of Position   69
     2-6   Interpreting and Understanding Standard Deviation   74
     2-7   The Art of Statistical Deception   78
       In Retrospect   81   Hands-On Problems   97
       Chapter Exercises   81

Chapter 3 Descriptive Analysis and Presentation of Bivariate Data — 100

Chapter Objectives — 102
3-1 Bivariate Data: Cross-Tabulation and Scatter Diagram — 103
3-2 Linear Correlation — 111
3-3 Linear Regression — 117
In Retrospect — 120 Hands-On Problems — 124
Chapter Exercises — 121

Chapter 4 Probability — 126

Chapter Objectives — 128

Unit 1 Concepts of Probability — 128
4-1 The Nature of Probability — 128
4-2 Probability of Events — 130
4-3 Simple Sample Spaces — 135
4-4 Rules of Probability — 141

Unit 2 Calculating Probabilities of Compound Events — 145
4-5 Complementary Events, Mutually Exclusive Events, and the Addition Rule — 145
4-6 Independence, Multiplication Rule, and Conditional Probability — 152
4-7 Combining the Rules of Probability — 159
4-8 Bayes's Rule — 164
In Retrospect — 167 Challenging Problem — 175
Chapter Exercises — 168 Hands-On Problems — 175

Chapter 5 Probability Distributions (Discrete Variables) — 178

Chapter Objectives — 180
5-1 Random Variables — 180
5-2 Probability Distributions of Discrete Random Variables — 182
5-3 Mean and Variance of a Discrete Probability Distribution — 185
5-4 The Binomial Probability Distribution — 189
5-5 Mean and Standard Deviation of the Binomial Distribution — 201
5-6 The Poisson Probability Distribution — 202
In Retrospect — 206 Challenging Problem — 212
Chapter Exercises — 207 Hands-On Problems — 213

Chapter 6 The Normal Probability Distribution — 214

Chapter Objectives — 216
6-1 The Normal Probability Distribution — 216
6-2 The Standard Normal Distribution — 218
6-3 Applications of the Normal Distribution — 224
6-4 Notation — 228

	6-5	Normal Approximation of the Binomial				232
		In Retrospect	237	Hands-On Problems	241	
		Chapter Exercises	237			

Chapter 7 Sample Variability 242

	Chapter Objectives		244
7-1	Sampling Distributions		244
7-2	The Central Limit Theorem		249
7-3	Application of the Central Limit Theorem		255
	In Retrospect	260	Hands-On Problems 263
	Chapter Exercises	261	

Chapter 8 Introduction to Statistical Inferences 266

	Chapter Objectives		268
8-1	The Nature of Hypothesis Testing		268
8-2	The Hypothesis Test (A Classical Approach)		275
8-3	The Hypothesis Test (A Probability-Value Approach)		286
8-4	Estimation		291
	In Retrospect	300	Hands-On Problems 305
	Chapter Exercises	300	

Chapter 9 Inferences Involving One Population 306

	Chapter Objectives		308
9-1	Inferences About the Population Mean		308
9-2	Inferences About Proportions		316
9-3	Inferences About Variance and Standard Deviation		323
	In Retrospect	330	Challenging Problem 336
	Chapter Exercises	331	Hands-On Problems 337

Chapter 10 Inferences Involving Two Populations 340

	Chapter Objectives		342
10-1	Independent and Dependent Samples		342
10-2	Inferences Concerning the Difference Between Two Independent Means (Variances Known or Large Samples)		344
10-3	Inferences Concerning Two Variances		351
10-4	Inferences Concerning the Difference Between Two Independent Means (Variances Unknown and Small Samples)		359
10-5	Inferences Concerning Two Dependent Means		367
10-6	Inferences Concerning Two Proportions		372
	In Retrospect	378	Challenging Problem 387
	Chapter Exercises	380	Hands-On Problems 388

Chapter 11 Additional Applications of Chi-Square — 390
Chapter Objectives — 392
- 11-1 Chi-Square Statistic — 392
- 11-2 Inferences Concerning Multinomial Experiments — 394
- 11-3 Inferences Concerning Contingency Tables — 399

In Retrospect — 408
Chapter Exercises — 409
Challenging Problem — 414
Hands-On Problem — 415

Chapter 12 Analysis of Variance — 416
Chapter Objectives — 418
- 12-1 The Logic Behind ANOVA — 418
- 12-2 Introduction to the Analysis of Variance Technique — 419
- 12-3 Applications of Single-Factor ANOVA — 426
- 12-4 Two-Factor ANOVA (Without Replication) — 432
- 12-5 Two-Factor ANOVA (with Replication) — 437

In Retrospect — 443
Chapter Exercises — 444
Challenging Problem — 452
Hands-On Problems — 452

Chapter 13 Linear Correlation and Simple Regression Analysis — 454
Chapter Objectives — 456
- 13-1 Linear Correlation Analysis — 456
- 13-2 Inferences About the Linear Correlation Coefficient — 461
- 13-3 Linear Regression: Fitting the Equation — 464
- 13-4 Coefficient of Determination: R Square — 472

In Retrospect — 476
Chapter Exercises — 477
Challenging Problem — 481
Hands-On Problems — 482

Chapter 14 Linear Regression Analysis — 484
Chapter Objectives — 486
- 14-1 Basic Concepts Needed for Inference in Regression — 486
- 14-2 Inferences Concerning the Slope — 491
- 14-3 Confidence Interval Estimates for Regression — 494
- 14-4 Common Mistakes Made in Using Regression — 500

In Retrospect — 501
Chapter Exercises — 501
Hands-On Problems — 507

Chapter 15 Multiple Regression — 510
Chapter Objectives — 512
- 15-1 Partial Correlation — 512
- 15-2 Model Building and Multiple Regression — 515
- 15-3 How Well Does the Model Fit: Multiple R^2 — 521

	15-4	Inferences Concerning the Goodness of Fit of a Multiple Regression Model	523
	15-5	Inferences Concerning Individual β's in the Multiple Regression Model	525
	15-6	Further Considerations in Modeling	529
		In Retrospect 531 Hands-On Problems 536	
		Chapter Exercises 531	

Chapter 16 Forecasting 538

		Chapter Objectives	540
	16-1	Forecasting	540
	16-2	Types of Models	542
	16-3	Basic Elements of a Forecast	543
	16-4	Measuring the Error of the Forecast: Mean Absolute Deviation and Mean Squared Error	547
	16-5	Defining What Is Meant by the Best Forecasting Technique	552
		In Retrospect 558 Hands-On Problems 562	
		Chapter Exercises 559	

Chapter 17 Classical Time Series Analysis 564

		Chapter Objectives	566
	17-1	Classical Time Series Decomposition	566
	17-2	Estimating Trend and Cyclical Components Together	571
	17-3	Separating the Trend and Cyclical Components: Trend	578
	17-4	Separating the Trend and Cyclical Components: Cyclical	584
	17-5	Seasonal and Irregular Components	586
		In Retrospect 590 Challenging Problems 594	
		Chapter Exercises 591 Hands-On Problems 594	

Chapter 18 Business Indicators and Index Numbers 596

		Chapter Objectives	598
	18-1	Business Indicators	598
	18-2	Indexes and Index Numbers	601
	18-3	Uses of Index Numbers	603
	18-4	Construction of a Weighted Price Index	610
	18-5	The Consumer Price Index	614
		In Retrospect 615 Challenging Problem 618	
		Chapter Exercises 615 Hands-On Problems 618	

Chapter 19 Elements of Nonparametric Statistics 620

		Chapter Objectives	622
	19-1	Nonparametric Statistics	622
	19-2	The Sign Test	623

19-3	The Mann-Whitney U Test		630
19-4	The Runs Test		636
19-5	Rank Correlation		640
19-6	Comparing Statistical Tests		645
	In Retrospect 646	Challenging Problem	649
	Chapter Exercises 647	Hands-On Problems	650

Chapter 20 Decision Theory — 652

Chapter Objectives — 654

20-1	Decision Making Under Certainty and Uncertainty	654
20-2	Payoff and Opportunity Loss Tables	656
20-3	Decision Strategies	662
20-4	Maximizing Expected Value of Payoff or Minimizing Expected Opportunity Loss	666
20-5	Expected Value of Perfect Information	671
20-6	Misuses of Maximizing Expected Value of Payoff	674

In Retrospect 675   Challenging Problem 679
Chapter Exercises 676

	Appendixes	**A-1**
A	**Summation Notation**	**A-2**
B	**Using the Random Number Table**	**A-8**
C	**Round-Off Procedure**	**A-10**
D	**Tables**	**A-12**
1	Random Numbers	A-12
2	Factorials	A-14
3	Binomial Coefficients	A-15
4	Binomial Probabilities	A-16
5	Poisson Probabilities	A-20
6	Areas of the Standard Normal Distribution	A-21
7	Critical Values of Student's t Distribution	A-22
8	Critical Values of the χ^2 Distribution	A-23
9a	Critical Values of the F Distribution ($\alpha = 0.05$)	A-24
9b	Critical Values of the F Distribution ($\alpha = 0.025$)	A-26
9c	Critical Values of the F Distribution ($\alpha = 0.01$)	A-28
10	Critical Values for the Sign Test	A-30
11	Critical Values of U in the Mann-Whitney Test	A-31
12	Critical Values for Total Number of Runs (V)	A-32
13	Critical Values of Spearman's Rank Correlation Coefficient	A-33
14	Critical Values of r When $\rho = 0$	A-34
15	Confidence Belts for the Correlation Coefficient ($1 - \alpha = 0.95$)	A-35

Answers to Selected Exercises	**A-37**
Index	**A-59**

Elementary Statistics for Business

1 Statistics

Chapter Outline

Unit 1: Introduction to Statistics

1-1 What Is Statistics?

1-2 Uses and Misuses of Statistics
Statistical analysis, like nuclear fission, can be applied either responsibly or irresponsibly—it is up to you.

1-3 Introduction to Basic Terms
Population, sample, variable, data, experiment, parameter, statistic, attribute data, and variable data.

1-4 Comparison of Probability and Statistics
Probability is related to statistics as the phrase "likelihood of rain today" is related to the phrase "actual amount of rainfall."

Unit 2: Sampling

1-5 Population and Sampling Frame
The sampling frame is a list of the items from which we select sample elements.

1-6 Errors in Sampling
The two types are random and systematic sampling errors.

1-7 Judgment Versus Probability Samples
Two distinct types of sampling designs.

1-8 Simple Random Sampling
One of the most common sample designs used to collect data.

1-9 Systematic Sampling
One of the easiest sample designs to use.

1-10 Stratified Sampling
A more complex design, which uses the characteristics of the population to improve the sample estimates.

1-11 Cluster Sampling
Uses the natural grouping of elements within the population.

The Current State of Statistical Sampling and Auditing

Statistical sampling could be used on more varied audit tests if the staffs of accounting firms were better trained in its use.... Specifically the experiences and views of a sample of CPA firm partners and sole practitioners are summarized....

TABLE I
Categorization of Respondents as Users or Nonusers of Statistical Sampling

	National Firm		Nonnational Firm		Total	
	Number	%	Number	%	Number	%
Use statistical sampling	89	89	37	36	126	62
Do not use statistical sampling	11	11	66	64	77	38
	100	100	103	100	203	100

From James P. Bedingfield, "The Current State of Statistical Sampling and Auditing," *Journal of Accountancy* (December 1975): 48–55. Copyright © 1975 by the American Institute of Certified Public Accountants, Inc. Reprinted by permission.

The Truth Is Often Lost in the Statistical Pollution

Daily Chuckle

The average man has 66 pounds of muscle and 3.2 pounds of brain, which explains a lot of things.

From *Listen*, vol. 24, no. 8 (1971):20. Reprinted by permission.

New American-made car rated at 40 mpg

4 out of 5 doctors recommend...

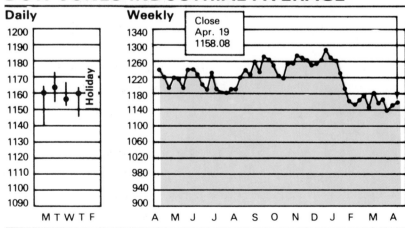

AVERAGE GAIN ON TODAY'S STOCKMARKET WAS 4.14

Reprinted with permission of the *Rochester Democrat and Chronicle*, Rochester, N.Y., April 22, 1984.

McDonald's restaurant chain has sold over 43,000,000,000 hamburgers!

Chapter Objectives *The purpose of this chapter is (1) to present an initial image of the field of statistics and (2) to explore the basic concepts and procedures of statistical data collection.*

Unit 1 Introduction to Statistics

Section 1-1 What Is Statistics?

The businessperson uses statistics in the same way the doctor uses blood tests. A laboratory report describes and summarizes the physical properties of a blood sample. The doctor then uses the results to diagnose the patient's condition and to make decisions as to the best course of treatment. Similarly, careful use of statistical methods enables us to (1) accurately describe and summarize information, (2) make estimations, and (3) make sound business decisions.

Statistics involves numbers, subjects, and the use of these numbers and subjects. The word "statistics" has different meanings to people of varied backgrounds and interests. To some it is hocus-pocus, whereby a person in the know overwhelms the layperson. To others it is a way of collecting and displaying large amounts of numerical information. And to still another group it is a way of "making decisions in the face of uncertainty." In the proper perspective each of these points of view is correct.

The field of statistics can be roughly subdivided into two areas: descriptive statistics and inferential statistics. **Descriptive statistics** is what most people think of when they hear the word "statistics." It includes the collection, presentation, and description of numerical data. The term **inferential statistics** refers to the technique of interpreting the values resulting from descriptive techniques and then using them to make decisions.

Statistics is more than just numbers—it is what is done to or with numbers. Let's use the following definition:

statistics

Statistics

The science of collecting, classifying, presenting, and interpreting numerical data.

Pick up the *Wall Street Journal, Business Week,* or any other major business publication and you will find many examples of statistics. The consumer price index summarizes the cost of living; the Dow-Jones average summarizes stock prices. A magazine tells its advertisers that the "average" reader is 35, married, with an income in excess of $20,000. To attract want ad customers, a newspaper tells us that the average used-car advertisement results in a sale within 3 days. Annual reports of companies present numerous statistics to describe the financial condition and outlook of the company.

These examples are just a few of the commonly used business statistics. Before we begin our detailed study, let's look at a few examples of how and when statistics can be applied.

Illustration 1-1

Suppose you own a 1980 Ford Mustang and want to sell it. How much should you ask for it? It would be extremely helpful if you had some statistics on the selling prices of 1980 Ford Mustangs. How could you collect such data? You could look in the "cars for sale" section of the newspaper and collect data on the sales prices asked. The prices will vary, but you can (1) look at the distribution of prices, (2) determine the high and the low values, and (3) determine the typical "asking" price. In setting your price, however, you will have to interpret these figures. What is the condition of your car, and how might the condition of the car be related to the spread of prices? How quickly must you sell your car, and how might the difference between your asking price and the average* asking price affect the speed of sale? □

Illustration 1-2

Wilson Corporation, a manufacturer of home television videotape machines, is considering opening a new manufacturing facility. To draw up an effective expansion plan, the firm needs the answer to the question: How many video units will it need to produce during the next 10 years? This question can be broken down into a number of smaller questions. How many television sets will be in use 10 years from now? What will be the total market for videotape machines at that time? What share of this market can Wilson expect to get? And so on. The corporate forecaster will need to (1) obtain data that will be useful in predicting the number of television sets 10 years from now; (2) relate the number of televisions to the number of videotape machines sold; and (3) project or predict what proportion of these sales will be Wilson's.

Consider this question: What will Wilson's share (percentage of total video set sales) of the market be? One answer to this question might be the share Wilson enjoys now. Another answer might be found by studying Wilson's share of the market over the past few years and looking for a trend to project into the future. For example, suppose that 4 years ago its share was 6%, 3 years ago 7%, 2 years ago 8%, last year 9%, and this year 10%. We might say its share is increasing by 1 percentage point per year;

---

* Many different types of average are used in statistics. Some of the more common types will be studied in chapter 2.

so that 10 years from now its share will be 20%. (This answer assumes that there is a relationship between past and future. This relationship does not always hold true, however. Wars, economic depressions, major technological changes, and other events alter the "natural" progression of events.)

As you can see, when accurate answers are desired, many problems must be overcome. One rather obvious problem is that of obtaining historical data. We can get Wilson's sales easily enough—but how about the total industry's sales? Do we look at just this year's figures, or do we look back over time; and if so, how far—1 year, 5 years, 10 years?

There are many other considerations: How accurate are our results? What is the probability that Wilson's market share will be less than we forecast? What is the probability that there will be more television sets than we predict 10 years from now? At this time we only wish to start you thinking about some of the problems involved in answering a question of this type. □

Illustration 1-3

How tall are sports car drivers? This question is posed by the owners of Custom Sport Coupe, a local manufacturer of the world's finest sports car. They want to design and build a new model that is truly comfortable for the driver. Their present model is designed for people between 5 feet 2 inches and 5 feet 8 inches tall. The manufacturers are concerned because they have heard rumors that their car is uncomfortable for a large proportion of sports car enthusiasts. (If you have ever ridden in a CSC, you will understand the rumor, "built for short people with very short necks.") How might you go about obtaining an answer to the original question? What special considerations would you give to the process of obtaining samples? □

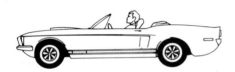

Each of these illustrations poses questions that should make you think about the situation and at the same time give you a feeling for statistics.

Section 1-2 Uses and Misuses of Statistics

The uses of statistics in business are unlimited. Statistics play an integral part in every area of business, from accounting and financial analysis to market research. It is difficult to find a company report that is not filled with statistics. The following are a few examples of how statistics are used in business:

1. Accountants, when auditing records of a company, often sample the records and, based on the sample results, estimate the correctness of all the records. (The importance of statistics in accounting is evidenced by the fact that approximately 25% of the questions on the national CPA exam are on statistics.)

2. In annual reports descriptive statistics are commonly used to describe the success (or failure) of the company during the year.
3. In production, samples of output are analyzed to determine their quality and to decide whether to stop production and make adjustments in the process.
4. In marketing, surveys are conducted and statistical information is analyzed to determine the best way to sell a product.
5. The U.S. government, the world's greatest collector and publisher of statistical data, constantly acquires and publishes data monitoring the well-being of the economy.

Misuses of statistics are often colorful and sometimes troublesome. Many people are concerned about the indifference of statistical descriptions; other people believe all statistics are lies. Most statistical lies are innocent, however, and result from using an inappropriate statistic; an open, nonspecific statement; or data derived from a faulty sample. All these errors lead to a common result—the consumer's misunderstanding of the information. Specific illustrations of the misuse of statistics are given in chapter 2.

Section 1-3 Introduction to Basic Terms

To study statistics, we need to be able to speak its language. Let's first define a few basic terms that will be used throughout the text. (These definitions are descriptive in nature and are not necessarily mathematically complete.)

population

Population

A collection, or set, of individuals, objects, or measurements whose properties are to be analyzed.

The population is the complete collection of individuals, objects, or measurements that are of interest to the sample collector. The concept of a population is the most fundamental idea in statistics. The population of concern must be carefully defined and is considered fully defined only when its membership list of elements is specified. The set of "all consumers who have ever purchased sports cars" is an example of a well-defined population (illustration 1-3).

Typically, we think of a population as a collection of people. However, in statistics a population can be a collection of documents, manufactured objects, or measurements. For example, the set of heights of all consumers who have ever purchased sports cars is a population.

sample   **Sample**

A subset of a population.

A sample consists of the individuals, objects, or measurements selected by the sample collector from the population.

response variable   **Response Variable (or simply, Variable)**

A characteristic of interest about each individual element of a population or sample.

The height of a sports car consumer is a variable.

data (singular)   **Data (singular)**

The value of the response variable associated with one element of a population.

For example, if we find that one sports car owner is 5 feet 8 inches tall, our data is "5 feet 8 inches."

data (plural)   **Data (plural)**

The set of values collected for the response variable from each of the elements belonging to the sample.

A set of 50 heights is an example of a set of data.

experiment   **Experiment**

A planned activity whose results yield a set of data.

parameter   **Parameter**

A numerical characteristic of an entire population.

The average height of all sports car consumers is an example of a parameter. A common practice in statistics is to use a Greek letter to symbolize the names of the various parameters. These symbols will be assigned as we study individual parameters.

statistic

Statistic

A numerical characteristic of a sample.

A statistic is a value that describes a sample. Most sample statistics are found with the aid of formulas and are assigned symbolic names that are letters of the English alphabet (e.g., $\bar{x}$, s, and r). The average height from the set of 50 heights is an example of a statistic.

Illustration 1-4

An airline is interested in finding out something about the age of the people who have flown between Philadelphia and Los Angeles this year. It feels that this information will help target its advertising efforts. Each of the eight terms previously described can be identified in this situation.

1. The *population* is composed of all persons who have flown between Philadelphia and Los Angeles this year. (Or it could be the collection of the ages of these people, since there is a one-to-one correspondence between people and their ages.)

2. A *sample* is any part of that population. The ages of passengers on this Friday's 11 A.M. flight from Philadelphia to Los Angeles is a sample.

3. The *response variable* is the age of each passenger.

4. Each response (each passenger's age) is one piece of *data*.

5. The *data* are the ages that correspond to a sample of the passengers.

6. An *experiment* is the method of determining (1) which passengers belong to the sample and (2) the age of each passenger in the sample. It could be accomplished by asking the age of each passenger on this Friday's 11 A.M. flight.

7. The *parameter* about which we are seeking information is the average age in the population.

8. The *statistic* that will be found is the average age of the sample. □

Basically, the two kinds of data are (1) data obtained from qualitative information and (2) data obtained from quantitative information.

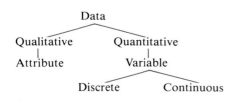

Introduction to Basic Terms   Section 1-3   **9**

attribute, or qualitative, data

Attribute, or Qualitative, Data

Result from a variable that asks for a quality type of description of the subject.

For example, color, brand name, and yes-no are attribute data. In general, the resulting data are a collection of word responses (as opposed to number responses). Qualitative data are generally referred to as *attribute data*. Some questions that the sample collector in illustration 1-2 might ask that would result in attribute data are the following: Do you plan to purchase a video set? (Answer: yes or no.) What brand of video set would you purchase?

variable, or quantitative, data

Variable, or Quantitative, Data

Result from obtaining quantities—counts (of how many) or measurements (length, weight, etc.)

discrete
continuous

Variable data can be subdivided into two classifications: (1) **discrete** variable data and (2) **continuous** variable data. In most cases the two can be distinguished by deciding whether the data result from a count or from a measurement. A count will always yield discrete variable data. The number of video sets sold per year will be 0, 1, 2,...;* it can never be 1.9 or 4.75. The idea of discontinuous numerical values is somewhat synonymous with that of discrete numerical values.

A measure of a quantity will usually be continuous. The heights of the owners of sports cars revealed that one owner was 68 inches (5 feet 8 inches) tall, to the nearest inch. All this number really tells us is that this person is between 67.5 and 68.5 inches tall. He or she could be 67.925 inches or exactly 68.00 inches. This concept is the basis of a continuous variable. There are many examples of continuous variables. Can you think of some?

The use of fractions or decimals does not necessarily imply that data are continuous. An example of a case where fractions appear in a discrete variable is stock prices. Stocks are traded only in units of $\frac{1}{8}$ dollars ($\frac{1}{8}, \frac{1}{4}, \frac{3}{8}$, etc.). These prices are actually discrete, because fractional values between $\frac{1}{8}$ and $\frac{2}{8}$ cannot occur. There are other illustrations of this situation, but in this text a count or a measurement is about the only distinction that we will need to make.

In some situations data are measured in variable form and reported and discussed in attribute form. Two examples are (1) the change in a stock

Note: The notation ... means that the listing of numbers continues on indefinitely.

10 Chapter 1 Statistics

price reported by a newspaper as advanced, declined, or unchanged, but measured in a specific amount and (2) a steel beam whose strength is measured in pounds per square inch, but reported as acceptable or unacceptable.

Let's explore the difference between these terms. If I were to go to the local supermarket and collect some data on customers, I might think of this data as a sample. However, it also could be considered as a population if I defined my area of concern to be the customers in the supermarket at the time I visited. The data would more likely be considered part of a larger collection of customers—for example, a sample of all the customers that have been in the supermarket this month, or a sample of all the customers in all supermarkets on this day. (You must define your population and your sample carefully. Generally, the purpose of collecting the data will lead you to the appropriate definition of the population and sample.) If I were to observe the sex of each customer, I would be collecting attribute data. If I were to record the number of items in the shopping cart of each customer as he or she checks out, I would be collecting discrete variable data. (This data clearly are a count of items purchased.) If I were to ask the age of each customer or weigh each filled shopping bag, continuous variable data would result.

Don't let the appearance of the data fool you in regard to their type. For example, suppose that after surveying the customers in the supermarket, I have a sample of 25 males and 75 females. These values appear to be the count of customers and might be considered discrete. This is not the case, however; they are attribute data. **You must look at each individual source to determine the kind of variable being used.** If one person is male, "male" constitutes the data and "male" is an attribute. Thus the previous collection (25 males and 75 females) is a *summary* of the attribute data. The definition of the variable must be relied on in making the distinction.

Let's consider a different situation. Suppose we are interested in the sex of customers in different types of retail stores. Our sample is 10 supermarkets and 10 gourmet food stores. We observe the number of males and the number of females that enter each store in a 1-hour time period. Here the individual source that determines each of the 40 data pieces is the *store*. We have discrete data, since each individual data piece represents a count of people.

Another example of a deceiving situation is the social security number of each customer. The numbers appear to be discrete variables, since only whole number values occur; however, these numbers are merely identification numbers. Identification numbers are not variables. Or consider the age of customers as recorded on a driver's license: 19, 22, 28, and so on. They are all whole numbers, but that does not make the variable discrete. The variable is "age," and these data are measured to the last birthday. In reality, age is continuous. Don't be misled into thinking a continuous variable is discrete because it is measured in rounded form (to the nearest inch, pound, minute, etc.). As you can see, the appearance of the data *after* they are recorded can be misleading in respect to their type. Remember to inspect an individual piece of data and you should have little trouble in distinguishing among attribute data and discrete and continuous variable data.

Exercises

1-1 Perform the "first-ace" experiment five times (instructions below) and observe the value of three different variables each time.

Variable 1: The color of the first ace to appear.

Variable 2: The longest distance across the dealt pile of cards.

Variable 3: The count of the card on which the first ace appears.

To perform the first-ace experiment, shuffle an ordinary deck of 52 playing cards containing four aces. Deal the cards one at a time onto a pile, stopping when the first ace appears. After the first ace has been placed on the pile, record the color of the ace. Now measure the longest distance across the scattered pile, before it is disturbed, and record it. Count and record the number of cards in the pile; repeat the experiment four more times.

(1)	Color of First Ace	(2)	Distance Across Pile	(3)	Count of Cards, Including First Ace
	$x_1 =$		$d_1 =$		$y_1 =$
	$x_2 =$		$d_2 =$		$y_2 =$
	$x_3 =$		$d_3 =$		$y_3 =$
	$x_4 =$		$d_4 =$		$y_4 =$
	$x_5 =$		$d_5 =$		$y_5 =$
			$\sum_{i=1}^{5} d_i =$		$\sum_{i=1}^{5} y_i =$

NOTE: The symbol $\sum$ represents "summation." See appendix A for information about the use of this symbol.

1-2 The variables identified in the first-ace experiment illustrate the three types of data discussed in this chapter. What type of data is (1) color, (2) distance, (3) count?

1-3 Extro Corporation's Board of Directors is interested in learning the expectations of its stockholders. At its annual meeting all stockholders who attend are asked three questions. In the next 3 years: (1) Do you expect a dividend increase? (2) What percentage of increase in stock value do you expect? (3) How many new products do you expect Extro to introduce?

(a) Carefully describe the population and the sample of concern.

(b) What type of data results from each variable—attribute, discrete, or continuous?

1-4 What is the average class size at your school this semester? To answer this question, consider all the classes in which you are personally enrolled as a sample of all the classes at your school this semester.

(a) Carefully describe the population and the sample involved.

(b) What is the variable being discussed? Is it attribute, discrete, or continuous?

(c) Describe the experiment; carry it out and record the data.

(d) Describe the parameter and the statistic involved and find the value of this statistic for your sample.

Section 1-4 Comparison of Probability and Statistics

Probability and statistics are two separate but related fields of mathematics. It has been said that "probability is the vehicle of statistics." That is, if it were not for the laws of probability, the theory of statistics would not be possible.

To illustrate the difference between probability and statistics, let's look at two boxes. The probability box contains five blue, five red, and five white poker chips. The subject of probability tries to answer such questions as this: If one chip is drawn from this box, what is the probability that it will be blue? On the other hand, in the statistics box we don't know what the combination of chips is. We draw a sample and make conjectures as to what we believe to be in the box. Note the difference: Probability asks about the chance that something (a sample) will happen when you know the possibilities (i.e., you know the population), whereas statistics asks what you believe the possibilities (population) to be when you know the results of a sample.

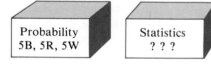

Unit 2 Sampling

Section 1-5 Population and Sampling Frame

The first problem faced by a statistician is the collection of data. The initial step in collecting data is to carefully define the population of interest.

Defining the population of interest means to clearly state those characteristics of the population that are of interest to the study. If an element has some of the characteristics but not all of them, it is not part of the population. An element in the population may have more characteristics than those in the definition; however, the additional characteristics are not of interest to the study and have no effect on the membership of the population.

The following illustrations provide examples of populations defined for specific investigations.

Illustration 1-5

As orders are received by a mail-order firm, they are sequentially numbered and the data are keypunched. The manager of the firm is interested in knowing what percentage of the orders had prices that were incorrectly keypunched during the past year. The population is the set of prices of every order keypunched during the past year. □

Illustration 1-6

To complete a form required by the Office of Federal Contract Compliance, a personnel manager must estimate the proportion of applicants for production and maintenance jobs during the past year who were members of a minority race. The population of interest is the set of all applicants for production or maintenance jobs last year. □

Illustration 1-7

A bank is studying the profitability of club accounts. It currently offers two types of club accounts, Vacation and Christmas. These accounts have fixed dollar deposits each week. For example, a member of the Vacation Club can deposit $5 per week for 52 weeks and receive $266.50 (the deposit plus interest) at the end of the period. The constant amount deposited and the amount paid out vary depending on the amount the customers decide to invest. An important factor in determining the profitability of club accounts is the average amount deposited each week. The population is the set of amounts deposited in all club accounts each week. □

Illustration 1-8

The Federal Reserve is analyzing the degree to which a certain questionable practice is being used by U.S. banks. To determine if the practice is widespread, a review of bank records is necessary. The population of interest is the set of records of all banks in the United States.

Illustration 1-9

The cost of a 1-minute radio advertisement varies drastically by station, day, and time of day. The rate is determined by the proportion of listeners who are tuned to a particular station on a given day of the week at a particular time. There are companies that compute and sell this information. The population of interest in computing this information is the set of daily radio listeners in a particular radio-listening area. □

census If each member or element in the population is listed, or enumerated, the compilation is called a *census*. Censuses are seldom taken because they are (1) expensive, (2) time consuming, or (3) difficult to compile. (Consider the task of contacting every person in the radio area where you live.) When compiling a census is unrealistic, part of the population, or a *sample*, is selected for analysis.

The first step in selecting a sample is to develop the sampling frame.

sampling frame

Sampling Frame

A list of the elements belonging to the population from which the sample will be drawn.

Ideally, the definition of the sampling frame should be identical to the definition of the population. However, this situation is not always possible. For instance, in illustration 1-5 the sampling frame could be the list of identification numbers for all orders keypunched; in illustration 1-7 it might be a list of all the club account numbers. In both these examples the sampling frame is the same as the population. In illustration 1-9, however, we cannot construct a sampling frame that is identical to the population because we would need a list of all people in a particular area who listen to the radio on a daily basis, and such a list is not likely to exist. One possible sampling frame might be all the names in the telephone directory; another might be a list of registered voters. Neither of these sampling frames contains the *complete* population of all radio listeners, and both will probably contain the names of some who are not radio listeners. Who from the population of all radio listeners would be excluded if (1) the telephone directory is used as the sampling frame or (2) the list of registered voters is used as the sampling frame? Since only the elements in the frame have a chance of being selected for the sample, it is important that the frame be representative of the population.

Exercises

1-5 For each of the following cases, define the population of interest and a possible sampling frame.

(a) A bank wants to determine the average daily balance of its checking account customers.

(b) An oil company wants to estimate the proportion of its current credit card customers it will lose if it eliminates its credit card system.

(c) The affirmative action office of a firm wants to estimate the proportion of people in the labor market who are female and possess the required skills to do a certain job.

1-6 For each of the illustrations 1-5 through 1-9, discuss whether a census is possible, and, if it is, why it still might be better to select a sample rather than use a census.

1-7 Select a problem of interest to you where survey data are useful. Define the population and specify a possible sampling frame.

1-8 Are the following sampling frames appropriate? Explain.

(a) The population is the set of all hourly employees of a firm. The sampling frame is the complete payroll list.

(b) The population is the set of all customers of a store. The sampling frame is the list of persons who maintain a charge account at that store.

Section 1-6 Errors in Sampling

Once the population and an appropriate sampling frame have been specified, the next step is to determine the procedure for selecting elements from this sampling frame. The concept of error in sampling must be understood so as to determine the best sampling procedure for the given situation.

sampling error

Sampling error occurs when the sample data do not represent the population. In analyzing such data, we would find that the statistic calculated from the sample does not equal the parameter or numerical value we would expect to get from measuring the entire population. The two basic causes of sampling error are *sampling variability* and *systematic bias* (also called nonsampling error).

sampling variability

The first type of error, **sampling variability** (or sample variability), occurs as a natural part of the sampling process. Sample results will vary depending upon the particular set of sample data drawn from the population. Regardless of what we do, when we sample, the possibility of error due to sample variability will be present. Because sampling variability is random, it may cause sample results to be either too high or too low. (Sampling variability is explored in chapter 7.)

systematic bias

Unlike sampling variability the second kind of sampling error, systematic bias, occurs because a mistake is made in the process of collecting the data. Because it is systematic and not random, systematic bias may cause the sample estimate to be consistently too high or consistently too low. Systematic bias can usually be eliminated and does not need to be present in a sampling problem. **You should always try to eliminate systematic bias from your sampling procedure.** Some of the more common causes of systematic bias are explored in the paragraphs that follow.

1. *Inappropriate sampling frame.* Systematic bias occurs when the objects in the sampling frame are not representative of the objects in the population.

2. *Natural bias in the reporting of data.* The population data may be incorrect. For example, gross incomes are probably underreported to the Internal Revenue Service. On the other hand, if an anonymous telephone survey were taken, more individuals might tend to overstate their gross income in order to appear more successful or affluent. This type of problem often occurs in opinion surveys, where persons tend to give what they think is the correct or socially acceptable answer rather than reveal their true feelings. This systematic bias is frequently easy to identify, but usually difficult to correct.

3. *"Indeterminancy" principle.* Individuals who know they are being observed often act differently than they normally would act. Two instances where this is true are time study analyses and quality control inspections. In a time study analysis, workers are studied on the job to determine the average length of time required to complete a task, usually so that quotas can be set up for piecework. When workers slow down during a time study, the results will lead to low quotas being set. This tendency has led to the design of elaborate methods, such as hidden cameras or one-way windows, for observing individuals without their being aware that they are under scrutiny.

Consider an example when a firm institutes a quality control system to monitor secretarial typing accuracy. When secretaries are aware they are being monitored, they tend to be more careful and accurate, but their typing speed suffers as a result. Once the monitoring system is removed or gradually becomes commonplace, their speed tends to increase and their accuracy returns to normal. Hence upward-biased estimates of accuracy are gathered from the initial monitoring system.

4. *Nonrespondents.* In surveys systematic bias often results from not being able to sample all the individuals initially selected from the sampling frame. This type of bias occurs because the likelihood of establishing contact or receiving cooperation from individuals is often related to the type of information collected. For example, in a survey of savings account balances, individuals with large savings account balances might be more cooperative than those with small balances. And those with very large accounts probably would not respond at all. In opinion polls people with strong opinions, either for or against, generally respond, whereas those with neutral opinions are much less likely to respond.

5. *Bias in the measuring device.* This type of systematic bias occurs when the physical measuring device consistently is in error, such as a scale calibrated an ounce too light. Bias due to physical measurement error is probably the easiest to avoid if all measuring devices are rigorously inspected and tested. In survey work this bias results if the questionnaire or interviewer is biased. Questionnaires can be inadvertently or deliberately worded to increase the probability of obtaining a certain response. A survey question such as "Do you think the small wage offering of Company X is a fair offer to its union employees?" will receive an inflated proportion of

"no" answers because of the use of the word "small." The use of leading, subjective adjectives such as "high," "disproportionate," or "small" biases the responses.

Exercises

1-9 Identify the type or types of systematic bias in the following situations:
(a) A television station asks viewers to phone in their votes on the question of whether the federal government should balance the budget.
(b) A public interest consumer group surveys doctors and asks them to state their net income per year.
(c) A public opinion survey evaluating unemployment insurance asks the question, "Do you think people should be paid not to work?"
(d) One thousand questionnaires are sent to various firms asking how much they spend per employee on safety. One hundred firms respond.

1-10 Do the following statements relate to systematic bias, sampling variability, or both?
(a) This error decreases as sample size increases.
(b) This error need not occur in a sample.
(c) This error is caused by a mistake in the method of collecting the data.
(d) This error may consistently cause a low estimate.
(e) This error is inherent in the concept of a sample.

1-11 Give examples of situations in which the procedure of collecting data could result in each of the five types of systematic bias.

Section 1-7 Judgment Versus Probability Samples

Now let us return to the problem of selecting the elements from the sampling frame. This selection process is defined by the sample design.

sample designs, or sampling plans

Sample Designs, or Sampling Plans

The procedures used to select the elements of a sample.

There are many different types of sample designs. Covering them all is beyond the scope of this introductory text. Thus we will restrict our attention to a few of the more common designs.

The two types of sample designs are *judgment samples* and *probability samples*.

judgment samples

Judgment Samples

Samples that are drawn on the basis of being "typical."

When a judgment sample is drawn, the person selecting the sample chooses items that he or she thinks are representative of the population. A marketing manager who wants to test-market a new product might select a certain geographical area that he or she feels is typical of all market areas. A quality control manager in conducting a stress test might select a piece of material he or she feels is "normal." A sales manager might select a consumer panel of what he or she believes are average consumers. The validity of the results from a judgment sample is a reflection of the soundness of the decision maker's judgment. Unfortunately, in this type of sample there is no objective way to measure sampling error or to assess how typical a judgment sample is of the population.

probability samples

Probability Samples

Samples in which the elements to be selected are drawn on the basis of probability. Each element in a population has a certain probability of being selected for a sample, and each possible sample has an equal probability of being selected.

In probability samples we can measure the sampling error. Measuring the sampling error allows us to make probability statements about how closely the sample results reflect the population.

NOTE: **Statistical inference requires that the sample design be a probability sample.** Throughout the rest of this book, when we discuss how to make statistical inferences, we will assume that the sample was a probability sample.

Section 1-8 Simple Random Sampling

One of the most common methods used to select a probability sample is the *simple random sample*.

simple random sample

Simple Random Sample

A sample in which every element in the population has an equal probability of being chosen and each of the possible samples has an equal chance to be selected.

When a simple random sample is drawn, an effort must be made to ensure that each element has an equal probability of being selected. Mistakes are frequently made because the term *random* is confused with *haphazard*. For example, suppose that 10 penlight batteries are to be drawn from a shipment of 100 batteries for the purpose of estimating the average life of the batteries in the lot. One method, seemingly random, would be to mix the batteries thoroughly and select 10 batteries. This procedure is haphazard, however, because it is extremely difficult to know when and if the items are thoroughly mixed. What if defective batteries were heavier and settled toward the bottom of the lot? The probability of their selection in that case would be less than if a true random selection design were used. The result would be that the defective batteries probably would not be in the sample and we would overestimate the average life of the 100 batteries.

A mixing procedure was used to select draft dates during the first year the draft lottery was used to conscript young men into the U.S. Army during the early 1970s. Afterward it was concluded that the selection was not a random process. The method used—placing the dates in a cylinder and mixing them—led to higher than chance probabilities that dates in the beginning and the end of the year would be selected and assigned either high or low draft numbers. The result was that 26 out of 31 December dates and 21 out of 30 November dates, but only 12 out of 31 January dates and 12 out of 29 February dates, were selected in the first half of the lottery and assigned low draft numbers.

The proper procedure for selecting a simple random sample is to use a random number generator or a table of random numbers. A random number generator is a device, usually a computer routine, that produces random numbers. A **random number table** is a collection of random digits, where each digit has an equal chance of appearing. The numbers in the random number table, table 1 in appendix D, can be thought of as single-digit numbers (0 to 9), two-digit numbers (00 to 99), or numbers of any other desired size (000 to 999, etc.).

To select a simple random sample, first assign a number to each element in the sampling frame. This assignment is usually done sequentially. Then go to a table of random numbers and select as many random numbers as are needed for the sample size desired (see appendix B). Each numbered element in the sampling frame that corresponds to a selected random number is chosen for the sample.

Illustration 1-10

G. Grant and Associates has been retained to audit the records of the Milson Company for the past fiscal year. The accounting firm wishes to do a statistical audit of 20 checks to see if they are recorded properly. The population of interest is the set of all the checks written last year. The sampling frame is identical to the population.

To select a simple random sample, each element in the sampling frame (the canceled checks) must first be numbered. Suppose there are 4143 checks; then these checks will be numbered sequentially from 0001 to 4143. To draw a sample, we turn to the random number table (table 1, appendix D) and, using four-digit numbers, we select 20 numbers that will represent the 20 checks to be drawn as the sample. We pick a starting number by arbitrarily pointing to the table while looking away. From this point we will follow the "path" of numbers until the 20 numbers have been selected.

Suppose that the starting point is 1585, which is located in the 38th line on the first page of the table. The check numbered 1585 would therefore be selected. Proceeding down the page, the next number, 4543, is discarded because it does not correspond to any numbered check in the sampling frame. If we continue down the column and then start at the top of the next column of four digits and work down, we would select the following 20 random numbers between 0001 and 4143:

1585 1553 3387 2591 2063 1595 2846 2940 1847 2111
0307 2625 1302 1874 0154 3909 2501 0545 3347 1168

The checks with these numbers would be the simple random sample of 20 checks.

Although the sampling pattern (path of numbers) is arbitrary, it must be specified before starting the selection process so that you are not influenced by the sight of the random numbers. □

Section 1-9 Systematic Sampling

systematic sampling

Systematic sampling is an easier method of selecting a sample than is simple random sampling. In systematic sampling every kth item in the sampling frame is selected. When a 10% sample is desired, we will want to select 1 out of every 10 items. This will be accomplished by selecting every 10th item. For a 5% sample, 1 out of every 20 is to be selected ($\frac{1}{20} = 0.05$). This will be accomplished by selecting every 20th item. In general, when an x% sample is desired, 1 out of every $100/x$ items will be selected by choosing every $(100/x)$th item. (If $100/x$ is not an integer, simply drop the fractional part and use only the integer part of $100/x$.)

Suppose a 3% systematic sample is desired. The first item (the *starting point*) is randomly selected (by using the random number table) from the first 33 elements ($100/3 = 33\frac{1}{3}$, which becomes 33) in the sampling frame, and thereafter every 33d element is included in the sample. This method of selection uses the random number table only once, to find the starting point. Moreover, it is not necessary to number the elements in the sampling frame or to know the total count of items in the sampling frame. Thus this procedure is good for sampling a percentage of a large population.

Using systematic sampling has inherent dangers, however. If repetitiveness exists in the population, systematic sampling should not be used. Suppose a sample is to be selected from a bottling assembly line. The filler machine has 10 heads, and every 10th bottle is filled by the same head. If one head is defective, every 10th bottle is incorrectly filled. If we select a 10% systematic sample, we either obtain a 100% defective sample or a 0% defective sample, depending on the random starting point.

Illustration 1-11

Suppose Grant and Associates audits purchase invoices. The invoices are filed by date of order in three file drawers. The accounting firm decides to audit 10% of the invoices. The population and sampling frame are the complete set of filed invoices. Rather than try to count and sequentially number the three file drawers of invoices, a systematic random sample design can be used. Since a 10% sample is desired, every 10th invoice can be selected. To begin, we randomly select 1 of the first 10 invoices. Using a randomly selected number from the random number table, we obtain a starting point of 9. Thus the 9th invoice in the file is selected, then the 19th, the 29th, and so forth. □

Section 1-10 Stratified Sampling

The accuracy of a simple random sample depends on the sample size and the similarity of the elements in the population. There are times, however, when a particular characteristic of the population that is of special interest to a study *may* be unevenly distributed throughout the population. In such cases it may be necessary to sacrifice some randomness to ensure that the full range of values for that characteristic are represented in the sample. To provide representation for the full range of values, the population is divided into subgroups that are known to include the needed values, and the elements of the sample are randomly selected from each of the subgroups. The different subgroups are called **strata**, and thus the term **stratified sample**.

    For example, suppose we are conducting a survey of the market potential of a new product line. If we survey consumers with a broad range of incomes, we can expect a wide diversity of opinions about the product. Rather than randomly selecting *n* potential customers to survey, it is better to divide the sample frame into strata on the basis of income and then sample from within each stratum. Stratified sampling results in more detailed information—in this case an estimate of the market potential of the new product line by income.

    Three basic questions must be answered before we can use a stratified sampling design:

1. How do we select the sample elements from each stratum in the sampling frame?
2. How many items do we select from each stratum?
3. How do we select the strata?

    The first question is easily answered. The usual method of selecting sample elements is to choose a simple random sample from each stratum. If systematic sampling is appropriate, it may be used instead of simple random sampling.

One method of allocating the sample among the strata—that is, determining the number of items to choose from each group—is to sample proportionately from each stratum. If p_i represents the proportion of the population included in stratum i, and n represents the total sample size, then the number of elements selected from stratum 1 is $n \cdot p_1$. That is, if stratum 1 contains 15% of the population, it should contribute 15% to the sample.

The third question is the most difficult to answer. The goal of stratifying the sampling frame is to select strata that have similar elements with respect to the response variable. Generally, the choice of strata is based on past experience and personal judgment about what factors in the population are related to the characteristic that is being estimated.

Illustration 1-12

Professional Services, Inc. is considering offering a new consulting service to corporations. To determine the market potential, it decides to survey 200 corporations. The sampling frame is the set of all corporations listed in Dun and Bradstreet. Professional Services feels that the amount of a firm's gross sales will affect its need for the new service.

The consulting firm decides to stratify the sampling frame into three strata: (1) corporations with under $1 million in gross sales, (2) corporations with from $1 to $10 million in gross sales, and (3) corporations with over $10 million in sales. The first group represents 25% of all the corporations listed in Dun and Bradstreet, the second group represents 60%, and the third group represents 15%. Thus Professional Services will survey 50 (0.25 × 200) firms from stratum 1, 120 (0.60 × 200) firms from stratum 2, and 30 (0.15 × 200) firms from stratum 3. To decide specifically which firms in each strata to survey, the corporations in each stratum are numbered sequentially. For example, the corporations with under $1 million in gross sales are numbered sequentially; the corporations with between $1 and $10 million in gross sales are numbered; and so forth. Finally, three sets of random numbers (50 numbers, 120 numbers, and 30 numbers) are selected from a random number table, and the corporations in each stratum that have identification numbers that match the random numbers selected for that stratum are surveyed. □

Section 1-11 Cluster Sampling

primary sampling units

The elements selected from the sampling frame are called the **primary sampling units**. In all the designs that we have discussed thus far, the sampling units have been the individual elements to be studied. To select the individual elements as primary sampling units, all individual elements in the sampling frame must be available or itemized. However, itemizing each element in the sampling frame or sampling from such a list is often extremely expensive, time consuming, or even impossible. It is sometimes easier to construct a sampling frame composed of primary sampling units that represent *groups* rather than individual elements in the population. All the elements clustered in each primary sampling unit (group) are then investigated. Because of the grouping of elements to be selected, often more elements can be selected for a given cost than in simple or stratified sampling. Such sample designs are called **cluster samples**. The clusters can be selected by either simple random sampling or systematic sampling. A stratified design, where the clustered primary sampling units are divided into strata, is also commonly used.

Illustration 1-13

Consider the problem of estimating the percentage of switches in a company's inventory that are defective. Assume that 10,000 switches are currently in the inventory, stored in 200 boxes of 50 each. Using a cluster sampling design, we can designate the 200 boxes as the sampling frame and randomly select n boxes, the primary sampling units, and examine all the switches clustered in each randomly selected box. □

Illustration 1-14

Cluster sampling could be used for an investigation of banks in the United States when the investigation requires a short interview with each bank manager. The individual element of interest is a bank. The population of all banks can be listed for the sampling frame, and simple random sampling, systematic sampling, or stratified sampling can be undertaken. However, if banks scattered all over the United States are selected, more time will be spent traveling to the banks than interviewing the managers. A more economical method is to construct a sampling frame based on geographic area, to randomly select areas, and then to interview all the banks in the selected areas. □

When the primary sampling units are clusters of elements based on geographic area, the design is called **area sampling**.

Exercises

1-12 Consider the set of checks drawn from the expense account (*A*) and the petty cash account (*B*) of a corporation, as shown in the accompanying table.

(a) Select a simple random sample of eight checks. List the elements.

(b) Select a systematic sample of 40% of the checks. List the elements.

(c) Select a stratified sample of eight checks. List the elements selected from each stratum (*A* and *B*).

Check Number	Amount (dollars)	Account Type	Check Number	Amount (dollars)	Account Type
1	35	B	11	137	A
2	139	A	12	16	B
3	20	B	13	162	A
4	8	B	14	12	B
5	105	A	15	15	B
6	160	A	16	160	A
7	12	B	17	135	A
8	130	A	18	15	B
9	18	B	19	139	A
10	25	B	20	152	A

1-13 (a) State the difference between a judgment sample and a probability sample.

(b) State the advantages of a probability sample in statistical analysis.

1-14 For each of the following situations, state which type of sample design might be appropriate and why.

(a) A sample of checks is audited to determine if they are entered properly in a journal.

(b) The proportion of defective products produced by a company that has two distinct lines of production is estimated.

(c) The racial and sexual composition of a group of applicants for a job is to be determined. Applications are currently filed by date of application in 10 file drawers.

(d) Physicians are personally interviewed about their position on proposed national health care legislation.

(e) You wish to estimate the number of broken bottles received at the R. T. Bottling Company. Bottles are received and stored in 25-bottle cartons.

1-15 Which type of sample design might be appropriate for each of the following sampling frames? Why?

(a) The sampling frame consists of a sequentially numbered list of elements.

(b) The sampling frame is large, not numbered, and not listed in any specific order.

(c) Every fifth item in the sampling frame was recorded by the same clerk.

(d) The sampling frame consists of male and female customers, and you believe customer's sex is related to the characteristics you are trying to estimate.

In Retrospect

You should now have a feeling of what statistics is about—an image that will grow and change as you work your way through this book. You know the distinction between attribute and variable data, and the difference between continuous and discrete variable data. You even know the difference between statistics and probability (although we will not study probability in detail until chapter 4). You know what a sample and a population are and how to solve the first problem in statistics—how to draw a sample.

Sampling consists of three steps: defining the population, developing a sampling frame, and selecting a method or sample design for choosing elements from the frame. You can use either a judgment sample or a probability sample as a method for selecting elements from the sampling frame. A probability sample is required if statistical inference techniques are to be used and sampling error is to be measured.

In addition to sampling error, another type of error, systematic bias, can occur because of the mechanics of the sampling process. Systematic bias has numerous causes in sampling, and you should be careful to avoid them as much as possible.

We have investigated four types of probability sample designs: simple random sampling, systematic sampling, stratified sampling, and cluster sampling. Each type of design offers certain advantages and disadvantages. The selection of a particular design depends on the specific situation.

You should realize that the news articles at the beginning of this chapter represent various aspects of statistics. The Daily Chuckle quotes two statistics and makes an implication (obviously a false implication in this case). The Dow Jones graph shows last week's daily results and the weekly average for each week during the last year. Can you relate these news articles to some of the statistical terms learned in this chapter?

As members of a computer-age society, we are constantly bombarded by "statistical" information. Information involving numbers can be found in nearly every daily newspaper. However, the fact that the information is numerical does not make it statistical, even though the terms "data" and "statistics" are sometimes used in referring to such information.

The results of a survey of CPA firms concerning their use of sampling in auditing show that 89% of the national firms use statistical sampling. After studying this chapter, you should have a better appreciation for what that survey shows. After some exercises using the material in this chapter we will begin to analyze sample data in detail in the next chapters.

Chapter Exercises

1-16 How would you try to arrive at a satisfactory answer to the question posed in illustration 1-3?

1-17 Indicate the type of data that the following variables would produce:
   (a) the current yield of a corporate bond (i.e., the dollar amount of interest paid divided by current bond price)

(b) the number of customers in a store during a given day
(c) the length of time between order and delivery
(d) the color preferred by consumers
(e) the quality of a product that is to be rated fair, good, or excellent
(f) the check number of an employee

1-18 You are interested in analyzing the steel industry and collecting the following information. State whether each item represents a statistic or a parameter.
  (a) The average number of people employed per steel-producing company for the whole industry is 26,405.
  (b) The average number of people employed per steel-producing company for those companies located near the East Coast of the United States is 23,570.
  (c) The average rate of return for all investments in the steel industry is 6.3%.
  (d) The average dividend rate for a sample of four steel-producing companies is 4.1%.

1-19 In each of the following cases, state what type of systematic error might be present.
  (a) The property values of an area are estimated based on a sample of assessment values recorded by the local tax office.
  (b) To determine if small-volume customers are happy with the service they receive from a stock brokerage house, each broker is asked to submit the names of 10 clients who purchased less than $10,000 in stocks last year. Each person whose name is submitted is sent a questionnaire.

1-20 State what type of sample design is described.
  (a) Every 10th check is audited.
  (b) Fifteen cases of cereal are randomly selected from inventory, and the boxes in each case are weighed.
  (c) An insurance company separates its whole-life policies into four categories: under $10,000, $10,000 to $49,999, $50,000 to $99,999, and $100,000 or more. It then surveys a simple random sample of each category to determine if the policyholders planned to increase their coverage within the next 2 years.

1-21 Use the data on bond spread presented in exercise 2-80 as the population.
  (a) Select a simple random sample of 10 bonds.
  (b) Select a systematic sample of 25% of the bonds.
  (c) Select a stratified sample of six grade Aaa bonds and four Aa bonds.

1-22 What type of sample design might be appropriate in the following situations and why?

(a) You want to survey major department stores to determine the amount of freedom that buyers have in ordering imported, rather than domestic, items. A short personal interview is needed, and you wish to cover the entire county geographically.

(b) You wish to sample employment applications to determine the distance that applicants live from a firm to estimate the relevant labor market for affirmative action. Applications are stored chronologically in six file drawers.

1-23 You have been retained by Nevada Calculators to do a marketing study of your college class. The company is interested in the answer to the following question: "If you were given a choice between two calculators that cost the same and had identical features, except calculator A had a square root key and calculator B had a memory, which calculator would you choose?"

(a) Define the population.
(b) Consider the four alternative sample designs discussed in this chapter: simple random sampling, systematic sampling, stratified sampling, and cluster sampling.
  (i) Discuss the sampling frame of each design.
  (ii) Discuss how you would obtain the sampling frame.
  (iii) Discuss how you would select the sample items from the frame.
(c) Discuss the pros and cons of each of the designs considered.
(d) Choose a design and conduct a survey of approximately 50 students.

Hands-On Problem

1-1 Find two articles from newspapers or magazines that exemplify the use of statistics. Select articles from two different areas of interest. Complete the following in reference to each of these articles:

(a) Describe the population of interest.
(b) Describe the response variable that was used.
(c) What type of data was used?

2 Descriptive Analysis and Presentation of Single-Variable Data

Chapter Outline

Unit 1: Graphic Presentation of Data

2-1 Graphs and Stem-and-Leaf Displays
A picture is often worth a thousand words.

2-2 Frequency Distributions, Histograms, Ogives
An increase in the amount of data requires us to modify our techniques.

Unit 2: Calculated Descriptive Statistics

2-3 Measures of Central Tendency
The four measures of central tendency—mean, median, mode, and midrange—are average values.

2-4 Measures of Dispersion
Measures of dispersion—range, variance, and standard deviation—assign a numerical value to the amount of spread in a set of data.

2-5 Measures of Position
Measures of position allow us to compare one piece of data with the set of data.

2-6 Interpreting and Understanding Standard Deviation
A standard deviation is the length of a standardized yardstick.

2-7 The Art of Statistical Deception
How the unwitting or the unscrupulous can use "tricky" graphs and insufficient information to mislead the unwary.

"Average" means different things

When it comes to convenience, few things can match that wonderful mathematical device called averaging.

How handy it is! With an average you can take a fistful of figures on any subject—temperatures, incomes, velocities, populations, light-years, hairbreadths, anything at all that can be measured—and compute one figure that will represent the whole fistful.

But there is one thing to remember. There are several kinds of measures ordinarily known as averages. And each gives a different picture of the figures it is called on to represent.

Take an example. Here are the annual incomes of ten families:

$44,000	$31,500
$39,000	$31,500
$37,500	$31,500
$36,750	$31,500
$35,250	$25,500

What would this group's "typical" income be? Averaging would provide the answer, so let's compute the typical income by the simpler and most frequently used kinds of averaging.

The arithmetic mean. When anyone cites an average without specifying which kind, you can probably assume that he has the arithmetic mean in mind.

It is the most common form of average, obtained by adding items in the series, then dividing by the number of items. In our example, the sum of the ten incomes divided by 10 is $35,400.

The mean is representative of the series in the sense that the sum of the amounts by which the higher figures exceed the mean is exactly the same as the sum of the amounts by which the lower figures fall short of the mean.

The median. As you may have observed, six families earn less than the mean, four earn more. You might very well wish to represent this varied group by the income of the family that is right smack dab in the middle of the whole bunch.

To do this, you need to find the median. It would be easy if there were 11 families in the group. The family sixth from highest (or sixth from lowest) would be in the middle and have the median income. But with ten families there is no middle family. So you add the two central incomes ($31,500 and $35,250 in this case) and divide by 2. The median works out to $33,375, less than the mean.

The midrange. The median, you will note, is the middle item in the series. Another number that might be used to represent the group is the midrange, computed by calculating the figure that lies halfway between the highest and lowest incomes. To find this figure, add the highest and lowest incomes ($54,000 and $25,500), divide by 2 and you have the amount that lies halfway between the extremes, $39,750.

The mode. So, three kinds of averages, and not one family actually has an income matching any of them.

Say you want to represent the group by stating the income that occurs most frequently. That kind of representativeness is called a mode. In this example $31,500 would be the modal income. More families earn that income than any other. If no two families earned the same income, there would be no mode.

Four different averages, each valid, correct and informative in its way. But how they differ!

arithmetic mean	$35,400
median	$33,375
midrange	$39,750
mode	$31,500

And they would differ still more if just one family in the group were a millionaire—or one were jobless!

So there are three lessons to take away from today's class in averages. First, when you see or hear an average, find out which average it is. Then you'll know what kind of picture you are being given.

Second, think about the figures being averaged so you can judge whether the average used is appropriate.

And third, don't assume that a literal mathematical quantification is intended every time somebody says "average." It isn't. All of us often say "the average person" with no thought of implying a mean, median or mode. All we intend to convey is the idea of other people who are in many ways a great deal like the rest of us.

Reprinted with permission from CHANGING TIMES Magazine, © 1980 Kiplinger Washington Editors, Inc., Mar. 1980. This reprint is not to be altered in any way, except with permission from CHANGING TIMES.

Chapter Objectives

In this chapter we will learn how to present and describe single-variable data. Single-variable means that we are going to deal with only one numerical response value from a given source. In chapter 3 we will work with two values from a common source.

The basic purpose of descriptive statistics is to describe a set of data. Primarily, we will try to describe a set of numerical values in a variety of abbreviated ways. Suppose your instructor just returned an exam paper to you. It carries a grade of 78, and you wish to interpret its value with respect to the rest of the class. Students usually are interested in three different things. What is the first thing that you would ask your instructor?

Most students immediately ask: What was the average exam grade? The information sought here is the location of the "middle" for the set of all the grades. "Measures of central tendency" are used to describe this concept. Your instructor replies that the class average was 68. So you know that your paper was above average by 10 points. Now you would like an indication of the "spread" of the grades. Since 78 is 10 points above the average, you ask: "How close to the highest is it?" Your instructor replies that the grades ranged from 42 to 87 points. Pictorially, the information we have so far looks like this:

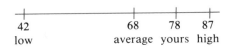

The third question is usually less important, since by now you have a fairly good idea how your grade compares with those of your classmates. However, the third concept still has some importance. It answers the question: How are the data distributed? You may have heard about the "normal curve," but what is it? As you will see later, the idea of distribution describes the data by telling you whether the values are evenly distributed or clustered (bunched) around a certain value. For example, your instructor says that half the class had grades between 65 and 75.

When very large sets of data are involved, measures of "position" are useful. When college board exams are taken, thousands of scores result. You are concerned with the position of your score with respect to all the others. Averages and ranges are not enough. You would like to say that your specific score is better than a certain percentage of scores. For example, your score is better than 75% of all the scores.

The four concepts necessary to describe sets of single-variable data are (1) measures of central tendency, (2) measures of dispersion (spread), (3) measures of position, and (4) types of distributions.

Unit 1 Graphic Presentation of Data

Section 2-1 Graphs and Stem-and-Leaf Displays

Once the sample of data has been collected, we must "get acquainted" with it. One of the most helpful ways to become acquainted is to employ an initial exploratory technique that will result in a pictorial representation of the data. The resulting displays will visually reveal patterns of the variable being studied. Data can be described in several graphic (pictorial) ways. The method used is determined by the type of data, the idea to be presented, and, in some cases, the preference of the investigator.

NOTE: You should realize from the start that there is no single correct answer when constructing a graphic display. The analyst's judgment and the circumstances surrounding the problem will play a major role in the development.

Circle and Bar Graphs

Circle graphs (pie diagrams) and bar graphs may be used. You have seen many examples of these graphs before. They are often used to summarize attribute data.

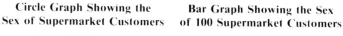

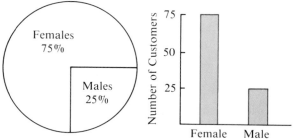

Stem-and-Leaf Displays

stem-and-leaf display

In recent years a technique known as the **stem-and-leaf display** has become very popular. This technique, a combination of a graphic and a sorting technique, is very simple to create and use. (To **sort** data is to make a listing of the data in rank order according to numerical value.) The data values themselves are used to do this sorting. The **stem** is the leading digit(s) of the data, whereas the **leaf** is the trailing digit(s). For example, the numerical data value 458 might be split 45—8 as shown in the following diagram:

```
LEADING DIGITS  | TRAILING DIGITS
            45  | 8
USED IN SORTING | SHOWN IN DISPLAY
```

Illustration 2-1

A corporation was considering changing the package of a certain cereal product. They asked a consumer panel of 20 persons to rate on a scale from 0 to 100 (100 being the best) the attractiveness of the new cereal box. The following data represent the results:

```
82  74  88  66  58  74  78  84  96  76
62  68  72  92  86  76  52  76  82  78
```

Let's construct a stem-and-leaf display for the ratings. At a quick glance we see that the scores are in the 50s, 60s, 70s, 80s, and 90s. Let's use the first digit of each score as the stem and the second digit as the leaf. Typically, we will construct the display in a vertical position. Draw a vertical line and locate the stems in order to the left of it.

```
5 |
6 |
7 |
8 |
9 |
```

The next step is to place each leaf on its stem. This task is accomplished by placing the trailing digit on the right side of the vertical line opposite its corresponding leading digit. Our first data value is 82; 8 is the stem and 2 is the leaf. Thus we place a 2 opposite the stem 8.

```
8 | 2
```

We continue by placing each of the other 19 leaves on the display. The resulting stem-and-leaf display is shown in figure 2-1.

The display in figure 2-1 places all scores with the same 10s digit on the same branch, which may not always be desired. Suppose we reconstruct the display, only this time instead of grouping 10 possible values on each stem, let's group it so that only 5 possible values could fall on each stem (figure 2-2).

20 RATINGS

```
5 | 8 2
6 | 6 2 8
7 | 4 4 8 6 2 6 6 8
8 | 2 8 4 6 2
9 | 6 2
```

FIGURE 2-1
Stem-and-Leaf Display with 10 Possible Values on Each Stem

FIGURE 2-2
Stem-and-Leaf Display with 5 Possible Values on Each Stem

```
        20 RATINGS
(50-54) 5 | 2
(55-59) 5 | 8
(60-64) 6 | 2
(65-69) 6 | 6 8
(70-74) 7 | 4 4 2
(75-79) 7 | 8 6 6 6 8
(80-84) 8 | 2 4 2
(85-89) 8 | 8 6
(90-94) 9 | 2
(95-99) 9 | 6
```

Do you notice a difference in appearance? The general shape is very much the same, approximately symmetrical about the 70s. A variable typically displays a distribution that is concentrated (mounded) about a central value and then, in some manner, dispersed in both directions. □

Often a graphic display will reveal something that the analyst may or may not have anticipated. Illustration 2-2 demonstrates what generally occurs when two populations are sampled together.

Illustration 2-2

A random sample of 50 college students was selected and their weights obtained from their medical records. The resulting data is listed in table 2-1.

TABLE 2-1
Data for Illustration 2-2

Student	1	2	3	4	5	6	7	8	9	10
Male/Female	F	M	F	M	M	F	F	M	M	F
Weight	98	150	108	158	162	112	118	167	170	120
Student	11	12	13	14	15	16	17	18	19	20
Male/Female	M	M	M	F	F	M	F	M	M	F
Weight	177	186	191	128	135	195	137	205	190	120
Student	21	22	23	24	25	26	27	28	29	30
Male/Female	M	M	F	M	F	F	M	M	M	M
Weight	188	176	118	168	115	115	162	157	154	148
Student	31	32	33	34	35	36	37	38	39	40
Male/Female	F	M	M	F	M	F	M	F	M	M
Weight	101	143	145	108	155	110	154	116	161	165
Student	41	42	43	44	45	46	47	48	49	50
Male/Female	F	M	F	M	M	F	F	M	M	M
Weight	142	184	120	170	195	132	129	215	176	183

Notice that the weights range from 98 to 215. Let's group the weights on stems of 10 units, using the 100s and the 10s digits as stems and the units digit as the leaf (figure 2-3).

FIGURE 2-3
Stem-and-Leaf Display

WEIGHTS OF 50 COLLEGE STUDENTS (LB)	
09	8
10	8 1 8
11	2 8 8 5 5 0 6
12	0 8 0 0 9
13	5 7 2
14	8 3 5 2
15	0 8 7 4 5 4
16	2 7 8 2 1 5
17	0 7 6 0 6
18	6 8 4 3
19	1 5 0 5
20	5
21	5

Close inspection of this display gives the impression that two overlapping distributions may be involved. That is exactly what we have, a distribution of female weights and a distribution of male weights. This distribution is shown in figure 2-4, a back-to-back stem-and-leaf display of this same set of data.

FIGURE 2-4
Back-to-Back Stem-and-Leaf Display

WEIGHTS OF 50 COLLEGE STUDENTS (LB)		
FEMALE		MALE
8	09	
8 1 8	10	
6 0 5 5 8 8 2	11	
9 0 0 8 0	12	
2 7 5	13	
2	14	8 3 5
	15	0 8 7 4 5 4
	16	2 7 8 2 1 5
	17	0 7 6 0 6
	18	6 8 4 3
	19	1 5 0 5
	20	5
	21	5

This graphic display makes it quite obvious that we do, in fact, have two distinct distributions within this set of data.

Exercises

2-1 Sales revenues for Gulf Resources & Chemical in 1981 were generated as follows:

Product	Sales Revenue (in millions of dollars)
Coal	$11.8
Lithium	7.5
Industrial explosives	6.6
Fertilizers and salt	1.5
Oil and gas	22.7
Other	0.1
Total	$50.2

Prepare a vertical bar graph of the sales revenue by product.

2-2 The Madison Gas & Electric Company uses various forms of energy for generating its electricity. The percentages of the sources of energy used last year and 10 years ago are shown as follows:

Source of Energy	10 Years Ago (%)	Last Year (%)
Coal	16.8%	62.6%
Nuclear	26.9	34.1
Gas	55.9	2.9
Other (including oil)	0.4	0.4
Total	100.0%	100.0%

Prepare a horizontal bar graph for this data by letting percent be along the horizontal axis and the sources of energy be along the vertical axis.

2-3 For Washington Gas Light Company, the distribution of number of shares owned by stockholders is as follows:

Number of Shares Owned	Proportion of Total
1– 99	0.43
100–249	0.41
250–999	0.14
1,000 or more	0.02
Total	1.00

Prepare a circle chart showing the distribution of number of shares owned.

2-4 Construct a stem-and-leaf display of the following data:

13.7 13.7 15.6 11.3 11.7 11.2 13.9
14.0 10.5 12.4 11.2 12.8 11.4 15.1

2-5 Construct a stem-and-leaf display of the following data:

39 52 39 56 87 40 73 38 48 55
44 65 51 63 60

2-6 The number of customers making deposits at an automatic teller machine was recorded for the past 22 days:

8   23 18 22 22 15 21 23 25 18 24
22 21 37 19 22 22 12 27 16 26 32

Construct a stem-and-leaf display of these data.

2-7 To study employee commuting patterns, 50 employees were asked for their one-way travel times from home to work (to the nearest 5 minutes). The resulting data were as follows:

20 20 30 25 20 25 30 15 10 40
35 25 15 25 25 40 25 30   5 25
25 30 15 20 45 25 35 25 10 10
15 20 20 20 20 25 20 20 15 20
 5 20 20 10   5 20 30 10 25 15

Construct a stem-and-leaf display of these data.

Section 2-2 Frequency Distributions, Histograms, Ogives

The entire listing of a large set of data does not present much of a picture to the reader. Sometimes we want to condense the data into a more manageable form. This objective can be accomplished with the aid of a **frequency distribution**.

frequency distribution

To demonstrate the concept of a frequency distribution, let's use the following set of data:

3 2 2 3 2
4 4 1 2 2
4 3 2 0 2
2 1 3 3 1

A frequency distribution is used to represent this set of data by listing the x values with their frequencies.

For example, the value 1 occurs in the sample three times; therefore the frequency for $x = 1$ is 3. The set of data is represented by the frequency distribution shown in table 2-2. The **frequency** f is the number of times the value x occurs in the sample. This table represents an **ungrouped frequency distribution**. We say "ungrouped" because each value of x in the distribution stands alone.

frequency
ungrouped frequency
distribution

TABLE 2-2
Frequency Distribution

x	f
0	1
1	3
2	8
3	5
4	3

classes

When a large set of data has many different x values instead of a few repeated values, as in the previous example, we can group the data into a set of **classes** and construct a frequency distribution. The stem-and-leaf display in figure 2-1 shows, in picture form, a grouped frequency distribution. Each stem represents a class, and the number of leaves on each stem is the same as the frequency for that same class. The data represented in figure 2-1 is listed as a frequency distribution in the following table:

Class	Frequency
50–59	2
60–69	3
70–79	8
80–89	5
90–99	2
Total	20

The stem-and-leaf process can be used to construct a frequency distribution; however, the stem representation is not compatible with all class widths. For example, class widths of 3, 4, or 7 are awkward to use. Thus sometimes we will find it advantageous to have a separate classifying procedure for constructing a grouped frequency distribution.

To illustrate this classifying procedure, let's use a sample of times elapsed between when orders are placed and when they are delivered at

Speciality Machine Works. Table 2-3 lists the 50 time intervals in ranked order.

TABLE 2-3
Time Intervals (days)

Ranked Data				
27	68	79	91	107
43	71	80	91	108
43	71	81	93	108
44	71	83	94	116
47	73	84	94	120
49	73	84	94	120
50	74	84	97	122
54	75	86	97	123
58	76	88	103	127
65	77	88	106	128

Three basic guidelines should be followed in constructing a grouped frequency distribution. They are as follows:

1. Five to twelve classes are most desirable.
2. Each class should be of the same width.
3. Classes should be set up so that they do not overlap and so that each piece of data belongs to exactly one class.

Two additional helpful (but not necessary) guidelines are:

4. An odd class width is often advantageous.
5. Use a system that takes advantage of a number pattern, to guarantee accuracy. (This procedure will be demonstrated in the following example.)

PROCEDURE:

1. Identify the high and the low values ($H = 128$, $L = 27$) and find the range. Range = $H - L = 128 - 27 = $ **101**.
2. Select a number of classes ($m = 10$) and a class width ($c = 11$) so that the product ($mc = 110$) is a bit larger than the range (range = 101).
3. Pick a starting point. The starting point should be a little smaller than the lowest score L. Suppose that we start at 22; counting from there by 11s, we get 22, 33, 44, 55, ..., 132. These values are called the *lower class limits*. (They are all multiples of 11, an easily recognized pattern.)

lower and upper class limits

The **lower class limit** is the smallest piece of data that can go into each class. The **upper class limits** are the largest values fitting into each class; in our example these values are 32, 43, 54, and so on. Our *classes* for this example are

22–32	77–87
33–43	88–98
44–54	99–109
55–65	110–120
66–76	121–131

NOTES:

1. At a glance you can check the number pattern to determine if the arithmetic used to form the classes was correct.

class width 2. The **class width** is the difference between the lower class limit and the next lower class limit. It is *not* the difference between the lower and upper limit of the same class.

class boundary 3. **Class boundaries** are numbers that do not necessarily occur in the sample data, but are halfway between the upper limit of one class and the lower limit of the next class. In the preceding example the class boundaries are 21.5, 32.5, 43.5, 54.5, ..., 120.5, and 131.5. The difference between the upper and lower class boundaries will also give you the class width.

When classifying data, it helps to use a **standard chart** (table 2-4). Once the classes are set up, we need to *tally* the data (table 2-4). If the data have been ranked, this tallying will be unnecessary; if not ranked, be careful as you tally them. The frequency f for each class is the number of pieces of

TABLE 2-4
Standard Chart or Frequency Distribution

Class Number	Class Limits	Tallies	Frequency, f
1	22–32	|	1
2	33–43	||	2
3	44–54	⦀⦀	5
4	55–65	||	2
5	66–76	⦀⦀ ||||	9
6	77–87	⦀⦀ ||||	9
7	88–98	⦀⦀ ⦀⦀	10
8	99–109	⦀⦀	5
9	110–120	|||	3
10	121–131	||||	4
			50

class mark

data that belong in that class. The sum of the frequencies should be exactly equal to the number of pieces of data n ($n = \sum f$). This summation serves as a good check.

In table 2-5 there is a column for the class mark x. The **class mark** is the numerical value that is exactly in the middle of each class. In table 2-5 the class marks are

$$x_1 = \frac{22 + 32}{2} = 27 \qquad x_2 = \frac{33 + 43}{2} = 38$$

and so on. As a check of your arithmetic, successive class marks should be a class width apart, which is 11 in this example.

TABLE 2-5
Frequency Table with Class Marks, a Grouped Frequency Distribution

Class Number	Class Limits	f	Class Mark, x
1	22–32	1	27
2	33–43	2	38
3	44–54	5	49
4	55–65	2	60
5	66–76	9	71
6	77–87	9	82
7	88–98	10	93
8	99–109	5	104
9	110–120	3	115
10	121–131	4	126
		50	

NOTES:

1. Now you can see why having an odd class width is helpful; the class marks are whole numbers.
2. The occasional use of subscripts is to identify the class being discussed.

grouped frequency distribution

Once the class marks are determined, we have a **grouped frequency distribution**. You should note that when we classify data in this fashion, we lose some information. Only when we have all the raw data before us do we know the exact values that were actually observed in each class. For example, we put a 58 and a 65 into class number 4, with class limits of 55 and 65. Once they are placed in the class, their values are lost to us and we use the class mark, 60, as their representative value. This loss of identity costs some accuracy, but the computations are made easier.

Histograms

histogram

frequency histogram

The **histogram** is a type of bar graph representing an entire set of data. The distribution of frequencies from table 2-5 appears in histogram form in figure 2-5a. It is therefore called a **frequency histogram**. A histogram is made up of the following components:

1. A title, which identifies the population of concern.
2. A vertical scale, which identifies the frequencies in the various classes.
3. A horizontal scale, which identifies the variable being studied.

FIGURE 2-5a
Frequency Histogram

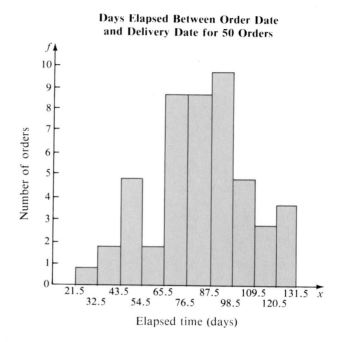

relative frequency histogram

relative frequency

Class marks may be used in place of class boundaries. (*Note*: Be sure to identify both scales so that the histogram tells the complete story.) Figure 2-5b shows the same histogram as in figure 2-5a, but generated by a computer.

By changing the vertical scale to $\frac{0}{50}$, $\frac{1}{50}$, $\frac{2}{50}$, and so on, the histogram would become a **relative frequency histogram**. The **relative frequency** is a proportional measure of the frequency of an occurrence. It is found by dividing the class frequency by the total number of observations. For example, in this illustration the frequency associated with the seventh class (88–98) is 10. The relative frequency is $\frac{10}{50}$ (or 0.20), since these values occurred 10 times in 50 observations. Relative frequency can often be useful in a presentation. Figure 2-6 is a relative frequency histogram of the sample of 50 observations. Compare and contrast figures 2-5a and 2-6.

FIGURE 2-5b
Computer-Generated Frequency Histogram

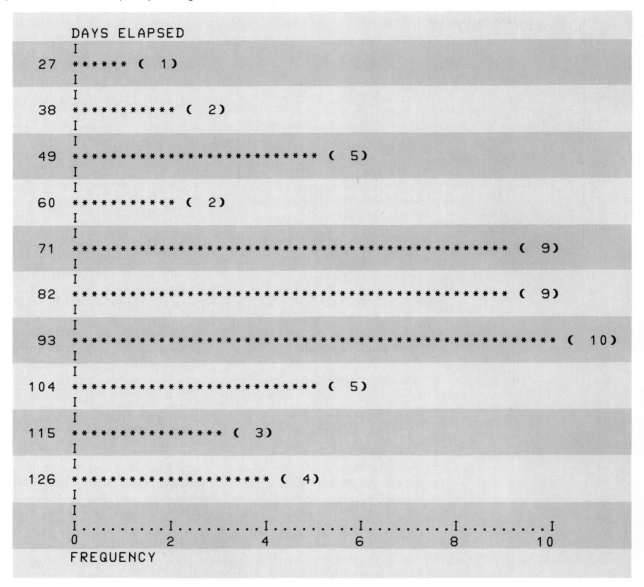

A stem-and-leaf display actually contains all the information of a histogram. Figure 2-7 shows that by rotating the stem-and-leaf display constructed in illustration 2-1, we produce a shape that conforms to that of the histogram.

Histograms are valuable tools. For example, the histogram of a sample should have a distribution shape very similar to that of the population from which the sample was drawn. If the reader of a histogram is at

FIGURE 2-6
Relative Frequency Histogram

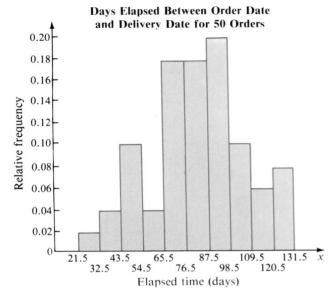

FIGURE 2-7
Rotated Stem-and-Leaf Display

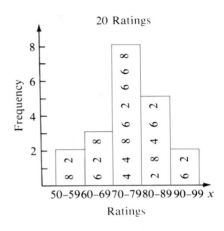

all familiar with the variable involved, he or she will usually be able to interpret several important facts from a histogram. Figure 2-8 presents histograms with descriptive labels resulting from their geometric shape. Can you think of populations whose samples might yield histograms like these?

Briefly, the terms used to describe histograms are as follows:

Symmetrical: Both sides of the distribution are identical.

Uniform (rectangular): Every value appears with equal frequency.

Skewed: One tail is stretched out longer than the other. The direction of skewness is on the side of the longer tail.

J-shaped: The side of the class with the highest frequency has no tail.

Bimodal: The two most populous classes are separated by one or more classes. This situation often implies that two populations are being sampled.

NOTES:

mode 1. The **mode** is the value of the *piece of data* that occurs with the greatest frequency. (The mode is discussed in section 2-3.)

modal class 2. The **modal class** is the class with the highest frequency.

bimodal frequency distribution 3. A **bimodal frequency distribution** has the two highest frequency classes separated by classes with lower frequencies.

FIGURE 2-8
Shapes of Histograms

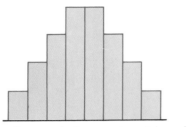
Symmetrical, normal, or triangular

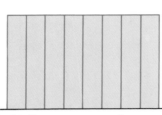
Uniform or rectangular

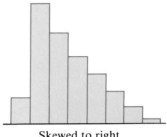
Skewed to right

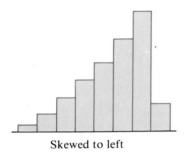
Skewed to left

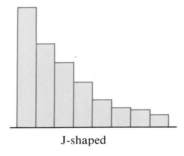
J-shaped

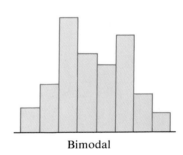
Bimodal

Ogive

A frequency distribution can be converted into a **cumulative frequency distribution** by replacing the frequencies with cumulative frequencies. This task is accomplished by placing a subtotal of the frequencies next to each class, as in table 2-6. The **cumulative frequency** for any given class is the sum of the frequency for that class and the frequencies of all classes of smaller values.

The same information can be presented by using a **cumulative relative frequency distribution** (table 2-7). This distribution combines the cumulative frequency idea and the relative frequency idea.

An **ogive** (pronounced "ō′ jīv") is a cumulative frequency or cumulative relative frequency graph. The components of an ogive are the following:

1. A title, which identifies the population.
2. A vertical scale, which identifies either the cumulative frequencies or the cumulative relative frequencies. (Figure 2-9 shows an ogive with cumulative relative frequencies.)
3. A horizontal scale, which identifies the upper class boundaries. Until the upper boundary of a class has been reached, you cannot be sure you have accumulated all the data in that class. Therefore **the horizontal scale for an ogive is always based on the upper boundaries**.

TABLE 2-6
Cumulative Frequency Distribution

Frequency Distribution		Cumulative Frequency Distribution		
Class Limits	Frequencies	Class Boundaries	Cumulative Frequencies	
22–32	1	21.5–32.5	1	
33–43	2	32.5–43.5	3	(1 + 2)
44–54	5	43.5–54.5	8	(1 + 2 + 5)
55–65	2	54.5–65.6	10	(1 + 2 + 5 + 2)
66–76	9	65.5–76.5	19	(1 + 2 + 5 + 2 + 9)
77–87	9	76.5–87.5	28	
88–98	10	87.5–98.5	38	
99–109	5	98.5–109.5	43	
110–120	3	109.5–120.5	46	
121–131	4	120.5–131.5	50	
	50			

TABLE 2-7
Cumulative Relative Frequency Distribution

Class Boundaries	Cumulative Relative Frequency
21.5–32.5	1/50 = 0.02
32.5–43.5	3/50 = 0.06
43.5–54.5	8/50 = 0.16
54.5–65.5	10/50 = 0.20
65.5–76.5	19/50 = 0.38
76.5–87.5	28/50 = 0.56
87.5–98.5	38/50 = 0.76
98.5–109.5	43/50 = 0.86
109.5–120.5	46/50 = 0.92
120.5–131.5	50/50 = 1.00

FIGURE 2-9
Ogive

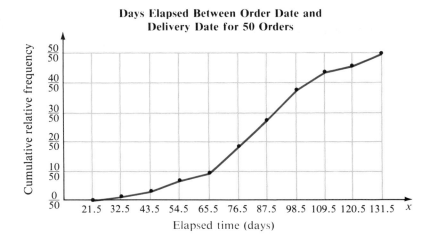

Days Elapsed Between Order Date and Delivery Date for 50 Orders

Elapsed time (days)

Frequency Distributions, Histograms, Ogives   Section 2-2

NOTE: Every ogive starts on the left with a cumulative relative frequency of 0 at the lower class boundary of the first class and ends on the right with a cumulative relative frequency of 100% at the upper class boundary of the last class.

Graphic representations of data should always be completely self-explanatory. That includes a descriptive and meaningful title and proper identification of the vertical and horizontal scales.

Exercises

2-8 The monthly percentage changes in the consumer price index for last year were as follows:

0.7 1.0 0.6 0.4 0.7 0.7 1.2 0.8 1.2 0.4 0.5 0.4

(a) Prepare a frequency distribution of these changes.
(b) Prepare a frequency histogram of these changes.

2-9 Hamburger Haven, Inc. retained an accounting firm to check the accuracy of the financial records kept by its franchises. The data were collected from an audit of 50 Hamburger Haven stores, and the number of errors in each store is listed as follows. One ledger was maintained in each store.

1	6	3	5	5	3	4	1	2	7
3	4	5	3	1	3	2	1	4	4
3	9	4	3	3	5	3	5	7	3
3	5	2	6	4	3	3	3	3	3
4	3	5	7	3	2	1	2	3	2

(a) Prepare a frequency distribution of the data.
(b) Prepare a frequency histogram of the data.
(Retain this solution for use in answering exercise 2-25.)

2-10 An inspection of 49 shipments of 1,000 glass bottles to Highlite Brewery yielded the following number of defective bottles per shipment:

11	8	13	14	11	8	14	7	11	13
7	4	9	11	10	12	6	12	5	10
12	10	3	6	9	10	10	13	12	9
9	5	12	7	11	9	4	8	13	12
13	11	12	8	6	13	11	6	12	

(a) Construct an ungrouped frequency distribution of the data.

(b) For the class containing eight defective bottles, find the (i) class mark, (ii) class width, and (iii) lower class boundary.

(c) Draw a frequency histogram that shows the number of defective bottles per shipment received at Highlite Brewery.

(d) Draw a relative frequency histogram that shows the same information.

2-11 The ages of 50 persons responding to an ad for counter help for a McDonald's are as follows:

```
21  19  22  19  18  20  23  19  19  20
19  20  21  22  21  20  22  20  21  20
21  19  21  21  19  19  20  19  19  19
20  20  19  21  21  22  19  19  21  19
18  21  19  18  22  21  24  20  24  17
```

(a) Prepare an ungrouped frequency distribution of these ages. (Retain this solution for answering exercises 2-34 and 2-55.)

(b) Prepare an ungrouped relative frequency distribution of the same data.

(c) Prepare a cumulative relative frequency distribution of the same data.

(d) Prepare a relative frequency histogram of this data.

(e) Prepare an ogive of this data.

2-12 To determine the optimum number of switchboard operators to hire, Elco Corporation recorded the number of phone calls received per hour for 50 hours.

```
27  23  22  38  43  24  35  26  28  18
25  23  22  52  31  30  41  45  29  27
29  28  27  25  29  28  24  37  28  29
26  33  25  27  25  34  32  36  22  32
21  23  24  18  48  23  16  38  26  21
```

(a) Classify the data into a grouped frequency distribution by using classes of 15–19, 20–24, ..., 50–54. (Retain this solution for use in answering exercises 2-27 and 2-54.)

(b) Find the class width.

(c) Find the set of class boundaries.

(d) For the class 20–24, find the (i) class mark and (ii) lower class limit.

(e) Prepare a frequency histogram of this data.

(f) Prepare an ogive of this data.

2-13 A salesperson made the following commissions (in dollars) on the last 35 sales:

$$
\begin{array}{ccccc}
176 & 690 & 32 & 140 & 143 \\
455 & 264 & 288 & 375 & 386 \\
325 & 148 & 654 & 83 & 466 \\
275 & 85 & 555 & 110 & 658 \\
108 & 410 & 434 & 542 & 473 \\
442 & 572 & 255 & 258 & 423 \\
21 & 688 & 409 & 61 & 112 \\
\end{array}
$$

(a) Construct a grouped frequency distribution of the data using class marks of 49.5, 149.5, 249.5,..., 649.5.

(b) Find the class width.

(c) What are the class limits of the classes whose class marks are 49.5 and 149.5?

(d) Name the boundary between the classes with class marks of 49.5 and 149.5.

(e) Prepare a frequency histogram of this data.

(f) Prepare an ogive of this data.

Unit 2 Calculated Descriptive Statistics

Section 2-3 Measures of Central Tendency

Measures of central tendency are numerical values that tend to locate, in some sense, the middle of a set of data. The term "average" is often associated with these measures. Each of the several measures of central tendency can be called the average value.

Mean

The average with which you are probably the most familiar is called the **mean**, $\bar{x}$ (read "x bar"). To find the mean, you add all the values of the variable x (this sum of the x values is symbolized by $\sum x$) and divide by the

mean

number of these values, n. We express this in formula form as

$$\text{Sample mean} = \bar{x} = \frac{\sum x}{n} \qquad (2\text{-}1)$$

NOTES:

1. See appendix A for information about the $\sum$ (read "sum") notation.
2. No indexes are shown in the formula in this chapter because the summations are over all the data.

Illustration 2-3

A set of data consists of the five values 6, 3, 8, 5, and 3. Find the mean. Using formula (2-1), we find

$$\bar{x} = \frac{\sum x}{n} = \frac{6 + 3 + 8 + 5 + 3}{5} = \frac{25}{5} = 5$$

Therefore the mean of this sample is 5. □

A physical representation of the mean can be constructed by thinking of a number line balanced on a fulcrum. A weight is placed on a number on the line corresponding to each number in the sample. In figure 2-10 the 6, 8, and 5 each contain one weight and the 3 contains two weights, since two 3s were in the sample (illustration 2-3). The mean is the value that balances the weights on the number line; for this illustration, $\bar{x} = 5$.

FIGURE 2-10
Physical Representation of the Mean

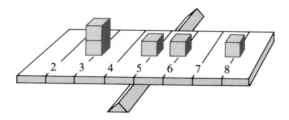

TABLE 2-8
Ungrouped Frequency Distribution

x	f
1	5
2	9
3	8
4	6
	28

Don't be misled by illustration 2-3 into believing that the mean value has to be a value in the data set. Consider the simple case where a salesperson sells merchandise worth $130 one week and $110 another week. The mean of these two values would be $120, which is not a value in our data set.

When the sample data is in the form of a frequency distribution, we will need to make a slight adaptation to our method of calculating $\bar{x}$. Consider the frequency distribution of table 2-8. This frequency distribution represents a sample of 28 values; five 1s, nine 2s, eight 3s, and six 4s. To calculate the mean $\bar{x}$ using formula (2-1), we need $\sum x$, the sum of the 28 x values.

$$\sum x = \underbrace{1 + 1 + \cdots + 1}_{5 \text{ of them}} + \underbrace{2 + 2 + \cdots + 2}_{9 \text{ of them}}$$

$$+ \underbrace{3 + 3 + \cdots + 3}_{8 \text{ of them}} + \underbrace{4 + 4 + \cdots + 4}_{6 \text{ of them}}$$

Five 1s equal $5 \times 1 = 5$, nine 2s equal $9 \times 2 = 18$, and so forth; therefore we write

$$\sum x = (5)(1) + (9)(2) + (8)(3) + (6)(4)$$
$$= 5 + 18 + 24 + 24 = 71$$

We can also obtain this sum by direct use of the frequency distribution. The extensions xf (see table 2-9) can be formed for each row and the extensions totaled to obtain $\sum xf$. The $\sum xf$ is the sum of the data. Therefore the mean of a frequency distribution may be found by dividing the sum of the data $\sum xf$ by the sample size $\sum f$. Formula (2-1) can be rewritten for use with a frequency distribution as

$$\bar{x} = \frac{\sum xf}{\sum f} \qquad (2\text{-}2)$$

TABLE 2-9
Extensions xf

x	f	xf
1	5	5
2	9	18
3	8	24
4	6	24
Total	28	71

The extensions for our illustration are shown in table 2-9. The mean for the sample is found by using formula (2-2).

$$\text{Mean:} \quad \bar{x} = \frac{\sum xf}{\sum f}$$

$$\bar{x} = \frac{71}{28} = 2.536 = 2.5$$

NOTES:
1. The values of xf are the same subtotals that we found previously.
2. The two column totals, $\sum f$ and $\sum xf$, are the same values that were previously known as n and $\sum x$. That is, $\sum f = n$; the sum of the frequencies is the number of pieces of data. The $\sum xf$ is the sum of the data.
3. The f in the symbol only indicates that the sum was obtained with the use of a frequency distribution.

Let's return now to the sample of 50 time periods elapsed between when orders were placed and filled. In table 2-10 you will find a grouped frequency distribution of the 50 times. We will calculate the mean for the sample by letting the class mark represent the value of x and using the frequencies and class marks in the same manner as in the preceding example. Table 2-10 shows the tabulations and totals.

TABLE 2-10
Calculations and Totals for Elapsed Time Periods

Class Number	Class Limits	f	Class Marks, x	xf
1	22–32	1	27	27
2	33–43	2	38	76
3	44–54	5	49	245
4	55–65	2	60	120
5	66–76	9	71	639
6	77–87	9	82	738
7	88–98	10	93	930
8	99–109	5	104	520
9	110–120	3	115	345
10	121–131	4	126	504
Total		50		4,144

The mean: Using formula (2-2), we get

$$\bar{x} = \frac{\sum xf}{\sum f} = \frac{4,144}{50} = 82.88 = 82.9$$

To illustrate again what the mean is, consider a balance beam (a real number line) with a cube placed on the value of each response. That is, nine cubes will be placed on 71, and so on. When the cubes are all placed, one cube for each piece of data, the beam will balance when the fulcrum is at 82.9, the mean. See figure 2-11.

FIGURE 2-11
Mean Is 82.9

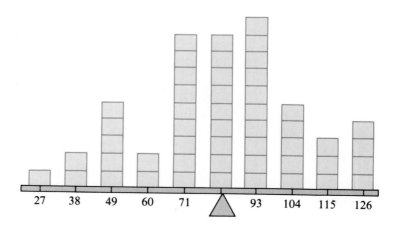

Measures of Central Tendency   Section 2-3

Because the value of the class mark is used to represent the value of each data that fell into that class interval, formula (2-2) gives an approximation of $\bar{x}$ when working with a grouped frequency distribution.

Median

median The **median**, $\tilde{x}$ (read "x tilde"), is the middle value of the sample when the data are ranked in order according to size. The data of illustration 2-3 ranked in order of size are 3, 3, 5, 6, and 8. The 5 is the third or middle value of the five numbers and thus is the median.

The **position** of the median, i, is determined by the following formula:

$$\text{Position of the median} = i = \frac{n + 1}{2} \qquad (2\text{-}3)$$

where n represents the number of pieces of data and i is the position that the median occupies in the ranked data.

To find the median $\tilde{x}$, you must first determine its position number i. Then count, starting from either end of the ranked data, to find the value that is in the ith position.

When n is an odd number, the median will be the exact middle piece of data. In our example, $n = 5$, and therefore the position of the median is

$$i = \frac{5 + 1}{2} = 3$$

That is, **the median is the third number** from either end of the ranked data, or $\tilde{x} = 5$.

However, if n is even, the position of the median will always be a half-number. For example, let's look at a sample whose ranked data are 6, 7, 8, 9, 9, and 10. Here $n = 6$, and therefore the median's position is

$$i = \frac{6 + 1}{2} = 3.5$$

This value says that the median is halfway between the third and fourth pieces of data. To find the number halfway between any two values, add the two values together and divide by 2. In this case add the 8 and the 9; then divide by 2. The **median is 8.5**, the number halfway between the middle two numbers. Notice that in figure 2.12 three numbers in the sample are smaller in value than 8.5 and three are larger.

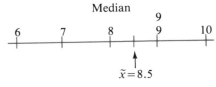

FIGURE 2-12
The Median

Other Averages

The **mode** is the value of x that occurs most frequently. In the preceding examples the mode in the sample of size 5 is 3. The mode in the sample of size 6 is 9. In both cases these values are the only numbers that occur more than once. If it happens that more than one of the values have the same greatest frequency (number of occurrences), we say there is no mode. For example, in the sample 3, 3, 4, 5, 5, and 7, both the 3 and the 5 appear an equal number of times. No one value appears most often. This sample has no mode.

midrange   Another measure of central tendency is the **midrange**. A set of data will always have a lowest value L and a highest value H. The midrange is the number exactly midway between them. It is found by averaging the low and the high values:

$$\text{Midrange} = \frac{L + H}{2} \qquad (2\text{-}4)$$

For the sample 3, 3, 5, 6, and 8, $L = 3$ and $H = 8$. Therefore the midrange is

$$\frac{3 + 8}{2} = 5.5$$

The midrange is the numerical value halfway between the two extreme values, the low L and the high H.

Exercises

2-14   Consider the sample 2, 3, 7, 8, 10. Find the following:
 (a) the mean $\bar{x}$ (b) the median $\tilde{x}$
 (c) the mode (d) the midrange

2-15   Consider the sample 5, 7, 6, 4, 2, 6. Find the following:
 (a) the mean $\bar{x}$ (b) the median $\tilde{x}$
 (c) the mode (d) the midrange

2-16   You are given the following sales figures (in billions of dollars) of the major food companies in fiscal year 1983:

Kraftco	4.5
Beatrice	4.2
General Foods	3.7
Consolidated Foods	2.4
General Mills	2.3

Find the following for these major food companies:
 (a) the mean sales (b) the median sales
 (c) the midrange

2-17 Welsher Dodge Agency's daily sales last week were (in number of cars sold) 10, 6, 15, 8, 13, 9, 12. Find the following:
(a) the mean $\bar{x}$
(b) the median $\tilde{x}$
(c) the mode
(d) the midrange

2-18 A sample of the volume of Xetra Corporation stock traded on the stock exchange for 10 days yields (in units of 100 shares) the data 6, 9, 8, 5, 3, 10, 6, 2, 6, 10. Find the following:
(a) the mean $\bar{x}$
(b) the median $\tilde{x}$
(c) the mode
(d) the midrange

2-19 In a brand loyalty study, 15 cigarette smokers were asked: "How many different brands of cigarettes have you smoked in the last year?" The responses were

5 7 3 2 2 5 4 3 1 8 5 2 7 3 4

Find the following:
(a) the mean, $\bar{x}$, number of brands smoked
(b) the median, $\tilde{x}$, number of brands smoked
(c) the mode number of brands smoked

2-20 The Elco personnel department is interested in studying employee absenteeism. During the month of August, time records show the following results for the number of production workers absent at the Elco Montreal facility:

13 14 9   17 21
10 15 22 19 13
22 13 19 23 17
21 10 9   20 18

Find the following:
(a) the mean $\bar{x}$
(b) the median $\tilde{x}$
(c) the mode
(d) the midrange

2-21 Use formula (2-2) to find the mean of the frequency distribution shown in the accompanying table.

x	f
0	1
1	3
2	8
3	5
4	3

2-22 Using the frequency distribution obtained in exercise 2-11, calculate the mean age of the job applicants. (Retain this solution for answering exercise 2-34.)

2-23 Berks Realty is interested in studying the activity of private-home sales in various offices. The accompanying table gives the frequency distribution of sales per office. Find the mean sales per office.

House Sales	f
3–5	1
6–8	7
9–11	8
12–14	11
15–17	7

2-24 INA Investments is analyzing the stock performance of the drug industry. It collected the price-earnings ratio on all listed drug stocks. Find the mean of the grouped frequency distribution of price-earnings ratios shown in the accompanying table.

Price-Earnings Ratio	f
2–5	6
6–9	12
10–13	18
14–17	16
18–21	8

2-25 Find the mean of the number of ledger errors given in exercise 2-9. Use the ungrouped frequency distribution found in the answer to exercise 2-9. (Retain this solution for use in answering exercise 2-41.)

2-26 The personnel department of Hayes Air West is interested in studying the monthly applicant flow for entry-level service jobs. Find the mean of the frequency distribution of job applicants shown in the accompanying table.

Job Applicants per Month	f
78–85	5
86–93	8
94–101	12
102–109	4
110–117	5
118–125	2

2-27 Find the mean for the number of phone calls given in exercise 2-12. Use the frequency distribution found in answering exercise 2-12. (Retain these solutions for use in answering exercise 2-43.)

2-28 Use the data presented in table 2-3 and calculate the mean. Compare this result with the result on page 58 obtained by using table 2-10 and formula (2-2). Which result is exact? Explain why there is a difference.

Section 2-4 Measures of Dispersion

Once the middle of a set of data has been determined, our search for information immediately turns to measures of dispersion (spread). The measures of dispersion include the range, variance, and standard deviation. These numerical values describe the amount of spread or variability that is found among the data. **Closely grouped data will have relatively small values, and more widely spread out data will have larger values for these measures of dispersion. The coefficient of variation allows us to compare standard deviations of different samples.**

Range

range The **range** is the simplest measure of dispersion. It is the difference between the highest- (largest-) valued (H) data and the lowest- (smallest-) valued (L) data:

$$\text{Range} = H - L \qquad (2\text{-}5)$$

For the sample 3, 3, 5, 6, and 8, the **range** is $8 - 3 = 5$. The range tells us that the five pieces of data all fall within a distance of 5 units on the number line.

The other two measures of dispersion, variance and standard deviation, are measures of dispersion about the mean. To develop a measure of

deviation from the mean dispersion about the mean, let's look first at the concept of **deviation from the mean**. An individual value x deviates, or is located, from the mean by an amount equal to $(x - \bar{x})$. The deviation $(x - \bar{x})$ is 0 when x is equal to the mean. The deviation $(x - \bar{x})$ will be positive if x is larger than $\bar{x}$, and negative if x is less than $\bar{x}$.

Consider the sample 6, 3, 8, 5, and 3. Find the deviation for each x. Using formula (2-1), $\bar{x} = \sum x/n$, we find that the mean is 5. Each deviation is then found by subtracting 5 from each x value, as follows:

x	6	3	8	5	3
$x - \bar{x}$	1	-2	3	0	-2

Figure 2-13 shows the deviation from the mean for each value. We might suspect that the sum of all these deviations, $\sum(x - \bar{x})$, would serve as

FIGURE 2-13
Deviations from the Mean

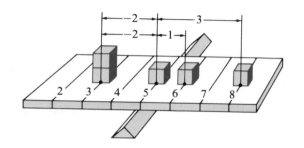

a measure of dispersion about the mean. However, that sum is exactly 0. As a matter of fact, it will always be 0. Why? Think back to the definition of the mean (page 50) and see if you can justify this statement.

If the sum of the deviations, $\sum (x - \bar{x})$, is always 0, it is not going to be of any value in describing a particular set of data. However, we want to be able to use the idea of deviation about the mean, since the mean is the most common average used. Yet there is a canceling effect between the deviations of x values smaller than the mean (negative) and those values larger than the mean (positive). This canceling effect can be removed if we make all the deviations positive. By using the absolute value of $x - \bar{x}$, that is, $|x - \bar{x}|$, we can accomplish this objective. Using the previous illustration with this definition, we now obtain the following absolute deviations:

x	6	3	8	5	3
$|x - \bar{x}|$	1	2	3	0	2

The sum of these deviations is 8. Thus we define a value known as the **mean absolute deviation**:

mean absolute deviation

$$\text{Mean absolute deviation} = \frac{\sum |x - \bar{x}|}{n} \qquad (2\text{-}6)$$

For our example the **mean absolute deviation is $\frac{8}{5}$, or 1.6**. Although this particular measure of spread is not used too frequently, it is a measure of dispersion. It tells us the average distance that a piece of data is from the mean.

The canceling effect can be eliminated in another way. Squaring the deviations will cause all these values to be nonnegative (positive or 0). When these values are totaled, the result is positive. The sum of the squares of the deviations from the mean, $\sum (x - \bar{x})^2$, is used to define the variance.

Variance

variance The **variance**, s^2, of a sample is the numerical value found by applying the following formula to the data:

$$\text{Sample variance} = s^2 = \frac{\sum (x - \bar{x})^2}{n - 1} \qquad (2\text{-}7)$$

where n is the sample size, that is, the number of items in the sample. The variance of a sample is a measure of the spread of the data about the mean. The variance of our sample 6, 3, 8, 5, and 3 is found as shown in table 2-11.

TABLE 2-11
Computing the Variance for 6, 3, 8, 5, 3

Step 1. x	Step 3. $x - \bar{x}$	Step 4. $(x - \bar{x})^2$
6	1	1
3	−2	4
8	3	9
5	0	0
3	−2	4
Total 25	0	18

Step 2. $\bar{x} = \dfrac{\sum x}{n} = \dfrac{25}{5} = 5$

Step 5. $s^2 = \dfrac{\sum (x - \bar{x})^2}{n - 1} = \dfrac{18}{4} = 4.5$

Let's look at another sample: 2, 3, 7, 8, and 10. The calculation of the totals, the mean, and the variance are shown in table 2-12.

TABLE 2-12
Computing the Variance for 2, 3, 7, 8, 10

1. x	3. $x - \bar{x}$	4. $(x - \bar{x})^2$
2	−4	16
3	−3	9
7	1	1
8	2	4
10	4	16
Total 30	0 ⓒⓚ	46

2. Using formula (2-1): $\bar{x} = \dfrac{\sum x}{n} = \dfrac{30}{5} = 6$

5. Using formula (2-7): $s^2 = \dfrac{46}{4} = 11.5$

Chapter 2 Descriptive Analysis and Presentation of Single-Variable Data

NOTES:

1. The sum of all the x's is used to find $\bar{x}$.
2. The sum of the deviations, $\sum (x - \bar{x})$, is always 0. Use this fact as a check in your calculations, as we did in table 2-12.

The last set of data is more dispersed than the previous set, and therefore its variance is larger. A comparison of these two samples is shown in figure 2-14. The two means are starred.

FIGURE 2-14
Comparison of Data

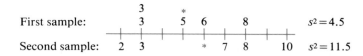

Standard Deviation

standard deviation   The **standard deviation** of a sample, s, is the positive square root of the sample variance:

$$\text{Sample standard deviation} = s = \sqrt{s^2} = \sqrt{\frac{\sum (x - \bar{x})^2}{n - 1}} \quad (2\text{-}8)$$

For the samples shown in figure 2-14, the **standard deviations** are $\sqrt{4.5}$, or 2.1, and $\sqrt{11.5}$, or 3.4.

Round-Off Rule

When rounding off an answer, let's agree to keep one more decimal place in our answer than was present in the original data. Round off only the final answer, not the intermediate steps. That is, do not use a rounded variance to obtain a standard deviation. In our previous examples the data were composed of whole numbers; therefore those answers that have decimal values should be rounded to the nearest 10th.

When rounding off a 5, always round to the even value. Thus 6.25 becomes 6.2, and 9.35 becomes 9.4. (See appendix C for further explanation.) If we use this rule, rounding up half the time and down half the time will, in the long run, balance out the round-off errors. The importance of a good round-off rule cannot be overemphasized.

NOTE: The numerator for sample variance, $\sum (x - \bar{x})^2$, is often called the *sum of squares for x* and symbolized by SS(x). Thus formula (2-7) can be expressed

$$s^2 = \frac{SS(x)}{n-1}$$

where $SS(x) = \sum (x - \bar{x})^2$.

The formula for variance can be modified into other forms for easier use in various situations. For example, suppose we have the sample 6, 3, 8, 5, and 2. Observe the computation of the variance shown in table 2-13.

TABLE 2-13
Computing the Variance for 6, 3, 8, 5, 2

1. x	3. $x - \bar{x}$	4. $(x - \bar{x})^2$
6	1.2	1.44
3	−1.8	3.24
8	3.2	10.24
5	0.2	0.04
2	−2.8	7.84
Total 24	0 Ⓚ	22.80

2. Using formula (2-1): $\bar{x} = \dfrac{\sum x}{n} = \dfrac{24}{5} = 4.8$

5. Using formula (2-7): $s^2 = \dfrac{\sum (x - \bar{x})^2}{n-1} = \dfrac{22.80}{4} = 5.70$

The arithmetic has become more complicated because the mean is in decimal form. The sum of squares for x can be found using an equivalent formula.

$$SS(x) = \sum (x - \bar{x})^2 = \sum x^2 - \frac{(\sum x)^2}{n} \qquad (2\text{-}9)$$

shortcut formula and is called the **shortcut formula** because it bypasses the calculation of $\bar{x}$. If we combine formulas (2-7) and (2-9),

$$s^2 = \frac{SS(x)}{n-1} = \frac{\sum x^2 - \dfrac{(\sum x)^2}{n}}{n-1} \qquad (2\text{-}10)$$

This shortcut formula can be shown to be equivalent to the original formula. The computations for s^2 using formula (2-10) are performed as shown in table 2-14.

The units of measure for the standard deviation are the same as the units of measure for the data. For example, if our data are in dollars, then the standard deviation s will also be in dollars. The unit of measure for variance might then be thought of as units squared. In our example of dollars, this unit would be dollars squared. As you can see, this unit has very little meaning.

The reason we have more than one formula to calculate variance is for convenience, not for confusion. As is so common in life, there is an easy way and a hard way to accomplish a goal. In statistics you can use several formulas for variance; and if you use the appropriate formula, your work will be kept to a minimum.

TABLE 2-14
Computing s^2 with the Shortcut Formula

x	x^2
6	36
3	9
8	64
5	25
2	4
Total 24	138

Using formula (2-9): $SS(x) = \sum x^2 - \dfrac{(\sum x)^2}{n} = 138 - \dfrac{(24)^2}{5}$

$$= 138 - 115.2 = \mathbf{22.8}$$

$$s^2 = \dfrac{22.8}{4} = \mathbf{5.7}$$

$$s = \sqrt{5.7} = \mathbf{2.4}$$

How do you decide which formula is applicable? With small samples only two formulas are used for variance:

$$\dfrac{\sum (x - \bar{x})^2}{n - 1} \quad \text{and} \quad \dfrac{\sum x^2 - \dfrac{(\sum x)^2}{n}}{n - 1}$$

Notice that the first formula involves the mean. If the mean is unknown or is a decimal (and thereby hard to work with), use the second formula. If the mean $\bar{x}$ is known and convenient to work with, use the first formula. If you need to find both the mean and the variance of a sample, always find the mean first.

To use formula (2-10) to calculate the variance when the sample is in the form of a frequency distribution, we will determine the values of n, $\sum x$,

TABLE 2-15
Ungrouped Frequency Distribution

x	f
1	5
2	9
3	8
4	6
Total	28

TABLE 2-16
$\sum x$ and $\sum x^2$

x	x^2	x	x^2
1	1	3	9
1	1	3	9
1	1	3	9
1	1	3	9
1	1	3	9
2	4	3	9
2	4	3	9
2	4	3	9
2	4	4	16
2	4	4	16
2	4	4	16
2	4	4	16
2	4	4	16
2	4	4	16
		71	209

TABLE 2-17
Subtotals of xf and x^2f

x	f	xf	x^2f
1	5	5	5
2	9	18	36
3	8	24	72
4	6	24	96
	28	71	209

and $\sum x^2$ using an extensions table. Recall that in section 2-3 the values of n and $\sum x$ were found on an extensions table as the values $\sum f$ and $\sum xf$. We need only add a column to our extensions table to obtain $\sum x^2f$ (it is the value of $\sum x^2$). Table 2-15 shows the same ungrouped frequency distribution used in section 2-3 (table 2-8). To calculate the sum $\sum x^2$ needed, we will find the extension x^2f for each class. These extensions will have the same value as the sum of the squares of the like values of x. For example, $x = 3$ has a frequency of 8, meaning that the value 3 occurs 8 times in the sample. Thus $x^2f = (3^2)(8) = 72$, the same value obtained by adding the squares of eight 3s. These extensions serve as subtotals. When the subtotals are added, we obtain the sum $\sum x^2f$, the replacement for $\sum x^2$.

These two sums can be found by subtotaling identical values of x and x^2 (table 2-16); that is,

$$\sum x = (5)(1) + (9)(2) + (8)(3) + (6)(4)$$
$$= 5 + 18 + 24 + 24 = 71$$

$$\sum x^2 = (5)(1) + (9)(4) + (8)(9) + (6)(16)$$
$$= 5 + 36 + 72 + 96 = 209$$

We can also obtain these sums by direct use of the frequency distribution. Subtotals of xf and x^2f are summed to obtain the totals $\sum x$ and $\sum x^2$ (table 2-17).

NOTES:

1. x^2f is found by multiplying xf by x; $x \cdot xf = x^2f$.
2. The three column totals, $\sum f$, $\sum xf$, and $\sum x^2f$, are the same values as were previously known as n, $\sum x$, and $\sum x^2$. That is, $\sum f = n$; the sum of the frequencies is the number of pieces of data. The $\sum xf = \sum x$ and $\sum x^2f = \sum x^2$. The f in the symbol only indicates that the sum was obtained with the use of a frequency distribution.

With these ideas in mind, the formula for the variance (2-10) becomes

$$s^2 = \frac{\sum x^2f - \frac{(\sum xf)^2}{\sum f}}{\sum f - 1} \qquad (2\text{-}11)$$

The standard deviation is the square root of the variance.
The variance and standard deviation for the example are found by using formula (2-11).

Variance:

$$SS(x) = 209 - \frac{(71)^2}{28}$$

$$= 209 - 180.036 = 28.964$$

$$s^2 = \frac{28.964}{27} = 1.073 = \mathbf{1.1}$$

Standard deviation:

$$s = \sqrt{s^2} = \sqrt{1.073} = 1.036 = \mathbf{1.0}$$

Let's return now to the sample of 50 time periods elapsed between when orders were placed and filled. In table 2-5 you will find a grouped frequency distribution of the 50 times. We will calculate the variance and standard deviation of the sample by using the frequencies and class marks in the same manner as in the preceding example. Table 2-18 shows the tabulations and totals. (*Note*: The class marks are now being used as representative values for the observed data. Hence the sum of the class marks is meaningless and is therefore not found.)

TABLE 2-18
Calculations and Totals
for Elapsed Time Periods

Class Number	Class Limits	f	Class Marks, x	xf	x²f
1	22–32	1	27	27	729
2	33–43	2	38	76	2,888
3	44–54	5	49	245	12,005
4	55–65	2	60	120	7,200
5	66–76	9	71	639	45,369
6	77–87	9	82	738	60,516
7	88–98	10	93	930	86,490
8	99–109	5	104	520	54,080
9	110–120	3	115	345	39,675
10	121–131	4	126	504	63,504
Total		50		4,144	372,456

The variance: Using formula (2-11), we get

$$s^2 = \frac{\sum x^2 f - \frac{(\sum xf)^2}{\sum f}}{\sum f - 1}$$

$$= \frac{372,456 - \frac{(4,144)^2}{50}}{49} = \mathbf{591.86}$$

The standard deviation is

$$s = \sqrt{s^2} = \sqrt{591.86} = 24.33 = \mathbf{24.3}$$

Coefficient of Variation

The variance and standard deviation are called measures of **absolute** dispersion. They deal with the difference between the observations and the mean, $x - \bar{x}$. If we want to compare the dispersion of two different samples from two different populations, the variance or standard deviation will not do.

For example, suppose we want to compare the dispersion of Eastman Kodak stock prices in a given month with the dispersion of Sprague Electronic stock. Since the standard deviation measures price movement away from the average price, it can be considered a measure of the swing in a stock's price, or a rough measure of risk. Suppose the standard deviation of Kodak is $10 and the standard deviation of Sprague is $2. Since Kodak has a greater standard deviation, can we conclude that its price is less stable? No; we must place each standard deviation in perspective with the average prices of its stock. If Kodak's average price for the month is $110, whereas the average of Sprague Electronics is $9, the $2 standard deviation of Sprague is *relatively* much greater (2/9, or 22%, of Sprague's average price) than Kodak's $10 standard deviation (10/110, or 9%, of Kodak's average price).

To make relative comparisons about the magnitude of the dispersion as measured by the standard deviation, we may express the standard deviation as a proportion of the mean. This proportion is called the **coefficient of variation**.

coefficient of variation

$$\text{Coefficient of variation} = \frac{s}{\bar{x}} \tag{2-12}$$

Thus the coefficient of variation is 0.09 (9%) for Kodak and 0.22 (22%) for Sprague. So we would say that Sprague's stock price has a greater dispersion (is less stable) than Kodak's.

Exercises

2-29 Consider the sample 2, 3, 7, 8, 10. Find the following:
   (a) the range
   (b) the variance s^2, using formula (2-7)

(c) the standard deviation s
(d) the coefficient of variation

2-30 Consider the sample 5, 7, 6, 4, 2, 6. Find the following:
(a) the range
(b) the variance s^2, using formula (2-7)
(c) the standard deviation s
(d) the coefficient of variation

2-31 Consider the following sample of the number of house closings processed per day by Provident Land Title Company:

$$10 \quad 4 \quad 10 \quad 9 \quad 11 \quad 7 \quad 10 \quad 9$$

Find the following:
(a) the variance s^2, using formula (2-7)
(b) the standard deviation s

2-32 Bud Company is trying to evaluate the assessment it received on a recently purchased commercially zoned parcel of land. It has obtained the assessment rates for similar parcels of land. The rates (in percentages) are

$$21 \quad 33 \quad 27 \quad 35 \quad 31 \quad 28 \quad 26 \quad 32 \quad 37 \quad 30$$

Find the following:
(a) the mean $\bar{x}$
(b) the variance s^2
(c) the standard deviation s

2-33 During the last few years, Portland General Electric Company has requested rate increases several times. The following revenues are the amounts to be earned as a result of the granted rate increases:

$34.5 million	$62.3 million
13.3	83.8
22.0	58.3
41.5	10.8

What is the variance for the revenue amounts realized?

2-34 Calculate the standard deviation for the set of ages given in exercise 2-11. (The mean was calculated in exercise 2-22.)

2-35 The accompanying table gives the rates of return of 12 mutual funds.

Fund	Rates of Return	Fund	Rates of Return
1	13.2	7	14.4
2	15.3	8	13.8
3	13.1	9	15.2
4	16.6	10	14.2
5	16.8	11	13.7
6	14.8	12	14.7

Find the following:

(a) the mean rate of return of these 12 funds

(b) the standard deviation of the rates of return of these 12 funds

2-36 Which sample has the higher dispersion relative to the size of its mean?

Sample A: $\bar{x} = 40$ and $s = 5$
Sample B: $\bar{x} = 106$ and $s = 11$

2-37 Consider the sample of prices of two bonds shown in the accompanying table. Which would you consider to have the more variable price? Explain your answer.

Ratco Electric	Warren Motors
88	9
74	5
83	8
95	12
78	11
93	13
84	6

2-38 Use formulas (2-2) and (2-11) to find the mean and variance of the frequency distribution shown in the accompanying table.

x	f
0	2
1	4
2	9
3	3
4	1

2-39 Fidelity Bank is interested in studying the activity of small-business loans in its various offices. The accompanying table gives the frequency distribution of loans per office. Find the mean number of loans per office and the variance of loans per office.

Number of Loans	f
3–5	2
6–8	6
9–11	9
12–14	13
15–17	8

2-40 A heating firm employs several troubleshooters to make emergency repairs of furnaces. The troubleshooters typically take many short trips. A sample of 20 travel expense vouchers related to troubleshooting was taken for the purpose of estimating travel expenses for the coming year. The following information resulted:

Dollar Amount on Voucher	Number of Vouchers
$0.01–$10.00	2
$0.01–$10.00	8
20.01–30.00	7
30.01–40.00	2
40.01–50.00	1
Total	20

Calculate the mean and the standard deviation for these travel account dollar amounts.

2-41 Find the standard deviation of the number of ledger errors given in exercise 2-9. Use the mean found in the answer to exercise 2-25. (Retain this solution for use in answering exercise 2-53.)

2-42 Find the standard deviation of the frequency distribution of job applicants given in exercise 2-26.

2-43 Find the standard deviation for the number of phone calls given in exercise 2-12. (Retain this solution for use in answering exercise 2-54.)

Section 2-5 Measures of Position

Measures of position are used to find the location of a specific piece of data in relation to the rest of the sample. Quartiles and percentiles are the two most popular measures of position.

Quartiles

quartile Quartiles are number values of the variable that divide a set of ranked data into quarters; each set of data has three quartiles. The **first quartile**, Q_1, is a number such that *at most* one-fourth of the data are smaller in value than Q_1 and *at most* three-fourths are larger. The **second quartile** is the median. The **third quartile**, Q_3, is a number such that *at most* three-fourths of the data are smaller in value than Q_3 and *at most* one-fourth are larger. See figure 2-15.

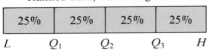

FIGURE 2-15
Quartiles

The procedure for determining the value of the quartiles is the same as that for percentiles and is shown after the following description of percentiles.

Percentiles

percentiles Percentiles are number values of the variable that divide a set of ranked data into 100 equal parts; each set of data has 99 percentiles. See figure 2-16.

The **kth percentile**, P_k, is a number value such that *at most* $k\%$ of the data are smaller in value than P_k and *at most* $(100-k)\%$ of the data are larger, as shown in figure 2-17.

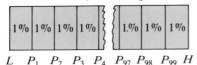

FIGURE 2-16
Percentiles

NOTES:

1. The 1st quartile and the 25th percentile are the same; that is, $Q_1 = P_{25}$. Also, $Q_3 = P_{75}$.

2. The median, the 2d quartile, and the 50th percentile are all the same ($\tilde{x} = Q_2 = P_{50}$). Therefore when asked to find P_{50} or Q_2, use the procedure for finding the median

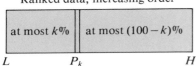

FIGURE 2-17
kth Percentile

The procedure for determining the value of any kth percentile P_k (or quartile) involves three basic steps.

First: The data must be ranked. All directions start at the lowest-valued piece of data and proceed toward the larger values.

Second: The position number i for the percentile in question must be determined. It is found by first calculating the value of $nk/100$. If $nk/100$ is *not* an integer (i.e., it contains a fraction), then i is equal to the next larger integer. For example, suppose $nk/100 = 17.2$; then $i = 18$. If $nk/100$ is an integer, then i will be equal to $nk/100 + 0.5$. For example, suppose $nk/100 = 23$; then $i = 23.5$.

Third: Locate the value of P_k. To locate this value, count from the lowest-valued piece of data, L, until the ith value is found if i is an integer. If i is not an integer, then it contains the fraction $\frac{1}{2}$. The value $\frac{1}{2}$ means that the value of P_k is halfway between the $(nk/100)$th and the $[(nk/100) + 1]$th pieces of data.

Illustration 2-4

A sample of 50 time periods (in days) that elapsed between when an order was taken and when it was delivered at Speciality Machine Works is listed in table 2-19 in order of collection. Find the 1st quartile, Q_1, and the 56th percentile, P_{56}.

TABLE 2-19
Times (days) Between When Order Taken and When Delivered

Raw Data				
75	97	74	120	71
97	58	73	94	106
71	94	68	79	86
65	43	54	80	108
84	116	50	83	84
27	123	49	71	93
108	91	81	88	77
91	120	128	88	107
122	94	103	47	44
84	43	76	73	127

The first step is to rank the data. The resulting ranked data are listed in table 2-20 in uniform columns.

TABLE 2-20
Time Intervals (days)

Ranked Data				
27	68	79	91	107
43	71	80	91	108
43	71	81	93	108
44	71	83	94	116
47	73	84	94	120
49	73	84	94	120
50	74	84	97	122
54	75	86	97	123
58	76	88	103	127
65	77	88	106	128

To find Q_1, we need to next determine the location of Q_1 by using $nk/100$.

$n = 50$ since there are 50 pieces of data

$k = 25$ since $Q_1 = P_{25}$

$$\frac{nk}{100} = \frac{(50)(25)}{100} = 12.5 \quad \text{which means that } i = 13$$

Q_1 is the 13th value, counting from L. Therefore $Q_1 = 71$.

To find P_{56}, we need to determine its location by using $nk/100$.

$n = 50$ since there are 50 pieces of data

$k = 56$ from P_{56}

$$\frac{nk}{100} = \frac{(50)(56)}{100} = 28 \quad \text{which means that } i = 28.5$$

P_{56} is the value halfway between the 28th and the 29th pieces of data. Therefore

$$P_{56} = \frac{86 + 88}{2} = 87 \qquad \square$$

z Score

z, or standard, score   The *z* score (often called the **standard score**) is the position of a value x in terms of the number of standard deviations it is located from the mean. The z score is found by the formula

$$z = \frac{\text{piece of data} - \text{mean of all data}}{\text{standard deviation of all data}} = \frac{x - \bar{x}}{s} \qquad (2\text{-}13)$$

Illustration 2-5

The youngest woman on the consumer panel (illustration 2-1) rated the new cereal box 92. Find the z score that corresponds to the 92 and describe what this value of z means.

The mean and standard deviation of the set of 20 ratings were found to be 75.9 and 11.1, respectively. Therefore the z score corresponding to $x = 92$ is found by using formula (2-13).

$$z = \frac{92 - 75.9}{11.1} = \frac{16.1}{11.1} = 1.45$$

This value means that the score of 92 is approximately $1\frac{1}{2}$ standard deviations above (larger than) the mean. $\square$

NOTE: The z score generally ranges between -3.0 and $+3.0$.

The z score, being a measure of relative position with respect to the mean, can be used to help make a comparison of two raw scores that come from different populations.

Illustration 2-6

The personnel manager of a department store wants you to compare the effectiveness of two salesworkers. One works in the men's clothing department and had gross sales of $150,000, whereas the other works in major appliances and had gross sales of $900,000. The second salesperson's sales were six times higher than those of the first salesperson. Is the second clerk's sales record better?

We need more information before we can draw a conclusion. Major appliance items cost more than men's clothing. Suppose the mean gross sales of all the salespeople in the men's clothing department is $100,000 and in the major appliance department it is $850,000. Both sales clerks are $50,000 above the mean, so we still can't draw any real conclusion. However, the standard deviation in the clothing department was $30,000 and in the appliance department it was $60,000. These values mean that the clothing salesperson's sales are $\frac{5}{3}$ standard deviations above the mean, whereas the appliance salesperson's sales are only $\frac{5}{6}$ standard deviation above the mean. Since the clothing salesperson's sales have the "better" relative position, we can conclude that this salesperson is the more effective of the two. (Again, we are speaking from a *relative* point of view.) □

Exercises

2-44 Find the first and third quartiles of the employee absentee data given in exercise 2-20.

2-45 Find the following percentiles for the set of absentee data given in exercise 2-20:
(a) 10th percentile, P_{10}
(b) 65th percentile, P_{65}
(c) 92d percentile, P_{92}

2-46 Find the 95th percentile P_{95} for the set of times elapsed between order and delivery shown in table 2-3.

2-47 Find the z score that corresponds to data $x = 35$ from a set of data whose mean $\bar{x}$ is 33 and whose standard deviation s is 2.0.

2-48 Find the z score that corresponds to the x value of 94.5 from a set of data whose mean is 105.0 and whose standard deviation is 5.0.

2-49 Find the x value that corresponds to a z score of 1.6 in a set of data whose mean is 10.0 and whose standard deviation is 4.2.

2-50 Find the value of the data whose z score is -2.1 in a set of data where the mean is 20.3 and the standard deviation is 5.3.

2-51 Which x value has the higher value relative to the set from which it comes?
A: $x = 42$, where $\bar{x} = 30$ and $s = 8$
B: $x = 65$, where $\bar{x} = 41$ and $s = 12$

2-52 Which x value has the lower value relative to the set from which it comes?
A: $x = 3.7$, where $\bar{x} = 5.2$ and $s = 0.6$
B: $x = 3.1$, where $\bar{x} = 4.7$ and $s = 0.8$

Section 2-6 Interpreting and Understanding Standard Deviation

Standard deviation is a measure of fluctuation (dispersion) in the data. It has been defined as a value calculated with the use of formulas. But you may wonder what it really is. It is a kind of yardstick by which we can compare one set of data with another. This particular "measure" can be understood further by examining the next two boxes.

Chebyshev's theorem

Chebyshev's Theorem

The proportion of any distribution that lies within k standard deviations of the mean is *at least* $1 - (1/k^2)$, where k is any positive number larger than 1. (This theorem applies to any distribution of data.)

This theorem says that within two standard deviations of the mean ($k = 2$), you will always find at least 75% (i.e., 75% *or more*) of the data.

$$1 - \frac{1}{k^2} = 1 - \frac{1}{2^2} = 1 - \frac{1}{4} = \frac{3}{4} = 0.75$$

Figure 2-18 shows a mounded distribution that illustrates this theorem.

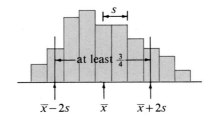

FIGURE 2-18
Chebyshev's Theorem with $k = 2$

If we were to consider the interval enclosed by three standard deviations on either side of the mean ($k = 3$), the theorem says that we will find at least 8/9, or 89%, of the data [$1 - (1/k^2) = 1 - (1/3^2) = 1 - (1/9) = 8/9$], as shown in figure 2-19.

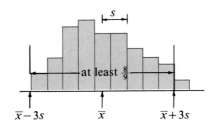

FIGURE 2-19
Chebyshev's Theorem with $k = 3$

empirical rule

Empirical Rule

If a variable is normally distributed, then within one standard deviation of the mean you will find approximately 68% of the data. Within two standard deviations of the mean there will be approximately 95% of the data, and within three standard deviations of the mean there will be approximately 99.7% of the data. [This rule applies specifically to a normal (bell-shaped) distribution, but it is frequently applied as an interpretation guide to any mounded distribution.]

Figure 2-20 shows the intervals of one, two, and three standard deviations about the mean of an approximately normal distribution. These proportions usually will not occur exactly in a sample, but your observed values will be close when a large sample is drawn from a normally distributed population.

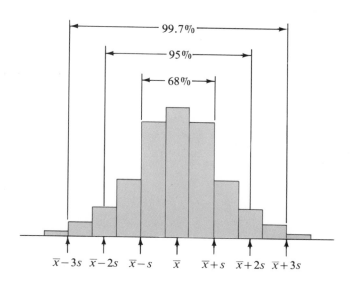

FIGURE 2-20
Empirical Rule

Normality may be tested crudely with the empirical rule. Let's demonstrate this test by working with the distribution of the time periods elapsed between order and delivery, which we have used throughout this chapter. The mean $\bar{x}$ was found to be 82.9, and the standard deviation s was 24.3. The interval from one standard deviation below the mean, $\bar{x} - s$, to one standard deviation above the mean, $\bar{x} + s$, is $82.9 - 24.3 = $ **58.6** to $82.9 + 24.3 = $ **107.2**. This interval includes 59, 60, 61, ..., 106, 107. Upon inspection of the ranked data (table 2-3, p. 40), we see that 32 of the 50 data pieces, or 64%, lie within one standard deviation of the mean. Further, $\bar{x} - 2s = 82.9 - (2)(24.3) = $ **34.3**, and $\bar{x} + 2s = 82.9 + (2)(24.3) = $ **131.5**. So the interval from 35 to 131, two standard deviations about the mean, includes 49 of the 50 data pieces, or 98%. All 50 data, or 100%, are included within three standard deviations of the mean (from 10.0 to 155.8). This information can be placed in a table for comparison with the values given by the empirical rule (table 2-21). These percentages are reasonably close to those found using the empirical rule. By combining this evidence with a histogram of the data, we can safely say that the sample data are approximately normally distributed.

TABLE 2-21
Observed Percentages Versus the Empirical Rule

Interval	Empirical Rule Percentage	Percentage Found
$\bar{x} - s$ to $\bar{x} + s$	Approximately 68	64
$\bar{x} - 2s$ to $\bar{x} + 2s$	Approximately 95	98
$\bar{x} - 3s$ to $\bar{x} + 3s$	Approximately 99.7	100

If a distribution is approximately normal, it will be nearly symmetrical and the mean (the mean and the median, therefore, are the same) will divide the distribution in half. This property allows us to refine the empirical rule. Figure 2-21 shows this refinement.

FIGURE 2-21
Refinement of Empirical Rule

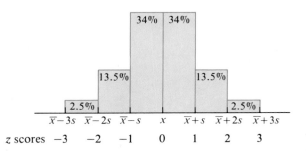

probability paper

test for normality

Normality can also be tested for graphically. This test is accomplished by drawing a relative frequency ogive of the grouped data on **probability paper** (which can be purchased at your college bookstore). On this paper the vertical scale is measured in percentages and is placed on the right side of the graph paper. All the directions and guidelines given on page 46 for drawing an ogive must be followed. An ogive of the time periods elapsed between order and delivery is drawn and labeled on a piece of probability paper in figure 2-22. The **test for normality** is to draw a straight line from the lower left corner to the upper right corner on the graph: if the ogive lies close to the straight line, the distribution is said to be approximately normal. The ogive for an exactly normal distribution will trace the straight line.

The dashed line in figure 2-22 is the straight-line test for normality. The ogive suggests that the distribution of delivery times is approximately normal. (*Warning*: This graphic technique is very sensitive to the scale used along the horizontal axis.)

FIGURE 2-22
Ogive on Probability Paper

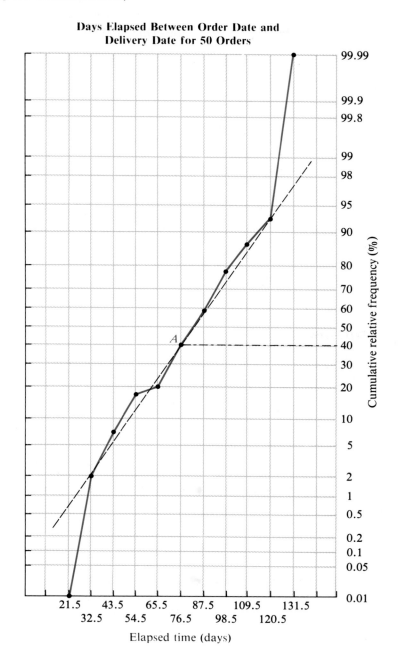

Exercises

2-53 Use the mean and standard deviation of the number of ledger errors data found in exercises 2-25 and 2-41.

(a) Find the values of $\bar{x} - s$ and $\bar{x} + s$.

(b) How many of the 50 pieces of data given in exercise 2-9 have values between $\bar{x} - s$ and $\bar{x} + s$? What percentage of the sample is this number?

(c) Find the values of $\bar{x} - 2s$ and $\bar{x} + 2s$.

(d) How many of the 50 pieces of data have values between $\bar{x} - 2s$ and $\bar{x} + 2s$? What percentage of the sample is this number?

(e) Find the values of $\bar{x} - 3s$ and $\bar{x} + 3s$.

(f) What percentage of the sample has values between $\bar{x} - 3s$ and $\bar{x} + 3s$?

(g) Compare the answers found in (d) and (f) with the results predicted by Chebyshev's theorem.

(h) Compare the answers in (b), (d), and (f) with the results predicted by the empirical rule. Does the result suggest an approximately normal distribution?

2-54 Answer the questions asked in exercise 2-53 (a) through (h) using the data given in exercise 2-12 and the values found in exercise 2-43.

2-55 Answer the questions asked in exercise 2-53 (a) through (h) using the data given in exercise 2-11 and the values found in exercise 2-34.

2-56 If a distribution is mounded so as to be approximately normal, how wide will an interval centered at the mean have to be to span the middle 95% of the data?

2-57 Can your sample data ever violate Chebyshev's theorem?

Section 2-7 The Art of Statistical Deception

"There are three kinds of lies—lies, damned lies, and statistics." These remarkable words spoken by Disraeli (a 19th-century British prime minister) represent the cynical view of statistics held by many people. Most people are on the consumer end of statistics and therefore have to "swallow" them.

Good Arithmetic, Bad Statistics

Let's explore an outright statistical lie. Suppose that a small business firm employs eight people who earn between $200 and $240 per week. The owner of the business pays himself $800 per week. He reports to the general public that the average wage paid to the employees of his firm is $280 per week. That may be an example of good arithmetic, but it is also an example of bad statistics. It is a misrepresentation of the situation, since only one employee, the owner, receives more than the mean salary. The public will think that most of the employees earn about $280 per week.

Tricky Graphs

Graphic representation can be tricky and misleading. The vertical scale (which is usually the frequency) should start at 0 in order to present a true picture. Graphs that do not start at 0 are used to save space. Nevertheless, these graphs can be, and usually are, deceptive.

The graph in figure 2-23 was presented in the annual report of a small business in connection with the amount of sales made by the company's

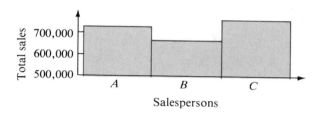

FIGURE 2-23
A Graph That Doesn't Start at 0

three salespersons. Without careful study the viewer of this graph would conclude that B only sold about half of what C sold. In reality the graph should look as shown in figure 2-24. The graph is not really different; it only creates a different impression.

FIGURE 2-24
The Same Graph, but Starting at 0

Insufficient Information

A realtor describes a mountainside plot of land as a perfect place for a summer home and tells you that the average high temperature during the months of July and August is 77 degrees. If the standard deviation is 5 degrees, this temperature is not too bad. Assuming that these daily highs are approximately normally distributed, the temperature will range from three standard deviations below to three standard deviations above the mean temperature. The range would be from 62 to 92 degrees, with only a few days above 87 and a few days below 67. However, what she didn't tell you was that the standard deviation of temperature highs is 10 degrees, which means the daily high temperatures range from 47 degrees to 107 degrees. These extremes are not too pleasant!

The familiar advertisement that shows the "higher level of pain reliever" is a tricky graph with no statistics. It has a statistical appearance due only to the clever use of a graphic presentation. There are no variables, no units—just implications.

Incorrect Interpretation of Percentages

Consider the following statements: "Conglom is the company of the small investor. Ninety percent of all Conglom stockholders own 100 shares or less." This statement makes it seem that a lot of small investors own Conglom Corporation. Is this true? Consider the following information:

	Number of Shareholders	Shares Owned
100 shares or less	9,000	225,000
More than 100 shares	1,000	1,000,000
Total	10,000	1,225,000

It is true that 90% of all Conglom stockholders own 100 shares or less, but the 10% who own more than 100 shares own 81.6% of the company.

Developed countries grow at an average rate of 4% per year, whereas underdeveloped countries have an average growth rate of 7% per year. Thus it is only a matter of time until the underdeveloped countries equal the developed countries. Wrong. The gross national product (GNP) of the United States is approximately $2.0 trillion ($2,000,000,000,000). Suppose that the gross national product of a certain underdeveloped Asian nation is $1 billion ($1,000,000,000). The United States GNP increases by 4% to $2.08 trillion [2.0 + (2.0)(0.04)] and the Asian country's GNP increases by 7% to $1.07 billion [1 + (1)(0.07)]. The difference now *increases* to $2.07893 trillion (2.08 − 0.00107). Rather than decreasing, the difference between the two GNPs actually increases. Remember that a small percentage of a large number can be greater than a large percentage of a small number. I would rather have 1% of IBM's total income than 99% of the corner grocery store's total income.

Statistics, like all languages, can be and is abused. In the hands of the careless, the unknowledgeable, or the unscrupulous, statistical information can be as false as "damned lies."

In Retrospect

You have been introduced to some of the more commonly used techniques of descriptive statistics. Far too many specific types of statistics are used in nearly every specialized field of study for us to review them here. We have outlined the uses of only the most universal statistics. Specifically, you have seen several basic graphic techniques (circle and bar graphs, stem-and-leaf displays, histograms, and ogives), which are used to present sample data in picture form. You have also been introduced to some of the more commonly used measures of central tendency (mean, median, mode, and midrange), measures of dispersion (range, variance, standard deviation, and coefficient of variation), and measures of position (quartiles, percentiles, and z score).

You should now be aware that an average can be any one of four different statistics, and you should understand the distinction between the different types of averages. The "Average Means Different Things" article at the beginning of the chapter discusses the four averages studied in this chapter. You might read it again; it should be more meaningful and interesting now. It will be time well spent!

You also should have obtained a feeling for and an understanding of the concept of a standard deviation. You were also introduced to Chebyshev's theorem and the empirical rule for this purpose.

The exercises in this chapter are extremely important; they will help you nail down the concepts studied before going on to learn how to use these ideas in later chapters.

Chapter Exercises

2-58 The percent earned on average shareholders' equity (the rate of return on common stock) for the Pacific Lighting Corporation for the last several years has been 8.6%, 9.8%, 10.9%, 11.8%, 13.7%, 13.9%, and 14.6%. Determine:
  (a) the median rate of return for these years
  (b) the mean rate of return for these years
  (c) the mean absolute deviation for this data

2-59 The average monthly occupancy rate for Metropolitan Hospital last year is given as follows:

Month	Jan.	Feb.	Mar.	Apr.	May	June	July	Aug.	Sept.	Oct.	Nov.	Dec.
Patients	265	259	258	242	245	249	234	222	226	254	215	190

  (a) Determine the median monthly occupancy rate for this information.
  (b) Determine the third quartile for this data.
  (c) What is the mean monthly occupancy rate?

2-60 A random sample of 10 state government employees showed the following number of months of service for each:

$$110 \quad 90 \quad 84 \quad 26 \quad 46 \quad 50 \quad 60 \quad 12 \quad 21 \quad 156$$

(a) What is the median length of service?

(b) Determine the 80th percentile for length of service.

2-61 The acreage of each offshore Louisiana lease acquired by Kerr-McGee last year is given as follows:

Location	Gross Acres
Main Pass	4,995
Ship Shoal	3,327
South Timbalier	3,772
Vermilion	5,000
West Cameron	5,000
West Cameron	2,500
West Cameron	5,000
West Delta	1,250

(a) What is the range of the acreage of these acquisitions?

(b) What is the mean?

(c) What is the median?

(d) What is the mode?

(e) Which measure do you think is most representative of the plots acquired? Explain.

2-62 The Crab Run Gas Company drills many gas wells each year. The depths of successful wells drilled last year are:

$$18{,}150 \quad 16{,}361 \quad 16{,}800 \quad 17{,}700$$
$$16{,}500 \quad 16{,}621 \quad 16{,}744$$

Determine the mean and the standard deviation for the depths of these wells.

2-63 Perks Plumbing, Inc. is considering advertising in a local reference directory. It surveyed 30 subscribers to the directory and asked, "How many times last year did you use the reference directory to find a plumbing contractor?" The responses were:

$$0 \quad 0 \quad 1 \quad 6 \quad 0 \quad 2 \quad 3 \quad 1 \quad 1 \quad 1 \quad 0 \quad 5 \quad 2 \quad 1 \quad 0$$
$$1 \quad 1 \quad 2 \quad 4 \quad 3 \quad 0 \quad 1 \quad 3 \quad 1 \quad 2 \quad 1 \quad 0 \quad 1 \quad 4 \quad 5$$

(a) Find the mean.
(b) Find the median.
(c) Find the mode.
(d) Find the midrange.
(e) Which one of the measures of central tendency in (a) through (e) would best represent the average subscriber's use of the directory if we were trying to portray the typical consumer? Explain.
(f) Which measure of central tendency would best describe the number of times a contractor was selected? Explain.
(g) Find the range.
(h) Find the variance.
(i) Find the standard deviation.

2-64 The total employment-population ratio for women (the percent of female population employed) for the years 1950 through 1982 were:

32.6 34.8 35.6 37.6 40.6 42.3
47.7 33.4 35.2 35.3 39.0 40.7
43.5 47.9 33.8 35.2 35.5 39.5
41.2 45.1 47.5 32.6 34.6 36.0
39.8 42.6 47.1 32.6 35.1 36.3
41.1 42.4 47.9

(a) Construct a stem-and-leaf display.
(b) Find the mean.
(c) Find the median.
(d) Find the standard deviation.

2-65 The following data represent the number of loss leader sales (sales where items are sold at cost to attract customers into a store) per month at Rex Brothers Department Stores:

25 22 17 31 15 19 22 26 22 32 12 16

(a) Find the mean.
(b) Find the median.
(c) Find the variance.
(d) Find the standard deviation.

2-66 The following data are the median household incomes of the readers of 20 major publications, according to a study by Simmons Corporation.

††Time	$18,097
Golf	19,803
Golf Digest	21,852
†Wall Street Journal	21,729
†Forbes	24,467
New York	19,791
New York Times Magazine	19,572
Psychology Today	17,689
Sports Illustrated	17,893
Scientific American	21,701
†Fortune	25,916
†Barrons	25,834
Harpers	18,725
National Geographic	18,437
††Newsweek	17,759
New Yorker	18,409
†Money	22,327
††U.S. News & World Report	18,433
†Business Week	22,877
Saturday Review	18,469

(a) Calculate the mean and standard deviation of this data set.

(b) Does the mean represent the average income of all readers of these publications? Explain.

(c) What are the z scores for the six financial publications (denoted with a †)?

(d) What are the z scores for the major weekly newsmagazines (denoted with a ††)?

(e) Based on your answers to parts (c) and (d), what can you conclude about the incomes of readers of financial publications compared with the incomes of readers of newsmagazines?

2-67 Samples A and B are shown in the following diagram. Notice that the two samples are the same with one exception, the 8 in A has been replaced by a 9 in B.

$$A: \underline{2 \quad . \quad 4 \quad \tfrac{5}{5} \quad . \quad 7 \quad 8}$$

$$B: \underline{2 \quad . \quad 4 \quad \tfrac{5}{5} \quad . \quad 7 \quad . \quad 9}$$

What effect does changing the 8 to a 9 have on each of the following statistics? Explain why.

(a) mean (b) median (c) mode
(d) midrange (e) range (f) variance
(g) standard deviation

2-68 The following figures are the "average family income needed to live moderately" in selected cities:

Atlanta	$20,797	Dallas–Fort Worth	$20,219
Baltimore	22,439	Houston	21,028
Boston	27,029	Los Angeles	21,954
Buffalo	23,393	San Diego	22,185
Chicago	22,717	Seattle–Everett	23,392

(a) Determine the median city on this list relative to family income.
(b) List the cities in which family incomes fall below the first quartile for this list.

2-69 The FLC accounting department is interested in the age of its delinquent accounts in order to determine when an account should be considered a bad debt. The accompanying table is a frequency distribution of age, x (in years), of delinquent accounts at FLC Finance Corporation.

x	1	2	3	4	5	6	7	8	9	10	11
f	4	9	11	8	7	5	4	2	2	0	1

(a) Draw a histogram of the data.
(b) Find the four measures of central tendency.
(c) Find Q_1 and Q_3.
(d) Find P_{15} and P_{12}.
(e) Find the three measures of dispersion (range, s^2, s).

2-70 A bank installed a single-line, multichannel service process whereby all the customers get in on one line, which feeds to all five tellers. To see how well the process is working, the bank collected data on the length of time 75 customers had to wait for service. The data are presented in the accompanying table.

(a) Draw a histogram that depicts the distribution.
(b) Find the mean.
(c) Find the standard deviation.
(d) An earlier method used by the bank, a separate line for each teller, had a mean waiting time of 8.4 minutes with a standard deviation of

5 minutes. How would you compare the effectiveness of the new process? Explain.

Time Waiting (minutes)	Frequency
1–3	10
4–6	14
7–9	27
10–12	10
13–15	8
16–18	6

2-71 The treasurer of Fillmore Corporation is interested in examining the accounts receivable of the corporation. To understand how long it generally takes for money to be received after billing, she randomly selected 100 invoices and recorded the time (in days) from invoice date to receipt of payment. The accompanying table presents the data.

Days	Frequency
0–8	17
9–17	12
18–26	8
27–35	12
36–44	6
45–53	16
54–62	8
63–71	6
72–80	4
81–89	11

(a) Construct a relative frequency histogram that shows the payment time of the 100 invoices.

(b) Construct an ogive that shows the cumulative distribution of the payment times.

(c) Calculate the mean of the payment times.

(d) Calculate the standard deviation of the payment times.

2-72 Forty students recently took an hour exam in statistics. The exam scores are as follows:

```
58  64  72  68  86  82  92  88  66  64
54  60  66  72  72  78  88  94  68  50
68  76  72  80  84  90  98  56  80  62
82  86  66  58  68  90  86  72  64  88
```

(a) Construct a stem-and-leaf display picturing the data.

(b) Find the mode using the 40 individual scores.

(c) Construct the frequency distribution of these scores using 50–54, 55–59, 60–64, and so forth as classes.

(d) Construct a frequency histogram of the 40 scores.

(e) Is the distribution shown on your histogram bimodal?

(f) If the distribution is bimodal, which two classes have the greatest frequencies?

(g) What does it mean for a class to be modal? A distribution to be bimodal?

(h) Compare your answers for (e), (f), and (g) with your answer to (b) and explain the difference between the "mode of a set of data" and the "modal class(es) in a frequency distribution."

2-73 Earnings per share for 40 firms in the radio and transmitting equipment industry are listed as follows:

4.62	0.25	1.07	5.56	0.10	1.34	2.50	1.62
1.29	2.11	2.14	1.36	7.25	5.39	3.46	1.93
6.04	0.84	1.91	2.05	3.20	−0.19	7.05	2.75
9.56	3.72	5.10	3.58	4.90	2.27	1.80	0.44
4.22	2.08	0.91	3.15	3.71	1.12	0.50	1.93

(a) Prepare a frequency distribution and a frequency histogram for this data.

(b) Which class of your frequency distribution contains the median?

(c) Find the mean, median, and mode for the earnings based on your frequency distribution.

(d) Find the variance and the standard deviation based on the frequency distribution.

(e) Find the mean, median, mode, variance, and standard deviation using the 40 ungrouped data.

(f) Compare the results of (e) with those obtained in (c) and (d).

2-74 The purchase price of some mutual funds includes what is called a *load*. A load, in reality, is the commission the sales agent receives for handling the mutual fund transaction. The following data represent the load for 100 mutual funds (in cents):

41	76	10	53	36	86	42	81	104	77
44	25	68	90	94	90	125	25	32	13
87	161	69	60	72	106	77	25	39	80
102	68	46	55	41	87	60	74	71	97
89	47	82	28	104	103	40	13	77	72
80	55	85	38	99	72	74	31	61	43
77	63	53	41	48	53	98	42	44	45
60	89	138	23	61	64	69	50	198	68
20	75	57	53	55	102	80	72	98	66
40	162	41	54	28	33	12	85	79	80

(a) Construct a stem-and-leaf display.

(b) Construct a grouped frequency distribution that uses a class width of 16. Start with 8.5 as the lower class boundary of the first class.

(c) Construct a histogram of the frequency distribution.

(d) Construct a cumulative relative frequency distribution.

(e) Construct an ogive of the cumulative relative frequency distribution.

(f) Find the mean of the frequency distribution.

(g) Find the standard deviation of the frequency distribution.

(h) Find the standard scores for loads of 124, 54, and 87.

(i) Find the loads that have z scores of 1.7, -2.3, and 0.7.

(j) What loads are equal to the following: $\bar{x} + s$, $\bar{x} - s$, $\bar{x} + 2s$, $\bar{x} - 2s$, $\bar{x} + 3s$, and $\bar{x} - 3s$?

(k) What percentage of the data lies between $\bar{x} - s$ and $\bar{x} + s$? Between $\bar{x} - 2s$ and $\bar{x} + 2s$? Between $\bar{x} - 3s$ and $\bar{x} + 3s$? (Refer back to the original data.)

(l) Compare the answers in part (k) with Chebyshev's theorem.

(m) Compare the answers in part (k) with the empirical rule.

(n) Is the distribution approximately normal? Explain.

2-75   The rates of return (in percentages) of 100 investments made by Pennline Investment Trust Company last year were as follows:

15.0	15.3	14.4	10.4	10.2	11.5	15.4	11.7	15.0	10.9
13.6	10.5	13.8	15.0	13.8	14.5	13.7	13.9	12.5	15.2
10.7	13.1	10.6	12.1	14.9	14.1	12.7	14.0	10.1	14.1
10.3	15.2	15.0	12.9	10.7	10.3	10.8	15.3	14.9	14.8
14.9	11.8	10.4	11.0	11.4	14.3	15.1	11.5	10.2	10.1
14.7	15.1	12.8	14.8	15.0	10.4	13.5	14.5	14.9	13.9
10.1	14.8	13.7	10.9	10.6	12.4	14.5	10.5	15.1	15.8
12.0	15.5	10.8	14.4	15.4	14.8	11.4	15.1	10.3	15.4
15.0	14.0	15.0	15.1	13.7	14.7	10.7	14.5	13.9	11.7
15.1	10.9	11.3	10.5	15.3	14.0	14.6	12.6	15.3	10.4

Answer the questions asked in exercise 2-74. Use 10.0–10.4 as the first class for your grouped distribution. Use 11.2, 12.4, and 15.2 to answer part (h).

2-76   The bad debt ratio of commercial loans of 250 banks is distributed as shown in the accompanying table.

Bad Debt Ratio	Frequency
0.03115–0.03117	2
0.03118–0.03120	6
0.03121–0.03123	8
0.03124–0.03126	15
0.03127–0.03129	42
0.03130–0.03132	68
0.03133–0.03135	49
0.03136–0.03138	25
0.03139–0.03141	18
0.03142–0.03144	12
0.03145–0.03147	4
0.03148–0.03150	1
Total	250

(a)   Construct a histogram of this distribution. Use relative frequency.
(b)   Find the mean bad debt ratio.
(c)   Find the standard deviation of the ratios.
(d)   Can we compute the bad debt ratio for all the commercial loans granted by the 250 banks? Explain.

2-77 In an attempt to study merchandise return patterns, a major men's clothing store chain collected the data shown in the accompanying table. The data included (1) the value of each customer's returned merchandise, (2) the number of days elapsed between the sale and the return, and (3) whether the sale was cash or charge.

Amount of Return (whole dollars)	Number of Days Elapsed	Cash (C) or Charge (X)	Amount of Return (whole dollars)	Number of Days Elapsed	Cash (C) or Charge (X)
66	9	X	96	12	X
71	4	C	97	13	X
71	12	X	97	20	X
72	6	C	97	12	X
73	13	X	97	4	C
74	10	X	98	17	X
76	8	X	98	17	X
77	19	X	98	5	C
80	10	X	98	11	X
81	1	C	98	6	C
81	16	X	99	11	X
83	13	X	99	8	C
83	9	X	99	10	X
84	14	X	99	13	X
85	11	C	100	9	C
85	7	X	100	16	X
85	10	C	101	5	X
86	5	X	101	10	X
86	8	C	101	13	C
88	8	X	102	11	X
88	5	X	102	13	X
89	15	C	102	7	C
89	8	X	102	12	X
90	9	C	103	11	C
91	12	X	103	1	X
92	3	X	103	6	C
92	10	X	104	15	X
92	16	C	105	15	X
92	17	X	106	5	X
93	7	C	106	4	C
93	18	X	107	9	X
95	4	X	107	7	X
95	13	C	107	11	X
96	10	X	108	11	X
96	1	X	110	4	C

(Continued)

Amount of Return (whole dollars)	Number of Days Elapsed	Cash (C) or Charge (X)	Amount of Return (whole dollars)	Number of Days Elapsed	Cash (C) or Charge (X)
110	14	X	118	15	X
111	10	X	118	10	C
111	21	X	119	21	X
112	21	X	119	11	C
112	13	X	121	16	C
112	8	C	122	12	X
113	5	X	123	10	X
113	14	X	126	14	X
114	9	C	126	18	X
114	2	C	127	7	C
115	12	X	130	14	X
117	1	C	136	1	C

For the amount of return, find each of the following:
(a) the mean
(b) the median
(c) the mode
(d) the midrange
(e) the range
(f) Q_1 and Q_3
(g) P_{35} and P_{64}

2-78 From the data in exercise 2-77, for the number of days elapsed between purchase and return, find the following:
(a) the mean
(b) the median
(c) the mode
(d) the range
(e) Q_1 and Q_3
(f) P_{10} and P_{95}

2-79 An advertising agency conducted a marketing survey of 50 randomly selected males at the request of an airline client. Each person was asked the following: (1) Do you prefer an "all news" radio station to a primarily music station? (2) Income (in thousands of dollars per year). (3) Age. (4) Number of nonbusiness, round-trip flights taken in the past year. The results of the survey are given in the accompanying table.

Person	Station Preference (N = news, M = music)	Income (rounded, thousands of dollars)	Age	Number of Flights
1	N	10	23	0
2	M	42	55	7
3	N	39	38	7
4	N	27	42	1
5	M	20	33	0
6	M	8	39	1
7	N	27	43	3
8	M	22	49	3
9	N	15	27	0
10	M	17	25	1
11	M	26	42	2
12	N	24	36	0
13	N	21	41	0
14	N	25	54	1
15	M	23	34	1
16	M	9	40	0
17	M	25	35	1
18	N	30	31	4
19	M	28	41	3
20	M	25	39	2
21	M	9	21	0
22	N	29	38	2
23	N	41	52	6
24	M	23	33	0
25	N	20	43	1
26	M	20	36	1
27	M	30	48	0
28	N	17	34	0
29	M	28	38	3
30	M	16	42	0
31	M	19	41	0
32	M	24	46	1
33	M	35	52	3
34	N	31	49	2
35	N	33	46	0
36	M	14	28	0
37	M	36	49	5
38	N	28	45	2
39	M	25	43	1
40	M	33	44	4
41	N	36	51	6
42	M	23	31	0

(Continued)

Person	Station Preference (N = news, M = music)	Income (rounded, thousands of dollars)	Age	Number of Flights
43	M	30	45	4
44	N	30	41	2
45	M	15	27	0
46	N	33	45	5
47	N	29	39	2
48	M	21	29	1
49	M	15	30	0
50	M	22	39	1

(a) Rank each of the three sets of variable data (income, age, and number of flights).
(b) Find the median of each of the three variables.
(c) Find the quartiles of each of the three variables.
(d) Find the 22d and 57th percentiles of each of the three variables.
(e) What kind of data is represented by each of the four responses?
(f) What proportion of the persons prefer all news? Music?

2-80 When an investment firm underwrites (guarantees the sale of) a new bond issue, it receives as commission a portion of the total amount received (gross proceeds) from the sale of the bonds. This portion is called the *spread*. The gross proceeds, the bond rating (Aaa is the best; Aa is the next highest rating, etc.), and the spread for 40 recent bond issues are given in the accompanying table.

Rating	Gross Proceeds (millions of dollars)	Spread (millions of dollars)	Rating	Gross Proceeds (millions of dollars)	Spread (millions of dollars)
Aaa	100	2.0	Aaa	105	2.1
Aaa	120	2.5	Aaa	100	2.1
Aaa	90	1.8	Aa	105	3.2
Aa	55	1.6	Aaa	65	1.4
Aaa	135	2.9	Aaa	50	1.0
Aaa	105	2.0	Aa	120	3.8
Aaa	110	2.2	Aaa	85	1.8
Aa	135	3.9	Aaa	100	2.0
Aaa	115	2.0	Aa	105	3.0
Aa	120	3.6	Aa	105	3.2
Aaa	60	1.0	Aaa	85	1.9
Aa	110	3.1	Aaa	95	1.9

(Continued)

Rating	Gross Proceeds (millions of dollars)	Spread (millions of dollars)	Rating	Gross Proceeds (millions of dollars)	Spread (millions of dollars)
Aa	80	2.4	Aaa	70	1.3
Aaa	105	2.1	Aa	70	2.1
Aaa	120	2.3	Aa	105	2.1
Aaa	95	2.0	Aaa	75	1.6
Aaa	100	2.0	Aa	90	2.8
Aa	75	2.4	Aa	100	3.0
Aaa	85	1.9	Aaa	130	2.6
Aa	130	3.6	Aa	95	3.9

(a) Find the mean gross proceeds per bond issue.
(b) Find the mean spread per bond issue.
(c) Find the mean gross proceeds per Aaa bond issue.
(d) Find the mean gross proceeds per Aa bond issue.
(e) Find the mean spread per Aaa bond issue.
(f) Find the mean spread per Aa bond issue.
(g) Compare the results of parts (c) through (f). Explain what conclusions can be made concerning the different types of bonds.

2-81 Convertible bonds can be exchanged for an equal dollar amount of common stock by their owners. A corporation conducted a study of the voluntary conversion of convertible bond issues. It collected data (shown in the accompanying table) on 15 bond issues. The data included (1) the percentage converted of the outstanding bonds, (2) the income difference between the dividends received from stock following conversion and the interest received on the bond, (3) the volatility of the market price of the company's common stock (a measure of the risk of holding the common stock), and (4) whether the bond had scheduled periodic reductions in its terms of conversion.

Percentage Converted of Outstanding Bonds	Difference	Index/Stock Price Volatility	Periodic Reduction in Conversion Terms
25	24.50	1.25	No
10	16.75	1.08	No
75	33.00	0.95	No
5	−12.50	1.12	No
55	18.50	0.87	No
42	7.75	1.04	Yes
80	18.25	0.91	Yes

(Continued)

Percentage Converted of Outstanding Bonds	Difference	Index/Stock Price Volatility	Periodic Reduction in Conversion Terms
1	−22.50	0.88	No
70	37.50	0.97	No
75	28.75	1.05	Yes
20	7.50	0.85	No
35	−16.25	1.09	Yes
40	18.00	0.93	No
18	24.25	1.18	No
28	−13.50	1.09	Yes

(a) Find the mean for each of the first three variables.

(b) Find the standard deviation for each of the first three variables.

(c) Calculate the mean percentage converted for stocks with a periodic reduction in conversion terms.

(d) Calculate the mean percentage converted for stocks without a periodic reduction in conversion terms.

(e) Calculate the standard deviation of percentage converted separately for those stocks with and without a periodic reduction in conversion terms.

(f) Calculate the coefficient of variation for percentage converted for those stocks with and without a periodic reduction in conversion terms.

(g) Based on your answers to parts (c) through (f), what can you say about the effect of a periodic reduction in conversion terms?

2-82 Use the ages from the market survey in exercise 2-79.
(a) Construct a stem-and-leaf display.
(b) Form a frequency distribution of all 50 ages using 19.5 to 25.5 as the first class.
(c) Construct a histogram showing the 50 ages.
(d) Calculate the mean and standard deviation of the ages using the frequency distribution of part (b). [Retain these solutions for use in answering exercises 2-88(a) and 2-89.]
(e) Separate the data into two subgroups on the basis of station preference (music or news). Form frequency distributions of the ages for each using the same set of classes.
(f) Construct a histogram of each subgroup on the same graph (they will overlap; use two colors).
(g) Calculate the mean and the standard deviation of the ages for each subgroup using the frequency distribution of part (e).

(h) Do the results of parts (a) through (e) indicate that the news and music stations are two distinct populations? Explain.

2-83 Use the sales return figures given in the study in exercise 2-77 to answer the questions in exercise 2-82. Use 60–70 for your first class.

2-84 Use the spread data given in the investment study of exercise 2-80.
(a) Construct a stem-and-leaf display.
(b) Construct a frequency distribution.
(c) Construct a frequency histogram.
(d) Calculate the mean and standard deviation of the data.

2-85 What proportion of a distribution is guaranteed, by Chebyshev's theorem, to fall within these intervals?
(a) $\bar{x} - 2s$ to $\bar{x} + 2s$
(b) $\bar{x} - 3s$ to $\bar{x} + 3s$

2-86 What proportion of a sample can be expected to be found within the following pairs of bounds when the sample is taken from a normal population?
(a) $\bar{x} - s$ and $\bar{x} + s$
(b) $\bar{x} - 2s$ and $\bar{x} + 2s$
(c) $\bar{x} - 3s$ and $\bar{x} + 3s$

2-87 Use the values calculated for the mean and standard deviation in exercise 2-69.
(a) Find $\bar{x} + s$, $\bar{x} + 2s$, $\bar{x} + 3s$, $\bar{x} - s$, $\bar{x} - 2s$, and $\bar{x} - 3s$.
(b) What proportion of the sample lies between $\bar{x} - s$ and $\bar{x} + s$? Between $\bar{x} - 2s$ and $\bar{x} + 2s$? Between $\bar{x} - 3s$ and $\bar{x} + 3s$?
(c) Compare the answers of part (b) with Chebyshev's theorem.
(d) Compare the answers of part (b) with the empirical rule.

2-88 Repeat exercise 2-87 with reference to each of the following sets of data:
(a) ages (exercise 2-82d, data in exercise 2-79)
(b) sales returns (exercise 2-83, data in exercise 2-77)
(c) spread (exercise 2-84, data in exercise 2-80)

2-89 Find the z score (standard score) associated with the value of each response for age obtained in the market survey in exercise 2-79. Use the values calculated in exercise 2-82d for the mean and standard deviation.

2-90 The number of units of a product sold by the salesmen of a firm fluctuates as the seasons change. For July the mean number of units sold per salesman has been 900 with a standard deviation of 94; whereas for December the number of units sold has a mean of 330 with a standard deviation of 40. John Scranton sold 1,080 units in July and 450 in December. In which month did he have the better relative performance?

2-91 The following information was generated from a study of stocks listed on the New York Stock Exchange and the American Stock Exchange:

	New York	American
Number of stocks	24	25
Mean 52-week price difference	$5.57	$11.79
Standard deviation	3.57	10.55

(a) At least 89% of the American Stock Exchange stocks have 52-week price differences within what range?

(b) At least 75% of the New York Stock Exchange stocks have 52-week price differences within what range?

2-92 The clean-up crew of a medium-sized firm has an average clean-up time of 84.0 total hours with a standard deviation of 6.8 hours. Assuming that the empirical rule is appropriate,

(a) What proportion of the time will it take the clean-up crew 97.6 or more hours for cleaning the plant?

(b) Ninety-five percent of the time the total clean-up time will fall within what interval?

Hands-On Problems

2-1 (a) Go to a newspaper and, beginning with the 5th stock listed on the New York Exchange and every 50th stock thereafter, record the closing price, rounded to the nearest dollar.

(b) Go to a newspaper and, beginning with the 3d stock listed on the American Exchange and every 25th stock thereafter, record the closing price, rounded to the nearest dollar.

2-2 Calculate separately for the stocks on each exchange the mean, median, range, 1st and 3d quartiles, 10th and 90th percentiles, the standard deviation, and the coefficient of variation.

2-3 Compare your results in problem 2-2 for each stock exchange. What can you say about the difference in the exchanges based on your sample data?

Use the sample stock price data collected in problem 2-1 to solve problems 2-4 through 2-8. Do each stock exchange separately.

2-4 Form a grouped frequency distribution.

2-5 Construct a histogram. Be sure to label it completely.

2-6 Construct a relative frequency ogive. Be sure to label it completely.

2-7 (a) Find the values of $\bar{x} + s$, $\bar{x} + 2s$, $\bar{x} + 3s$, $\bar{x} - s$, $\bar{x} - 2s$, and $\bar{x} - 3s$.

(b) Use the set of raw data collected in problem 2-1 to determine the percentage of your sample that falls within each of the following intervals: $\bar{x} - 2s$ to $\bar{x} + 2s$, and $\bar{x} - 3s$ to $\bar{x} + 3s$.

(c) Compare these values with values obtained by using Chebyshev's theorem.

(d) Use the empirical rule to check for closeness to a normal distribution.

2-8 Concluding statements: Comment interpretatively on these questions:

(a) What type of distribution does your histogram suggest?

(b) Does the histogram suggest the type of distribution that you would expect your population to have? Explain.

2-9 As an alternative to problems 2-1 through 2-8, read Darrell Huff and Irving Geis, *How to Lie with Statistics* (New York: Norton, 1954). Write a report describing how some aspect of the book relates to some specific personal experience. Be specific.

The calculations for this problem set can most easily be accomplished with the assistance of an electronic calculator or a packaged program on a computer. A list of the available programs can be obtained from your computer center. There are a variety of packaged programs available: Minitab, Biomed (Biomedical Programs), SAS (Statistical Analysis System), IBM Scientific Subroutine Packages, and SPSS (Statistical Package for the Social Sciences) program libraries. Your local computer center will assist you.

3 Descriptive Analysis and Presentation of Bivariate Data

Chapter Outline

3-1 Bivariate Data: Cross-Tabulation and Scatter Diagrams

Bivariate data—two sets of data individually paired together for analysis. Can be displayed by using cross-tabulation tables and scatter diagrams.

3-2 Linear Correlation

Does an increase in the value of one variable indicate a change in the value of a second variable?

3-3 Linear Regression

The statistician seeks a mathematical expression or equation for the relationship between two variables.

Agricultural Imports Related to Income, by Regions of the World

It is sometimes said that when we help foreign countries to become more efficient, they expand their own output and import less. Thus, it is said that some of our efforts to help developing countries tend to undercut our own markets. Also, when our own imports of food increase, many people say that it is doing great damage to our farmers.

Probably no sweeping dicta would hold good for all situations of this kind. But there is good evidence that most countries tend to increase their imports as their incomes go up, and to decrease their imports as their incomes go down. This is true of agricultural imports as well as of total imports.

The chart is patterned after one ... which shows the average per capita agricultural imports for each of 11 major regions of the world.

Looking at the 11 regions plotted on the diagram, it is clear that there is a close and positive relationship between agricultural imports and income. The relationship seems to be almost linear....

Thus, if we can effectively help developing countries to modernize their agriculture and industry and to increase their incomes, we can expect them to import more goods, including more agricultural products....

The diagram shows a high correlation between agricultural imports and income. This is indicated by the fact that the observed data lie very close to the regression line....

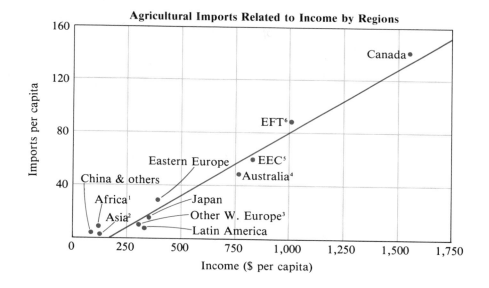

[1] Africa excludes Republic of South Africa.
[2] Asia excludes Japan, USSR, China, N. Korea, and N. Viet Nam.
[3] Other W. Europe includes Finland, Greece, Ireland, Iceland, Spain, Turkey, and Yugoslavia.
[4] Australia includes New Zealand and Republic of S. Africa.
[5] EEC includes Belgium, Luxembourg, France, Italy, W. Germany, and the Netherlands.
[6] EFT includes United Kingdom, Denmark, Sweden, Switzerland, Austria, and Portugal.

From Frederick V. Waugh, *Graphic Analysis: Applications in Agricultural Economics*, Agriculture Handbook No. 326, prepared for U.S. Department of Agriculture, Economic Research Service (Washington, D.C.: Government Printing Office, 1966), pp. 36–37.

Chapter Objectives

In the field of business, many problems require the analysis of more than one variable. We often try to obtain answers to the following questions: Are these two variables related? If so, how are they related? Are these variables correlated? The relationships discussed are not cause-and-effect relationships. They are only mathematical relationships that predict the behavior of one variable from knowledge about a second variable.

Now let's look at a few specific illustrations.

Illustration 3-1

As earnings per share of a stock increase, the stock price usually advances. The question can be asked: Is there a relationship between earnings per share and stock prices? ☐

Illustration 3-2

As the price of a consumer good increases, sales normally decrease. Total revenue, however, is equal to price times sales. Since the price goes up and sales go down, unless we know the exact amount of the decrease in sales, we cannot tell whether total revenue will decrease or increase when the price is raised. ☐

Illustration 3-3

Marketing departments often aim their advertising at certain income groups. Is it true that certain income groups are more likely to purchase some products? ☐

Illustration 3-4

A personnel department would like to predict the future success of a sales applicant. The predicted value of sales volume is based on personal traits such as ambition, intelligence, prior sales experience, and so on. ☐

These illustrations all require an analysis of the relationship between two or more variables. In this chapter we will explore the graphic and tabular presentation of data when two variables are studied. Scatter diagrams will replace histograms, and contingency, or cross-tabulation, tables will replace frequency distributions. We will also take a first look at the techniques of linear correlation and regression analysis, which are used to measure the relationship between variables. Since regression and linear correlation analyses are so important in business, we will spend more time with these techniques in chapters 13, 14, and 15.

The basic objectives of this chapter are (1) to become familiar with several forms of descriptive presentation of two variables, (2) to *understand the basic idea of regression and linear correlation, and (3) to be able to distinguish between the basic purposes of linear correlation and of regression analysis.*

Section 3-1 Bivariate Data: Cross-Tabulation and Scatter Diagrams

bivariate data
ordered pair

The term **bivariate data** is used to describe two different pieces of data that are paired. Expressed mathematically, bivariate data comprise **ordered pairs**—let's call them x and y, where x is the value of the first variable and y is the value of the second variable. We write an ordered pair as (x, y). The data are said to be *ordered* because one value, x, is always written first. They are said to be *paired* because for each x value there is a corresponding y value. For example, if x is height and y is weight, a height and corresponding weight are recorded for each person. If y is monthly sales and x is advertising expenditures, a sales figure and corresponding advertising expenditure are reported for each month.

independent and dependent variable

It is customary to call the **independent variable** (sometimes called *predictor* variable) x and the **dependent variable** (sometimes called the *response* variable) y. The independent variable, x, is measured or controlled to predict the dependent variable, y. For example, in illustration 3-2 the company controls the price; thus price is the independent variable x. In illustration 3-1 we would expect earnings to affect stock prices, not stock prices to affect earnings; earnings per share is the independent variable x, and stock price is the dependent variable y. In the case of height and weight, either variable could be the independent or dependent variable, depending on the question being asked. If we want to predict a person's height on the basis of his or her weight, weight is the independent variable and height is the dependent variable; if we want to predict weight on the basis of height, the reverse is the case.

Cross-Tabulations

cross-tabulation, or contingency, table

Bivariate data can be either variable or attribute data, or a mixture of both. Bivariate data are presented in tabular form in **cross-tabulation**, or **contingency, tables**. These tables are the bivariate analogy, or counterpart, to the single-variable frequency distribution. The contingency table format is illustrated in table 3-1.

In a contingency table the choice of whether the rows or columns are labeled x or y is up to you. The 1 through 5 under the x and 1 through 6 under the y in table 3-1 represent the classes of each variable. If the variable is an attribute, the classes are categories. For example, if a variable is a

TABLE 3-1
Contingency, or Cross-Tabulation, Table

	x					
y	1	2	3	4	5	Sum
1						
2						
3						
4						
5						
6						
Sum						Total number of observations

product's quality, the classes might be "good" and "defective." If the variable is stock movement, the classes might be "up," "down," and "unchanged."

If one or both of the variables are quantitative and there are many pieces of data taking on a wide range of values, we can group the data into classes, similar to the grouped frequency distributions we discussed in chapter 2. The classes are found by the rules for frequency distributions outlined in chapter 2.

cell Each set of (x, y) classes is referred to as a **cell**. If there are a classes of x and b classes of y, the table will have $a \times b$ cells, or $a \times b$ combinations, of x and y values.

The far right column and last row are labeled *sum*. These categories contain the total number of observations in each column and row and are **marginal total** referred to as the **marginal totals** of x and y. The value at the bottom far right is the total number of paired observations.

Illustration 3-5

The marketing department of Vita Vitamins, Inc. is interested in determining if there is a relationship between the age of a prospective customer and the likelihood that he or she will purchase vitamins. People in a random sample of 100 persons were asked their ages and whether they had ever purchased Vita Vitamins products. The raw data are given in table 3-2, and the contingency table is shown in table 3-3.

TABLE 3-2
Raw Data for Vita Vitamins, Inc. Survey

Person	Age	Purchased	Person	Age	Purchased	Person	Age	Purchased
1	43	No	35	19	No	68	66	Yes
2	58	Yes	36	59	Yes	69	51	Yes
3	34	No	37	32	No	70	44	No
4	66	No	38	46	Yes	71	25	No
5	46	Yes	39	53	No	72	62	Yes
6	52	Yes	40	31	No	73	30	No
7	18	No	41	50	Yes	74	36	No
8	54	No	42	52	No	75	35	Yes
9	39	Yes	43	68	Yes	76	24	No
10	46	No	44	35	No	77	65	Yes
11	62	Yes	45	32	Yes	78	48	No
12	40	No	46	37	No	79	46	Yes
13	47	Yes	47	39	No	80	28	No
14	64	No	48	28	No	81	41	No
15	21	No	49	68	No	82	67	Yes
16	42	No	50	64	Yes	83	26	No
17	19	No	51	19	No	84	40	No
18	69	Yes	52	57	Yes	85	57	No
19	59	No	53	45	No	86	43	No
20	54	Yes	54	22	No	87	71	Yes
21	48	No	55	63	Yes	88	47	No
22	49	Yes	56	35	No	89	26	Yes
23	42	Yes	57	29	No	90	36	No
24	49	No	58	51	No	91	60	No
25	61	Yes	59	43	Yes	92	55	Yes
26	56	Yes	60	30	No	93	44	Yes
27	44	No	61	70	Yes	94	23	No
28	60	Yes	62	21	No	95	55	No
29	48	Yes	63	47	No	96	24	No
30	50	No	64	59	No	97	68	Yes
31	37	Yes	65	34	No	98	38	No
32	35	No	66	31	No	99	45	Yes
33	70	Yes	67	27	No	100	53	Yes
34	43	No						

TABLE 3-3
Contingency Table for
Vita Vitamins Data

	Age of Potential Customer							
Purchased Product?	18–25	26–33	34–41	42–49	50–57	58–65	66–73	Sum
Yes	0	2	3	10	8	9	8	40
No	11	10	13	13	7	4	2	60
Sum	11	12	16	23	15	13	10	100

Instead of reporting the actual number of occurrences, we can present the relative frequencies of occurrence. Recall that the relative frequency is simply the frequency divided by the number of observations. To convert from frequency to relative frequency, divide each cell count by the total number of observations. The sum of the relative frequencies must equal 1, since the frequencies of occurrence of all events are totaled.

Illustration 3-6

A contingency table of the relative frequencies of customer age by purchase pattern given in illustration 3-5 is presented in table 3-4.

TABLE 3-4
Relative Frequency Contingency
Table for Vita Vitamins Data

	Age of Potential Customer							
Purchased Product?	18–25	26–33	34–41	42–49	50–57	58–65	66–73	Sum
Yes	0.00	0.02	0.03	0.10	0.08	0.09	0.08	0.40
No	0.11	0.10	0.13	0.13	0.07	0.04	0.02	0.60
Sum	0.11	0.12	0.16	0.23	0.15	0.13	0.10	1.00

Scatter Diagrams

scatter diagram

In situations where both x and y are *variable data*, the sample data can be displayed pictorially in a **scatter diagram**. A scatter diagram plots the ordered pairs of bivariate data on a coordinate axis system. The independent variable x is plotted on the horizontal axis, and the dependent variable y is plotted on the vertical axis.

NOTE: When constructing a scatter diagram, it is convenient to construct scales so that the range of data along the vertical axis is equal to or slightly shorter than the range along the horizontal axis.

Illustration 3-7

A marketing department selected a sample of 10 salespeople and recorded their sales y and entertainment expenses x. The entertainment expenditures, measured in hundreds of dollars, and sales, measured in thousands of dollars, for each salesperson were

$$(27, 30) \quad (22, 26) \quad (15, 25) \quad (35, 36) \quad (33, 33)$$
$$(52, 36) \quad (35, 32) \quad (40, 54) \quad (40, 50) \quad (40, 43)$$

The data (see the following table) have been plotted in the scatter diagram in figure 3-1a.

Salesperson	1	2	3	4	5	6	7	8	9	10
Entertainment Expense, x	27	22	15	35	33	52	35	40	40	40
Sales, y	30	26	25	36	33	36	32	54	50	43

FIGURE 3-1a
Scatter Diagram

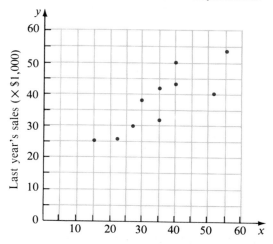

Sales Versus Entertainment Expenditures

Figure 3-1b shows the same scatter diagram as in figure 3-1a, except that it was computer generated.

FIGURE 3-1b
Computer-Generated Scatter Diagram

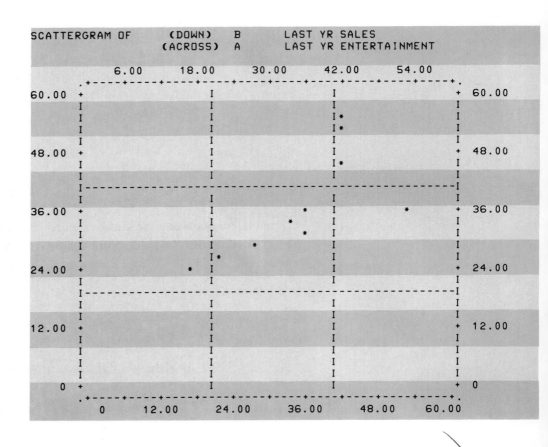

Exercises

3-1 Which of the following represents bivariate sets of data?

(a) the price of 10 stocks and the price of 10 bonds

(b) the sex and age of 100 customers

(c) a sample of 100 cars, where the quality of the car and the shift it was produced in is recorded

(d) a broker's stock transactions (in dollars) and percentage of commission on each transaction

3-2 Consider the accompanying contingency table, which presents the results of an advertising survey about the use of credit by Martan Oil Company customers.

	Number of Purchases at Gasoline Station Last Year					
Preferred Method of Payment	0–4	5–9	10–14	15–19	20 and Over	Sum
Cash	150	100	25	0	0	275
Oil company card	50	35	115	80	70	350
National or bank credit card	50	60	65	45	5	225
Sum	250	195	205	125	75	850

(a) How many customers were surveyed?
(b) How many customers preferred to use an oil company credit card?
(c) How many customers made 20 or more purchases last year?
(d) How many customers preferred to use an oil company credit card and made only between five and nine purchases last year?
(e) What does the 80 in the fourth cell in the second row mean?

3-3 Use the data in exercise 2-80 to construct a contingency table of the rating and spread bivariate data.

3-4 Use the data in exercise 2-80 to construct a contingency table of the gross proceeds and spread bivariate data.

3-5 Consider the data in the accompanying table, which give the views of 20 managers and laborers concerning whether there will be a strike when the firm's labor contract expires. Construct a contingency table of these data.

Person	Management (M) or Labor (L)	Strike? (yes/no)	Person	Management (M) or Labor (L)	Strike? (yes/no)
1	M	No	11	L	Yes
2	M	No	12	L	Yes
3	M	No	13	L	No
4	M	Yes	14	L	No
5	M	Yes	15	L	Yes
6	M	No	16	L	Yes
7	M	No	17	L	Yes
8	M	No	18	L	Yes
9	M	Yes	19	L	Yes
10	M	No	20	L	No

3-6 Plot a scatter diagram of the following data. (Retain for use in answering exercise 3-16.)

x	2	12	4	6	9	4	11	3	10	11	3	1	13	12	14	7	2	8
y	4	8	10	9	10	8	8	5	10	9	8	3	9	8	8	11	6	9

3-7 A personnel department wants to determine whether the size of help-wanted advertisements for senior executives is related to the number of applicants. It collected the data shown in the accompanying table on the number of lines of advertising copy, x, and the number of applicants, y. Construct a scatter diagram using these data. (Retain this solution for use in answering exercise 3-11.)

x	2	3	3	4	5	5	6	6	7	7
y	7	7	8	8	9	8	9	10	9	10

3-8 A credit department wants to know if there is a relationship between the length of time a customer has had a credit card and payment delinquency. It collected the data shown in the accompanying table. Construct a scatter diagram of these data. (Retain this solution for use in answering exercise 3-12.)

Age of Account (years)	2	3	4	5	6	7	8	9	10	11
Percentage Delinquent (2 or more months)	14	11	11	9	8	8	4	5	2	1

3-9 Clark Corporation's economist collected the data shown in the accompanying table on the price and sales of 10 different brands of pens to study the price elasticity of pens (i.e., the relationship of price and sales). Construct a scatter diagram using these data. (Retain this solution for use in answering exercise 3-13.)

Price, x	0.39	0.59	0.79	0.99	1.29	1.59	2.29	2.59	2.99	3.99
Sales, y (× $1,000)	800	785	755	730	720	680	600	585	570	470

3-10 The sample data shown in the accompanying table were taken to determine if the average cost of production is related to the amount of product manufactured in a week. Construct a scatter diagram using these data. (Retain this solution for use in answering exercise 3-17.)

Week	1	2	3	4	5	6	7	8	9	10
Amount Produced	21	25	20	31	24	36	33	28	26	29
Average Production Cost per Item	52	59	48	70	55	85	78	66	59	69

Section 3-2 Linear Correlation

correlation analysis

The primary purpose of linear correlation analysis is to measure the strength of a linear relationship between two variables. Let's examine some scatter diagrams demonstrating different relationships between independent variables x and dependent variables y. If, as x increases there is no definite shift in the values of y, we say there is **no correlation**, or no relationship, between x and y. If as x increases there is a shift in the values of y, there is a **correlation**. The correlation is positive when y tends to increase and negative when y tends to decrease. If the values of x and y tend to follow a straight-line path, there is a **linear** correlation. The preciseness of the shift in y as x increases determines the strength of the linear correlation. The scatter diagrams in figure 3-2 demonstrate this idea.

linear correlation

FIGURE 3-2
Scatter Diagrams and Correlation

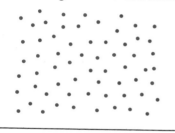
No correlation

Positive correlation

High positive correlation

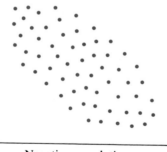
Negative correlation

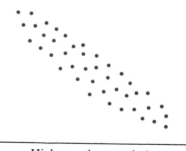
High negative correlation

Perfect linear correlation occurs when all the points fall exactly along a straight line, as shown in figure 3-3. This correlation can be either positive or negative, depending on whether y increases or decreases as x increases. If the data form a straight horizontal or vertical line, there is no correlation, since one variable has no effect on the other.

Scatter diagrams do not always appear in one of the five forms shown in figure 3-2. Sometimes they suggest relationships that are not linear. The

FIGURE 3-3
Perfect Positive Correlation

two variables in figure 3-4 are not related linearly, although a definite pattern is formed when they are plotted.

FIGURE 3-4
No Linear Correlation

coefficient of linear correlation

The **coefficient of linear correlation**, r, is a measure of the strength of the linear relationship between two variables. The coefficient reflects the consistency of the effect that a change in one variable has on the other.

The value of the linear correlation coefficient helps us to answer the question: Is there a linear correlation between the two variables under consideration? The linear correlation coefficient r always has a value between -1 and $+1$. A value of $+1$ signifies a perfect **positive correlation**, and a value of -1 shows a perfect **negative correlation**.

positive and negative correlation

If as x increases there is a general increase in the value of y, then r will indicate a positive linear correlation. For example, a positive value of r would be expected for age and height of children, because as children grow older, they grow taller. Also consider the age x and relative value y of an automobile. As the car ages, its value decreases. Since as x increases, y decreases, the correlation coefficient will have a negative value.

The value of r for a sample is obtained from the formula

$$r = \frac{\sum (x - \bar{x})(y - \bar{y})}{(n-1)s_x s_y} \qquad (3\text{-}1)$$

NOTE: s_x and s_y are the standard deviations of the x and y variables.

Pearson's product moment r

This formula is called **Pearson's product moment** r. The development of this formula is discussed in chapter 13.

To calculate r, we will use an alternative formula, which is equivalent to (3-1). We will calculate three separate sums of squares and then substitute them into formula (3-2) to obtain r.

$$r = \frac{SS(xy)}{\sqrt{SS(x)SS(y)}} \qquad (3\text{-}2)$$

where $SS(x) = \sum x^2 - [(\sum x)^2/n]$; from formula (2-9) on page 62,

$$SS(y) = \sum y^2 - \frac{(\sum y)^2}{n} \qquad (3\text{-}3)$$

and

$$SS(xy) = \sum xy - \frac{(\sum x)(\sum y)}{n} \qquad (3\text{-}4)$$

Illustration 3-8

Find the linear correlation coefficient of the data in illustration 3-7.

Solution First construct a table, as shown in table 3-5, listing all the pairs of values (x, y), and compute their extensions, x^2, xy, and y^2, and the column totals.

TABLE 3-5 Computations Needed to Calculate r for the Sales-Entertainment Expenses Data

Salesperson	Entertainment Expenses, x	x^2	Sales, y	y^2	xy
1	27	729	30	900	810
2	22	484	26	676	572
3	15	225	25	625	375
4	35	1,225	36	1,296	1,260
5	33	1,089	33	1,089	1,089
6	52	2,704	36	1,296	1,872
7	35	1,225	32	1,024	1,120
8	40	1,600	54	2,916	2,160
9	40	1,600	50	2,500	2,000
10	40	1,600	43	1,849	1,720
Total	339	12,481	365	14,171	12,978

Now the totals from the table are used to calculate r.
Using formula (2-9),

$$SS(x) = 12,481 - \frac{(339)^2}{10} = 988.9$$

Using formula (3-3),

$$SS(y) = 14,171 - \frac{(365)^2}{10} = 848.5$$

Using formula (3-4),

$$SS(xy) = 12,978 - \frac{(339)(365)}{10} = 604.5$$

Using formula (3-2),

$$r = \frac{604.5}{\sqrt{(988.9)(848.5)}} = 0.65992 = \mathbf{0.66}$$

NOTE: Typically, r is rounded to the nearest hundredth.

The value of the calculated linear correlation coefficient is supposed to help us answer the question: Is there a linear correlation between the two variables under consideration in the population from which the sample was drawn? When the calculated value of r is close to 0, we conclude that there is no linear correlation. When r is close to $+1$ or -1, we suspect that there is a linear correlation between the two variables.

Between 0 and 1.0 there is a value that marks the division point between concluding that there is or there is not a linear correlation. This point is called a **decision point**. A similar point is between 0 and -1.0; see figure 3-5. The value of the decision point is determined by the size of the

decision point

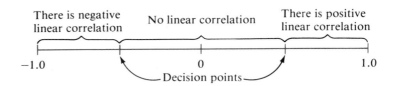

FIGURE 3-5
Decision Points for r

sample. Table 3-6 presents the positive decision points of various sample sizes from 5 to 100. Notice that the values for the decision points decrease as n, the sample size, increases. If r lies between the negative and the positive value of a decision point, we can assume there is no evidence of a linear relationship between the two variables. If r is either less than the negative value or more than the positive value of a decision point, we can conclude there is evidence of a linear relationship between the two variables.

TABLE 3-6
Decision Points in Determining
Linear Correlation for
Different Sample Sizes

n	Decision Point	n	Decision Point	n	Decision Point
5	0.878	14	0.532	26	0.388
6	0.811	15	0.514	28	0.374
7	0.754	16	0.497	30	0.361
8	0.707	17	0.482	40	0.312
9	0.666	18	0.468	50	0.279
10	0.632	19	0.456	60	0.254
11	0.602	20	0.444	80	0.220
12	0.576	22	0.423	100	0.196
13	0.553	24	0.404		

Returning to the solution for illustration 3-8 and the data in table 3-5, we find that $r = 0.66$ with a sample size of 10. Using the decision point from table 3-6, we conclude that there is a linear relationship between the variables.

To conclude that there is a relationship between two variables does not mean that it is a cause-and-effect relationship. This relationship must be determined by your understanding of the situation. For example, let x be installment credit interest rates and y the amount of installment credit over the last 10 years. A sample of such data would show a positive correlation. Yet high interest rates logically would decrease the amount of credit. So what we have here is simply an increase in both interest rates and the amount of credit over time. That is, an increase in one variable is not causing an increase in the other. Or consider the price of steel, x, and the price of corn, y, over the past 5 years. A sample of these data would yield a strong positive correlation. But clearly the price of corn does not affect the price of steel or vice versa. A third variable, inflation, has increased both x and y.

With a formula as complex as formula (3-2), it would be very convenient to inspect the scatter diagram of the data and be able to estimate the calculated value of r from the diagram. This procedure would serve as a check on the calculations. The following **method for estimating r** is quick and generally yields a reasonable estimate.

1. Draw an oval around the plot of points; see figure 3-6.

FIGURE 3-6
An Oval Is Drawn Around the Plotted Points

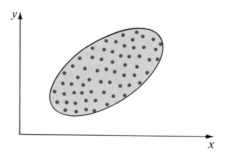

2. Measure the length of the maximum diameter D; see figure 3-7, where $D = 57$ millimeters.

FIGURE 3-7
The Maximum Diameter D

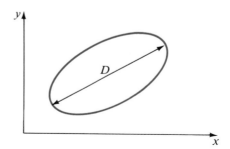

Linear Correlation Section 3-2

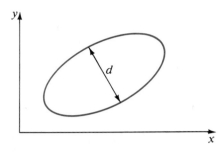

FIGURE 3-8
The Minimum Diameter d

3. Measure the length of the minimum diameter d; see figure 3-8, where $d = 28$ millimeters.

The value of r may be estimated as

$$\pm\left(1 - \frac{d}{D}\right)$$

The sign assigned to r may be determined by observing the general position of the maximum diameter. If the diameter lies in an increasing position (see figure 3-9), r will be positive; if it lies in a decreasing position, r will be negative. If the diameters are in horizontal or vertical positions, r will be 0.

Note that this estimating technique is very crude. It is very sensitive to the "spread" of the diagram. However, if the range of the x values and the range of the y values are approximately equal, the estimation will be helpful. **This technique should be used only as a mental check.**

FIGURE 3-9
Position of the Maximum Diameter

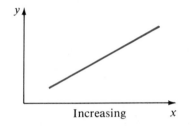
Increasing

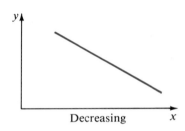
Decreasing

Let's estimate the value of the linear correlation coefficient r for the relationship between last year's sales and entertainment expenses from illustration 3-7. From figure 3-10 we find that $d = 28$ millimeters and $D = 57$ millimeters. Therefore our estimate for r is 0.5. It is positive because of the increasing position of the maximum diameter. The value is 0.5 because

$$1 - \frac{28}{57} \approx 0.5$$

FIGURE 3-10
Estimating r for the Sales-Entertainment Expense Data

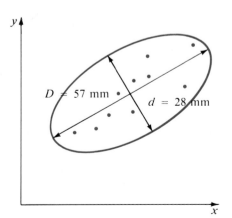

Exercises

3-11 (a) Use the scatter diagram constructed in answering exercise 3-7 to estimate r for the sample data relating number of lines of advertising copy and number of applicants.

    (b) Calculate r.

    (c) Does there seem to be a linear correlation? Explain.

3-12 (a) Use the scatter diagram constructed in answering exercise 3-8 to estimate r for the data involving the age and delinquency of credit accounts.

    (b) Calculate r.

3-13 (a) Use the scatter diagram constructed in answering exercise 3-9 to estimate r for the sample data.

    (b) Calculate the value of Pearson's product moment r.

3-14 (a) Calculate the Pearson product moment r for the data on production and costs in exercise 3-10.

    (b) Would you say, based on the calculation in part (a), that there is a relationship between the amount of production and average cost per item? Explain.

3-15 A marketing firm wished to determine whether the number of broadcasted television commercials is linearly correlated to the sales of its product. The data given in the accompanying table were obtained for each of several cities.

City	A	B	C	D	E	F	G	H
No. of TV Commercials, x	11	6	9	15	11	15	8	16
Sales Units, y	8	5	10	14	12	9	7	11

(a) Draw a scatter diagram.

(b) Estimate r.

(c) Calculate r.

(d) Is there evidence that the number of commercials is linearly correlated to sales? Explain.

Section 3-3 Linear Regression

Whereas correlation tells us the strength of a linear relationship, it does not tell us the exact numerical relationship. For example, by applying the decision point table (3-6) to the correlation coefficient calculated for the data in table 3-5, we concluded that there is a linear relationship between a salesperson's sales and entertainment expenses. That means we should be

regression analysis

able to use the amount of entertainment expenses to predict sales. But correlation analysis does not show us how to determine a y value given an x value. This calculation is done by regression analysis. **Regression analysis calculates an equation that provides values of y for given values of x.** One of the primary objectives of regression analysis is to **make predictions**—for example, predicting the productivity of an employee on the basis of a preemployment test, or predicting sales based on the characteristics of the economy.

Generally, the exact value of y is not predicted. We are usually satisfied if the predictions are reasonably close. The statistician seeks an equation to express the relationship between the two variables. **The equation he or she chooses is the one that best fits the scatter diagram.**

prediction equations

Here are some examples of various relationships; they are called **prediction equations**:

$$y = b_0 + b_1 x \quad \text{(linear)} \tag{3-5}$$

$$y = a + bx + cx^2 \quad \text{(quadratic)} \tag{3-6}$$

$$y = a(b^x) \quad \text{(exponential, or logarithmic)} \tag{3-7}$$

$$y = \frac{a}{1 + 10^{c-bx}} \quad \text{(S-shaped, or logistic)} \tag{3-8}$$

Figures 3-11 through 3-15 illustrate some of these relationships. The graphs of these relationships might result if we were to sample the following bivariate populations:

Figure 3-11: The number of man-hours worked x and output y

Figure 3-12: The price of an item x and the sales of the item y

Figure 3-13: The number of items produced x and the average cost per item y

Figure 3-14: The time since introduction x and the market share of a product y

Figure 3-15: The height of a salesman x and his sales volume y

In each of figures 3-11, 3-12, 3-13, and 3-14, there appears to be a relationship. In figure 3-15, however, the values of x and y do not seem to be related.

The graphic relationships between variables can be represented by algebraic expressions such as those in formulas (3-5), (3-6), (3-7), and (3-8).

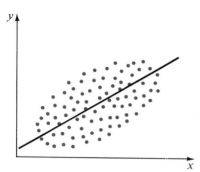

FIGURE 3-11
Linear Regression with Positive Slope

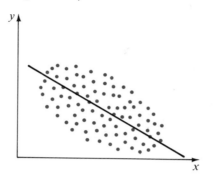

FIGURE 3-12
Linear Regression with Negative Slope

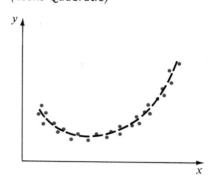

FIGURE 3-13
Curvilinear Regression (Looks Quadratic)

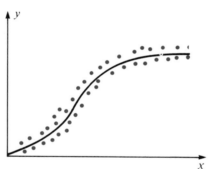

FIGURE 3-14
S-Shaped, or Logistic, Curve

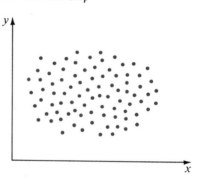

FIGURE 3-15
No Relationship

Regression analysis provides the equation and the a, b, c, b_0, and b_1 values that best fit a set of data. At this point we do not know how to find a regression equation. Since regression is probably the most frequently used statistical technique in business, it is important that we cover it in depth. More details must, by necessity, wait until additional development work has been completed. It is sufficient for the present that you know the purpose of regression analysis. In later chapters (13, 14, and 15) we will return to this technique.

Exercises

3-16 Do the data in exercise 3-6 appear to fit one of the patterns described in the preceding figures?

3-17 Which of the predictive equations might you fit to the data in exercise 3-10?

3-18 If you put $1,000 in a bank at 6% compounded interest, which predictive equation would you use to calculate the total amount of money in the bank, given a specified number of years the money has been in your account?

3-19 If the correlation coefficient r between the age of a machine x and yearly maintenance cost y is $+0.96$, which predictive equation would you use? Which figure in the text, 3-11, 3-12, 3-13, 3-14, or 3-15, most appropriately describes the data?

3-20 The data in the accompanying table were collected on the sales of a new tranquilizer drug.

Years Since Introduction	Market Share (%)
1	2
2	5
3	10
4	13
5	15
6	16
7	16

(a) Draw a scatter diagram of these data.

(b) What predictive curve might you want to fit?

In Retrospect

You have learned that bivariate data can be presented in cross-tabulation, or contingency, tables and that two quantitative variables can be presented pictorially in a scatter diagram. These displays are the bivariate analogies to frequency distributions and histograms.

Thus far you have been introduced to the basic concepts of bivariate data—correlation and regression—which have been presented for the purpose of a first (descriptive) look. After reading this chapter, you should have a basic understanding of bivariate data and its presentation and of regression analysis and correlation analysis and when each is used. You also should be able to calculate a linear correlation coefficient and explain what it signifies.

Reread the article at the beginning of this chapter. You should now have a better understanding of the article, which uses correlation and linear regression to study the relationship between a country's income and its agricultural imports per capita.

Chapter Exercises

3-21 Determine whether each of these questions requires regression analysis or correlation analysis to obtain an answer.

(a) Is the amount of advertising expenditures linearly related to sales?

(b) How much should sales increase if advertising expenditures are increased from $1 million to $1.2 million?

(c) Is the score on the Wunderlich Intelligence Test related to the job performance of a welder?

(d) Is the amount of overtime worked in a week related to the average output per hour?

(e) The amount of overtime is linearly related to average output per hour; if overtime is increased, will the average output tend to increase or decrease?

(f) The amount of overtime is linearly related to average output per hour; if overtime is increased by 3 hours a week, how much should average hourly output change?

3-22 A market survey to help select advertisement media was conducted. People in the sample were asked their income last year and the number of times they ate dinner in a restaurant. Why would correlation analysis be conducted on the survey data, and what would be the nature of the results?

3-23 Estimate the value of the correlation coefficient for each of the following pairs of variables.

(a) number of man-hours worked and number of items produced

(b) the percentage of cars sold in the United States that were foreign made and the percentage of cars sold in the United States that were made in the United States over the past 10 years

(c) the price of a pen and the numbers of pens sold

(d) the interest rate of new-car loans and the number of new-car loans made

3-24 A study of beef prices showed that as the demand for beef increased by 1,000 units, the price tended to increase by 10¢ a pound.

(a) What type of analysis would produce this result?

(b) Is the correlation coefficient for the demand for beef and the price of beef positive or negative?

(c) If total beef sales is equal to price times demand, what type of predictive equation would you use to determine the mathematical relationship between sales volume and demand?

3-25 PSFS 5-Year Summary (in thousands)

	1982	1981	1980	1979	1978
Assets	$10,105,013	$7,424,435	$7,062,797	$6,742,069	$5,941,674
Interest income	$917,462	$695,568	$586,998	$516,065	$417,439
Interest expense	$934,699	$703,552	$541,963	$431,042	$329,870
Net operating income (loss)	$(114,353)	$(38,720)	$12,249	$42,523	$48,661
Net income (loss)	$(73,960)	$(33,337)	$21,422	$42,125	$44,685

The above table represents data concerning a large savings and loan. Draw a scatter diagram and calculate the linear correlation coefficient between:

(a) assets and interest income
(b) interest income and interest expense
(c) assets and interest income
(d) net income and the difference between interest income and interest expense

3-26 To determine when copiers should be replaced, the maintenance costs of a sample of copiers used nationwide by a firm were collected. The following table gives data for a sample of 12 copiers.

Age, x (years)	Maintenance Cost, y (dollars)
4	$1,274
5	1,568
3	825
1	75
1	0
3	428
2	610
2	559
3	720
1	25
4	1,325
5	1,622

(a) Construct a scatter diagram.
(b) Calculate the linear correlation coefficient.

3-27 (a) Construct a scatter diagram and calculate r for the data in exercise 2-81 on percentage converted y and income difference x.
(b) Construct a scatter diagram and calculate r for the data in exercise 2-81 on percentage converted y and index of stock volatility x.
(c) Which variable is a better predictor of percentage converted?

3-28 Construct a scatter diagram and calculate r for the data in exercise 2-80 on gross proceeds y and spread x.

3-29 Construct a scatter diagram and calculate r for the data in exercise 2-77 on amount of return y and days elapsed x.

3-30 According to Engel's economic law, "The poorer a family is, the greater the proportion of its expenditures that must be used to provide nourishment." Consider the data in the accompanying table collected from nine countries in Western Europe.

Country	Income per Capita (U.S. dollars)	Percentage of Income for Food Expenditures
Austria	$588	35.0
Denmark	985	24.9
France	876	31.3
Greece	304	46.3
Iceland	421	45.2
Italy	467	45.2
Netherlands	775	31.7
Norway	900	30.3
United Kingdom	1,337	30.3

(a) Compute the correlation between income and percentage spent on food.

(b) Do these data support Engel's law? Explain.

3-31 The data in the accompanying table report the price paid to farmers for beef cattle and the production of beef and veal for a 15-year period.

Year	Price of Beef Cattle per 100 Pounds	Beef and Veal Production (billion pounds)
1	$23.30	9.53
2	28.70	8.84
3	24.30	9.65
4	16.30	12.41
5	16.00	12.96
6	15.60	13.57
7	14.90	14.46
8	17.20	14.20
9	21.90	13.33
10	22.60	13.58
11	20.40	14.75
12	20.20	15.30
13	21.30	15.30
14	19.80	16.42
15	18.00	18.42

(a) Construct a scatter diagram of price and years.
(b) Construct a scatter diagram of price and production.
(c) Compute r for price and year.
(d) Compute r for price and production.
(e) What does the preceding analysis tell us about beef prices and production?

3-32 Use the data in exercise 2-79 to construct a contingency table for the following:
(a) station preference and income
(b) station preference and number of flights

3-33 Use the data in exercise 2-79 and the results from exercise 3-32.
(a) Construct a relative frequency contingency table for station preference and income.
(b) Construct a relative frequency contingency table for station preference and number of flights.
(c) Does station preference appear to correlate to income and/or number of flights?

3-34 (a) Use the data in exercise 2-79 to construct a contingency table for income and number of flights.
(b) Use the data in exercise 2-79 to construct a contingency table for age and number of flights.
(c) Does number of flights appear to correlate to income and/or age of person?
(d) Compute the correlation coefficient for number of flights and age.
(e) Compute the correlation coefficient for number of flights and income.
(f) Reevaluate your answer to (c) in light of the answers to (d) and (e).

Hands-On Problems

3-1 Collect a sample of 15 pieces of bivariate data from a population of your choice. Make a brief statement defining the population, the predictor (independent) variable, and the response (dependent) variable. Include a list of your original data. Try to obtain a representative sample.

3-2 Plot a scatter diagram of the data you collected for problem 3-1. Be sure to label the axes and give a title. (Use graph paper.)

3-3 (a) Estimate r, the linear correlation coefficient.
(b) Draw the line of best fit by eye on your scatter diagram.

3-4 Construct and use a table to calculate the summations for x, y, x^2, xy, and y^2 for your data.

3-5 Calculate the linear correlation coefficient, r, for your data.

3-6 Comment on your results (problems 3-2 through 3-5). Some of the points to cover would be the following:

(a) The strength of r as compared with what you expected before the data were analyzed.

(b) If your results are different from what you expected, can you explain the difference? Were your expectations wrong or do you think the sample isn't representative?

(c) If your results are similar to what you expected, why did you expect these results?

Each population and sample presents its own set of circumstances. Try to be as specific as you can. (*Warning*: Do not try to draw any earth-shattering conclusions from your results. Samples of size 15 are relatively small and inconclusive.)

The calculations for this problem set can most easily be accomplished with the assistance of an electronic calculator or a packaged program on a computer. A list of the available programs can be obtained from your computer center. There are a variety of packaged programs available: Minitab, Biomed (Biomedical Programs), SAS (Statistical Analysis System), IBM Scientific Subroutine Packages, and SPSS (Statistical Package for the Social Sciences) program libraries. Your local computer center will assist you.

4 Probability

Chapter Outline

Unit 1: Concepts of Probability

4-1 The Nature of Probability

*Probability is the **relative frequency** with which an event is expected to occur.*

4-2 Probability of Events

*The three methods for assigning probabilities to an event are **empirical, theoretical, or subjective**.*

4-3 Simple Sample Spaces

*Listing of **all possible outcomes** for an experiment.*

4-4 Rules of Probability

*Probabilities for the outcomes of a sample space **total exactly 1**.*

Unit 2: Calculating Probabilities of Compound Events

4-5 Complementary Events, Mutually Exclusive Events, and the Addition Rule

*Events are mutually exclusive if they **cannot both occur** at the **same time**.*

4-6 Independence, Multiplication Rule, and Conditional Probability

*Events are independent if the **occurrence** of one event does **not change** the **probability** of the other event.*

4-7 Combining the Rules of Probability

*The addition and multiplication rules are often used together to calculate the probability of **compound events**.*

4-8 Bayes's Rule

*A method for finding **conditional probabilities**.*

Still a Good Year

What changes in monetary policy should we expect now that G. William Miller has replaced Arthur Burns as chairman of the Fed? How should we second-guess the consensus forecasters in light of this news? Is there a widening uncertainty about all our estimates?

• • •

The initial reaction to Burns's termination has been favorable. Common stocks recovered in a matter of hours. Foreign-exchange and gold markets moved less than had been feared.

• • •

When we economists put our computer models through new runs based upon a new chairman at the Fed, it is surprising how minute are the resulting changes in estimates of inflation. Employing a variety of new reasonable assumptions, one comes out with less than half of 1 per cent of change in the 1980 level of consumer prices.

• • •

The other side of the coin is that vast improvements in the unemployment rate are also unlikely. A new Supreme Court has more genuine autonomy than a new Federal Reserve Board, for the reason that the Court does not have to report quarterly to Congress. The Fed under Miller can be expected to continue its uneasy compromising between holding down excessive monetary growth and excessive increases in interest rates.

All that being said, it should be a good year for the American economy. The evidence suggests these odds in 1978:

- 2 to 1 against a growth recession in which unemployment worsens.
- 5 to 1 against a recession in which the economy shows negative real growth.
- 100 to 1 against a 1930s-like depression.
- 10 to 1 against another bout of two-digit price inflation.
- 100 to 1 against real stability of the price level.

This is not good enough to realize Jimmy Carter's promises to create jobs for youths, the unskilled and minority workers. Those anxious about continuing inflation and possible weakness in the dollar exchange rate will continue to sleep uneasily. The stock market will still nervously contend with rising interest rates and fears for the future.

Still, for what is the fourth year of an American recovery, the scenario envisaged by 30 out of 40 of the leading consensus forecasters is not at all a bad one.

Condensed from Paul A. Samuelson, "Still a Good Year," *Newsweek*, 9 January 1978, pp. 52–53. Copyright 1978 by Newsweek, Inc. All rights reserved. Reprinted by permission.

Chapter Objectives

Before continuing our study of statistics, we must make a slight detour and study basic probability. Probability is often referred to as the "vehicle" of statistics; that is, the probability associated with chance occurrences is the underlying theory of statistics. Recall that in chapter 1 we described probability as the science of making statements about what will occur when samples are drawn from known populations. Statistics was described as the science of selecting a sample and making inferences about the unknown population from which it is drawn. To make these inferences, we need to study sample results in situations where the population is known, so that we will be able to understand the behavior of chance occurrences.

You may already be familiar with some ideas of probability, because probability is part of our everyday culture. We constantly hear people making probability-oriented statements:

"There is a 40% chance of rain tonight."

"Xerox will most likely introduce a new copier next year."

"I have a 50–50 chance of passing today's accounting exam."

"If I buy that stock, its price will probably fall."

Everyone has made or heard these kinds of statements. What exactly do they mean? Do they, in fact, mean what they say? Some statements are based on scientific information and others on subjective prejudice. Whatever the case may be, they are probabilistic inferences—not fact, but conjectures.

In this chapter we will learn about the basic concept of probability and the rules that apply to the probability of both simple and compound events.

Unit 1 Concepts of Probability

Section 4-1 The Nature of Probability

Let's consider an experiment in which we toss two coins simultaneously and record the number of heads that occur: $0H$ (zero heads), $1H$ (one head), and $2H$ (two heads) are the only possible outcomes. Let's toss the two coins 10 times and record our findings.

$$2H \quad 1H \quad 1H \quad 2H \quad 1H \quad 0H \quad 1H \quad 1H \quad 1H \quad 2H$$

Totals: $0H$, 1 $1H$, 6 $2H$, 3

Suppose the experiment is repeated 19 more times. Table 4-1 shows the totals for the 20 sets of 10 tosses.

TABLE 4-1
Experimental Results of Tossing Two Coins

Outcome	\										Trial											Total
		1	2	3	4	5	6	7	8	9	10	11	12	13	14	15	16	17	18	19	20	
2H		3	3	5	1	4	2	4	3	1	1	2	5	6	3	1	4	1	0	3	1	53
1H		6	5	5	5	5	7	5	5	5	5	8	4	3	7	5	1	5	4	5	9	104
0H		1	2	0	4	1	1	1	2	4	4	0	1	1	0	4	5	4	6	2	0	43

The total of 200 tosses of the pair of coins resulted in 2H on 53 occasions, 1H on 104 occasions, and 0H on 43 occasions. We can express these results in terms of relative frequencies and show the results with the aid of a histogram, as is done in figure 4-1.

FIGURE 4-1
Relative Frequency Histogram

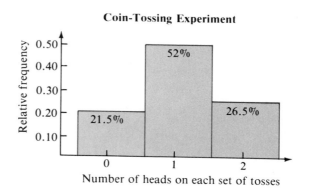

What would happen if this experiment were repeated or continued? Would the relative frequencies change? If so, by how much? If we look at the individual sets of 10 tosses, we notice a large variation in the number of times each of the events (2H, 1H, and 0H) occurred. In both the 0H and 2H categories, there were as many as six occurrences and as few as zero occurrences in a given set of 10 tosses. In the 1H category there were as few as one occurrence and as many as nine occurrences.

If we were to continue this experiment for several hundred tosses, what would you expect to happen in terms of the relative frequency of these three events? It looks as if we have approximately a 1:2:1 ratio in the totals of table 4-1. We might therefore expect to find the relative frequency of 0H to be approximately $\frac{1}{4}$, or 25%; the relative frequency of 1H to be approximately $\frac{1}{2}$, or 50%; and the relative frequency of 2H to be approximately $\frac{1}{4}$, or 25%. These relative frequencies accurately reflect the concept of probability.

Section 4-2 Probability of Events

We are now ready to define what is meant by probability. Specifically, we talk about the *probability that a certain event will occur*.

probability of an event

Probability That an Event Will Occur

The relative frequency with which that event can be expected to occur.

The probability of an event may be obtained in three different ways: (1) empirically, (2) theoretically, and (3) subjectively. The first method was illustrated in the experiment in section 4-1 and might be called **experimental, or empirical, probability**. This method is nothing more than the **observed relative frequency with which an event occurs**. In the coin-tossing illustration we observed exactly one head ($1H$) on 104 of the 200 tosses of a pair of coins. The observed empirical probability for the occurrence of $1H$ was 104/200, or 0.52.

experimental, or empirical, probability

When the value assigned to the probability of an event results from experimental data, we will identify the probability of the event with the symbol $P'(\ )$. The name of the event is placed in parentheses: $P'(1H) = 0.52$, $P'(0H) = 0.215$, and $P'(2H) = 0.265$. **Thus the prime notation is used to denote observed probabilities.**

The value assigned to the probability of event A as a result of experimentation is found by means of the formula

$$P'(A) = \frac{\text{No.}(A)}{n} \qquad (4\text{-}1)$$

where No. (A) is the number of times event A actually is observed and n is the number of times the experiment is attempted.

$P'(A)$ represents the relative frequency with which event A occurred. This value may or may not be the relative frequency with which that event "can be expected" to occur. $P'(A)$, however, represents our best *estimate* of what can be expected to occur. But what is meant by the relative frequency that can be expected to occur?

Consider the rolling of a single die. Define event A as the occurrence of a 1. In a single roll of a die, six outcomes are possible. Assuming the die is symmetrical, each number should have an equal likelihood of occurring. Intuitively, we see that the probability of A, or the expected relative frequency of a 1, is $\frac{1}{6}$. (Later we will formalize this calculation.)

What does this mean? Does it mean that once in every six rolls a 1 will occur? No, it does not. Saying that the probability of a 1, $P(1)$, is $\frac{1}{6}$

means that in the long run the proportion of times that a 1 occurs is approximately $\frac{1}{6}$. How close to $\frac{1}{6}$ can we expect the observed relative frequency to be?

Table 4-2 shows the number of 1s observed in each set of six rolls of a single die [column (1)], an observed relative frequency for each set of six rolls [column (2)], and a cumulative relative frequency [column (3)]. Each trial is a set of six rolls.

Figure 4-2(a) shows the fluctuation of the observed probability for event A on each of the 20 trials [column (2), table 4-2]. Figure 4-2b shows the fluctuation of the cumulative relative frequency [column (3), table 4-2]. Notice that the observed relative frequency on each trial of six rolls of a die tends to fluctuate about $\frac{1}{6}$. Notice also that the observed values on the cumulative graph seem to become more stable; in fact, they become relatively close to the expected $\frac{1}{6}$.

FIGURE 4-2
Fluctuations Found in the Die-Tossing Experiment

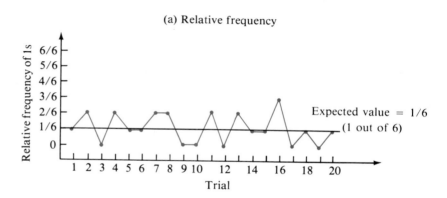

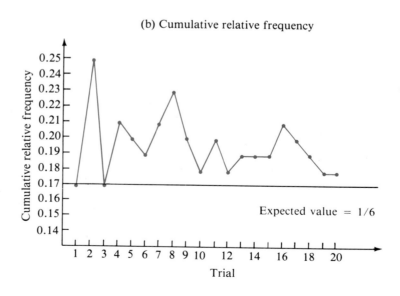

Probability of Events   Section 4-2   **131**

TABLE 4-2
Experimental Results of Rolling a Die Six Times in Each Trial

Trial	(1) Number of 1s Observed	(2) Relative Frequency	(3) Cumulative Relative Frequency
1	1	1/6	1/6 = 0.17
2	2	2/6	3/12 = 0.25
3	0	0/6	3/18 = 0.17
4	2	2/6	5/24 = 0.21
5	1	1/6	6/30 = 0.20
6	1	1/6	7/36 = 0.19
7	2	2/6	9/42 = 0.21
8	2	2/6	11/48 = 0.23
9	0	0/6	11/54 = 0.20
10	0	0/6	11/60 = 0.18
11	2	2/6	13/66 = 0.20
12	0	0/6	13/72 = 0.18
13	2	2/6	15/78 = 0.19
14	1	1/6	16/84 = 0.19
15	1	1/6	17/90 = 0.19
16	3	3/6	20/96 = 0.21
17	0	0/6	20/102 = 0.20
18	1	1/6	21/108 = 0.19
19	0	0/6	21/114 = 0.18
20	1	1/6	22/120 = 0.18

A cumulative graph such as figure 4-2b demonstrates the idea of long-range average. When only a few rolls were observed (as on each trial), the probability $P'(A)$ fluctuated between 0 and $\frac{1}{2}$ [see the relative frequency of each trial, column (2), table 4-2]. As the experiment was repeated, however, the cumulative graph suggests a stabilizing effect on the observed cumulative probability. This stabilizing effect, or **long-term average** value, is often referred to as the *law of large numbers*.

long-term average

law of large numbers

Law of Large Numbers

If the number of times that an experiment is repeated is increased, the ratio of the number of successful occurrences to the number of trials will tend to approach the theoretical probability of the outcome for an individual trial.

The law of large numbers can be applied in a different manner. Consider the following experiment. A thumbtack is tossed, and the result is observed and recorded as a point up (⊥) or point down (⨀). What is the probability that the tack will land point up? That is, $P(⚇) = ?$ When we worked with the die in the preceding example, we could easily see the true probability. With the tack we have no idea. Is $P(⚇)$ equal to $\frac{1}{2}$? Since the tack is not symmetrical, we would suspect not. More than $\frac{1}{2}$? Less than $\frac{1}{2}$? About the only way we can tell is to conduct an experiment. And the law of large numbers can help us here. After many trials of the experiment, $P(⚇)$ should indicate a value that is very close to the true value of $P(⚇)$, even though the true value cannot be proven mathematically.

Table 4-3 lists the data recorded after repeated tosses of a thumbtack. After each 10 tosses the number of times the tack landed point up was recorded. The cumulative value of the number of times it landed point up was computed and reported as a fraction of the total tosses. Data on 500

TABLE 4-3
Experimental Results of Tossing a Thumbtack

Multiples of 10 Tosses	$n(⚇)$ in Each Set of 10 Tosses	Cumulative $P'(⚇)$	Multiples of 10 Tosses	$n(⚇)$ in Each Set of 10 Tosses	Cumulative $P'(⚇)$
1	5	5/10 = 0.500	26	3	156/260 = 0.600
2	7	12/20 = 0.600	27	5	161/270 = 0.596
3	10	22/30 = 0.733	28	7	168/280 = 0.600
4	6	28/40 = 0.700	29	7	175/290 = 0.603
5	9	37/50 = 0.740	30	9	184/300 = 0.613
6	5	42/60 = 0.700	31	7	191/310 = 0.616
7	6	48/70 = 0.686	32	7	198/320 = 0.619
8	6	54/80 = 0.675	33	7	205/330 = 0.621
9	2	56/90 = 0.622	34	6	211/340 = 0.621
10	4	60/100 = 0.600	35	7	218/350 = 0.623
11	4	64/110 = 0.582	36	6	224/360 = 0.622
12	7	71/120 = 0.592	37	4	228/370 = 0.616
13	6	77/130 = 0.592	38	6	234/380 = 0.616
14	8	85/140 = 0.607	39	4	238/390 = 0.610
15	6	91/150 = 0.607	40	5	243/400 = 0.608
16	3	94/160 = 0.588	41	5	248/410 = 0.605
17	7	101/170 = 0.594	42	7	255/420 = 0.607
18	9	110/180 = 0.611	43	7	262/430 = 0.609
19	7	117/190 = 0.616	44	6	268/440 = 0.609
20	3	120/200 = 0.600	45	4	272/450 = 0.604
21	5	125/210 = 0.595	46	7	279/460 = 0.607
22	8	133/220 = 0.605	47	5	284/470 = 0.604
23	8	141/230 = 0.613	48	8	292/480 = 0.608
24	6	147/240 = 0.612	49	5	297/490 = 0.606
25	6	153/250 = 0.612	50	6	303/500 = 0.606

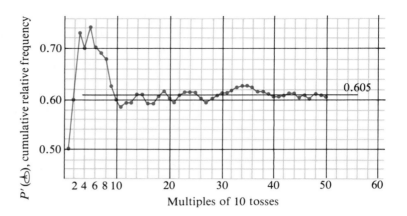

FIGURE 4-3
Experimental Probability of Thumbtack Pointing Up

tosses are recorded. Figure 4-3 illustrates the value of $P(⚲)$ that we are attempting to estimate. The figure illustrates the tendency for the observed probability $P'(⚲)$ to be zeroing in on the value 0.605. We can conclude that $P(⚲)$ is very close to 0.605.

An important point is illustrated in figure 4-3 and table 4-3.

The larger the number of experimental trials n, the closer the experimental probability $P'(A)$ is expected to be to the true probability $P(A)$.

Many business situations are analogous to the thumbtack experiment. We'll look at one example in illustration 4-1.

Illustration 4-1

The key to establishing proper life insurance rates is the use of the probability that insured people will live 1, 2, or 3 years, and so forth, from the time they purchase policies. These probabilities are derived from actual life and death statistics and hence are experimental probabilities. They are published by the government and are extremely important to the life insurance industry. □

Exercises

4-1 Explain what it means to say that the probability of a "1" when a single die is rolled is $\frac{1}{6}$.

4-2 Toss a single coin 20 times, recording H (head) or T (tail) after each toss. Using your results, find the observed probabilities:

(a) $P'(H)$ (b) $P'(T)$

4-3 Roll a single die 20 times, recording a 1, 2, 3, 4, 5, or 6 after each roll. Using your results, find the observed probabilities:

(a) $P'(1)$ (b) $P'(2)$ (c) $P'(3)$
(d) $P'(4)$ (e) $P'(5)$ (f) $P'(6)$

4-4 Place three coins in a cup, shake them, and dump them out, observing the number of heads showing. Record $0H$, $1H$, $2H$, or $3H$ after each trial. Repeat the process 25 times. Using your results, find:

(a) $P'(0H)$ (b) $P'(1H)$ (c) $P'(2H)$ (d) $P'(3H)$

4-5 Take two dice (one white and one black) and roll them 50 times and record the results as an ordered pair [e.g., white 3, black 5, (3, 5)]. Calculate these observed probabilities:

(a) P'(black die is odd)
(b) P'(sum is 8)
(c) P'(same number appears on both dice)
(d) P'(number on black die exceeds number on white die)

4-6 Place 10 pennies in a cup, shake, and dump them in a box. Count and record the number of heads observed. Repeat 19 more times. Display your data in a table similar to table 4-2 and graphically similar to figures 4-2a and 4-2b. Do your data seem to support the claim that $P(\text{head}) = \frac{1}{2}$? If not, what do you think caused the difference?

Section 4-3 Simple Sample Spaces

Let's return to a question that was posed earlier: What values might be expected to be assigned to the three events $(0H, 1H, 2H)$ associated with the coin-tossing experiment? As we inspect these three events, we see that they do not tend to happen with the same relative frequency. Why? Suppose the experiment of tossing two pennies and observing the number of heads had actually been carried out using a penny and a nickel: two distinct coins. Would this have changed our results? No, it would have had no effect on the experiment. However, it does show that more than three outcomes are possible.

When a penny is tossed, it may land as heads or tails. When a nickel is tossed, it may also land as heads or tails. If we toss them simultaneously, we see that each toss actually has four different possible outcomes. Each observation would be one of the following possibilities: (1) heads on the penny and heads on the nickel, (2) heads on the penny and tails on the nickel, (3) tails on the penny and heads on the nickel, or (4) tails on the penny and tails on the nickel. Notice that the previous events $(0H, 1H, 2H)$ pair up with these four in the following manner: event 1 is the same as $2H$, event 4 is the same as $0H$, and events 2 and 3 together make up what was previously called $1H$.

In this experiment with the penny and the nickel, let's use an ordered-pair notation. The first listing will correspond to the penny, and the second will correspond to the nickel. Thus (H, T) represents the event that a head occurs on the penny and a tail occurs on the nickel. Our listing of events for the tossing of a penny and a nickel looks like this:

$$(H, H) \quad (H, T) \quad (T, H) \quad (T, T)$$

What we have accomplished here is a listing of what is known as the *sample space* for this experiment.

Before we define sample space, let's define a few of the terms that have been used intuitively on the previous pages.

experiment

Experiment

Any process that yields a result or an observation.

outcome

Outcome

A particular result of an experiment.

event

Event

Any set of outcomes of an experiment.

sample space

Sample space

The set of all possible outcomes of an experiment. The sample space is typically called S and may take any number of forms: a list, a tree diagram, a lattice grid system, and so on. The individual outcomes in a sample space are called **sample points**.

sample points

Regardless of the form in which they are presented, the **outcomes in a sample space can never overlap**. Also, **all possible outcomes must be represented**. These characteristics are called **mutually exclusive** and **all-inclusive**, respectively. A more detailed explanation of these characteristics will be presented later; for the moment, however, an intuitive grasp of their meaning is sufficient.

mutually exclusive
all-inclusive

Now let's look at some illustrations of probability experiments and their associated sample spaces.

Experiment 4-1   A single coin is tossed once, and the outcome—a head (H) or a tail (T)—is recorded.

$$\text{Sample space:} \quad S = \{H, T\}$$

Experiment 4-2   Half of all checks deposited in a bank clear and become available as cash the next day. The second half of all checks clear 2 days after deposit. A check is deposited and the number of days it takes to clear is recorded.

Sample space: $S = \{1 \text{ day}, 2 \text{ days}\}$

Experiment 4-3 Two coins, one penny and one nickel, are tossed simultaneously, and the outcome for each coin is recorded using ordered notation: (penny, nickel).

The sample space is shown here in three different ways.

Tree diagram representation*:

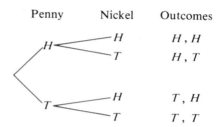

(four branches, each branch shows a possible outcome)

Listing:

$$S = \{(H, H), (H, T), (T, H), (T, T)\}$$

Lattice grid representation:

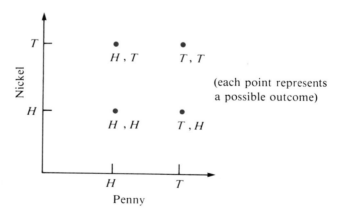

(each point represents a possible outcome)

Notice that all three representations show the same four possible outcomes. For example, the top branch of the tree diagram shows heads on both coins, as does the first ordered pair in the listing and the lower left point in the lattice grid diagram.

* See the *Study Guide* for information about tree diagrams.

Experiment 4-4 Enco Products purchases the same amount of chemical from three manufacturers. The quality control department randomly selects one batch of chemicals for testing and records the manufacturer.

Sample space: $S = \{\text{company } A, \text{ company } B, \text{ company } C\}$

Experiment 4-5 A box contains three relay switches: one is in working order, one needs repairs, and one is beyond repair. Two relays are drawn with replacement. Replacement means that one is selected, its condition is observed, and then it is replaced in the box. The relays are scrambled before the second relay is selected and its condition is observed. Each observed relay is classified as good (G), repairable (R), or scrap (S).

The sample space is shown in three different ways.

Tree diagram representation:

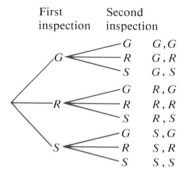

Listing:

$S = \{(G, G), (G, R), (G, S), (R, G), (R, R), (R, S), (S, G), (S, R), (S, S)\}$

(The order of the pairs corresponds to the drawing.)

Lattice grid representation:

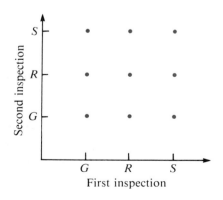

Experiment 4-6   This experiment is the same as experiment 4-5 except that the first relay is not replaced before the second selection is made.

The sample space is shown in two ways.

Tree diagram representation:

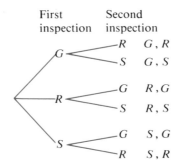

Listing:

$$S = \{(G, R), (G, S), (R, G), (R, S), (S, G), (S, R)\}$$

Experiment 4-7   A white die and a black die are each rolled one time, and the number of spots showing on each die is observed.

The sample space is shown by a chart representation.

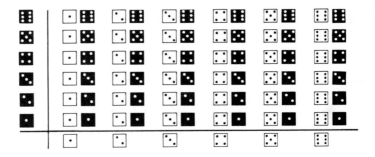

Experiment 4-8   Two dice are rolled, and the sum of their spots are observed.

Sample space: $S = \{2, 3, 4, 5, 6, 7, 8, 9, 10, 11, 12\}$

(Or the 36-point sample space listed in experiment 4-7.)

You will notice that two different sample spaces were suggested for experiment 4-8. Both of these sets satisfy our definition of a sample space, and thus either can be used. We will learn later why the 36-point sample space is more useful than the other.

Experiment 4-9 Two common stocks (A, B) are purchased, and their prices 1 year later are compared with their purchase price. (Assume for simplicity that the price of each stock changes.)

Sample space:
$S = \{(A$ up, B up), $(A$ up, B down), $(A$ down, B up), $(A$ down, B down)$\}$

NOTE: The list $S = \{A$ up, A down, B up, B down$\}$ is incorrect because the elements listed do not identify the specific outcomes that can occur. For instance, A up and B up can both occur.

Experiment 4-10 A new product is marketed. One year later the company observes whether or not it has been profitable.

Sample space: $S = \{$profitable, not profitable$\}$

Experiment 4-11 Four ledger entries are audited, and the number of errors determined.

Sample space: $S = \{0, 1, 2, 3, 4\}$

Special attention should always be given to the sample space. Like the statistical population, the sample space must be well defined. Once the sample space is defined, you will find the remaining work much easier.

Exercises

4-7 An investment counselor is asked to rate a stock as a poor, average, or good investment. List the sample space.

4-8 An investment counselor is asked to rate two stocks (see exercise 4-7). What is the sample space?

4-9 An experiment consists of two trials. On the first trial a penny is tossed and the outcome is recorded as a head or a tail. On the second trial a die is rolled and the outcome is recorded as a 1, 2, 3, 4, 5, or 6. Construct the sample space.

4-10 A salesperson calls on four clients and either makes a sale or does not make a sale each time. List the sample space.

4-11 Two printed circuits are selected from a production lot of ten circuits. Three circuits in the lot are known to be defective. Give two alternative sample spaces showing the selected circuits' quality.

4-12 (a) A balanced coin is tossed twice. List a sample space showing the possible outcomes.

    (b) A biased coin (it favors heads in a ratio of 3 to 1) is tossed twice. List a sample space showing the possible outcomes.

Section 4-4 Rules of Probability

Let's return now to the concept of probability and relate the sample space to it. Recall that the probability of an event was defined as the relative frequency with which the event could be expected to occur.

In the sample space associated with experiment 4-1, the tossing of one coin, we find two possible outcomes: heads (H) and tails (T). We have an "intuitive feeling" that these two events will occur with approximately the same frequency. The coin is a symmetrical object and therefore would not be expected to favor either of the two outcomes. We would expect heads to occur one-half of the time. Thus the probability that a head will occur on a single toss of a coin is thought to be $\frac{1}{2}$.

The preceding description is the basis for the second technique for assigning the probability of an event. In a sample space containing **sample points that are equally likely to occur**, the probability of an event A, $P(A)$, is the ratio of the number of points satisfying the definition of event A, $n(A)$, to the number of sample points in the entire sample space, $n(S)$. That is,

equally likely events

$$P(A) = \frac{n(A)}{n(S)} \qquad (4\text{-}2)$$

theoretical probability

This formula gives a **theoretical probability** value of event A's occurrence. The *prime symbol* of formula (4-1) *is not used* with theoretical probabilities.

The probability of event A is still a relative frequency, but it is now based on all the possible outcomes, regardless of whether they actually occur. We can now see that the probability of obtaining two heads, $P(2H)$, when the penny and nickel are tossed is $\frac{1}{4}$. The probability of exactly one head, $P(1H)$ [(H, T) or (T, H)], is $\frac{2}{4}$, or $\frac{1}{2}$. And $P(0H) = \frac{1}{4}$. These values are those that were indicated by the experiment.

The use of formula (4-2) requires the existence of a sample space in which each outcome is equally likely. Thus when dealing with experiments that have more than one possible sample space, it is helpful to construct a sample space in which the sample points are equally likely.

Consider experiment 4-8, where two dice were rolled. If you list the sample space as the 11 sums, the sample points are not equally likely. If you use the 36-point sample space, all the sample points are equally likely. The 11 sums in the first sample space represent combinations of the 36 equally likely sample points. For example, the sum of 2 represents $\{(1, 1)\}$; the sum of 3 represents $\{(2, 1), (1, 2)\}$; and the sum of 4 represents $\{(1, 3), (3, 1), (2, 2)\}$. Thus we can use formula (4-2) and the 36-point sample space to obtain the probabilities for the 11 sums.

$$P(2) = \frac{1}{36} \qquad P(3) = \frac{2}{36} \qquad P(4) = \frac{3}{36}$$

and so forth.

In many cases the assumption of equally likely events does not make sense. Experiments 4-1 through 4-7 are examples where the sample space elements are all equally likely. The sample points in experiment 4-8 need not be equally likely, and we have no reason to believe that the sample points in experiments 4-9 through 4-11 are equally likely.

What do we do when the sample space elements are not equally likely or not a combination of equally likely events? We could use empirical probabilities. But what do we do when no experiment has been done or can be performed?

Let's look again at experiment 4-10. Every new product does not have a 50-50 chance of success. Moreover, truly new products have no past track record. In such cases the only method available is personal judgment. These probability assignments are called **subjective probabilities**. The accuracy of subjective probabilities depends on the individual's ability to correctly assess the situation.

subjective probability

Often personal judgment of the probability of the possible outcomes of an experiment is expressed by comparing the likelihood between the various outcomes. For example, an investment counselor's personal assessment of a stock's price next year might be that "it is nine times more likely to increase (I) than not increase (NI)," or $P(I) = 9 \cdot P(NI)$. But what values should be assigned to $P(I)$ and $P(NI)$? To answer this question, we need to examine the two basic properties of probability. But first let's review some of the ideas about probability we've already discussed:

1. Probability represents a relative frequency.
2. $P(A)$ is the ratio of the number of times an event A can be expected to occur divided by the number of trials.
3. The numerator of the probability ratio must be a positive number or 0.
4. The denominator of the probability ratio must be a positive number (greater than 0).
5. The number of times an event can be expected to occur in n trials is always less than or equal to the total number of trials.

Thus we can reasonably conclude that a probability is always a numerical value between 0 and 1.

Property I

$$0 \leq P(A) \leq 1$$

NOTES:

1. The probability is 0 if the event cannot occur.
2. The probability is 1 if the event occurs every time.

Property 2

$$\sum_{\text{all outcomes}} P(A) = 1$$

Property 2 states that if we add up the probabilities of each of the sample points in the sample space, the sum must equal 1. This follows, since when we sum up all the probabilities, we are really asking, "What is the probability the experiment will yield an outcome?" and this will happen every time.

We are now ready to assign probabilities to $P(I)$ and $P(NI)$. The events I and NI represent the sample space for the experiment (stock price movement a year from now). The investment counselor's personal judgment was

$$P(I) = 9 \cdot P(NI)$$

From property 2 we know

$$P(I) + P(NI) = 1$$

Substituting $9 \cdot P(NI)$ for $P(I)$, we get

$$9 \cdot P(NI) + P(NI) = 1$$
$$10 \cdot P(NI) = 1$$
$$P(NI) = \frac{1}{10} = \mathbf{0.1}$$
$$P(I) = 1.0 - P(NI) = \mathbf{0.9}$$

odds

The statement "it is nine times more likely to increase (I) than not increase (NI)" is often expressed as "the odds are 9 to 1 in favor of increase." (It is also written as 9:1.) **Odds** are simply another way of expressing probabilities. To convert odds for or against an event to probability statements, we can use the following rules:

1. If the **odds in favor** of event A are a to b, then

$$P(A) = \frac{a}{a + b}$$

2. If the **odds against** event A are c to d, then

$$P(A) = \frac{d}{c + d}$$

To illustrate these rules, consider the preceding statement, "the odds favoring I are 9 to 1." Then $P(I)$, found by use of rule 1, is $9/(9 + 1) = \mathbf{0.9}$. Now consider the statement of odds "5 to 1 against a recession" in the news article at the beginning of the chapter. The probability P (recession) is then $1/(1 + 5)$, or $\frac{1}{6}$.

Exercises

4-13 The probability that an interview with a job candidate will result in a job offer is $\frac{1}{5}$.
   (a) Explain what this statement means.
   (b) Express this statement in terms of odds.

4-14 The probability that a direct mail advertisement will result in a sale is 0.01.
   (a) Explain what this statement means.
   (b) Express this statement in terms of odds.

4-15 Three managers of a company were asked to assign probabilities to the chances of success of the new advertising campaign. Their respective subjective probability assignments were as follows:

Possible Outcome of the Advertising Campaign	Manager's Probability Assignments		
	Manager A	Manager B	Manager C
Highly successful	0.5	0.6	0.3
Successful	0.4	0.3	0.3
Not successful	0.3	−0.1	0.3

Do any of the manager's assignments represent valid probability assignments? If not, why not?

4-16 Suppose on Monday you have $100 in your checking account and write four checks to pay four bills. The amounts of the checks are $5, $10, $20, and $25. The first two are to local stores, the latter two to out-of-state stores. Two checks are presented to your bank for payment on Thursday. List the sample space representing your checking account balance at the end of the day on Thursday. If each check is equally likely to be processed in 4 days, what probabilities would you assign to each outcome? Is this probability reasonable?

4-17 Find the probability that event A will happen if the odds are:
   (a) 7 to 3 in favor of event A
   (b) 1 to 3 in favor of event A
   (c) 3 to 2 against event A
   (d) 3 to 5 against event A

4-18 State whether theoretical, empirical, or subjective probability assignments are appropriate for the sample space described in:
(a) exercise 4-7
(b) exercise 4-9
(c) exercise 4-10
(d) experiment 4-1
(e) experiment 4-2
(f) experiment 4-10

Unit 2 Calculating Probabilities of Compound Events

compound events

Compound events are formed by combining several simple events. The probabilities of the following four compound events will be studied in the remainder of this chapter:

1. the probability of the complementary event, $P(\bar{A})$
2. the probability that either event A or event B will occur, $P(A \text{ or } B)$
3. the probability that both events A and B will occur, $P(A \text{ and } B)$
4. the probability that event A will occur given that event B has occurred, $P(A|B)$

NOTE: When we determine which compound probability we are seeking, it is not enough to look for words "either/or," "and," or "given" in the question. We must carefully examine the question to determine what combinations of events are called for.

Section 4-5 Complementary Events, Mutually Exclusive Events, and the Addition Rule

complementary events

Complement of an Event A

The set of all sample points in the sample space that do not belong to event A. The complement of event A is denoted by $\bar{A}$ (read "A complement").

For example, the complement of the event "success" is "failure"; the complement of "heads" is "tails" for the tossing of one coin; the complement of "at least one head" in 10 tosses of a coin is "no heads."

By combining the information in the definition of complement with property 2, we can say

$$P(A) + P(\bar{A}) = 1.0 \quad \text{for any event } A$$

It then follows that

$$P(\bar{A}) = 1 - P(A) \tag{4-3}$$

NOTE: Every event A has a complementary event $\bar{A}$. Complementary probabilities are very useful when the question asks for the probability of "at least one." Generally, this probability represents a combination of several events, but the complementary event "none" is a single outcome. It is easier to solve for the complementary event's probability and get the answer by using formula (4-3).

Illustration 4-2

Two coins are tossed. What is the probability that at least one head appears?

Solution Let event A be the occurrence of no heads; then $\bar{A}$ represents the occurrence of one or more heads, that is, at least one head. The sample space is $\{(H, H), (H, T), (T, H), (T, T)\}$.

$$P(A) = \frac{1}{4} = 0.25 \quad \text{[using formula (4-2)]}$$

$$P(\bar{A}) = 1 - P(A) = 1 - \frac{1}{4} = \frac{3}{4} = 0.75 \quad \text{[using formula (4-3)]} \quad \square$$

Mutually Exclusive Events

mutually exclusive events

Mutually Exclusive Events

Events defined in such a way that the occurrence of one event precludes the occurrence of any of the other events. (In short, if one of them happens, the other cannot happen.)

The following illustrations give examples of events that are mutually exclusive.

Illustration 4-3

In the experiment where two coins were tossed and the events of no heads (0H), one head (1H), and two heads (2H) were observed, the events 1H and

2H are examples of mutually exclusive events.

$$S = \{(H, H), (H, T), (T, H), (T, T)\}$$
$$1H = \{(H, T), (T, H)\} \quad \text{and} \quad 2H = \{(H, H)\}$$

The events $1H$ and $2H$ have no common sample points (i.e., no intersection); therefore they are mutually exclusive events. □

Illustration 4-4

Consider an experiment in which two dice are rolled. Three events are defined:

A: The sum of the numbers on the two dice is 7

B: The sum of the numbers on the two dice is 10

C: Each of the two dice shows the same number

Let's determine if these three events are mutually exclusive.

Are events A and B mutually exclusive? Yes, they are, since the sum on the two dice cannot be both 7 and 10 at the same time. If a sum of 7 occurs, the sum cannot possibly be 10.

Figure 4-4 presents the sample space for this experiment. This sample space is the same as that shown in experiment 4-7, except that ordered pairs are used in place of the pictures. The ovals, diamonds, and rectangles show the ordered pairs that are in events A, B, and C, respectively. We can see that events A and B do not **intersect** at a common sample point. Therefore they are mutually exclusive. Point (5, 5) satisfies both events B and C. Therefore B and C are not mutually exclusive. Two dice can each show a 5, which satisfies C; and the total of the two 5s satisfies B.

intersection

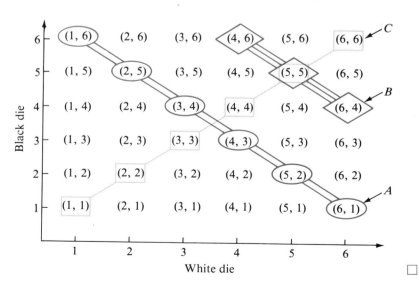

FIGURE 4-4
Sample Space for the Roll of Two Dice

□

Venn diagrams

Figures 4-5, 4-6, and 4-7 are **Venn diagrams** that show the relationships between the events defined in illustration 4-4. Venn diagrams are simple "area" or "region" diagrams. The basic concept of a Venn diagram is that those points (and only those points) belonging to a given event are shown inside a circle. (For further background information, see the *Study Guide*.)

Figure 4-5 shows that events A and B are mutually exclusive. Six sample points belong to event A, and three sample points belong to event B. The two events do not have a point in common. Therefore the sets represented in the Venn diagram do not intersect; that is, they are mutually exclusive.

Figure 4-6 presents a similar situation. Events A and C do not intersect; therefore they are mutually exclusive.

In figure 4-7 events C and B share a common sample point (5, 5). Notice that the point (5, 5) is in the intersection of the two Venn diagram circles. Consequently, events C and B are not mutually exclusive.

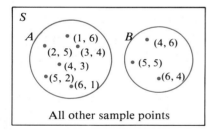

FIGURE 4-5
Venn Diagram: Events A and B Are Mutually Exclusive

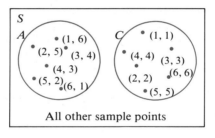

FIGURE 4-6
Events A and C Are Mutually Exclusive

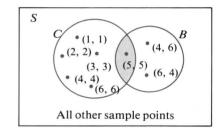

FIGURE 4-7
Events B and C Are Not Mutually Exclusive

Addition Rule

Let us now consider the compound probability $P(A \text{ or } B)$, where A and B are mutually exclusive events.

Illustration 4-5

Earlier in this chapter we considered an experiment where two coins were tossed and we observed either no heads ($0H$), one head ($1H$), or two heads ($2H$). We found the probabilities of these events to be $P(0H) = \frac{1}{4}$, $P(1H) = \frac{1}{2}$, and $P(2H) = \frac{1}{4}$, as shown in the histogram of figure 4-8. The area in this histogram represents probability.

Let's find the probability of at least one head. This event, "at least one head," can be restated as "either one head or two heads occurred." These two events are represented on the histogram by the two shaded bars. We can reasonably say that the probability of at least one head is $\frac{3}{4}$, the sum of $P(1H) = \frac{1}{2}$ and $P(2H) = \frac{1}{4}$. This value can be justified by looking at the sample space, where three of the four sample points satisfy the event. □

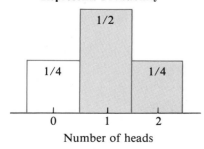

The Toss of Two Coins; the Area Represents Probability

FIGURE 4-8
Histogram

When events are **not mutually exclusive**, we cannot find the probability that one or the other occurs by simply adding the individual probabilities, as shown previously. Why not? Let's look at an illustration and see what happens when events are not mutually exclusive.

Illustration 4-6

Find the probability that a sum of 10 or a pair of double numbers is observed when a pair of dice are rolled. [Double numbers are (1, 1), (2, 2), (3, 3), and so on; this event is the same as event C defined in illustration 4-4.]

Solution $P(10) = \frac{1}{12}$ and $P(\text{double}) = \frac{1}{6}$. If we add the two probabilities, we obtain $P(10 \text{ or double}) = \frac{1}{12} + \frac{1}{6} = \frac{1}{4}$. However, if we go to the sample space (figure 4-4) and count the number of points that satisfy the statement "a sum of 10 or a double was observed," we find eight points. Using an adaptation of formula (4-2), we obtain

$$P(10 \text{ or double}) = \frac{n(10 \text{ or double})}{n(S)} = \frac{8}{36} = \frac{2}{9}$$

But $\frac{2}{9}$ does not equal $\frac{1}{4}$. The probability of an event cannot have two values. What caused the difference?

Look at the sample space and you will find that the sample point (5, 5) satisfies both "sum is 10" and "double." In adding the two probabilities, we "double counted" the probability of point (5, 5). However, this practice is not allowed. ☐

We can add probabilities to find the probability of an "or" compound event, but we must make an adjustment in situations like the previous example.

general addition rule

General Addition Rule

Let A and B be two events defined in a sample space S.

$$P(A \text{ or } B) = P(A) + P(B) - P(A \text{ and } B) \quad (4\text{-}4a)$$

special addition rule

Special Addition Rule

Let A and B be two events defined in a sample space. If A and B are **mutually exclusive** events, then

$$P(A \text{ or } B) = P(A) + P(B) \quad (4\text{-}4b)$$

This equation can be expanded to consider **more than two** mutually exclusive events:

$$P(A \text{ or } B \text{ or } C \text{ or} \ldots \text{or } E) = P(A) + P(B) + P(C) + \cdots + P(E)$$
$$(4\text{-}4c)$$

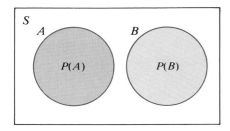

FIGURE 4-9
Mutually Exclusive Events

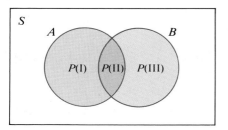

FIGURE 4-10
Nonmutually Exclusive Events

The key to this formula is the property "mutually exclusive." If two events are mutually exclusive, double counting of sample points will not occur. If events are not mutually exclusive, then when probabilities are added, the double counting will occur.

Let's look at some examples. In figure 4-9 events A and B are mutually exclusive. Simple addition is justified, since the total probability of the shaded regions is sought. [In a Venn diagram, probabilities are often represented by enclosed areas. In figure 4-9, for example, we've shaded the areas representing the probabilities $P(A)$ and $P(B)$.]

In figure 4-10 events A and B are not mutually exclusive. The probability of the event "A and B," $P(A$ and $B)$, is represented by the area contained in region II. The probability $P(A)$ is represented by the area of circle A. That is, $P(A) = P(\text{region I}) + P(\text{region II})$. In addition, $P(B) = P(\text{region II}) + P(\text{region III})$. And $P(A$ or $B)$ is the sum of the probabilities associated with the three regions: $P(A$ or $B) = P(\text{I}) + P(\text{II}) + P(\text{III})$. However, if $P(A)$ is added to $P(B)$, we have

$$P(A) + P(B) = [P(\text{I}) + P(\text{II})] + [P(\text{II}) + P(\text{III})]$$
$$= P(\text{I}) + 2P(\text{II}) + P(\text{III})$$

This result is the double count previously mentioned. However, if we subtract one measure of region II from this total, we will be left with the correct value.

The addition formula (4-4b) is a special case of the more general rule stated in formula (4-4a). If A and B are mutually exclusive events, $P(A$ and $B) = 0$. (They cannot both happen at the same time.) Thus the last term in formula (4-4a) is 0 when events are mutually exclusive.

Illustration 4-7

One white die and one black die are rolled. Find the probability that the white die shows a number smaller than 3 or the sum of the dice is greater than 9.

Solution I A = white die shows a 1 or a 2; B = sum of both dice is 10, 11, or 12.

$$P(A) = \frac{12}{36} = \frac{1}{3} \quad \text{and} \quad P(B) = \frac{6}{36} = \frac{1}{6}$$

$$P(A \text{ or } B) = P(A) + P(B) - P(A \text{ and } B)$$

$$= \frac{1}{3} + \frac{1}{6} - 0 = \frac{1}{2}$$

[$P(A$ and $B) = 0$, since the events do not intersect.] □

Solution 2

$$P(A \text{ or } B) = \frac{n(A \text{ or } B)}{n(S)} = \frac{18}{36} = \frac{1}{2}$$

(Look at the sample space, figure 4-4, and count.) □

Exercises

4-19 Determine whether or not each of the following pairs of events are mutually exclusive. An investment manager makes decisions concerning the investment of $100,000:

(a) She invests $50,000 in bonds; she invests $70,000 in a money market fund.

(b) She invests $50,000 in bonds; she invests $50,000 in a money market fund.

(c) She invests $10,000 in bonds; she buys no bonds.

(d) She invests $70,000 in bonds; she invests no money in the money market fund.

4-20 Determine whether or not each of the following pairs of events are mutually exclusive.

(a) You make a sale; the sale exceeds $100.

(b) You don't make a sale; the sale exceeds $100.

(c) You pass this course; you receive a B.

(d) You make a ledger entry; the ledger entry is correct.

(e) The customer pays cash; the customer charges the purchase.

(f) The stock price goes up; the dividend rate of the stock goes up.

4-21 $P(A) = 0.4$, $P(B) = 0.2$, and A and B are mutually exclusive. Find the following:

(a) $P(A)$

(b) $P(A \text{ or } B)$

(c) $P(A \text{ and } B)$

4-22 An investor has $10,000 to invest. If the probability that she will buy common stock is 0.6 and the probability that she will buy bonds is 0.5, are the two events mutually exclusive? Explain.

4-23 An investor buys a straddle on stock. A straddle is a bet that the stock price will either increase or decrease significantly within a certain time period. For the straddle to make money, Xerox's stock price must increase or decrease by $10 within the next 30 days. If the probability it will increase by $10 is 0.4 and the probability it will decrease by $10 is 0.2, what is the probability the straddle will make money?

4-24 If $P(A) = 0.4$, $P(B) = 0.4$, and $P(A \text{ and } B) = 0.2$, find $P(A \text{ or } B)$.

4-25 If $P(A) = 0.1$, $P(B) = 0.4$, and $P(A \text{ and } B) = 0.1$, find $P(A \text{ or } B)$.

4-26 Consider the following two events:

A: A customer reads *Newsweek*

B: A customer reads *Time*

If you know that 30% of your customers read *Newsweek*, 30% read *Time*, and 55% read at least one of two newsmagazines, what is the probability a customer will read both magazines?

Section 4-6 Independence, Multiplication Rule, and Conditional Probability

We will study next the compound event that A and B both occur. For example, what is the probability that two heads occur when two coins (one penny and one nickel) are tossed? Let A represent the occurrence of a head on the penny and B represent the occurrence of a head on the nickel. What is $P(A \text{ and } B)$? $P(A \text{ and } B)$ is found by using the definition $n(A \text{ and } B)/n(S)$. The sample space for experiment 4-3 suggests that this value should be $\frac{1}{4}$. How can we obtain this value by using $P(A)$ and $P(B)$? Both $P(A)$ and $P(B)$ have values of $\frac{1}{2}$. If $P(A)$ is multiplied by $P(B)$, we obtain $\frac{1}{4}$. Thus we might suspect that $P(A \text{ and } B)$ equals $P(A) \cdot P(B)$.

Consider this example. A is the event that 2 shows on a white die, and B is the event that 2 shows on a black die. If both dice are rolled once, what is the probability that two 2s occur?

$$P(A) = \frac{1}{6} \quad \text{and} \quad P(B) = \frac{1}{6}$$

$$P(A \text{ and } B) = \frac{n(A \text{ and } B)}{n(S)} = \frac{1}{36}$$

However, note that $\frac{1}{6}$ multiplied by $\frac{1}{6}$ is also $\frac{1}{36}$. In this case multiplication yields the correct answer. Multiplication does not always work, however. For example, $P(\text{sum of 7 and double})$ when two dice are rolled is 0 (as seen in figure 4-4). However, if $P(7)$ is multiplied by $P(\text{double})$, we obtain $(\frac{1}{6})(\frac{1}{6}) = \frac{1}{36}$.

Multiplication does not work for $P(\text{sum of 10 and double})$ either. By definition and by inspection of the sample space, we know that $P(10 \text{ and double}) = \frac{1}{36}$ [the point (5, 5) is the only element]. However, if we multiply $P(10)$ by $P(\text{double})$, we obtain $(\frac{3}{36}) \cdot (\frac{6}{36}) = \frac{1}{72}$. The probability of this event cannot be both values.

The property that is required for multiplying probabilities is **independence**. Multiplication worked in the two examples cited previously because the events were independent. In the other two cases, the events were not independent.

Independence and Conditional Probabilities

independent events

Independent Events

Two events A and B are independent events if the occurrence (or nonoccurrence) of one does not affect the probability assigned to the occurrence of the other.

Consider again the experiment where one penny and one nickel are tossed. Intuitively, it appears reasonable that a result of a head or a tail on a nickel has no effect on the experiment of tossing a penny. The two events can be thought of as separate events. The same is true with rolling two dice and looking for 2s on both.

dependent events

Lack of independence, called *dependence*, is demonstrated by the following illustration. Reconsider the experiment of rolling two dice and observing the two events "sum of 10" and "double." As stated previously, $P(10) = \frac{3}{36} = \frac{1}{12}$ and $P(\text{double}) = \frac{6}{36} = \frac{1}{6}$. Does the occurrence of 10 affect the probability of a double? Think of it this way. A sum of 10 has occurred; it must be one of the following: $\{(4, 6), (5, 5), (6, 4)\}$. One of these three possibilities is a double. Therefore we must conclude that the $P(\text{double},$ *knowing* 10 has occurred), written $P(\text{double}|10)$, is $\frac{1}{3}$. Since $\frac{1}{3}$ does not equal the original probability of a double, $\frac{1}{6}$, we can conclude that the event 10 has an effect on the probability of a double. Therefore "double" and "10" are dependent events.

Whether events are independent is often clear from examination of the events in question. Rolling one die does not affect the outcomes of a second roll. However, in many cases independence is not self-evident, and the question of independence may itself be of special interest. Consider the events "having a checking account at a bank" and "having a loan account at the same bank." The possession of a checking account at a bank may increase the probability that a person has a loan account. This relationship has practical implications. For example, advertising loan programs to checking account clients would make sense if they are more likely to apply for loans than are people who are not customers of the bank.

One approach to the problem is to *assume* independence or dependence. The correctness of the probability analysis depends on the truth of the assumption. In practice, we often assume independence and compare *calculated* probabilities with *actual frequencies* of outcomes to infer whether the assumption of independence is warranted.

conditional probability

Conditional Probability

The probability that A will occur given that B has occurred. This probability is represented by the symbol $P(A|B)$.

The previous definition of independent events can now be written in a more formal manner.

Independent Events

Two events A and B are independent events if

$$P(A|B) = P(A) \quad \text{or} \quad P(B|A) = P(B) \tag{4-5}$$

Let's become more familiar with conditional probability. Consider the experiment where a single die is rolled: $S = \{1, 2, 3, 4, 5, 6\}$. Two events that can be defined for this experiment are $B =$ an even number occurs and $A =$ a 4 occurs. Then $P(A) = \frac{1}{6}$. Event A is satisfied by exactly one of the six equally likely sample points in S. The conditional probability of A given B, $P(A|B)$, is found in a similar manner, but S is no longer the sample space. Think of it this way. A die is rolled out of your sight, and you are told that the number showing is even. That is the given condition. Knowing this condition, you are asked to assign a probability to the event that the even number is a 4. The new (or reduced) sample space contains only three possibilities: $\{2, 4, 6\}$. Each of the three outcomes is equally likely; thus $P(A|B) = \frac{1}{3}$.

What we have done can be written as:

$$P(A|B) = \frac{P(A \text{ and } B)}{P(B)} \tag{4-6}$$

Thus for our example

$$P(A|B) = \frac{1/6}{1/2} = \frac{1}{3}$$

Illustration 4-8

Age discrimination laws prohibit employers from considering the fact that a job applicant is 45 or more years of age as a negative factor in hiring. Suppose a firm has 150 applicants for 30 job openings. Some of the applicants are over 45. Table 4-4 shows the breakdown by age of who was hired. Suppose one person is selected at random from these 150 applicants.

TABLE 4-4
Probabilities of Who Is Hired, by Age

	Age of Applicant		
Employment Decision	A, Under 45	B, 45 or Older	Total
Not hired, NH	80	40	120
Hired, H	20	10	30
Total	100	50	150

(a) What is the probability that the person selected was hired knowing that he or she is 45 or over, $P(H|B)$?

(b) What is the probability that the person selected was hired knowing that he or she is under 45, $P(H|A)$?

(c) What is the probability that the person selected was hired, $P(H)$?

(d) Is there any evidence of age discrimination?

Solution

(a) First we consider $P(H|B)$. Given that the applicant is 45 or older, the sample space is reduced to those applicants in column B of table 4-4. (The 100 applicants in column A are under 45 and have been eliminated from this problem.) Of the remaining 50, 10 are hired and 40 are not hired. Thus we have

$$P(H|B) = \frac{10}{50} = 0.20$$

(b) Now let's solve for $P(H|A)$. Of the 100 applicants under 45 (column A only), 20 are hired. Thus we have

$$P(H|A) = \frac{20}{100} = 0.20$$

(c) Since 30 of the 150 applicants were hired, the probability of being hired is

$$P(H) = \frac{30}{150} = 0.20$$

(d) When we examine the results, we find that $P(H|B) = P(H|A) = P(H)$. That is, the probability of having been hired given that the applicant is 45 or older is the same as the probability of having been hired given that the applicant is under 45, which is the same as the probability of being hired for all applicants. So the data do not indicate age discrimination. □

Multiplication Rule

general multiplication rule

General Multiplication Rule

A and B are two events defined in sample space S. Then

$$P(A \text{ and } B) = P(A) \cdot P(B|A) \quad (4\text{-}7a)$$

or

$$P(A \text{ and } B) = P(B) \cdot P(A|B) \quad (4\text{-}7b)$$

special multiplication rule

Special Multiplication Rule

Let A and B be two events defined in sample space S. If A and B are **independent events**, then

$$P(A \text{ and } B) = P(A) \cdot P(B) \quad (4\text{-}8a)$$

This formula can be expanded. If $A, B, C, \ldots, G$ are independent events, then

$$P(A \text{ and } B \text{ and } C \text{ and} \ldots G) = P(A) \cdot P(B) \cdot P(C) \ldots P(G) \quad (4\text{-}8b)$$

If events A and B are independent, then the general multiplication rule [formulas (4-7a) and (4-7b)] reduces to the special multiplication rule, shown above as formulas (4-8a) and (4-8b)

Illustration 4-9

One white and one black die are rolled. Find the probability that the sum of their numbers is 7 and that the number on the black die is larger than the number on the white die.

Solution $A = $ sum is 7; $B = $ black number larger than white number. The "and" requires the use of the multiplication rule. However, we do not yet know if events A and B are independent. Refer to figure 4-4 for the sample space of this experiment. We see that $P(A) = \frac{6}{36} = \frac{1}{6}$. Also, $P(A|B)$ is obtained from the reduced sample space, which includes 15 points above the gray diagonal line. Of the 15 equally likely points, 3 of them—(1, 6), (2, 5), and (3, 4)—satisfy event A. Therefore $P(A|B) = \frac{3}{15} = \frac{1}{5}$. Since this value is different from $P(A)$, the events are dependent. So we must use formula (4-7b) to obtain $P(A$ and $B)$.

$$P(A \text{ and } B) = P(B) \cdot P(A|B) = \frac{15}{36} \cdot \frac{3}{15} = \frac{3}{36} = \frac{1}{12}$$

Since the two events in illustration 4-9 are defined for an equally likely sample space, we also could use an adaptation of formula (4-2) to find the probability:

$$P(A \text{ and } B) = \frac{n(A \text{ and } B)}{n(S)} \qquad (4\text{-}9)$$

Thus for our example

$$P(A \text{ and } B) = \frac{3}{36} = \frac{1}{12} \qquad \square$$

NOTE: This formula is a modification of formula (4-2) and requires the sample points within the sample space to be equally likely.

NOTES:

1. Independence and mutual exclusiveness are two completely separate concepts.
2. The term *mutually exclusive* refers to whether the events can occur together, whereas *independence* refers to the effect that one event has on the probability of the other event's occurrence.
3. If two events are *not* mutually exclusive, then their corresponding sets of sample points intersect.
4. The relationship between independence (or dependence) and mutual exclusiveness (or nonmutual exclusiveness) is summarized by the following four statements:

   (a) If two events are known to be mutually exclusive, then they are also dependent events. For example: If $P(A) = 0.3$ and $P(B) = 0.4$ and if A and B are mutually exclusive, then $P(A|B) = 0$. Since $P(A) \neq P(A|B)$, A and B are dependent events.

   (b) If two events are known to be independent, then they are *not* mutually exclusive events. For example: If $P(C) = 0.2$ and $P(D) = 0.5$ and if C and D are independent, then $P(C \text{ and } D) = 0.2 \times 0.5 = 0.10$. Since $P(C \text{ and } D)$ is greater than 0, C and D intersect. Therefore C and D are nonmutually exclusive events.

   (c) If two events are known to be nonmutually exclusive, then they may be either independent *or* dependent events. For example: (i) If $P(C) = 0.2$, $P(D) = 0.5$, and $P(C \text{ and } D) = 0.1$ (C and D intersect; that is, they are nonmutually exclusive), then $P(C|D) = 0.1/0.5 = 0.2$. Since $P(C) = P(C|D)$, C and D are independent events. (ii) If $P(E) = 0.2$, $P(F) = 0.5$, and $P(E \text{ and } F) = 0.05$ (E and F intersect; that is, they are nonmutually exclusive), then $P(E|F) = 0.05/0.5 = 0.1$. Since $P(E) \neq P(E|F)$, E and F are dependent events.

   (d) If two events are known to be dependent, then they may be *either* mutually exclusive *or* nonmutually exclusive. For example: (i) If $P(E) = 0.2$, $P(F) = 0.5$, and $P(E|F) = 0.1$ [E and F are dependent, since $P(E) \neq P(E|F)$], then $P(E \text{ and } F) = 0.5 \times 0.1 = 0.05$.

Since $P(E$ and $F)$ is greater than 0, E and F are intersecting, or nonmutually exclusive, events. (ii) If $P(A) = 0.3$, $P(B) = 0.4$, and $P(B|A) = 0$ [A and B are dependent, since $P(B) \neq P(B|A)$], then $P(A$ and $B) = 0.3 \times 0 = 0$. Since $P(A$ and $B) = 0$, A and B do not intersect, or are mutually exclusive events.

Exercises

4-27 If $P(A) = 0.5$, $P(B) = 0.2$, and A and B are independent, what is $P(A$ and $B)$? $P(B|A)$?

4-28 $P(A) = 0.4$, $P(B) = 0.2$, and $P(A$ and $B) = 0.06$.
   (a) What is $P(A|B)$?
   (b) Are A and B independent?

4-29 Determine whether or not each of the following pairs of events are independent:
   (a) rolling a die and observing a 1 and then rolling it again and observing another 1.
   (b) drawing a heart from a regular deck of playing cards and then drawing another heart from the same deck without replacing the first card
   (c) same as (b) except the first card drawn is replaced before the second drawing
   (d) owning a red automobile and having blonde hair
   (e) bad weather and being in an automobile accident

4-30 A manufacturing process historically produces, on the average, 5 defective items for every 100 items produced.
   (a) If two items are randomly selected, what is the probability both are defective?
   (b) If three items are randomly selected, what is the probability all are defective?

4-31 Last year Value Worth Investment Advisor chose IBM as its best-listed stock buy and Glaxo as its best over-the-counter stock buy. Both increased in price over the year. If 80% of all listed stocks and 70% of all over-the-counter stocks increased in price last year, what is the probability that by pure random selection the investment service's best-listed stock buy and best over-the-counter stock buy would have both increased in price?

4-32 Suppose that when a job candidate comes to interview for a job at K Industries, the probability that the candidate will want the job (event A) after the interview is 0.88. Also, the probability that K Industries will want the candidate (event B) is 0.45.
   (a) If $P(A|B) = 0.92$, find the probability $P(A$ and $B)$.
   (b) Find the probability $P(B|A)$.

4-33 In a survey of high school students and their attitudes toward business, 0.85 said "honesty is the best policy" and 0.28 agreed that "success in business requires some dishonesty." Assuming that the survey results accurately reflect the attitudes of all high school students and further assuming that the two attitudes are independent of one another, what is the probability that:

(a) a student did not say "honesty is the best policy" and agreed that "success in business requires some dishonesty"

(b) a student said "honesty is the best policy" or agreed that "success in business requires some dishonesty"

Section 4-7 Combining the Rules of Probability

Many probability problems can be presented by tree diagrams. When these diagrams are possible, the addition and multiplication rules can be applied quite readily.

To illustrate the use of tree diagrams in the solution of probability problems, consider the following: The vice-president of New Product Research and Development is considering three proposed new product ideas. Let's denote them as R, B, and W. She has staff and funds to pursue only one idea. She sends the proposals to the engineering department and asks it to eliminate the proposal least likely to be technically feasible. She then forwards the remaining two to the marketing department and asks it to eliminate the proposal that is least marketable. The tree diagram of figure 4-11 represents all the possible combinations of proposals that can be eliminated during the two decision steps.

If we assume that any project is equally likely to be eliminated at each stage, we can assign a probability to each branch segment of the tree, as shown in figure 4-12. Notice that the set of branches that initiate from a

FIGURE 4-11
All Possible Combinations That Can Be Eliminated

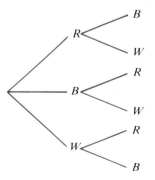

FIGURE 4-12
Probabilities of All Possible Combinations

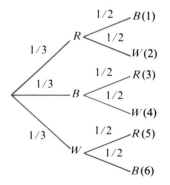

single point have a total probability of 1. Four such sets of branch segments are in this diagram. The tree diagram shows six distinct outcomes. Reading down: path (1) shows (R, B), path (2) shows (R, W), and so on. (*Note*: Each outcome of the experiment is represented by a path that begins at the common starting point and ends at a terminal point on the right.)

The probability associated with the outcome (R, B)—that is, R eliminated by engineering and B eliminated by marketing—is found by multiplying P(R eliminated by engineering) by P(B eliminated by marketing|R eliminated by engineering). The two probabilities $\frac{1}{3}$ and $\frac{1}{2}$ are shown on the two branches of path (1) in figure 4-12. The $\frac{1}{2}$ is the conditional probability required by the multiplication rule. Thus along the branches that form a path, we will multiply.

Some events may be composed of more than one outcome. For example, suppose we ask for the probability that project R is selected. You will find two outcomes that satisfy this event, path (4) or path (6). With "or" we will use the addition rule, formula (4-4b). Since the branches of a tree diagram represent mutually exclusive events, we have

$$P(R \text{ selected}) = P[(W, B) \text{ or } (B, W)]$$

$$= \left(\frac{1}{3}\right)\left(\frac{1}{2}\right) + \left(\frac{1}{3}\right)\left(\frac{1}{2}\right)$$

$$= \frac{1}{6} + \frac{1}{6} = \frac{1}{3}$$

NOTES:

1. Multiply along the branches.
2. Add across the branches.

Let us now consider an example where all the rules are placed in perspective.

Illustration 4-10

A firm plans to test a new product. It plans to randomly select one market area to test the product. The market areas can be categorized on the basis of location and population density. The number of markets in each category is presented in table 4-5.

TABLE 4-5
Number of Markets, by Location and by Population Density

	Population Density		
Location	Urban, U	Rural, R	Total
East, E	25	50	75
West, W	20	30	50
Total	45	80	125

What is the probability that the test market selected is in the East, $P(E)$? In the West, $P(W)$? What is the probability that the test market is in an urban area, $P(U)$? In a rural area, $P(R)$? What is the probability that the market is a western rural area, $P(W \text{ and } R)$? What is the probability it is an eastern or urban area, $P(E \text{ or } U)$? What is the probability that if it is in the East, it is an urban area, $P(U|E)$? Are location and population density independent? (What do we mean by independence or dependence in this situation?)

Solution   The first four probabilities, $P(E)$, $P(W)$, $P(U)$, and $P(R)$, represent "or" questions. For example, $P(E)$ means that the area is an eastern urban area or an eastern rural area. Since in this and the other three cases the two components are mutually exclusive (an area can't be both urban and rural), the desired probabilities can be found by simply adding. Since in each case the probabilities are added across all the rows or columns of the table, the totals are found in the total column or row.

$$P(E) = \frac{75}{125} \quad \text{(total for East divided by total number of markets)}$$

$$P(W) = \frac{50}{125} \quad \text{(total for West divided by total number of markets)}$$

$$P(U) = \frac{45}{125} \quad \text{(total for urban divided by total number of markets)}$$

$$P(R) = \frac{80}{125} \quad \text{(total for rural divided by total number of markets)}$$

Now we solve for $P(W \text{ and } R)$. There are 30 western rural markets and 125 markets in all. Thus

$$P(W \text{ and } R) = \frac{30}{125}$$

Note that $P(W) \cdot P(R)$ does *not* give the right answer.

$$\left[\left(\frac{50}{125}\right) \cdot \left(\frac{80}{125}\right) = \frac{32}{125} \right]$$

Therefore location and population density are dependent events.

$P(E \text{ or } U)$ can be solved in several different ways. The most direct way is to simply examine the table and count the number of markets that satisfy the condition that they are in the East or they are urban. We find the number of markets to be 95 (25 + 50 + 20). Thus

$$P(E \text{ or } U) = \frac{95}{125}$$

Note that the first 25 markets were both in the East and urban; thus E and U are not mutually exclusive events.

Another way to solve for $P(E \text{ or } U)$ is to use the addition formula:

$$P(E \text{ or } U) = P(E) + P(U) - P(E \text{ and } U)$$

which yields

$$\frac{75}{125} + \frac{45}{125} - \frac{25}{125} = \frac{95}{125}$$

A third way to solve the problem is to recognize that the complement of $(E \text{ or } U)$ is $(W \text{ and } R)$. Thus $P(E \text{ or } U) = 1 - P(W \text{ and } R)$. Using the previous calculation, we get $1 - 30/125 = \mathbf{95/125}$.

Finally, we solve for $P(U|E)$. Looking at the table, we see that 75 markets are in the East. Of the 75 eastern markets, 25 are urban. Thus

$$P(U|E) = \frac{25}{75}$$

The conditional probability formula also could be used:

$$P(U|E) = \frac{P(U \text{ and } E)}{P(E)} = \frac{(25/125)}{(75/125)} = \frac{25}{75} \qquad \square$$

Although each rule for computing compound probabilities has been discussed separately, you should not think that they are only used separately. In many cases they are combined to solve problems. Consider the following two illustrations.

Illustration 4-11

A production process produces light bulbs. On the average, 20% of all bulbs produced are defective. Each item is inspected before being shipped. The inspector misclasses an item 10% of the time; that is

$$P(\text{classified good}|\text{defective item}) = P(\text{classified defective}|\text{good item})$$
$$= 0.10$$

What proportion of the items will be classified good?

Solution What do we mean by the event "classified good"?

$$G = \text{item good}$$
$$D = \text{item defective}$$
$$CG = \text{item called good by inspector}$$
$$CD = \text{item called defective by inspector}$$

CG consists of two possibilities: "the item is good and is correctly classified good" or "the item is defective and is misclassified good." Thus

$$P(CG) = P[(CG \text{ and } G) \text{ or } (CG \text{ and } D)]$$

Since the two possibilities are mutually exclusive, we can start by using the addition rule (formula 4-4b).

$$P(CG) = P(CG \text{ and } G) + P(CG \text{ and } D)$$

The condition of a bulb and its classification by the inspector are not independent. The general multiplication rule must be used. Therefore

$$P(CG) = [P(G) \cdot P(CG|G)] + [P(D) \cdot P(CG|D)]$$

Substituting the known probabilities, we get

$$P(CG) = [(0.8)(0.9)] + [(0.2)(0.1)] = 0.72 + 0.02 = \mathbf{0.74}$$

That is, 74% of the items are classified good. □

Illustration 4-12

Let's reconsider the previous example. Suppose only items that pass inspection are shipped. Items not classified good are scrapped. What is the quality of the shipped items? That is, what percentage of the items shipped are good, $P(G|CG)$?

Solution   Using the conditional probability formula (4-6), we get

$$P(G|CG) = \frac{P(G \text{ and } CG)}{P(CG)}$$

In illustration 4-11 we found $P(CG)$, and $P(G \text{ and } CG)$ was also found as the first term in the "or" probability statement. Thus

$$P(G|CG) = \frac{P(G) \cdot P(CG|G)}{P(CG)}$$

$$= \frac{(0.8)(0.9)}{0.74} = 0.9729 = \mathbf{0.973}$$

In other words, 97.3% of all items shipped will be good. Inspection increases the quality of items sold from 80% good to 97.3% good. □

Exercises

4-34 If $P(A) = 0.7$ and $P(B|A) = 0.2$, find $P(A \text{ and } B)$.

4-35 Whenever a new home is purchased, Dirk Insurance Company mails an advertisement for mortgage insurance to the new homeowner. The company encloses a coupon for the homeowner to return if more information is desired. If the coupon is returned, a salesperson will call on the customer. Records show that 1 out of 100 advertisements results in a response and that a salesperson can sell insurance to 1 out of every 10 customers visited. What is the probability that a new homeowner will purchase mortgage insurance from Dirk Insurance Company?

4-36 A market survey found that 90% of the individuals questioned own a color television set. The survey also found that 70% of the individuals own a stereo set of some kind. Of those who own a color television set, 68% own a stereo.

(a) What is the probability that a randomly selected individual will own both a color television and a stereo set?

(b) What is the probability that a randomly selected individual will own a color television or a stereo set?

4-37 A recent market survey found that if a business executive is enrolled in a frequent flyer bonus program, the probability that he or she will specify a particular airline when requesting travel arrangements is 60%, whereas if he or she is not enrolled, it is only 20%. Forty percent of all business executives are enrolled in a frequent flyer bonus program.

(a) What is the probability that a randomly selected business executive will specify a particular airline?

(b) Given that an executive has specified a particular airline, what is the probability that he or she is enrolled in a frequent flyer bonus program?

Section 4-8 Bayes's Rule

The Reverend Thomas Bayes (1702–1761), an English Presbyterian minister and mathematician, developed an expanded form for conditional probabilities. This expanded rule, called Bayes's rule, allows us to revise (or adjust) the probabilities assigned to events in accordance with new information.

Bayes's rule

Bayes's Rule

$$P(A_i|B) = \frac{P(A_i) \cdot P(B|A_i)}{\sum [P(A_i) \cdot P(B|A_i)]} \qquad (4\text{-}10)$$

where $A_1, \ldots, A_n$ is an all-inclusive set of possible outcomes given B.

Although Bayes's formula looks difficult, if you use a tabular approach, you will find it easy to use.

Illustration 4-13

Let's solve illustration 4-12 by using Bayes's rule.

Solution The first step is to set up a table that shows all the possible outcomes given event B, that is, "shipped." These outcomes are listed in the first column of table 4-6.

In the second column we list the probabilities for each of the A_i outcomes in the first column. In the third column we list the conditional probability that B happens for each A_i, $P(B|A_i)$. [For our illustration, $P(B|A_1) = P(CG|G)$ and $P(B|A_2) = P(CG|D)$.] These first three columns represent the information obtained from the problem.

TABLE 4-6
Tabular Presentation of Given Information

(1) A_i, Possible Outcomes	(2) $P(A_i)$	(3) $P(B|A_i)$
A_1, item good	0.8	0.9
A_2, item defective	0.2	0.1
Total	1.0 ck	

To solve for the conditional probabilities $P(A_i|B)$, the first calculation is to multiply each number in the row of column (2) by the number in the same row of column (3). This product is placed in column (4) of the table (see table 4-7). The column is labeled $P(A_i$ and $B)$. The values calculated

TABLE 4-7
Tabular Solution of Bayes's Rule

(1) A_i, Possible Outcomes	(2) $P(A_i)$	(3) $P(B|A_i)$	(4) $P(A_i$ and $B)$ $= P(A_i)P(B|A_i)$	(5) $P(A_i|B)$
A_1, item good	0.8	0.9	0.72	0.72/0.74 = **0.973** $= P(G|\text{shipped})$
A_2, item defective	0.2	0.1	0.02	0.02/0.74 = **0.027** $= P(D|\text{shipped})$
Total	1.0 ck		0.74 = $P(B)$	1.000 ck

represent the probability that both A_i and B will occur. Thus 72% of the items produced will be good and classified good; 2% of the items will be defective and classified good.

The second step is to add up column (4). The sum represents $P(B)$. Thus 74% of the items produced will be classified good.

Finally, the answers we are looking for, the conditional probabilities $P(A_i|B)$, are obtained by dividing each number in column (4) by the total of column (4). The results are placed in column (5) and are the answers. Thus 0.973 is the proportion of items classified good that are good. And 0.027 is the proportion of items classified good that are actually defective.

NOTE: The total of columns (2) and (5) must equal 1. The total of column (3) need not equal 1.

Bayes's rule is of special interest because it gives us a mechanism to revise initial probability estimates when new information is learned, as we see in the next illustration.

Illustration 4-14

prior

revised, or posterior

Consider the situation where we feel that the probability that a stock is a good buy is 0.4. That is, our **prior** (before new information) probabilities are $P(\text{good buy}) = 0.4$ and $P(\text{bad buy}) = 0.6$. Now we find out that an investment service that has a record of being right 80% of the time recommends the stock. What should be our **revised, or posterior** (after new information), probability that the stock is a good buy, that is, $P(\text{good buy}|\text{investment service recommends it})$, or $P(A_i|B)$?

Solution Using the Bayesian tabular analysis, we find that the revised, or posterior, probability is 0.727 that the stock is a good buy and 0.273 that it is a bad buy (see table 4-8).

TABLE 4-8
Tabular Analysis for Illustration 4-14

A_i	$P(A_i)$	$P(B\|A_i)$	$P(A_i)P(B\|A_i)$	$P(A_i\|B)$
A_1, good buy	0.4	0.8	0.32	0.32/0.44 = 0.727
A_2, bad buy	0.6	0.2	0.12	0.12/0.44 = 0.273
Total	1.0 ck		0.44 = $P(B)$	1.000 ck

In analyzing the use of Bayes's rule to revise prior probability estimates in light of new information, we note the following relationship: The stronger the prior probability, the less the effect of the new information on changing the probability. Also, the more conclusive the new information, the greater the impact on the revised probability.

Exercises

4-38

	Male		Female		
	Skilled	Unskilled	Skilled	Unskilled	Total
Satisfied	350	150	25	100	625
Unsatisfied	150	100	75	50	375
Total	500	250	100	150	1,000

Use the accompanying table on worker satisfaction in the Russell Microprocessor Company.

(a) Find the probability that an unskilled employee is satisfied with the work.

(b) Find the probability that a skilled woman employee is satisfied with the work.

(c) Is satisfaction for women employees independent of their being skilled or unskilled?

4-39 Given the information in the accompanying table, compute $P(A_1|UF)$ and $P(A_2|UF)$ by filling in the rest of the table. (UF = unfavorable survey results.)

	$P(A_i)$	$P[(UF)|A_i]$	$P(A_i$ and $UF)$	$P(A_i|UF)$
A_1, profitable	0.6	0.4		
A_2, not profitable	0.4	0.7		

4-40 Given the following:

$$P(A_1) = 0.2 \quad P(A_2) = 0.4 \quad P(A_3) = 0.3 \quad P(A_4) = 0.1$$

$$P(B|A_1) = 0.5 \quad P(B|A_2) = 0.4 \quad P(B|A_3) = 0.2 \quad P(B|A_4) = 0.1$$

Find $P(A_1|B)$, $P(A_2|B)$, $P(A_3|B)$, and $P(A_4|B)$.

In Retrospect

You now have studied the basic concepts of probability. These fundamentals need to be understood to allow us to continue our study of statistics. Probability is the vehicle of statistics, and we have begun to see how probabilistic events occur. We have explored theoretical and experimental probabilities of the same event. Does the experimental probability turn out to have the same value as the theoretical probability? Not exactly, but over the long run we have seen that it does have approximately the same value.

You must, of course, know and understand the basic definition of probability. The properties of mutual exclusiveness and independence as they apply to the concepts presented in this chapter must also be well understood.

Professor Samuelson, in the news article at the beginning of this chapter, uses odds to assess the likelihood of certain economic conditions occurring. You should recognize that odds are simply a way of expressing probabilities. What type of probabilities do you think these odds are: theoretical, empirical, or subjective?

From reading this chapter you are expected to be able to relate the "and" and the "or" of compound events to the multiplication and addition rules. You should also be able to calculate conditional probabilities and use Bayes's rule.

In the next three chapters, we will look at distributions associated with probabilistic events. These chapters will prepare us for the statistics that will follow. We must be able to predict the variability that the sample will show with respect to the population before we will be successful at inferential statistics, in which we describe the population based on the sample statistics available.

Chapter Exercises

4-41 If $P(A) = 0.6$, $P(B) = 0.3$, and $P(A \text{ and } B) = 0.18$, find the following:
(a) $P(A|B)$ (b) $P(\bar{A})$ (c) $P(A \text{ or } B)$

4-42 Gerco has raised the price of pocket butane lighters 4 out of the past 6 years. When Gerco has raised its price, its chief competitor, Lito, has correspondingly raised its price three out of the four times. Lito has never raised its price unless Gerco first increased its price. Assuming future events follow past trends, what are the following probabilities?
(a) that Gerco will raise its price next year
(b) that Gerco and Lito will raise their prices next year
(c) that Gerco or Lito will raise its price next year
(d) that neither Gerco nor Lito will raise its price next year

4-43 An inventory number at Retra Corporation consists of three digits. The first two digits indicate product line, and the third indicates style. Inventory is maintained on a computer system. When an item is sold, its inventory number is keypunched on a card and entered into the computer. If the keypunch operator randomly mispunches 1 out of every 100 digits, what are the following probabilities?
(a) that when an item is sold, it is incorrectly recorded
(b) that when an item is sold, the wrong product line is recorded regardless of style correctness
(c) that when an item is sold, the wrong product line and correct style are recorded
(d) that when an item is sold, the correct product line and wrong style are recorded

4-44 Two out of every ten manuscripts sent to a publisher would be profitable if published. Each manuscript is reviewed. The reviewer responds favorably to five out of every ten manuscripts that would be profitable and to one out of every ten manuscripts that would be unprofitable.
(a) If a reviewer likes a manuscript, what is the probability it will be profitable?
(b) If a reviewer does not like a manuscript, what is the probability it will be profitable?

4-45 A testing organization wishes to rate a particular brand of television. Six TVs are selected at random from stock, and the brand is judged to be satisfactory if nothing is found wrong with any of the six.
(a) What is the probability that the brand will be rated satisfactory if 10% of the TVs actually are defective?
(b) What is the probability that the brand will be rated satisfactory if 20% of the TVs actually are defective?
(c) What is the probability that the brand will be rated satisfactory if 40% of the TVs actually are defective?

4-46 State why each of the following pairs of events is or is not independent.
(a) A potential customer earns less than $10,000, and the potential customer decides to purchase a yacht.
(b) A student gets an A in statistics, and the student has an A average.
(c) A consumer earns less than $10,000, and the consumer purchases a pack of chewing gum.
(d) A car is blue, and the car is in an accident.
(e) A person reads *Fortune* magazine, and the person is a business executive.

4-47 (a) Find the probability of each of the various "sums" when two dice are rolled: $P(2), P(3), \ldots, P(12)$.
(b) Use the data recorded in exercise 4-5 (or use 50 rolls of a pair of dice) and find $P(2), P(3), P(4), \ldots, P(12)$. Compare the probabilities with the values found in part (a). Are they reasonably close in value?

4-48 A two-page advertising copy contains an error on one page. Two proofreaders review the copy. Each has an 80% chance of catching the error.
(a) What is the probability the error will be identified if each reads a different page?
(b) What is the probability the error will be identified if they both read both pages?
(c) What is the probability the error will be identified if the first person randomly selects a page to read and then the second person randomly selects a page, unaware of which page the first selected?

4-49 Events R and S are defined on the same sample space. If $P(R) = 0.2$ and $P(S) = 0.5$, explain why each of the following statements is either true or false.
(a) If R and S are mutually exclusive, then $P(R$ or $S) = 0.10$.
(b) If R and S are independent, then $P(R$ or $S) = 0.6$.
(c) If R and S are mutually exclusive, then $P(R$ and $S) = 0.7$.
(d) If R and S are mutually exclusive, then $P(R$ or $S) = 0.6$.

4-50 A supermarket is planning a sale of unmarked boxes of cookies, pretzels, and crackers. The 300 boxes of cookies, 300 boxes of pretzels, and 400 boxes of crackers all look identical. Consider the first box selected by a shopper.
(a) What is the probability it is a box of pretzels?
(b) What is the probability it is not a box of cookies?
(c) Are the two events "pretzels" and "cookies" mutually exclusive?
(d) Are the two events "pretzels" and "not cookies" mutually exclusive?
(e) Are the two events "pretzels" and "cookies" independent?
(f) Are the two events "pretzels" and "not cookies" independent?

4-51 The Getrich Tire Company is having a tire sale on tires salvaged from a train wreck. Of the 15 tires offered in the sale, 5 tires have suffered internal damage and the remaining 10 are damage free. You are to randomly select and purchase two of these tires.
(a) What is the probability that the tires you purchase are both damage free?
(b) What is the probability that exactly one of the tires you purchase is damage free?
(c) What is the probability that at least one of the tires you purchase is damage free?

4-52 C and D are mutually exclusive events defined on a common sample space. If $P(C) = 0.3$ and $P(D) = 0.2$, find the following:
(a) $P(C$ and $D)$ (b) $P(C$ or $D)$
(c) $P(\bar{C}$ and $D)$ (d) $P(C$ or $\bar{D})$
(e) $P(\bar{C}$ or $D)$ (f) $P(D|\bar{C})$

4-53 Consider the experiment in which a box contains five items, four of which are good and the other is defective. You are to randomly take two of these items from the box.
(a) Describe the sample space by a tree diagram.
(b) Assign probabilities to each branch of your tree diagram.

Consider these three events:

A: Draw a good item on first drawing
B: Draw a defective item on second drawing
C: Draw a good item on second drawing

Find the following probabilities:

(c) $P(A)$ (d) $P(B)$
(e) $P(C)$ (f) $P(\bar{A})$
(g) $P(\bar{B})$ (h) $P(\bar{C})$
(i) $P(A$ and $B)$ (j) $P(A$ or $B)$
(k) $P(A$ and $C)$ (l) $P(B$ and $C)$
(m) $P(A$ or $C)$ (n) $P(B$ or $C)$

4-54 Given $P(A) = 0.3$, $P(B) = 0.5$, and $P(A$ and $B) = 0.1$, find the following:

(a) $P(\bar{A})$ (b) $P(A$ or $B)$
(c) $P(A$ and $\bar{B})$ (d) $P(\bar{A}$ and $\bar{B})$
(e) $P(\overline{A \text{ and } B})$

4-55 Over the last several years, the distribution of grades A, B, C, D, and F in statistics classes has been 0.08, 0.26, 0.50, 0.10, and 0.06, respectively.

(a) What is the probability a student will receive a grade of C or better?
(b) Are the events "passing with a C or better" and "passing with an A" mutually exclusive?

4-56 Consider an experiment in which a die is weighted so that each even number occurs twice as often as each odd number.

(a) List the sample space for one roll of this die.
(b) Assign probabilities to each outcome.
(c) Find P (even number).
(d) Find P (number is 4 or more).
(e) Find $P(4|\text{even})$. Are the events "4" and "even" independent? Explain.
(f) Find $P(4|\text{odd})$, $P(\text{even}|4)$, and $P(\text{even}|\text{not } 4)$. Do these values agree with your answer in (e)?

4-57 When a plant is spot checked by the Environmental Protection Agency for three pollutants, the level of smoke exhaust is unsatisfactory 20% of the time, the level of chemical toxins is unsatisfactory 15% of the time, and the level of water pollution is unsatisfactory 25% of the time. Assume independence (perhaps unreasonable) among the three events and find these probabilities:

(a) that the plant has unsatisfactory levels of all three items
(b) that the plant has unsatisfactory levels of exactly two items
(c) that the plant has an unsatisfactory level of exactly one item
(d) that the plant has satisfactory levels of all three items

4-58 According to automobile accident statistics, one out of every six accidents results in an insurance claim of $100 or less in property damage. Three cars insured by an insurance company are involved in different accidents. Consider the following two events:

A: The majority of claims exceed $100

B: Exactly two claims are $100 or less

(a) List the sample points for this experiment.
(b) Are the sample points equally likely?
(c) Find $P(A)$ and $P(B)$.
(d) Are A and B independent? Justify your answer.

4-59 A slot machine consists of three reels with digits 0, 1, 2, 3, and 4 and a flower on each reel. When a coin is inserted and the lever pulled, each of the three reels spins independently and then stops at one of the six positions. Find these probabilities:

(a) that a flower shows on all three reels
(b) that a flower shows on exactly one reel
(c) that a flower shows on exactly two reels
(d) that no flowers appear
(e) that a three-digit sequence appears in order (e.g., 2, 3, 4)
(f) that one flower and two odd integers appear

4-60 The probabilities of employed females working in certain occupational groups are listed as follows:

Occupational Group	Probability
Professional-technical	0.16
Managerial-administrative	0.06
Sales	0.07
Clerical	0.34
Craft	0.02
Operatives, including transport	0.12
Nonfarm laborers	0.01
Service, except private households	0.18
Private households	0.03
Farm	0.01
Total	1.00

(a) What is the probability that a randomly selected employed female is either a professional or a managerial employee?

(b) What is the probability that two randomly selected employed females will both be clerical employees?

(c) What is the probability of a randomly selected employed female being both a sales and a clerical employee?

4-61 From a recent survey of females employed by a firm that predominantly employs females, the following information was generated in answer to the question, "Is college more important for a male than for a female?"

Age	Yes	No	Total
18–24	0.17	0.43	0.60
Over 24	0.26	0.14	0.40
Total	0.43	0.57	1.00

Using this information:

(a) What is the probability that a randomly selected female employee felt college was more important for a male than for a female?

(b) What is the probability that a randomly selected female answered "yes" or was "over 24?"

(c) What is the probability that a randomly selected female was "over 24" given that she answered "no" to the question?

4-62 The accompanying table shows the sentiments of 3,000 wage employees of the J and J Company on a proposal to emphasize fringe benefits rather

	Opinion			
Employee	Favor	Neutral	Opposed	Total
Male	700	300	200	1,200
Female	200	1,100	500	1,800
Total	900	1,400	700	3,000

than wage increases during an impending contract discussion. Calculate the probability that an employee selected at random from this group will be:

(a) neutral

(b) a female

(c) a female, given the person is opposed

4-63 A lobbyist for the Openpit Mining Corporation entertained 180 members of a state legislature, which was considering three bills of interest to Openpit. For each legislator, x = the number of votes cast by the legislator that were favorable to Openpit and y = the amount of money spent by the lobbyist in entertaining the legislator. A statistician at Openpit analyzed the legislator's votes and obtained the accompanying table.

	x				
y	0	1	2	3	Total
$ 0	2	2	2	2	8
10	2	2	2	6	12
20	0	24	24	22	70
50	10	16	24	40	90
Total	14	44	52	70	180

(a) Find $P(x = 0)$ and $P(x = 3)$.
(b) Find $P(x = 0 | y = 0)$; $P(x = 0 | y = 50)$.
(c) Find $P(x = 3 | y = 0)$; $P(x = 3 | y = 50)$.
(d) The company comptroller claimed the entertainment expenditure was a waste of money, since the votes and expenditures were independent. Are they independent?

4-64 A company is considering five alternative investments. Each will result in either a 20% profit or a 20% loss. The probability of success for each investment is 0.6; their outcomes are independent. The firm makes three investments.

(a) What is the probability that all three will be successful?
(b) What is the probability that none will be successful?
(c) What is the probability that at least two will be successful?

4-65 Reconsider exercise 4-64. The firm has $3 million to invest. Each investment alternative costs $1 million. The firm invests all $3 million.

(a) What is the probability it will make $600,000?
(b) What is the probability it will lose $600,000?
(c) What is the probability it will make at least $200,000?

4-66 Reconsider exercise 4-64. A firm decides to invest $3 million. It can invest $1, $2, or $3 million in a single investment, $1 million in each of three investment alternatives, or $2 million in one and $1 million in another.

(a) If it invests in one investment, what is the probability that it will (i) make $600,000, (ii) lose $600,000, and (iii) make at least $200,000?
(b) If it invests in two investments, what is the probability that it will (i) make $600,000, (ii) lose $600,000, and (iii) make at least $200,000?
(c) Compare your results for exercises 4-64, 4-65, and parts (a) and (b) above. What can you say about risk diversification of investments?

4-67 Solve this exercise by using Bayes's formula in tabular form. The treasurer's initial opinion is that there is a 30% chance that an investment will exceed expectations, a 50% chance that it will equal expectations, and a 20% chance that it will return less than expected. A private investment consulting service reviews the investment and reports that it should equal expectations. In the past the consultants were correct 60% of the time, underestimated the return 10% of the time, and overestimated the return 30% of the time. What should be the treasurer's revised probabilities?

4-68 Ninety percent of the insulators produced by Superior Insulator Company are satisfactory. The firm hires an inspector. The inspector inspects all the insulators and correctly classifies an item 90% of the time; that is, P(classify good|good) = P(classify defective|defective) = 0.9. Items classified good are shipped and those classified defective are scrapped.

(a) What percentage of items shipped can be expected to be good?
(b) What percentage of items scrapped can be expected to be good?

4-69 The firm in exercise 4-68 hires a second inspector, who has the same accuracy record. The second inspector inspects all insulators independently of the first inspector. What percentage of items shipped and what percentage of items scrapped can be expected to be good if items are shipped only if

(a) both inspectors independently say they are good
(b) at least one inspector says they are good

Challenging Problem

4-70 In sports, championships are often decided by two teams playing each other in a championship series. Often the fans of the losing team claim they were unlucky and their team is actually the better team. Suppose team A is the better team, and the probability it will beat team B in any given game is 0.6. What is the probability that the better team A will lose the series if it is:

(a) a one-game series?
(b) a best out of three series?
(c) a best out of seven series?
(d) Suppose the probability that A would beat B in any given game were actually 0.7. Recompute (a) through (c).
(e) Suppose the probability that A would beat B in a given game were actually 0.9. Recompute (a) through (c).
(f) What is the relationship between the "best" team winning and the number of games played? The best team winning and the probabilities that each will win?

Hands-On Problems

Design and conduct a probability experiment using two dice. Modify each die so it has three faces with one dot, two faces with two dots, and one face with three dots; or let the 1, 2, and 3 on each die represent a "1," the 4 and

5 a "2," and the 6 a "3." The dice should be two different colors, such as white and black.

4-1 Construct the sample space for the experiment. (*Hint*: It has nine sample points.)

4-2 Assign a theoretical probability to each sample point.

4-3 Find the theoretical probability of the first die showing
  (a) one dot
  (b) two dots
  (c) three dots

4-4 Roll the dice 100 times and record the results as ordered pairs.

4-5 Find the observed probability of each of the nine possible outcomes.

4-6 Find the observed probability of each of the possible outcomes (one, two, or three dots) on the first die.

4-7 In problems 4-2 and 4-3, 12 theoretical probabilities were calculated. In problems 4-5 and 4-6, 12 observed probabilities for the same set of events were calculated. Pair the 24 probability values. How well do the experimental probabilities compare with the theoretical values you calculated?

5 Probability Distributions (Discrete Variables)

Chapter Outline

5-1 Random Variables
To study probability distributions, a *numerical value* will be assigned to the random variable for each event.

5-2 Probability Distributions of Discrete Random Variables
The probability of any event may be expressed in the form of a *probability function*.

5-3 Mean and Variance of a Discrete Probability Distribution
Population parameters are used to measure probability distributions.

5-4 The Binomial Probability Distribution
Binomial probability distributions occur in situations where each *trial* of an experiment has *two possible outcomes*.

5-5 Mean and Standard Deviation of the Binomial Distribution
Two simple *formulas* are used to measure the binomial distribution.

5-6 The Poisson Probability Distribution
Poisson probability distributions occur in situations where *repeated observations* are made within a *unit of measure*.

Alphanumeric Terminal Market Overview

PERIPHERALS DIGEST

The alphanumeric terminal market had a relatively calm year. Competition was tougher than 1981, but technical innovations were fewer and less spectacular than in previous years. The definitions that have characterized terminal types are becoming increasing blurred. With distribution channels becoming more numerous, thanks to retailers, mail-order suppliers and industrial distributors, market analysts have a tougher time, but buyers have more options.

The terms "dumb terminal" and "low-end terminal" used to mean the same thing, but because of rampant price cutting, the less-than-$1000 low end now encompasses a number of smart or editing terminals. At the high-end, distinctions between intelligent terminals and desk-top microcomputers blur as more micros gain flexible communications capabilities and as micro sales induce terminal vendors to re-label their high-end terminals as microcomputer systems.

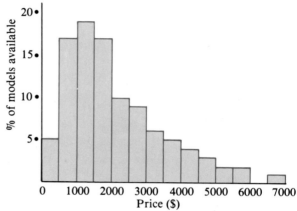

Data: Mini-Micro Systems

Terminal prices are falling and becoming more homogeneous. *Taken from the product tables that follow, these data mirror the intense competition in the dumb and low-end editing markets in which features are standardized, and price differentials of only a few percentage points affect large- and small-volume buying decisions.*

Copyright 1982 by Cahners Publishing Company. Reprinted by permission from *Mini-Micro Systems*, November 1982.

Chapter Objectives

Chapter 2 dealt with frequency distributions of data sets, and chapter 4 with the fundamentals of probability. Now we are ready to combine these ideas to form probability distributions, which are much like relative frequency distributions. The basic difference between probability and relative frequency distributions is the use of the random variable. The random variable of a probability distribution corresponds to the response variable of a frequency distribution.

In this chapter we will investigate discrete probability distributions and study measures of central tendency and dispersion for such distributions. Special emphasis will be given to the binomial random variable and its probability distribution, since it is the most important discrete random variable encountered in most fields of application. A discussion of the Poisson probability distribution is also included.

Section 5-1 Random Variables

The events in a probability experiment must be defined in such a manner that each event can be assigned a numerical value. This numerical value will be the value of the *random variable* under study.

random variable

Random Variable

A variable that assumes a unique numerical value for each of the outcomes in the sample space of a probability experiment.

In other words, a random variable is used to denote the outcome of a probability experiment. It can take on any of the numerical values that belong to the set of all possible outcomes of the experiment. (It is called "random" because the value it assumes is the result of a chance or random event.)

Each event in a probability experiment must also be defined in such a way that only one value of the random variable is assigned to it, and each event must have a value assigned to it. Typically, the **discrete random variable** is a count of something.

discrete random variable

The following illustrations demonstrate what we mean by random variable.

Illustration 5-1

An accountant audits five transactions and determines whether each transaction has been properly recorded. The random variable x is the number of correct transactions. It can take on integer values from 0 to 5.   □

Illustration 5-2

A salesman calls on 15 customers. The random variable x is the number of sales made. It can take on integer values from 0 to 15.   □

Illustration 5-3

Let the number of customers per day serviced by a company be a random variable. The random variable can take on integer values ranging from 0 to some very large number. □

Illustration 5-4

The length of time it takes a deposited check to clear through the banking process can be a random variable. An observed probability distribution would result if the data were presented as a relative frequency distribution.

For example, 0.20 of the checks are processed in 1 day, 0.25 in 2 days, 0.28 in 3 days, 0.15 in 4 days, and so on, as shown in figure 5-1. The random variable "length of time to process a check" is actually a continuous random variable, since it measures time. It only appears to be discrete because of the context in which it is presented. Continuous random variables and their distributions are the topics of chapter 6. □

FIGURE 5-1
Histogram

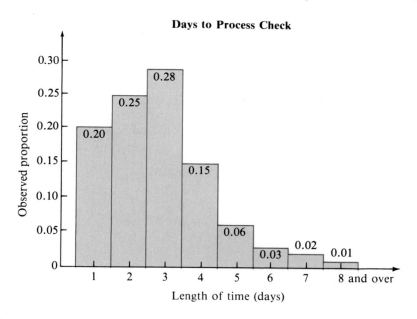

Exercises

5-1 Consider an experiment where three stocks listed on the New York Stock Exchange are randomly selected. Suppose we define the random variable x to be the number of stocks selected that are also included in the Dow-Jones index.

(a) Make a list of the possible values of x.

(b) List the sample space and assign a value of the random variable to each of the eight outcomes in your sample space.

(c) Is x a discrete or a continuous variable?

5-2 Two dice are rolled, and the random variable of interest is the total number of dots seen on the two dice.

(a) Make a list of the values that the random variable can assume.

(b) Use the sample space shown in figure 4-4 (p. 147) to assign a value of the random variable to each sample point.

(c) Is this random variable discrete or continuous?

5-3 Five television sets are inspected by a quality control department before they are shipped. Identify the random variable most likely to be of interest here and list its possible values.

5-4 Only a few of the records produced by a record company become hits. Ray Company releases new records in lots of 10. Identify a random variable that would be of interest to this company and list its possible values.

Section 5-2 Probability Distributions of Discrete Random Variables

Recall the experiment of tossing two coins that was used at the beginning of section 4-1. Two coins were tossed, and no heads, one head, or two heads were observed. If we define the random variable x to be the number of heads that are observed when two coins are tossed, x can take on the values 0, 1, or 2. The probability of each of these three events is the same as the probability we calculated in chapter 4.

$$P(x = 0) = P(0H) = \frac{1}{4}$$

$$P(x = 1) = P(1H) = \frac{1}{2}$$

$$P(x = 2) = P(2H) = \frac{1}{4}$$

TABLE 5-1
Probability Distribution, Tossing a Coin

x	P(x)
0	1/4
1	1/2
2	1/4

These probabilities can be listed in any number of ways, but they are best displayed in the form illustrated by table 5-1. Can you see why the name *probability distribution* is used?

probability distribution

Probability Distribution

A distribution of the probabilities associated with each of the values of a random variable.

In the experiment in which a single die is rolled and the number of dots on the top surface is observed, the random variable is the number observed. The probability distribution for this random variable is shown in table 5-2.

Sometimes it is convenient to write a rule that expresses the probability of an event in terms of the value of the random variable. This expression is typically written in formula form and is called a *probability function*.

probability function

Probability Function

A rule that assigns probabilities to the values of the random variables.

TABLE 5-2
Probability Distribution, Rolling a Die

x	P(x)
1	1/6
2	1/6
3	1/6
4	1/6
5	1/6
6	1/6

A probability function can be as simple as a list pairing the values of a random variable with their probabilities. Tables 5-1 and 5-2 show two such listings. However, a probability function is most often expressed in formula form.

Consider a die that has been modified to have one face with one dot, two faces with two dots, and three faces with three dots. Let x be the number of dots observed when the die is rolled. The probability distribution for this experiment is presented in table 5-3. Each of the probabilities can be represented by the value of x divided by 6. That is, each $P(x)$ is equal to the value of x divided by 6, where $x = 1$, 2, and 3. Thus $P(x) = x/6$ for $x = 1, 2, 3$ is the formula expression of the probability function of this experiment.

TABLE 5-3
Probability Distribution, Rolling the Modified Die

x	P(x)
1	1/6
2	2/6
3	3/6

The probability function of the experiment of rolling one ordinary die is $P(x) = \frac{1}{6}$ for $x = 1, 2, 3, 4, 5,$ and 6. This particular function is called a **constant function** because the value of $P(x)$ does not change as x changes.

Every probability function must display the two basic properties of probability. These two properties are (1) the probability assigned to each value of the random variable must be between 0 and 1 inclusive, that is,

$$0 \leq \text{each } P(x) \leq 1$$

and (2) the sum of the probabilities assigned to all the values of the random variable must equal 1, that is,

$$\sum_{\text{all } x} P(x) = 1$$

TABLE 5-4
Probability Distribution for $P(x) = x/10$

x	P(x)
1	1/10
2	2/10
3	3/10
4	4/10
	10/10 = 1

Is $P(x) = x/10$ for $x = 1, 2, 3,$ and 4 a probability function? To answer this question, we need only test the function in terms of the two basic properties. The probability distribution is shown in table 5-4. Property 1 is satisfied, since $\frac{1}{10}, \frac{2}{10}, \frac{3}{10},$ and $\frac{4}{10}$ are all numerical values between 0 and 1. Property 2 is also satisfied, since the sum of all four probabilities is exactly 1. Since both properties are satisfied, we can conclude that $P(x) = x/10$ for $x = 1, 2, 3, 4$ is a probability function. What about $x = 5$ (or any value other than 1, 2, 3, or 4) in the function $P(x) = x/10$ for $x = 1, 2, 3,$ and 4?

$P(x = 5)$ is considered to be 0. That is, the probability function provides a probability of 0 for all values of x other than the values specified.

Figure 5-2 presents a probability distribution graphically. Regardless of the specific representation, the values of the random variable are plotted on the horizontal scale, and the probability associated with each value of the random variable is plotted on the vertical scale. A discrete random variable should really be presented by a **line histogram**, since the variable can assume only discrete values. Figure 5-2 shows the probability distribution of $P(x) = x/10$ for $x = 1, 2, 3,$ and 4. This line representation makes sense because the variable x can assume only the values 1, 2, 3, and 4. The length of the vertical line represents the measure of the probability.

However, a regular **histogram** is more frequently used to present probability distributions. Figure 5-3 shows the same probability distribution of figure 5-2, but in histogram form. This representation suggests

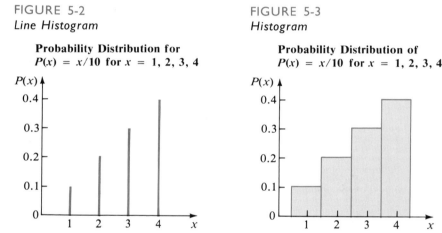

FIGURE 5-2
Line Histogram

FIGURE 5-3
Histogram

that x can assume all numerical values (fractions, decimals, etc.). The histogram probability distribution uses the physical area of each "bar" to represent its assigned probability. The bar for $x = 2$ is 1 unit wide (from 1.5 to 2.5) and is 0.2 units high. Therefore its area is 0.2 (1×0.2), the probability assigned to $x = 2$. The areas of the other bars can be determined in similar fashion. This area representation will be an important concept in chapter 6, when we begin work with continuous random variables.

Exercises

5-5 Test the following function to determine if it is a probability function. If it is not, try to make it a probability function. List the distribution of probabilities and sketch a histogram.

$$P(x) = \frac{4 - x}{6} \quad \text{for } x = 1, 2, 3$$

5-6 Test the following function to determine if it is a probability function. If it is not, try to make it a probability function. List the distribution of probabilities and sketch a histogram.

$$P(x) = \frac{x}{18} \quad \text{for } x = 2, 3, 4, 5$$

5-7 Test the following function to determine if it is a probability function. If it is not, try to make it a probability function.

$$P(x) = \frac{1}{6} \quad \text{for } x = 1, 2, 3, 4, 5, 6$$

(a) List the distribution of probabilities and sketch a histogram.
(b) Do you recognize $P(x)$? If so, identify it.

5-8 Test the following function to determine if it is a probability function. If it is not, try to make it a probability function.

$$P(x) = \frac{6 - |x - 7|}{36} \quad \text{for } x = 2, 3, 4, 5, 6, 7, \ldots, 11, 12$$

(a) List the distribution of probabilities and sketch a histogram.
(b) Do you recognize $P(x)$? If so, identify it.

Section 5-3 Mean and Variance of a Discrete Probability Distribution

Recall that the mean of a frequency distribution for a sample is

$$\bar{x} = \frac{\sum xf}{\sum f} = \frac{\sum xf}{n}$$

which also can be written as

$$\bar{x} = \sum \left(x \cdot \frac{f}{n} \right)$$

We learned in our study of probability that the probability of an event is the expected relative frequency of its occurrence. Thus if we replace f/n with $P(x)$, we can find the mean of a theoretical probability distribution:

$$\text{Mean value of } x = \sum [x \cdot P(x)] \qquad (5\text{-}1)$$

mean

Recall that the expected relative frequency represents what would occur in the long run. The mean of x is therefore the mean of the entire population of experimental outcomes. The symbol for the **mean** value of x in a probability distribution is μ, and formula (5-1) can be written as

$$\mu = \sum [x \cdot P(x)] \qquad (5\text{-}2)$$

NOTES:
1. $\bar{x}$ is the mean of a sample.
2. s is the standard deviation of the individual elements of the sample.
3. $\bar{x}$ and s are called **sample statistics**.
4. μ (Greek letter mu) is the mean of the population under consideration.
5. σ (Greek letter sigma) is the standard deviation of the individual elements of the population under consideration.
6. μ and σ are called **population parameters**. (A parameter is a constant. μ and σ are typically unknown values.)

Illustration 5-5

Let's return to the previous probability function: $P(x) = x/10$ for $x = 1, 2, 3,$ and 4. We can find its mean by use of formula (5-2) after we compile a probability distribution table (see table 5-5). The mean μ, or the sum of the products of x times $P(x)$, is $\frac{30}{10}$, or **3.0**. [The sum of the probabilities, $\sum P(x)$, must be 1.0. Use this as a check.]

TABLE 5-5
Probability Distribution for $P(x) = x/10$

x	$P(x)$	$x \cdot P(x)$
1	1/10	1/10
2	2/10	4/10
3	3/10	9/10
4	4/10	16/10
	10/10 = 1.0	$\mu = 30/10 = 3.0$

variance of probability distribution

The **variance** of a discrete probability distribution is defined in much the same way as is the variance of sample data.

$$\sigma^2 = \sum [(x - \mu)^2 \cdot P(x)] \qquad (5\text{-}3)$$

The variance of x of the probability distribution discussed in illustration 5-5 is found in table 5-6. The variance σ^2 is $\frac{10}{10}$, or 1.0. ($\mu = 3$ was found in table 5-5.) □

TABLE 5-6
Finding the Variance for $P(x) = x/10$

x	$x - \mu$	$(x - \mu)^2$	$P(x)$	$(x - \mu)^2 P(x)$
1	-2	4	1/10	4/10
2	-1	1	2/10	2/10
3	0	0	3/10	0/10
4	1	1	4/10	4/10
				$\sigma^2 = 10/10$

Formula (5-3) is often inconvenient to use; fortunately, it can be reworked to appear in the following form:

$$\sigma^2 = \sum [x^2 \cdot P(x)] - \{\sum [x \cdot P(x)]\}^2 \qquad (5\text{-}4a)$$

or

$$\sigma^2 = \sum [x^2 \cdot P(x)] - \mu^2 \qquad (5\text{-}4b)$$

It is left for you (exercise 5-13) to verify that formulas (5-4a) and (5-4b) are equivalent to formula (5-3).

Illustration 5-6

To find the variance of the probability distribution in illustration 5-5 by using formula (5-4a), we will need to add two additional columns to table 5-5. These columns are shown in table 5-7.

TABLE 5-7
Calculations Needed to Find the Variance for $P(x)$

x	$P(x)$	$x \cdot P(x)$	x^2	$x^2 \cdot P(x)$
1	1/10	1/10	1	1/10
2	2/10	4/10	4	8/10
3	3/10	9/10	9	27/10
4	4/10	16/10	16	64/10
	10/10	30/10		100/10

The variance is

$$\sigma^2 = \sum [x^2 \cdot P(x)] - \{\sum [x \cdot P(x)]\}^2$$
$$= \frac{100}{10} - \left(\frac{30}{10}\right)^2 = 10 - (3)^2 = 10 - 9 = 1.0$$

standard deviation of probability distribution   The **standard deviation** is the square root of the variance; therefore $\sigma = 1.0$. □

Illustration 5-7

United Electric has found from past experience that the probability of submitting a low electrical-contracting bid and thus winning a job is $\frac{1}{2}$. Last month the firm submitted three bids. Let the number of successful bids be the random variable x, which can assume values of 0, 1, 2, or 3. The probability distribution and extensions are presented in table 5-8. [In section 5-4 we will show you how the probability distribution $P(x)$ can be calculated. For now you can verify that the distribution shown in table 5-8 is correct by using the methods of chapter 4—say, by constructing a sample space and counting.]

TABLE 5-8
Probability Distribution and Extensions, Illustration 5-7

x	$P(x)$	$x \cdot P(x)$	x^2	$x^2 \cdot P(x)$
0	1/8	0/8	0	0/8
1	3/8	3/8	1	3/8
2	3/8	6/8	4	12/8
3	1/8	3/8	9	9/8
	8/8 = 1 ✓	$\mu = 12/8 = \mathbf{1.5}$		24/8 = 3.0

The mean μ is the mean number of successful bids expected per three bids. The mean is calculated in table 5-8 and is found to be **1.5**.

The variance is found with the aid of formula (5-4a):

$$\sigma^2 = \sum [x^2 \cdot P(x)] - \{\sum [x \cdot P(x)]\}^2$$

$$= 3.0 - (1.5)^2 = 3.0 - 2.25 = \mathbf{0.75}$$

The standard deviation is the positive square root of the variance:

$$\sigma = \sqrt{\sigma^2} = \sqrt{0.75} = 0.866 = \mathbf{0.87}$$

That is, 0.87 is the standard deviation expected among the number of successful bids when the bids are submitted in lots of three. □

Exercises

5-9 Given the probability function $P(x) = (4 - x)/6$ for $x = 1, 2, 3$, find the mean and standard deviation.

5-10 Given the probability function $P(x) = \frac{1}{6}$ for $x = 1, 2, 3, 4, 5, 6$, find the mean and standard deviation.

5-11 Given the probability function $P(x) = 6 - |x - 7|/36$ for $x = 2, 3, 4, 5, \ldots, 12$, find the mean and standard deviation.

5-12 (a) Draw a histogram of the probability distribution of the single-digit random numbers (0, 1, 2, ..., 9).

(b) Calculate the mean and standard deviation associated with the population of single-digit random numbers.

5-13 Verify that formulas (5-4a) and (5-4b) are equivalent to formula (5-3).

5-14 Sales of 1-pound cartons of cottage cheese at a convenience food store have the following distribution based on past history:

Number of Cartons Sold per Day, x	Probability
10	0.2
11	0.4
12	0.2
13	0.1
14	0.1
	1.0

Determine the mean and the standard deviation of x, the number of cartons sold per day.

Section 5-4 The Binomial Probability Distribution

Consider the following probability experiment. A six-question, multiple-choice quiz is to be given. You did not study the material to be quizzed and therefore decide to answer the six questions by randomly guessing the answers without reading the questions or the answers.

Answer Page to Quiz

Directions: Circle the best answer to each question.

1.   A   B   C
2.   A   B   C
3.   A   B   C
4.   A   B   C
5.   A   B   C
6.   A   B   C

Circle your answers before continuing.

Before we look at the correct answers to the quiz, let's think about some of the things we might consider when a quiz is answered in this way.

1. How many of the six questions do you think you answered correctly?
2. If an entire class answered the quiz by guessing, what do you think the average number of correct answers would be?
3. What is the probability that you selected the correct answer to all six questions?
4. What is the probability that you selected wrong answers to all six questions?

Let's construct the probability distribution associated with this experiment. Let x be the number of correct answers on your paper; x may then take on the values 0, 1, 2, ..., 6. Since each individual question has three possible answers and since only one answer is correct, the probability of selecting the correct answer to any particular question is $\frac{1}{3}$. The probability that a wrong answer is selected is $\frac{2}{3}$. $P(x = 0)$ is the probability that all questions were answered incorrectly.

$$P(x = 0) = \left(\frac{2}{3}\right)\left(\frac{2}{3}\right)\left(\frac{2}{3}\right)\left(\frac{2}{3}\right)\left(\frac{2}{3}\right)\left(\frac{2}{3}\right) = \frac{64}{729} \approx 0.09$$

These events are independent because each question is a separate event. We can therefore multiply the probabilities according to formula (4-8b). Also,

$$P(x = 6) = \left(\frac{1}{3}\right)^6 = \frac{1}{729} \approx 0.001$$

Before we find the other five probabilities, let's look at the events on a tree diagram (figure 5-4). Notice that event $x = 0$, "zero correct answers," is

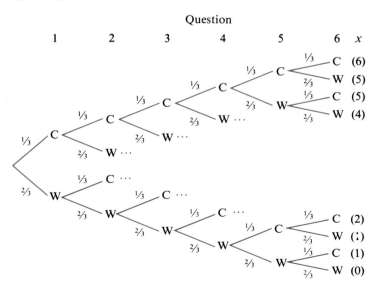

FIGURE 5-4
Tree Diagram for the Multiple-Choice Quiz

shown by the bottom branch, and that event $x = 6$, "six correct answers," is shown by the top branch. The other events, "one correct answer," "two correct answers," and so on, are represented by several branches of the tree.

Although not all branches are shown in figure 5-4, if we did show all branches, we would find that event $x = 1$ occurs on 6 different branches, event $x = 2$ occurs on 15 branches, $x = 3$ occurs on 20 branches, $x = 4$ on 15 branches, and $x = 5$ on 6 branches. Thus the remaining probabilities can be calculated by finding the probability of a single branch representing the event and multiplying it by the number of branches that represent that event.

$$P(x = 1) = \left[\left(\frac{1}{3}\right)\left(\frac{2}{3}\right)\left(\frac{2}{3}\right)\left(\frac{2}{3}\right)\left(\frac{2}{3}\right)\left(\frac{2}{3}\right)\right](6)$$

$$= \left(\frac{1}{3}\right)^1 \left(\frac{2}{3}\right)^5 (6) = \frac{192}{729} \approx 0.26$$

$$P(x = 2) = \left(\frac{1}{3}\right)^2 \left(\frac{2}{3}\right)^4 (15) = \frac{240}{729} \approx 0.33$$

$$P(x = 3) = \left(\frac{1}{3}\right)^3 \left(\frac{2}{3}\right)^3 (20) = \frac{160}{729} \approx 0.22$$

$$P(x = 4) = \left(\frac{1}{3}\right)^4 \left(\frac{2}{3}\right)^2 (15) = \frac{60}{729} \approx 0.08$$

$$P(x = 5) = \left(\frac{1}{3}\right)^5 \left(\frac{2}{3}\right)^1 (6) = \frac{12}{729} \approx 0.02$$

The answer to question 1 on page 190 is suggested by the probability distribution in table 5-9. The most likely occurrence would be to get two answers correct. One, two, or three correct answers are expected to result approximately 80% of the time $(0.26 + 0.33 + 0.22 = 0.81)$. We might also argue that $\frac{1}{3}$ (i.e., $\frac{2}{6}$) of the questions would be expected to be answered correctly. This information provides a reason for answering both questions

TABLE 5-9
Probability Distribution for the Quiz Example

x	P(x)
0	0.09
1	0.26
2	0.33
3	0.22
4	0.08
5	0.02
6	0.001
	1.001 Ⓒⓚ (because of round-off, the sum is not exactly 1)

1 and 2 on page 190 with the value of 2. Questions 3 and 4 were asked with the expectation that you would feel that the chance of "all correct" was very slight, 0.001. "All wrong" is not too likely, but it is not as rare an event as "all correct." "All wrong" has a probability of 0.09.

The correct answers to the quiz are B, C, B, A, C, and C. How many correct answers did you have? You might ask several people to answer the quiz by guessing at the answers. Then construct an observed relative frequency distribution and compare it with the distribution shown in table 5-9.

success or failure Many experiments result in outcomes that can be classified in one of two categories, **success or failure**. Examples of such experiments include the previous experiments of tossing coins or thumbtacks and other more practical experiments, such as making or not making a sale, determining the success or failure of a product, and determining the success or failure of a new business venture. There are experiments that have many outcomes that, under the right conditions, may fit this general description of being classified in one of two categories. For example, when we buy or sell a stock, we usually consider the random variable x to be the dollar profit or loss. However, if we are only interested in knowing whether or not we made money, only two outcomes are important: we made money or we didn't make money.

The experiments just described are usually referred to as *binomial probability experiments*.

---

binomial experiment

Binomial Probability Experiment

An experiment that is made up of repeated trials of a basic experimental event. The binomial experiment must possess the following properties:

trial
1. Each **trial** has two possible outcomes (success, failure).
2. There are n repeated independent and identical trials.
3. $P(\text{success}) = p$, $P(\text{failure}) = q$, and $p + q = 1$.

binomial random variable
4. The **binomial random variable** x is the count of the number of successful trials that occur; x may take on any integer value from 0 to n.

NOTE: The probability of success, p, remains constant for all the trials. The independence of the repeated trials gives us this property.

The two basic properties that a binomial experiment *must* demonstrate are properties 1 and 2. Properties 3 and 4 concern notation and the identification of the variables involved.

NOTE: It is of utmost importance that a probability p be assigned to the particular outcome that is considered to be the "success," since the binomial random variable indicates the number of successes that occur.

Illustration 5-8

Consider a salesman's success in selling new cars, where the number of sales is counted. The random variable x is the number of cars sold to n customers. Assume customers buy either one or no cars. Each customer is considered a trial, and P(selling a car) $= p$. □

Illustration 5-9

If you were an inspector on a production line in a plant that manufactures flashbulbs, you would be concerned with identifying the number of defective flashbulbs. You would probably define success as the occurrence of a defective bulb. This definition is not what we normally think of as success, but if we count defective bulbs in a binomial experiment, we must define success as a defective bulb. The random variable x indicates the number of defective bulbs found per lot of n bulbs. p represents P(bulb is defective), and q is P(bulb is good). □

NOTE: Independent trials means that the result of one trial does not affect the probability of success of any other trial in the experiment. In other words, the probability of success remains constant throughout the entire experiment.

Each binomial probability experiment has its own specific probability function. However, all such functions have a common format, and all such distributions display some similarities. Let's look at a relatively simple binomial experiment and discuss its probability function.

Let's reconsider illustration 5-7; three bids are made, and we observe the number of successful bids. This example is a binomial experiment because it displays all the properties of a binomial experiment:

1. Each trial (one bid) has two possible outcomes: success (low bid, what we are counting) and failure (not low bid).
2. n (three) repeated trials are independent (each is a separate bid, and the outcome of any one bid does not affect the outcome of any other bid).
3. The probability of success (low bid) is p ($\frac{1}{2}$), the probability of failure (not low bid) is q ($\frac{1}{2}$), and $p + q = 1$.
4. The random variable x is the number of successes (winning bids) that occurs in n (three) trials. x can assume the values 0, 1, 2, or 3.

Let's consider first the probability of $x = 1$:

$P(x = 1) = P$(exactly one low bid is observed in three bids)

We want to find this probability. What does it mean to say "exactly one low bid in three?" The probability that one bid is low is $\frac{1}{2}$. The probability that the other two bids are not low bids is $\frac{1}{2} \times \frac{1}{2} = \frac{1}{4}$. The probability that exactly one low bid will occur in three bids, then, must be related to $\frac{1}{2} \times \frac{1}{4} = \frac{1}{8}$ [from the multiplication rule, formula (4-8a)]. But we have one other consideration: In how many different ways can exactly one of three bids be low?

1. The first bid can be low and the next two not low.
2. The second bid can be low and the first and last not low.
3. The first two bids can be not low and the last low.

These combinations are the three ways in which three trials can result in exactly one success. This means that the $\frac{1}{8}$ found earlier can occur three different ways, resulting in $P(x = 1) = 3(\frac{1}{8}) = \frac{3}{8}$.

If we look at the case $x = 2$, we find a similar situation: there are three ways to obtain exactly two low bids. Thus $P(x = 2) = \frac{3}{8}$, also. We can find only one way for $x = 0$ and $x = 3$ to occur: each bid loses or each bid wins. In both cases the probability is equal to $(\frac{1}{2})^3 = \frac{1}{8}$. The rest of the probability distribution is exactly as shown in table 5-8.

If we take a careful look at the probability of each case shown in the tree diagram in figure 5-4, we can argue that the probability that a random variable x takes on a particular value in a binomial experiment is always the product of three basic factors. These three factors are as follows:

1. the probability of exactly x successes, p^x
2. the probability that failure will occur on the remaining $(n - x)$ trials, q^{n-x}
3. the number of ways that exactly x successes can occur in n trials

The number of ways that exactly x successes can occur in a set of n trials is represented by the symbol

$$\binom{n}{x}$$

binomial coefficient   which must always be a positive integer. This integer is termed the **binomial coefficient**. The binomial coefficient is found with the following formula:

$$\binom{n}{x} = \frac{n!}{x!(n-x)!} \qquad (5\text{-}5)$$

If the factorial notation, $n!$, is unfamiliar, see the *Study Guide*. See the *Study Guide* for general information on the binomial coefficient. The values for

$$\binom{n}{x}$$

when n is equal to or smaller than 20 are found in table 3 of appendix D.

This information allows us to form a general *binomial probability function*.

binomial probability function

Binomial Probability Function

$$P(x) = \binom{n}{x} \cdot p^x \cdot q^{n-x} \quad \text{for } x = 0, 1, 2, \ldots, n \quad (5\text{-}6)$$

where $P(x)$ is the probability of obtaining x successes in n independent trials. p is the probability of success on any particular trial; q is the probability of failure on any particular trial; and $q = 1 - p$.

When this general binomial probability function is applied to the illustration with three bids, we find

$$P(x) = \binom{3}{x} \cdot \left(\frac{1}{2}\right)^x \cdot \left(\frac{1}{2}\right)^{3-x} \quad \text{for } x = 0, 1, 2, 3$$

Let's calculate each of the probabilities and see if they form a probability distribution.

$$P(x = 0) = \binom{3}{0} \cdot \left(\frac{1}{2}\right)^0 \cdot \left(\frac{1}{2}\right)^3 = 1 \cdot 1 \cdot \frac{1}{8} = \frac{1}{8}$$

$P(x = 0) = \frac{1}{8}$ is the probability that no low bids will occur in three bids. (*Note:* $(\frac{1}{2})^0 = 1$.)

$$P(x = 1) = \binom{3}{1} \cdot \left(\frac{1}{2}\right)^1 \cdot \left(\frac{1}{2}\right)^2 = 3 \cdot \frac{1}{2} \cdot \frac{1}{4} = \frac{3}{8}$$

$P(x = 1) = \frac{3}{8}$ is the probability that exactly one winning bid will occur in three bids.

$$P(x = 2) = \binom{3}{2} \cdot \left(\frac{1}{2}\right)^2 \cdot \left(\frac{1}{2}\right)^1 = 3 \cdot \frac{1}{4} \cdot \frac{1}{2} = \frac{3}{8}$$

$P(x = 2) = \frac{3}{8}$ is the probability that two winning bids will occur in three bids.

$$P(x = 3) = \binom{3}{3} \cdot \left(\frac{1}{2}\right)^3 \cdot \left(\frac{1}{2}\right)^0 = 1 \cdot \frac{1}{8} \cdot 1 = \frac{1}{8}$$

$P(x = 3) = \frac{1}{8}$ is the probability that three winning bids will occur in three bids. Since each of these probabilities is between 0 and 1, and the sum of all the probabilities is exactly 1, this is a probability distribution.

Let's look at another example. Consider an experiment in which five customers in a department store are observed and the number who charge their purchases rather than pay cash (or by check) is recorded. In addition, we know that, historically, 75% of all persons charge their purchases. The random variable x is the number of persons who charged their purchases when they checked out. This is a binomial experiment. Let's identify the various properties.

1. Each person's purchase (trial) has two outcomes: "charged" or "not charged."

2. Five persons are being observed (trials), so $n = 5$. These individual trials are independent, since the way one customer pays should not affect the way the other customers pay for their purchases.

3. $p = 0.75$ and $q = 0.25$.

4. x is the number of charged sales recorded for five separate purchases.

The binomial function is

$$P(x) = \binom{n}{x} \cdot p^x \cdot q^{n-x} = \binom{5}{x} \cdot (0.75)^x \cdot (0.25)^{5-x}$$

for $x = 0, 1, 2, 3, 4,$ and 5. The probabilities are

$$P(x = 0) = \binom{5}{0} \cdot (0.75)^0 \cdot (0.25)^5 = 1 \cdot 1 \cdot (0.25)^5 = 0.0010$$

$$P(x = 1) = \binom{5}{1} \cdot (0.75)^1 \cdot (0.25)^4 = 5 \cdot (0.75)^1 \cdot (0.25)^4 = 0.0146$$

$$P(x = 2) = \binom{5}{2} \cdot (0.75)^2 \cdot (0.25)^3 = 10 \cdot (0.75)^2 \cdot (0.25)^3 = 0.0879$$

$$P(x = 3) = \binom{5}{3} \cdot (0.75)^3 \cdot (0.25)^2 = 10 \cdot (0.75)^3 \cdot (0.25)^2 = 0.2637$$

$$P(x = 4) = \binom{5}{4} \cdot (0.75)^4 \cdot (0.25)^1 = 5 \cdot (0.75)^4 \cdot (0.25)^1 = 0.3955$$

$$P(x = 5) = \binom{5}{5} \cdot (0.75)^5 \cdot (0.25)^0 = 1 \cdot (0.75)^5 \cdot 1 = 0.2373$$

$$\sum P(x) = 1.0000$$

The preceding distribution of probabilities indicated that the single most likely value of x is 4, the event of observing exactly four charges in five sales. What is the least likely number of charges that would probably be observed?

Let's consider one more binomial probability problem.

Illustration 5-10

The manager of Steve's Food Market guarantees that none of his cartons of one dozen eggs will contain more than one bad egg. If a carton contains more than one bad egg, he will replace the whole dozen and allow the customer to keep the original eggs. The probability of an individual egg being bad is 0.05. What is the probability that Steve will have to replace a given carton of eggs?

Solution Is this a binomial experiment? To find out, let x be the number of bad eggs found in a carton of a dozen eggs. $p = 0.05$, and let the inspection of each egg be a trial resulting in finding a "bad" or "not bad" egg. To find the probability that the manager will have to make good on his guarantee, we need the probability function associated with this experiment:

$$P(x) = \binom{12}{x} \cdot (0.05)^x \cdot (0.95)^{12-x} \qquad \text{for } x = 0, 1, 2, \ldots, 12$$

The probability that Steve will replace a dozen eggs is the probability that $x = 2, 3, 4, 5, \ldots$, or 12. Recall that $\sum P(x) = 1$, that is,

$$P(x = 0) + P(x = 1) + P(x = 2) + \cdots + P(x = 12) = 1$$

Therefore

$$P(x = 2) + P(x = 3) + \cdots + P(x = 12) = 1 - [P(x = 0) + P(x = 1)]$$

Finding $P(x = 0)$ and $P(x = 1)$ and subtracting them from 1 is easier than finding each of the other probabilities.

$$P(x) = \binom{12}{x} \cdot (0.05)^x \cdot (0.95)^{12-x}$$

$$P(0) = \binom{12}{0} \cdot (0.05)^0 \cdot (0.95)^{12} = \mathbf{0.540}$$

$$P(1) = \binom{12}{1} \cdot (0.05)^1 \cdot (0.95)^{11} = \mathbf{0.341}$$

NOTE: The value of many binomial probabilities, for small values of n and common values of p, are found in table 4 of appendix D. In this example we have $n = 12$ and $p = 0.05$, and we want the probabilities for $x = 0$ and 1. We need to locate the section of table 4 where $n = 12$; find the column marked $p = 0.05$ and read the numbers opposite $x = 0$ and 1. We find 540 and 341, as shown in table 5-10. (Look these values up in table 4.) The decimal point was left out of the table to save space, and we must replace it. It belongs at the front of each entry. Therefore our values are 0.540 and 0.341.

TABLE 5-10

Abbreviated Portion of Table 4 in Appendix D, Binomial Probabilities

n	x	0.01	0.05	0.10	0.20	0.30	0.40	*p* 0.50	0.60	0.70	0.80	0.90	0.95	0.99	x
⋮	⋮	⋮	⋮												⋮
→12	0	886	540	282	069	014	002	0+	0+	0+	0+	0+	0+	0+	0
→	1	107	341	377	206	071	017	003	0+	0+	0+	0+	0+	0+	1
	2	006	099	230	283	168	064	016	002	0+	0+	0+	0+	0+	2
	3	0+	017	085	236	240	142	054	012	001	0+	0+	0+	0+	3
	4	0+	002	021	133	231	213	121	042	008	001	0+	0+	0+	4
⋮	⋮	⋮													⋮

Now let's return to our illustration:

$$P(\text{replacement}) = 1 - (0.540 + 0.341) = \mathbf{0.119}$$

If $p = 0.05$ is correct, Steve will be busy replacing cartons of eggs. If he replaces 11.9% of all the cartons of eggs he sells, he is not likely to make a profit. This situation suggests that he should adjust his guarantee. For example, if he were to replace a carton of eggs only when four or more were found bad, he would expect to replace only 0.003 cartons [$1.0 - (0.540 + 0.341 + 0.099 + 0.017)$], or 0.3% of the cartons sold. Notice that he will be able to control his *risk* (probability of replacement) if he adjusts the value of the random variable stated in his guarantee.

Exercises

5-15 Evaluate each of the following:

(a) 4!

(b) 7!

(c) 0!

(d) $\dfrac{6!}{2!}$

(e) $\dfrac{5!}{3!\,2!}$

(f) $\dfrac{6!}{4!(6-4)!}$

(g) $(0.3)^4$

(h) $\binom{7}{3}$

(i) $\binom{5}{2}$

(j) $\binom{3}{0}$

(k) $\binom{4}{1}(0.2)^1(0.8)^3$

(l) $\binom{5}{0}(0.3)^0(0.7)^5$

5-16 Evaluate each of the following:

(a) 5!

(b) 6!

(c) $\dfrac{6!}{3!}$

(d) $\dfrac{6!}{4!\,2!}$

(e) $\dfrac{5!}{3!(5-3)!}$

(f) $(0.4)^3$

(g) $\binom{6}{2}$ (h) $\binom{6}{4}$ (i) $\binom{5}{4}$

(j) $\binom{6}{4}(0.4)^4(0.6)^2$ (k) $\binom{5}{1}(0.2)^1(0.8)^4$

5-17 If x is a binomial random variable, calculate the probability of x for each case.

(a) $n = 4$, $x = 1$, $p = 0.3$ (b) $n = 3$, $x = 2$, $p = 0.8$
(c) $n = 2$, $x = 0$, $p = \frac{1}{4}$ (d) $n = 5$, $x = 2$, $p = \frac{1}{3}$
(e) $n = 4$, $x = 2$, $p = 0.5$ (f) $n = 3$, $x = 3$, $p = \frac{1}{6}$

5-18 If x is a binomial random variable, use table 4 in appendix D and determine the probability of x for each case.

(a) $n = 10$, $x = 8$, $p = 0.3$ (b) $n = 8$, $x = 7$, $p = 0.95$
(c) $n = 15$, $x = 3$, $p = 0.05$ (d) $n = 12$, $x = 12$, $p = 0.99$
(e) $n = 9$, $x = 0$, $p = 0.5$ (f) $n = 6$, $x = 1$, $p = 0.01$
(g) Explain the meaning of the symbol 0+ that appears in table 4.

5-19 A new car undergoes three road tests before being shipped to a dealer. Each test has outcomes of success S and failure F, where $P(S) = p$ and $P(F) = q$.

(a) Complete the accompanying tree diagram. Label all branches completely.
(b) In column (b) of the tree diagram, express the probability of each outcome represented by the branches as a product of p and q.
(c) Let x be the random variable, the number of successes observed. In column (c) identify the value of x for each branch of the tree diagram.
(d) Write the equation of the binomial probability function of this situation.

5-20 Show that guessing the answers to the quiz of six multiple-choice questions (see p. 190) is actually a binomial experiment.
  (a) Specify how this experiment satisfies the four properties of a binomial experiment.
  (b) Complete the tree diagram, showing all possible outcomes of the quiz. A total of 64 outcomes should be shown. How many outcomes are there for each of the events "no correct answers," "one correct answer," "two correct answers,"..., "five correct answers," "six correct answers"?
  (c) Write the equation of the binomial probability function of this experiment.

5-21 Test the following function to determine whether or not it is a binomial probability function. List the probability distribution and sketch a histogram.

$$P(x) = \binom{4}{x} \cdot \left(\frac{1}{2}\right)^x \cdot \left(\frac{1}{2}\right)^{4-x} \quad \text{for } x = 0, 1, 2, 3, 4$$

(Retain this solution for use in answering exercise 5-28.)

5-22 State a very practical reason why a defective item in an industrial situation would be defined as a success in a binomial experiment.

5-23 Consider an experiment in which four stocks listed on the New York Stock Exchange are randomly selected with replacement and the number of those stocks that are also included in the Dow–Jones index is called the random variable x. (Assume that 5% of the stocks listed on the New York Exchange are in the Dow–Jones index.)
  (a) Is this a binomial experiment? Describe all the properties and the variable.
  (b) Write the equation of the binomial probability function.
  (c) Find the various binomial probabilities that are in the probability distribution.

5-24 According to studies in 1982, 1 out of every 15 individuals entering a department store will attempt to shoplift something before leaving the store. Assuming that the binomial distribution is appropriate, what is the probability that exactly one out of three randomly selected individuals who have entered the store will attempt to shoplift something before leaving the store?

5-25 A multiple-choice apprentice exam has 12 questions, each with 5 possible choices and only 1 correct answer. What is the probability that from 5 to 10 questions are answered correctly by random guessing? (Use table 4 of appendix D.)

5-26 A salesperson averages 0.40 sales per customer contact. What is the probability that this salesperson makes exactly two sales in the next five customer contacts?

5-27 Seventy percent of all audits result in finding at least three journal entries misclassified. What is the probability that two of the next three audits do not find at least three journal entries misclassified?

Section 5-5 Mean and Standard Deviation of the Binomial Distribution

mean and standard deviation of binomial distribution

The **mean** and **standard deviation** of a theoretical binomial probability distribution can be found by using these two formulas:

$$\mu = np \qquad (5\text{-}7)$$

$$\sigma = \sqrt{npq} \qquad (5\text{-}8)$$

The formula for the mean seems appropriate as the number of trials multiplied by the probability that a success will occur. (We previously learned that the mean number of correct answers on the binomial quiz was expected to be $\frac{1}{3}$ of 6, $6 \cdot \frac{1}{3}$, or np.) The formula for the standard deviation is not as easily understood. Thus at this point it is appropriate to look at an example that demonstrates that formulas (5-7) and (5-8) yield the same results as formulas (5-2) and either (5-4a) or (5-4b).

Referring to the case of the three bids in illustration 5-7, $n = 3$ and $p = \frac{1}{2}$. Using formulas (5-7) and (5-8), we find

$$\mu = np = (3)\left(\frac{1}{2}\right) = 1.5$$

$\mu = 1.5$ is the mean value of the random variable x.

$$\sigma = \sqrt{npq} = \sqrt{(3)\left(\frac{1}{2}\right)\left(\frac{1}{2}\right)} = \sqrt{\frac{3}{4}} = \sqrt{0.75} = 0.866 = \mathbf{0.87}$$

$\sigma = 0.87$ is the standard deviation of the random variable x. Look back at the solution for illustration 5-7 and compare the use of formulas (5-2) and (5-4a) with formulas (5-7) and (5-8). Note that the results are the same, regardless of the formula you use. However, formulas (5-7) and (5-8) are much easier to use when x is a binomial random variable.

Illustration 5-11

Find the mean and standard deviation of the binomial distribution where $n = 20$ and $p = \frac{1}{5}$.

$$\mu = np = (20)\left(\frac{1}{5}\right) = 4.0$$

$$\sigma = \sqrt{npq} = \sqrt{(20)\left(\frac{1}{5}\right)\left(\frac{4}{5}\right)} = \sqrt{\frac{80}{25}}$$

$$= \frac{(4\sqrt{5})}{5} = 1.79 \qquad \square$$

Exercises

5-28 (a) Calculate the mean and standard deviation of the random variable in exercise 5-21 by using formulas (5-2) and (5-4a).

(b) Check the values found in part (a) by using formulas (5-7) and (5-8).

5-29 (a) Calculate the mean and the standard deviation of x in exercise 5-23 by using formulas (5-2) and (5-4a).

(b) Calculate the mean and the standard deviation of x in exercise 5-23 by using formulas (5-7) and (5-8).

(c) How do the answers to parts (a) and (b) compare?

5-30 Find the mean and standard deviation of each of the following binomial random variables:

(a) The number of leaky valves determined in testing 200 industrial valves. The probability of a leaky valve is $\frac{1}{6}$.

(b) The number of wine bottles with improper caps observed in 500 bottles, where $P(\text{improper cap}) = 0.06$.

(c) The number of delinquent charge accounts among 500 accounts. Assume that the probability of an account being delinquent is 0.12.

(d) The number of brokerage accounts that will be inactive this week out of 300 accounts. Assume that the probability that an account will be inactive in a given week is 0.60.

Section 5-6 The Poisson Probability Distribution

Consider the probability experiment where the random variable x is the number of accident claims filed against an insurance company on a given day. x is a count; therefore it is a discrete random variable. However, it is not a binomial random variable. The binomial probability experiment conditions as outlined on page 192 are not met, since this experiment contains no trials. Instead of asking for the number of successes that occur in a set of

n trials, we are asked about the number of successes that occur in a given period of time. The number of successes is no longer limited by a fixed number of trials and can therefore take on any value 0, 1, 2, 3, and so on. In situations where the random variable is in the form of a count of the **number of successes per unit of measure**, the probability function that is often used is the **Poisson probability function** (named after the French mathematician, Simeon Denis Poisson (1781–1840), who developed it).

Poisson probability function

Poisson probability experiment

Poisson Probability Experiment

An experiment where repeated observations are made for a unit of measure. The Poisson experiment must possess the following properties:

1. The Poisson random variable x is the count of the number of successes that occur in a given unit of measure; x may take on any integer value from 0 to infinity.

2. If the unit of measure is subdivided into very small subunits (e.g., if x is the number of accident claims filed per day, then consider the number of claims filed per second), then:

   (a) The probability of success in each subunit would be very small.

   (b) The probability of two or more successes per subunit is so small that it may be considered 0.

   (c) The probability of success in each subunit is the same as in all other subunits.

   (d) The number of successes that occur in each subunit does not depend on the subunit being sampled.

[The statements (c) and (d) describe the independence of the occurrences.]

Illustration 5-2

Consider the number of telephone calls that pass through a particular switchboard per hour. Let x be the number of wrong numbers (successes) in a given hour (unit of measure—time). If we were to consider a small subunit of an hour, say a second, we would expect: (a) only a small probability for receiving a wrong number telephone call within a given second, (b) it to be nearly impossible to receive two or more wrong number telephone calls within the same second, (c) the probability of receiving a wrong number call would be the same for every second, and (d) the fact that a wrong number call was received during one 1-second interval should not affect the probability of receiving a wrong number call in any other second. ☐

NOTE: When using the Poisson probability distribution, we often must simply assume that the conditions hold true. We are not able to inspect the experiment and verify the conditions as we can when the binomial probability distribution is being used.

Many variables have a Poisson distribution. Some of the random variables that are often assumed to have a Poisson distribution are: the number of customers arriving per day at a store, the number of accident claims filed per month, the number of customers arriving for service at a checkout counter per hour in a large department store, the number of defects in a yard of material, and the number of machine failures per shift at a manufacturing plant.

The Poisson probability function used to calculate the probability of a given x value is expressed as

$$P(x) = \frac{\mu^x \cdot e^{-\mu}}{x!} \quad \text{for } x = 0, 1, 2, \ldots \quad (5\text{-}9)$$

where $P(x)$ is the probability of obtaining exactly x successes in a given unit of measure, μ is the mean number of successes that occur in a given unit of measure, and e is the constant $2.71828\ldots$. The probabilities for key values of $\mu(0.1, 0.5, 1.0, 1.5, \ldots, 5.0)$ and $x(0, 1, 2, \ldots, 15)$ have been calculated and are given in table 5. Table 5 may be used to find the Poisson probabilities for all questions asked in this textbook.

Illustration 5-13
Find the probability that $x = 3$ for the Poisson random variable x given that $\mu = 2$.

Solution To find the probability $P(x = 3 | x \text{ is Poisson and } \mu = 2)$, we find the column identified $x = 3$ and the row identified $\mu = 2$ on table 5. The value for the probability is found at the intersection. The answer is $P(x = 3) = 0.1804$. □

Illustration 5-14
If x represents a Poisson random variable where $\mu = 1$, find:

(a) $P(x = 0)$ (b) $P(x = 3)$ (c) $P(x \leq 2)$

Solution Since x is a Poisson random variable, the probabilities are found in table 5.

(a) $P(x = 0) = 0.3679$
(b) $P(x = 3) = 0.0613$
(c) $P(x \leq 2) = P(x = 0, 1, \text{ or } 2)$
$= P(x = 0) + P(x = 1) + P(x = 2)$
$= 0.3679 + 0.3679 + 0.1839$
$= \mathbf{0.9197}$ □

Illustration 5-15

Assume that the number of customers arriving at JC's Hair Village for haircuts per hour is a Poisson probability experiment. By reviewing past records, we can determine that a mean of two customers arrives per hour.

(a) What is the probability that in a given hour exactly four customers will arrive?

(b) What is the probability that in a given 2-hour period exactly eight customers will arrive?

(c) What is the probability that in a given half-hour period exactly two customers will arrive?

Solution

(a) The probability that exactly four customers arrive in a 1-hour time period given that x is Poisson and that $\mu = 2$ for that 1-hour time period is found in table 5:

$$P(x = 4) = 0.0902$$

(b) Notice that the time period has changed. The given information was in terms of 1-hour time periods; the question asks about a 2-hour time period. We must adjust the value of μ, the mean number of arrivals per time period. If the mean number of arrivals per hour is 2, then the mean number of arrivals per 2-hour period must be double that, or $\mu = 4$. Thus we are looking for the probability

$$P(x = 8) = 0.0298$$

(c) Again the time period has been changed, this time to half-hour periods. The mean number of arrivals for a half-hour is one-half the mean number of arrivals for an hour. Thus μ becomes 1 for this question.

$$P(x = 2) = 0.1839$$

Exercises

5-31 If x is a Poisson random variable with $\mu = 3$, find:
(a) $P(x = 2)$ (b) $P(x \leq 2)$ (c) $P(x > 2)$

5-32 Which variables would not be Poisson random variables? Explain.
(a) the value of a stock on a given day
(b) the number of planes arriving at an airport per hour
(c) the number of wrong numbers in 20 calls
(d) the number of wrong numbers received per hour

5-33 A telephone magazine salesperson averages two sales per hour of work. What is the probability that in a given hour he will:

(a) make exactly two sales

(b) make at least two sales

(c) make more than two sales

5-34 The number of car accident insurance claims received by an insurance broker per day follows the Poisson probability distribution with a mean number of claims per day of 1.

(a) On any given day, what is the probability that no claims would be made?

(b) In any given week (work week is 5 days), what is the probability that no claims would be made?

(c) In a 2-day period, what is the probability that exactly one claim will be made?

5-35 Using table 5 draw a histogram of the Poisson probability distribution when:

(a) $\mu = 1$ (b) $\mu = 5$

In Retrospect

In this chapter we combined concepts of probability with some of the ideas presented in chapter 2. We are now able to deal with distributions of probability values and find means, standard deviations, and so on.

In chapter 4 we explored the concepts of mutually exclusive and independent events. The addition and multiplication rules were used on several occasions in this chapter, but very little was said about mutual exclusiveness or independence. Recall that every time we add probabilities together, as we did in each of the probability distributions, we need to know that the associated events are mutually exclusive. If you look back over the chapter, you will notice that the random variable actually requires events to be mutually exclusive; therefore no real emphasis was placed on this concept. The same basic comment can be made in reference to the multiplication of probabilities and the concept of independent events. Throughout the chapter, probabilities were multiplied together and occasionally independence was mentioned. Independence, of course, is necessary to be able to multiply probabilities together.

If we now look at the news article at the beginning of the chapter, we can see that the percentage of models available per unit of price could be reorganized to form a probability distribution. Some of the sets of data in chapter 2 could also be used to form a probability distribution. For example, we could let the random variable x be the number of days elapsed between order and delivery in the illustration used throughout chapter 2.

We are now ready to extend these concepts to continuous random variables, which we will do in chapter 6.

Chapter Exercises

5-36 What are the two basic properties of every probability distribution?

5-37 Verify whether or not the following is a probability function. State your conclusion and explain.

$$f(x) = \frac{1/2}{x!(2-x)!} \quad \text{for } x = 0, 1, 2$$

5-38 Calculate the mean and standard deviation, to the nearest 10th, for this probability distribution of x:

x	P(x)
0	0.1
2	0.2
4	0.4
6	0.2
8	0.1

5-39 Two vice-presidents rank a prospective director of personnel on a scale from 1 to 4. All possible rankings are shown in the accompanying table. Define x to be the random variable "total of the two rankings." (Assume each pair of rankings is equally likely.)

Second Vice-President	First Vice-President			
	1	2	3	4
1	1, 1	2, 1	3, 1	4, 1
2	1, 2	2, 2	3, 2	4, 2
3	1, 3	2, 3	3, 3	4, 3
4	1, 4	2, 4	3, 4	4, 4

(a) Construct a probability distribution for x.

(b) Calculate the mean value of x.

(c) Calculate the standard deviation of x.

5-40 Seventy-five percent of the foreign-made autos that were sold in the United States in 1978 are now falling apart.

(a) Determine the probability distribution of x, the number of these autos that are falling apart in a random sample of size 5.

(b) Draw a histogram of the distribution.

(c) Calculate the mean and the standard deviation of this distribution.

5-41 A business firm is considering two investments, of which it will choose the one that promises the greater average profit. Which of the investments should it accept?

Invest in Tool Shop		Invest in Book Store	
Profit	Probability	Profit	Probability
$100,000	0.10	$400,000	0.20
50,000	0.30	90,000	0.10
20,000	0.30	−20,000	0.40
−80,000	0.30	−250,000	0.30
	1.00		1.00

5-42 One alphanumeric terminal is to be selected at random from the product tables discussed in the news article at the beginning of the chapter. What is the probability that the terminal selected has the following prices:
(a) between $500 and $1,000
(b) between $1,000 and $2,500

5-43 Let x be a random variable with the following probability distribution:

$$x = 0 \quad 1 \quad 2 \quad 3$$
$$P(x) = 0.4 \quad 0.3 \quad 0.2 \quad 0.1$$

Does x have a binomial distribution? Justify your answer.

5-44 As a quality control inspector, you have observed that wooden wheels that are bored off-center occur about 3% of the time. If six of these wheels are to be used on each toy truck produced, what is the probability that a randomly selected set of wheels has no wheels off-center?

5-45 In a recent marketing survey, 900 out of 1,000 females admitted that they had never looked at a copy of *Vogue* magazine. Assuming that this information is accurate, what is the probability that a random sample of three females will show fewer than two who have looked at the magazine?

5-46 One-fourth of all orders received by a supplier are delivered on time. If five firms each place an order, what is the probability that exactly three will be delivered on time? (Find the answer by using a formula.)

5-47 One-third of all persons who have medical insurance with the Stover Agency make at least one claim during the year. What is the probability that three randomly selected accounts will have no claims this year? (Find the answer by using a formula.)

5-48 According to a college placement service, one out of every five job interviews on campus leads to a follow-up interview at the firm. If five students each have an interview on campus, what is the probability that fewer than three will have follow-up interviews? (Use a formula to find the answer.)

5-49 Records show that 10% of a certain model of computer terminals fail before 2,000 hours of use. If a firm has four terminals, what is the probability that none fail before 2,000 hours? (Find the answer by using a formula.)

5-50 Chicken Spice, Inc. has found that 30% of all prospective franchises have adequate capital. What is the probability that exactly 5 of the next 15 prospective franchises will have adequate capital? (Find the answer by using a table.)

5-51 On the average, one out of every five personal computers purchased by small firms is primarily for word-processing capabilities. What is the probability that between 8 and 10 of the 14 personal computers sold to a small firm this week were for word processing? (Find the answer by using a table.)

5-52 A typist makes, on the average, 2.5 errors per 5 minutes of typing. Assume the number of errors follows a Poisson distribution. What is the probability that:
 (a) in 5 minutes the typist makes exactly five errors
 (b) in 5 minutes the typist makes less than five errors
 (c) in 1 minute the typist makes exactly one error
 (d) in 1 minute the typist makes no errors

5-53 The average number of defects per yard of material is four. If the number of defects per yard follows the Poisson distribution, what is the probability that:
 (a) exactly three defects are in a given yard of material
 (b) at least three defects are in a given yard of material
 (c) at most three defects are in a given yard of material

5-54 The probability that an hourly employee will refuse to work overtime (allowed by the union contract) is 0.30. Let x be the number of employees refusing to work overtime. If 10 are asked, what is the probability that at least 5 will work overtime?

5-55 Consider the variable in exercise 5-54. The probability of x consecutive employees refusing to work overtime in a sequence of n requests depends on n.
 (a) What is the probability of $x = 3$ when $n = 3$?
 (b) What is the probability of $x = 5$ when $n = 5$?
 (c) What is an "unusual" number of consecutive refusals, if an "unusual event" is something that has a probability of less than 0.02?

5-56 A large shipment of radios is accepted upon delivery if an inspection of eight randomly selected radios yields no more than one defective radio.
  (a) Find the probability that a shipment is accepted if 5% of the total shipment is defective.
  (b) Find the probability that a shipment is not accepted if 20% of the shipment is defective.

5-57 Experience has shown that 30% of the applicants for work at U.S. Anvil Works fail the initial physical exam. The company is concerned with the number of applicants who fail the physical on a day when 10 people apply.
  (a) Describe how this experiment exhibits the properties of a binomial experiment.
  (b) Find the probability that at least half of the applicants will fail the physical.

5-58 The American Rifle Company must test each gun sold to the U.S. Army. Bullets are shot at a target in sets of eight; the probability of a bull's-eye is 0.3 from a given distance for an expert shot.
  (a) Find the probability distribution for the random variable "number of bull's-eyes per set of eight shots."
  (b) What is the probability that a set of eight shots will result in more than half hitting the bull's-eye?
  (c) Find the mean number of bullets expected to hit the bull's-eye per set of eight shots.
  (d) Find the standard deviation of the number of shots that hit the bull's-eye per set.

5-59 An experiment consists of placing each of eight numbered objects into one of two containers, A or B. After the objects have been placed in the containers, the results are recorded and the set of objects in container A is listed.
  (a) How many different sets can be found in container A at the conclusion of the experiment?
  (b) How many of the sets found in container A contain exactly no elements? One element? Two elements? Three elements?...Seven elements? Eight elements?
  (c) If the objects are placed into the containers randomly, what is the probability of the occurrence of each of the sets in part (a)?
  (d) Define a random variable x to be the number of objects in container A and construct the probability distribution of x.
  (e) Find the mean and standard deviation of x.

5-60 A box contains eight identical balls numbered from 1 to 8. An experiment consists of selecting two of these balls, without replacement, and identifying their numbers.

(a) How many different sets of two balls can be selected?
(b) List the sample space for the experiment.
(c) If the balls are selected randomly, what is the probability of each of the sets in the sample space?
(d) Define the random variable x to be the total of the numbers on the two balls. Construct a probability distribution of x.
(e) Find the mean and standard deviation of x.

5-61 An advertising agency prepared six new ads for one of its customers. The agency ranked each of the ads from 1 to 6 according to its opinion of the ad's effectiveness in selling the product. The customer is to select two of the ads for use in promoting the product.
(a) Construct the sample space, showing the possible selections that can be made. (Use the agency's rank number to identify each ad.)
(b) Assume that the customer selects the ads in a random manner. Find the probability distribution of the random variable x, the sum of the two rank numbers of the ads selected.
(c) Find the mean and the standard deviation of x.

5-62 The construction of Reliable Brand CB radios uses a "series" process for the major electronic component. The process consists of components A and B, which are linked so that both must work for the system to work. Both parts work independently of each other. The probability of either part failing is 0.02. See the following diagram:

A newly hired reliability engineer suggests that the system can be built in "parallel." See the following diagram:

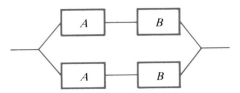

If a system is built in parallel, the system works as long as either set of A and B works. Both sets, as well as each A and B component, work independently.
(a) The engineer claims that the cost of the second system (an extra A and B) can be absorbed by reducing quality so that $P(A \text{ fails}) = P(B \text{ fails}) = 0.10$, and yet the reliability of the system $P(\text{works})$ will increase. Do you agree? (Support your answer with probabilities.)
(b) $P(A \text{ failing}) = P(B \text{ failing})$ is increased to 0.40. How many parallel systems are needed if the overall probability of the system not failing is 0.95 or better? (What is the actual probability that the system will work?)

5-63 The chief economist of a trust fund wants to evaluate the performance of a stock recommendation service. Last year 50% of all stocks increased at least 6% in value. Of the 12 stocks the firm recommended as best buy of the month, 8 increased at least 6% in value. Would you conclude that the service did better than if the stocks were picked out of a hat? Suppose 11 had increased. What would you conclude?

Challenging Problem

5-64 Some stock analysts believe that when the Dow-Jones index drops, the stocks with the greatest activity (highest volume) will be the stocks most likely to decline in price that day. The following table summarizes the activity on the New York Stock Exchange for January 27, 1983. How likely is it that, of the 15 most active stocks, 9 or more would decline when overall only 36% of all stocks declined? Based on this data, would you agree with the analyst? Why or why not?

Market Diary

	Wed.	Tues.	Mon.	Fri.	Thurs.	Wed.
Issues traded	1,963	1,965	1,993	1,947	1,967	1,987
Advances	877	1,095	173	389	827	554
Declines	709	534	1,631	1,185	749	1,076
Unchanged	377	336	189	367	391	352
New highs	53	34	25	72	106	129
New lows	1	3	4	3	0	1

Most Active Stocks

	Open	High	Low	Close	Chg.	Volume
Natomas	$14\frac{1}{4}$	$14\frac{3}{8}$	$13\frac{5}{8}$	$13\frac{7}{8}$	$-2\frac{5}{8}$	1,820,400
Exxon	30	$30\frac{1}{8}$	$29\frac{1}{8}$	$29\frac{3}{8}$	$-\frac{1}{4}$	1,306,400
Amer T & T	68	$68\frac{1}{2}$	$67\frac{1}{4}$	$68\frac{1}{8}$	$+\frac{1}{8}$	1,016,900
MesaOffsh	$1\frac{1}{2}$	$1\frac{5}{8}$	$1\frac{1}{2}$	$1\frac{1}{2}$		921,200
ArchDnM	$22\frac{1}{2}$	$22\frac{1}{2}$	21	$21\frac{3}{4}$	$-\frac{3}{4}$	910,800
SperryCo	34	$36\frac{1}{2}$	34	$35\frac{1}{2}$	$+1\frac{7}{8}$	892,000
IBM	96	$96\frac{1}{8}$	$94\frac{7}{8}$	$95\frac{1}{8}$	$-\frac{7}{8}$	842,500
SuperOil	$30\frac{1}{4}$	$31\frac{3}{4}$	$30\frac{1}{4}$	$31\frac{1}{4}$	$+\frac{3}{4}$	735,300
GenMotors	$58\frac{1}{4}$	$58\frac{3}{4}$	$57\frac{1}{4}$	$57\frac{1}{4}$	$-\frac{3}{4}$	724,600
Rowan	$13\frac{1}{8}$	$13\frac{1}{4}$	$12\frac{3}{4}$	$12\frac{7}{8}$	$-\frac{1}{8}$	716,200
AMR Corp	21	21	$19\frac{3}{4}$	$19\frac{3}{4}$	$-\frac{3}{4}$	689,200
StaOilCal	$31\frac{3}{4}$	$32\frac{5}{8}$	$31\frac{1}{2}$	$32\frac{1}{4}$	$+\frac{3}{4}$	669,600
StaOilInd	40	$40\frac{1}{4}$	$39\frac{1}{4}$	$39\frac{1}{4}$	$-\frac{1}{4}$	639,000
Sterl Drug	23	23	$22\frac{1}{4}$	$22\frac{3}{4}$	$-\frac{1}{2}$	629,900
DigitalEq	114	$115\frac{7}{8}$	$113\frac{1}{2}$	$115\frac{7}{8}$	$+\frac{3}{8}$	626,100

Hands-On Problems

The probability experiment of rolling two dice is to be performed. Each die is modified so that it has three faces with one dot, two faces with two dots, and one face with three dots. One die is white and one die is black. (This probability experiment is the same as the one used in the chapter 4 hands-on problems.)

5-1 Determine the theoretical probability distribution for the sum of the dots shown on the two dice. Construct the sample space for this experiment and find the probabilities associated with each point of the sample space.

5-2 Draw a histogram of this probability distribution. (Be sure to label it completely.)

5-3 Calculate the mean and the standard deviation of this probability distribution [formulas (5-2) and (5-4)].

5-4 Perform the experiment 100 times, recording the sum of the number of dots shown on the two dice. (See the hands-on problems in chapter 4 for directions on how this experiment might be carried out.)

5-5 Find the relative frequency distribution for the variable "sum of the number of dots," and draw a relative frequency histogram showing your results.

5-6 Calculate the mean and the standard deviation for the observed data by using a frequency distribution [formulas (2-2) and (2-11)].

5-7 Compare the value of the theoretical mean and standard deviation with the values found for the observed data. Compare the two histograms; do they seem to be related? Do the experimental results support the information calculated in problem 5-3? Do the observed probabilities seem to agree with those calculated as theoretical values?

6 The Normal Probability Distribution

Chapter Outline

6-1 The Normal Probability Distribution
*The **bell-shaped distribution** whose domain is the set of all real numbers.*

6-2 The Standard Normal Distribution
*To work with the normal distribution, we need the **standard score**.*

6-3 Applications of the Normal Distribution
*The normal distribution can help us to determine **probabilities**.*

6-4 Notation
*The **z notation** is critical in the use of the normal distribution.*

6-5 Normal Approximation of the Binomial
***Binomial probabilities** can be **estimated** by using the normal distribution.*

The Valuation of Human Resources

Do people have a quantifiable value to a firm? The authors believe they do. Experimental work with a valuation model at the Bank of America shows promise in human resource management.

It should be recognized that HRA [human resource accounting] is a generic term, including diverse techniques for evaluating human resources that yield quite different kinds of information. HRA theory is based primarily on the cost approach and the value approach....

Expected Tenure

If a person is expected to stay with an organization until retirement, value can be measured by summing discounted salaries for every year until retirement. However, if a person separates, the expected future contributions will never be realized. Therefore, expected future salary should be reduced by the probability of separation.

Although separations may be explained by a variety of factors, our research has shown that length of service has the highest correlation to separations. Using historical separation rates, we determined the probability of terminating after specified years of service and used this probability to reduce the expected future salary. An example of the kind of functional relationship we found is ... the figure [shown here]....

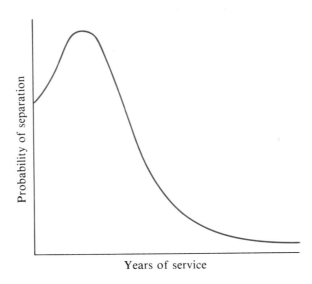

From Richard B. Frantzreb, Linda L. T. Landau, and Donald P. Lundberg, "The Valuation of Human Resources," *Business Horizons* (Indiana University Graduate School of Business), vol. 17, no. 3 (June 1974): 73–80. Reprinted by permission.

Chapter Objectives

*Until now we have considered distributions of discrete variables only. In this chapter we will examine one particular **continuous** probability distribution of major importance, whose domain is the set of all real numbers. This distribution is called the "normal," the "bell-shaped," or the "Gaussian" distribution. "Normal" is simply the traditional title of this particular distribution and is not a descriptive name meaning "typical." Although there are many other types of continuous distributions (rectangular, triangular, skewed, etc.), many variables have an approximately normal distribution. For example, several of the histograms drawn in chapter 2 suggested a normal distribution. A mounded histogram that is approximately symmetric is an indication of such a distribution.*

In addition to learning what a normal distribution is, we will consider (1) how probabilities are found, (2) how they are represented, and (3) how the normal distribution is used. Although continuous variables have other distributions, the normal distribution is the most important.

Section 6-1 The Normal Probability Distribution

normal distribution

As in all other probability distributions, there are formulas that give us information about the **normal distribution**. Previously, the probability function was the only function of interest. However, two functions define and describe the normal distribution. Each formula uses the random variable as an independent variable. Following this paragraph you will find two formulas. Formula (6-1) gives an ordinate (y value) for each point on the graph of the normal distribution for each given abscissa (x value). Formula (6-2) yields the probability associated with x when x is in the interval between the values a and b. Note that the random variable x is a continuous variable. For continuous variables we will discuss the probability that x has a value between the two extreme values of an interval, and we will depict the probability as an area under the graph of the probability distribution.

$$y = f(x) = \frac{1}{\sigma\sqrt{2\pi}} \cdot \exp\left[-\frac{1}{2}\left(\frac{x-\mu}{\sigma}\right)^2\right] \quad \text{for all real } x \quad (6\text{-}1)$$

$$P(a \leq x \leq b) = \int_a^b f(x)\, dx \quad (6\text{-}2)$$

Don't worry if you don't understand these formulas. We will not be using them in this book. However, a few comments about their meaning will be helpful if you ever use a probability table in any of the standard reference books. The directions for the use of these tables are often described with the aid of mathematical formulas such as (6-1) and (6-2).

normal, or bell-shaped, curve

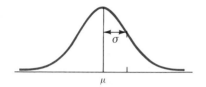

FIGURE 6-1
The Normal Distribution

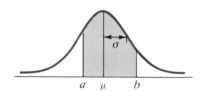

FIGURE 6-2
Shaded Area Under the
Circle Is $P(a \leq x \leq b)$

Formula (6-1) and its associated table can be used to graph a normal distribution with a given mean and standard deviation. [The table associated with formula (6-1) is not included in this text, but it is available in many other textbooks and in all standard books of statistical tables.] In a more practical sense, we can draw a **normal**, or **bell-shaped, curve** and then label it to approximate scale, as indicated in figure 6-1.

To talk intelligently about formula (6-2), we need to know what the symbols used in the formula represent. The $\int_a^b f(x)\,dx$ is called a *definite integral* and it comes from calculus. The definite integral is a number that is the measure of the area under the curve of $f(x)$. This area is bounded by a on the left, b on the right, $f(x)$ at the top, and the x-axis at the bottom. The equation of the normal curve in formula (6-2) gives the measure of the shaded area shown in figure 6-2.

Since $P(a < x < b)$ (read "the probability that x is between a and b") is given by $\int_a^b f(x)\,dx$, the measure of this probability is also the measure of the probability that a continuous random variable will assume a value within a specified interval. As a student of elementary statistics, you are not expected to know anything about calculus, so we will say nothing more about the definite integral in this text.

The values calculated by using these formulas [(6-1) and (6-2)] are also found in a table, with the probabilities expressed to four decimal places. Like all other tables, this table must be read in the manner in which it was constructed.

However, before we show you how to read the table, we must point out that the table is expressed in a "standardized" form. It is standardized so that it is unnecessary to have a different table for every different pair of values of the mean and standard deviation. For example, the area under the normal curve of a distribution with a mean of 15 and a standard deviation of 3 must be somehow related to the area under the curve of a normal distribution with a mean of 113 and standard deviation of 38.5. If you will recall, the empirical rule concerns the percentage of a distribution within a certain number of standard deviations of the mean (see p. 75).

percentage—proportion

NOTE: Percentage and probability are the same relative frequency concept. **Percentage** is usually used when talking about a **proportion** of a population. **Probability** is usually used when talking about the **chance** that the next individual item will display a particular characteristic.

The empirical rule is a fairly crude measuring device. With it we are able to find probabilities associated only with whole-number multiples of the standard deviation (e.g., within one standard deviation of the mean). We will often be interested in probabilities associated with fractional parts of the standard deviation. For example, we might want to know the probability that x is within 1.85 standard deviations of the mean. Therefore we must refine the empirical rule so that we can deal with more precise measurements. This refinement is discussed in the next section.

Section 6-2 The Standard Normal Distribution

The key to working with the normal distribution is the **standard score** z. The measures of probabilities associated with the independent variable x are determined by the relative position of x with respect to the mean and the standard deviation of the distribution. The empirical rule told us that approximately 68% of the data is within one standard deviation of the mean. The value of the mean and the size of the standard deviation do not change this fact.

Recall that z was defined in chapter 2 to be

$$\frac{x - (\text{mean of } x)}{(\text{standard deviation of } x)}$$

Therefore we can write the formula for z as

$$z = \frac{x - \mu}{\sigma} \qquad (6\text{-}3)$$

(Note that if $x = \mu$, then $z = 0$.)

The z score is known as a "standardized" variable because its units are standard deviations. The normal probability distribution associated with a standard score z is called the **standard normal distribution**. Table 6 in appendix D lists the probabilities associated with intervals of standard deviation from the mean for specific positive values of z. Other probabilities may be found by addition, subtraction, and so on, based on the concept of symmetry that exists in the normal distribution.

Let's look at an illustration to see how we read table 6. The z scores are in the margins: the left margin has the units digit and the tenths digit; the top has the hundredths digit. Let's look up a z score of 1.52. We find 0.4357, as shown in table 6-1. Now, exactly what is 0.4357? It is the measure of the area under the standard normal curve between $z = 0$ (which locates the mean) and $z = 1.52$ (a number 1.52 standard deviations larger than the mean). See figure 6-3. **This area is also the measure of the probability associated with the same interval**, that is,

$$P(0 < z < 1.52) = 0.4357$$

Read this result as "the probability that a value of the variable picked at random will fall between the mean and 1.52 standard deviations above the mean is 0.4357" or, equivalently, "the probability that a z score picked at random will fall between 0 and 1.52 is 0.4357."

standard normal distribution

are representation for probability

TABLE 6-1
A Portion of Table 6, Appendix D

z	0.00	0.01	0.02	...
⋮				
1.5			0.4357	
⋮				

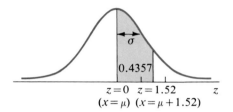

FIGURE 6-3
Area Under Normal Curve from z = 0 to z = 1.52 Is 0.4357

Recall that one of the basic properties of probability is that the sum of all probabilities is exactly 1.0. Since the area under the normal curve represents the measure of probability, the total area under the bell-shaped curve is exactly 1 unit. This distribution is symmetric with respect to the vertical line drawn through $z = 0$, which cuts the area exactly in half at the mean. Can you verify this fact by inspecting formula (6-1)? That is, the area under the curve to the right of the mean is exactly $\frac{1}{2}$ unit (0.5), and the area to the left is also $\frac{1}{2}$ unit (0.5) of probability. Areas (probabilities) not given directly in the table can be found by relying on these facts.

Now let's look at some illustrations.

Illustration 6-1

Find the area under the normal curve to the right of $z = 1.52$.

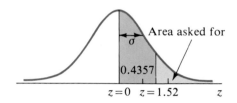

Solution The area to the right of the mean (all the shading in the figure) is exactly 0.5000. The question asks for the shaded area that is not included in the 0.4357. Therefore subtract 0.4357 from 0.5000.

$$0.5000 - 0.4357 = 0.0643 \qquad \square$$

SUGGESTION: As we have done here, always draw and label a sketch. It is most helpful.

Illustration 6-2

Find the area to the left of $z = 1.52$.

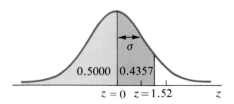

Solution The total shaded area is made up of the 0.4357 found in the table and the 0.5000 that is to the left of the mean. Therefore add 0.4357 and 0.5000.

$$0.4357 + 0.5000 = 0.9357 \qquad \square$$

NOTE: The addition and subtraction done in illustrations 6-1 and 6-2 are correct because the "areas" represent mutually exclusive events (discussed in section 4-5).

The symmetry of the normal distribution is a key factor in determining probabilities associated with values below (to the left of) the mean. The area between the mean ($z = 0$) and $z = -1.52$ is exactly the same as the area between $z = 0$ and $z = +1.52$. This fact allows us to look up values related to the left side of the distribution.

Illustration 6-3

The area between the mean ($z = 0$) and $z = -2.1$ is the same as the area between $z = 0$ and $z = +2.1$.

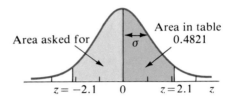

Thus we have

$$P(-2.1 < z < 0) = \mathbf{0.4821}$$

Illustration 6-4

The area to the left of $z = -1.35$ is found by subtracting 0.4115 from 0.5000.

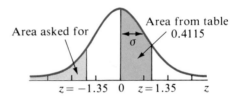

Therefore we obtain

$$P(z < -1.35) = \mathbf{0.0885}$$

Illustration 6-5

The area between $z = -1.5$ and $z = 2.1$ is found by adding the two areas together.

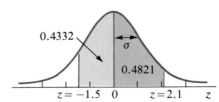

Therefore we obtain
$$P(-1.5 < z < 2.1) = P(-1.5 < z < 0) + P(0 < z < 2.1)$$
$$= 0.4332 + 0.4821 = \mathbf{0.9153}$$

Illustration 6-6

The area between $z = 0.7$ and $z = 2.1$ is found by subtracting. The area between $z = 0$ and $z = 2.1$ contains all the area between $z = 0$ and $z = 0.7$. The area between $z = 0$ and $z = 0.7$ is therefore subtracted from the area between $z = 0$ and $z = 2.1$.

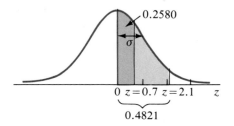

Therefore
$$P(0.7 < z < 2.1) = 0.4821 - 0.2580 = \mathbf{0.2241}$$

The normal distribution table can also be used to determine a z score if we are given an area. The next illustrations consider this idea.

Illustration 6-7

What is the z score associated with the 75th percentile? (Assume the distribution is normal.)

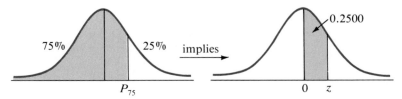

Solution To find this z score, look in table 6, appendix D and find the "area" entry that is closest to 0.2500 (this area entry is 0.2486). Now read the z score that corresponds to this area. From the table the z score is found to be $z = \mathbf{0.67}$. This value says that the 75th percentile in a normal distribution is 0.67 (approximately $\frac{2}{3}$) standard deviation above the mean.

z	...	0.07	0.08	...
⋮				
0.6		0.2486	0.2517	
⋮				

Illustration 6-8

What z scores bound the middle 95% of a normal distribution?

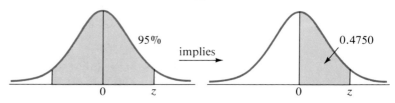

z	...	0.06	...
:			
1.9		0.4750	
:			

Solution The 95% is split into two equal parts by the mean; 0.4750 is the area (percentage) between $z = 0$, the mean, and the z score at the right boundary. Since we have the area, we look for the entry in table 6 closest to 0.4750 (it happens to be exactly 0.4750) and read the z score in the margin. We obtain $z = 1.96$. Therefore, $z = -1.96$ and $z = 1.96$ bound the middle 95% of a normal distribution. □

Exercises

6-1 Describe the distribution of the standard normal score z.

6-2 Find the area under the normal curve that lies between the following pairs of z values:
(a) $z = 0$ to $z = 2.30$
(b) $z = 0$ to $z = 1.14$
(c) $z = 0$ to $z = -3.45$
(d) $z = 0$ to $z = -2.98$

6-3 Find the probability that a piece of data picked at random from a normal population will have a standard score (z) that lies between the following pairs of z values:
(a) $z = 0$ to $z = 1.80$
(b) $z = 0$ to $z = 3.47$
(c) $z = 0$ to $z = -2.20$
(d) $z = 0$ to $z = -2.77$

6-4 Find the area under the normal curve that lies between the following pairs of z values:
(a) $z = -1.30$ to $z = 2.22$
(b) $z = -2.51$ to $z = 1.04$
(c) $z = -3.45$ to $z = -1.20$
(d) $z = -4.0$ to $z = -0.53$

6-5 Find the probability that a piece of data picked at random from a normal population will have a standard score (z) that lies between the following pairs of z values:
(a) $z = -3.20$ to $z = 0.64$
(b) $z = 0.12$ to $z = 3.05$
(c) $z = -1.95$ to $z = -1.52$

6-6 Find the following areas under the normal curve.
(a) to the right of $z = 2.10$
(b) to the right of $z = 0.00$
(c) to the right of $z = -3.05$
(d) to the left of $z = 1.10$
(e) to the left of $z = -1.40$

6-7 Find the probability that a piece of data picked at random from a normally distributed population will have the following standard score:
(a) less than 2.00
(b) greater than −2.50
(c) less than −1.25
(d) less than 0.75
(e) greater than −1.72

6-8 Find these probabilities:
(a) $P(0.00 < z < 1.65)$
(b) $P(-1.10 < z < 3.10)$
(c) $P(z > 1.95)$
(d) $P(z < 2.28)$

6-9 Find the z score for the standard normal distribution shown in each of the following diagrams:

(a)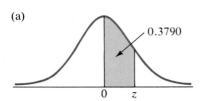
(b)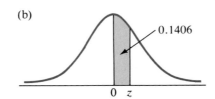
(c)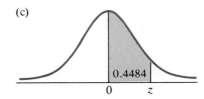
(d)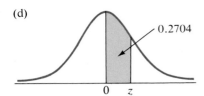
(e)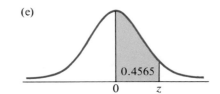
(f)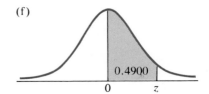

6-10 Find the z score for the standard normal distribution shown in each of the following diagrams:

(a)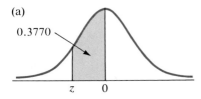
(b)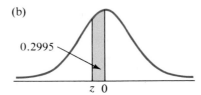
(c)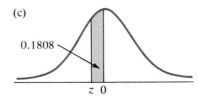
(d)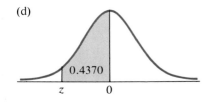
(e)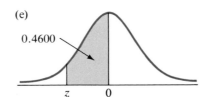
(f)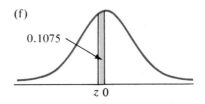

6-11 Find a value of z such that 30% of a distribution lies between it and the mean. (There are two possible answers.)

6-12 Find the standard score z such that:
(a) 60% of the distribution is below (to the left of) this value
(b) the area to the right of this value is 0.05

6-13 Find the two standard scores z such that:
(a) the middle 75% of a normal distribution is bounded by them
(b) the middle 90% of a normal distribution is bounded by them

Section 6-3 Applications of the Normal Distribution

The probabilities associated with any normal distribution can be found by applying the techniques discussed in section 6-2. First, however, we must "standardize" the given information. When dealing with a normal distribution, we need to know its mean μ and its standard deviation σ. Once these values are known, any value of a random variable x can be easily converted to the standard score z by use of formula (6-3):

$$z = \frac{x - \mu}{\sigma}$$

Illustration 6-9

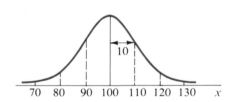

Superior Electronics sells and installs CB radios. The installation is free for its customers. Records show that the installation time, in minutes, is an approximately normally distributed random variable with a mean time of 100 minutes and a standard deviation of 10 minutes (see the accompanying figure). Mr. West just bought a CB radio from Superior. What is the probability that the installation will take between 100 and 115 minutes?

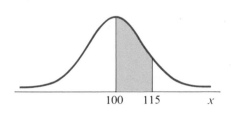

Solution $P(100 < x < 115)$ is represented by the shaded area in the figure on the left.

The variable x must be standardized by using formula (6-3). The z values are shown on the next diagram on the following page.

$$z = \frac{x - \mu}{\sigma}$$

When $x = 100$: $z = \dfrac{100 - 100}{10} = 0.0$

When $x = 115$: $z = \dfrac{115 - 100}{10} = 1.5$

Therefore

$$P(100 < x < 115) = P(0.0 < z < 1.5) = \mathbf{0.4332}$$

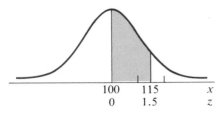

(The value 0.4332 is found by using table 6.) Thus the probability is 0.4332 that the installation will take between 100 and 115 minutes. □

Illustration 6-10

Ms. North has purchased a CB radio for her van, and the service department at Superior Electronics closes in 90 minutes. What is the probability that her CB radio can be completely installed within the 90 minutes before closing?

Solution

$$z = \frac{x - \mu}{\sigma}$$

When $x = 90$: $z = \dfrac{90 - 100}{10} = \dfrac{-10}{10} = -1.0$

$P(x < 90) = P(z < -1.0) = 0.5000 - 0.3413 = \mathbf{0.1587}$

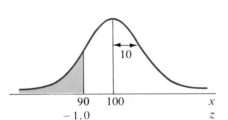

Thus the probability is 0.1587 that her radio can be installed within the 90 minutes before closing. □

The normal table can be used to answer many kinds of questions that involve the normal distribution. Many times a problem will call for the location of a "cutoff point," that is, a particular value of x such that there is exactly a certain percentage in a specified area. The following illustrations concern some of these problems.

Illustration 6-11

Rosen Appliances is the local distributor for a certain brand of television set. They receive monthly shipments from the factory, inventory them, and distribute them on demand to area retailers. Their profit margin depends on ordering the proper number of televisions each month. Too large an order results in excessive inventory costs; too small an order may cost more per set and in addition may mean that the retailer will not be able to fill an order (lost sales). Prior sales figures show that June sales orders for 19-inch color portables, x, are approximately normally distributed with a mean of 72 and a standard deviation of 13. Management will not accept a probability of a lost sale (due to sets being out of stock) greater than 1%. How many 19-inch color portable models should be ordered?

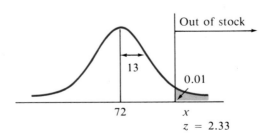

Solution x is the number ordered, and the 1% chance of being out of stock means that sales exceeded x. The area of the figure that relates to table 6 is 0.4900 (0.5000 − 0.0100). The closest z score entry is 2.33.

$$z = \frac{x - \mu}{\sigma}$$

When $z = 2.33$: $\quad 2.33 = \dfrac{x - 72}{13}$

$$x - 72 = (13)(2.33) = 30.29$$
$$x = 72 + 30.29 = 102.29$$
$$x = \mathbf{103}$$

Thus if Rosen Appliances orders 103 sets, there is less than a 1% chance that they will be out of stock on this 19-inch portable model. □

Illustration 6-12

Referring back to the CB radio installation problem of illustration 6-9, find the amount of time T such that 67% of all installations take more than T minutes.

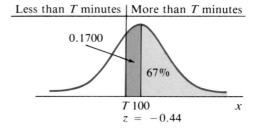

Solution Since the area to the right of the mean represents 50% of the cases, and those take 100 minutes or more, the additional 17% representing "more than T" must come from the left of the mean; therefore the value 0.67 is represented by the entire shaded portion in the diagram, but only 0.17 (0.67 − 0.50) relates to table 6. The 0.1700 entry in table 6 tells us that $z = -0.44$ (0.44 is read from the table; we use a negative value because it is below the mean value).

$$z = \frac{x - \mu}{\sigma}$$

When $z = -0.44$: $\quad -0.44 = \dfrac{T - 100}{10}$

$$T - 100 = (-0.44)(10) = -4.4$$
$$T = 100 - 4.4 = \mathbf{95.6 \text{ minutes}}$$

That is, 67% of the installations take more than 95.6 minutes to complete. □

Illustration 6-13

The incomes of junior executives in a large corporation are normally distributed with a standard deviation of $2,200. A cutback is pending, at which time those who earn less than $19,000 are to be discharged. If such a cut represents 10% of the junior executives, what is the current mean salary of the group of junior executives?

Solution If 10% of the salaries are below $19,000, then 40% (or 0.4000) are between $19,000 and the mean μ. Table 6 indicates that $z = -1.28$ is the standard score that occurs at $x = \$19,000$. Using formula (6-3), we can find the value of μ:

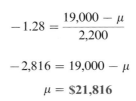

$$-1.28 = \frac{19{,}000 - \mu}{2{,}200}$$

$$-2{,}816 = 19{,}000 - \mu$$

$$\mu = \$21{,}816$$

That is, the current mean salary of junior executives is $21,816. □

Referring again to the installation of CB radios, what is the probability that an installation will require 125 minutes, $P(x = 125)$? This situation has two interpretations: theoretical and practical. Let's look at the theoretical interpretation first. Recall that the probability associated with a continuous random variable is represented by the area under the curve. That is, $P(a < x < b)$ is equal to the area between a and b under the curve. $P(x = 125)$—that is, x is exactly 125—is then $P(125 < x < 125)$, or the area of a vertical line segment at $x = 125$. This area is 0. However, this interpretation is not the *practical* meaning of $x = 125$. It generally means 125 minutes to the nearest minute. Thus $P(x = 125)$ would most likely be interpreted as $P(124.5 < x < 125.5)$. The interval 124.5 to 125.5 under the curve has a measurable area and is then nonzero. In situations of this nature, you must be sure of the meaning being used.

Exercises

6-14 Given that x is a normally distributed random variable with a mean of 50 and a standard deviation of 8, find the following probabilities:
(a) $P(x > 50)$ (b) $P(50 < x < 62)$
(c) $P(48 < x < 72)$ (d) $P(54 < x < 58)$
(e) $P(30 < x < 78)$ (f) $P(x < 30)$

6-15 Let x be a normally distributed random variable with a mean of 6 and a standard deviation of 2. If a single value of x is taken from this population at random, what is the probability that it will be between 7 and 9?

6-16 Let x be a normally distributed random variable with a mean of 15.5 and a standard deviation of 2.6. Find the probability that an individual value of x, selected at random, will fall in the following intervals:
- (a) between 15.5 and 19.2
- (b) between 16.0 and 20.0
- (c) between 10.0 and 14.0
- (d) between 10.0 and 20.0

6-17 The distribution of the lengths of useful life of a fluorescent tube used for indoor gardening has a mean of 600 hours and a standard deviation of 40 hours. The useful lifetimes are normally distributed. Determine:
- (a) the probability that a tube chosen at random will last between 620 and 680 hours
- (b) the probability that such a tube will last more than 740 hours

6-18 Railroad stock grows at an average of 4.6% a year with a standard deviation of 2.3%. (Assume normality.)
- (a) Find the probability that a given railroad stock will grow at least 3% next year.
- (b) Find the percentage of railroads whose stock grows less than 2% a year.

6-19 The number of checks returned for insufficient funds on a given day by Lincoln Bank averages 46 with a standard deviation of 13. Assume that the number of checks that bounce is approximately normally distributed.
- (a) Find the probability that at least 40 checks are returned for insufficient funds on a given day.
- (b) Find the proportion of days in a year that no more than 20 checks are returned for insufficient funds.

Section 6-4 Notation

When working with the standard score z, it is often helpful and necessary to identify the z score with an area under the normal curve. **The standard procedure is to use the area under the curve and to the right of the z.** We will write this area within parentheses following the z.

Illustration 6-14

$z(0.05)$ is the value of z such that exactly 0.05 of the area under the curve lies to its right, as shown in the diagram.

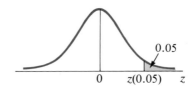

$z(0.60)$ is that value of z such that 0.60 of the area lies to its right, as shown in the next figure.

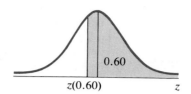

Now let's find the value of $z(0.05)$. We must convert this information into a value that can be read from table 6; see the areas shown in the next figure.

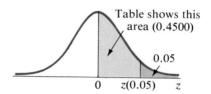

Now we look in table 6 in appendix D and find an area as close as possible to 0.4500.

z	...	0.04	0.05	...
⋮				
1.6	...	0.4495	0.4505	
⋮				

Therefore $z(0.05) = \mathbf{1.65}$. (*Note*: We always round up when finding the z score; this procedure is an exception to the round-off rule.) □

Illustration 6-15

Find the value of $z(0.60)$.

Solution The value 0.60 is related to table 6 by use of the area 0.1000, as shown in the diagram. The closest values in table 6 are 0.0987 and 0.1026.

z	...	0.05	0.06	...
⋮				
0.2		0.0987	0.1026	
⋮				

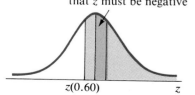

Look for 0.1000; remember that z must be negative

Therefore $z(0.60)$ is related to 0.26. Since $z(0.60)$ is below the mean, we conclude that $z(0.60) = -\mathbf{0.26}$. □

In later chapters the notation just discussed will be used on a regular basis. Only a few values of z will be used regularly. These values come from one of two situations: (1) the z score that cuts off a specified area in one tail of the normal distribution or (2) the z scores that bound a specified middle proportion of the normal distribution. Illustration 6-14 showed a commonly used one-tail situation; z(0.05) = 1.65 is located so that 0.05 of the area under the normal distribution curve is in the tail to the right.

Illustration 6-16

Find z(0.95).

Solution z(0.95) is located on the left-hand side of the normal distribution, since the area to the right is 0.95. The area in the tail to the left then contains the other 0.05, as shown in the accompanying figure. Because of the symmetrical nature of the normal distribution, z(0.95) is −z(0.05), that is, z(0.05) with its sign changed. Thus z(0.95) = −z(0.05) = **−1.65**. □

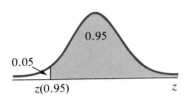

When the middle proportion is specified, we still can use the "area to the right" notation to identify the specific z score involved.

Illustration 6-17

Find the z scores that bound the middle 0.95 of the normal distribution.

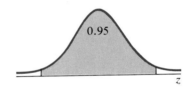

Solution With 0.95 as the area in the middle, the two tails must contain a total of 0.05, and each therefore contains 0.025, as shown in the accompanying figure.

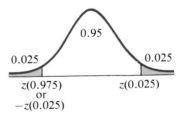

To find z(0.025) in table 6, we must determine the area between the mean and z(0.025). It is 0.4750, as shown in the figure. From table 6,

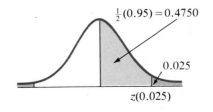

	...	0.06	...
⋮			
1.9		0.4750	

Therefore $z(0.025) = 1.96$ and $z(0.975) = -z(0.025) = -1.96$, and -1.96 and 1.96 bound the middle 0.95 of the normal distribution. □

Exercises

6-20 Use the notation defined in this section to identify the z score shown in each of the diagrams.

(a)

(b)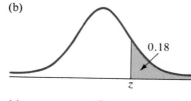

(c)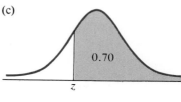

(d)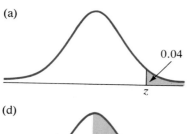

(e)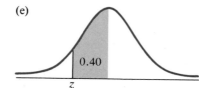

(f)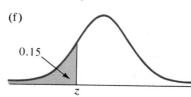

6-21 Use table 6 to find the following values of z:
(a) $z(0.01)$ (b) $z(0.02)$ (c) $z(0.025)$
(d) $z(0.95)$ (e) $z(0.90)$

6-22 Complete the following charts of z scores. The area A that is given is the area to the right under the normal distribution.

(a) z scores associated with the right-hand tail: Given the area A, find $z(A)$.

A =	0.10	0.05	0.025	0.02	0.01	0.005
$z(A) =$						

(b) z scores associated with the left-hand tail: Given the area A, find $z(A)$.

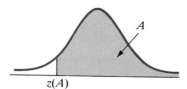

A =	0.995	0.99	0.98	0.975	0.95	0.90
$z(A) =$						

Section 6-5 Normal Approximation of the Binomial

binomial probability

In chapter 5 we introduced the binomial distribution. Recall that the binomial distribution is a probability distribution of the discrete random variable x, the number of successes observed in n repeated independent trials. We will now see how **binomial probabilities**—that is, probabilities associated with the binomial distribution—can be reasonably estimated by use of the normal probability distribution.

Let's look first at a few specific binomial distributions. Figures 6-4a, 6-4b, and 6-4c show the probabilities of x for 0 to n for three situations: $n = 4$, $n = 8$, and $n = 24$. For each of these distributions, the probability of success in one trial is 0.5. Notice that as n becomes larger, the distribution appears more and more like the normal distribution.

FIGURE 6-4
Binomial Distributions

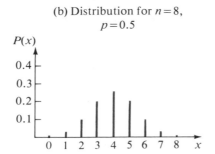
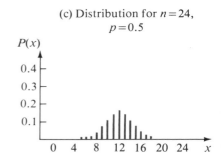

To make the desired approximation, we need to take into account one major difference between the binomial and the normal probability distributions. The binomial random variable is discrete, whereas the normal random variable is continuous. Recall that in chapter 5 we demonstrated that the probability assigned to a particular value of x should be shown on a diagram by means of a straight-line segment whose length represents the probability (as in figure 6-4). We suggested, however, that we also can use a histogram in which the area of each bar is equal to the probability of x.

Let's look at the distribution of the binomial variable x, where $n = 14$ and $p = 0.5$. The probabilities for each x value can be obtained from table 4 in appendix D. This distribution of x is shown in figure 6-5. In histogram form we see the very same distribution in figure 6-6.

Let's examine $P(x = 4)$ for $n = 14$ and $p = 0.5$ to study the approximation technique. $P(x = 4)$ is equal to 0.061 (see table 4 of appendix D), the area of the bar above $x = 4$ in figure 6-7. Area is the product of width and height. In this case the height is 0.061 and the width is 1.0; thus the area is 0.061. Let's take a closer look at the width. For $x = 4$, the bar starts at 3.5

Distribution of x When $n = 14$, $p = 0.5$

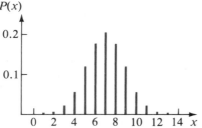

FIGURE 6-5
Line Histogram

FIGURE 6-6
Histogram

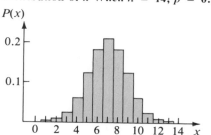

FIGURE 6-7
Histogram

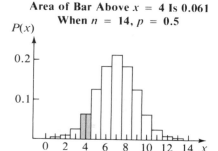

and ends at 4.5, so we are looking at an area bounded by $x = 3.5$ and $x = 4.5$. The addition and subtraction of 0.5 to the x value is commonly referred to as the **continuity correction factor**. It is our method of adjustment so that a continuous variable can be used to approximate a discrete variable.

continuity correction factor

Now let's look at the normal distribution related to this situation. We will first need a normal distribution with a mean and a standard deviation equal to those of the binomial distribution we are discussing. Formulas (5-7) and (5-8) give us these values.

$$\mu = np = (14)(0.5) = 7.0$$

$$\sigma = \sqrt{npq} = \sqrt{(14)(0.5)(0.5)} = \sqrt{3.5} = 1.87$$

The probability that $x = 4$ is approximated by the area under the normal curve between $x = 3.5$ and $x = 4.5$, as shown in figure 6-8. Figure 6-9 shows the entire distribution of the binomial variable x with a normal distribution of the same mean and standard deviation superimposed. Notice that the "bars" and the "interval areas" under the curve cover nearly the same area.

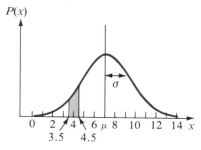

FIGURE 6-8
Probability That $x = 4$ Is Approximated by Shaded Area

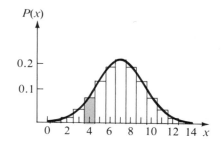

FIGURE 6-9
Normal Distribution Superimposed over Distribution for Binomial Variable x

The probability that x is between 3.5 and 4.5 under this normal curve is found by using table 6 and the methods outlined in section 6-3.

$$P(3.5 < x < 4.5) = P\left(\frac{3.5 - 7.0}{1.87} < z < \frac{4.5 - 7.0}{1.87}\right)$$
$$= P(-1.87 < z < -1.34)$$
$$= 0.4693 - 0.4099 = \mathbf{0.0594}$$

Since the binomial probability of 0.061 and the normal probability of 0.0594 are reasonably close in value, the normal probability distribution seems to be a reasonable approximation of the binomial distribution.

By now you may be thinking, "So what? I will just use the binomial table and find the probabilities directly and avoid all the extra work." But consider for a moment a situation such as that presented in illustration 6-18.

Illustration 6-18

An unnoticed mechanical failure has caused a machine shop's output of 5,000 rifle firing pins to be one-third defective. What is the probability that an inspector will find no more than 3 defective firing pins in a random sample of 25?

Solution In this illustration of a binomial experiment, x is the number of defectives found in the sample, $n = 25$, and $p = P(\text{defective}) = \frac{1}{3}$. To answer the question by using the binomial distribution, we will need to use the binomial probability function [formula (5-6)]:

$$P(x) = \binom{25}{x} \cdot \left(\frac{1}{3}\right)^x \cdot \left(\frac{2}{3}\right)^{25-x} \quad \text{for } x = 0, 1, 2, \ldots, 25$$

We must calculate the values for $P(0)$, $P(1)$, $P(2)$, and $P(3)$, since they do not appear in table 4, appendix D. This job is very tedious because of the size of the exponent. In situations such as this, we can use the normal approximation method.

Now let's find $P(x \leq 3)$ by using the normal approximation method. We first need to find the mean and standard deviation of x [formulas (5-7) and (5-8)]:

$$\mu = np = (25)\left(\frac{1}{3}\right) = 8.333$$

$$\sigma = \sqrt{npq} = \sqrt{(25)\left(\frac{1}{3}\right)\left(\frac{2}{3}\right)} = 2.357$$

These values are shown in the diagram. The measure of the shaded area $(x < 3.5)$ in the diagram represents the probability of $x = 0, 1, 2,$ or 3. Remember that $x = 3$, the discrete binomial variable, covers the continuous interval from 2.5 to 3.5.

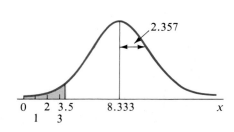

Chapter 6 The Normal Probability Distribution

$$P(x \text{ is no more than } 3) = P(x \leq 3) \quad \text{(for a discrete variable } x\text{)}$$
$$= P(x < 3.5) \quad \text{(using a continuous variable } x\text{)}$$

$$P(x < 3.5) = P\left(z < \frac{3.5 - 8.333}{2.357}\right) = P(z < -2.05)$$
$$= 0.5000 - 0.4798 = \mathbf{0.0202}$$

Thus P(no more than three defectives) is approximately 0.02. ☐

The normal approximation of the binomial distribution is also useful for values of p that are not close to 0.5. The binomial probability distributions shown in figures 6-10 and 6-11 suggest that binomial probabilities can be approximated by use of the normal distribution. Notice that as n increases in size, the binomial distribution begins to look like the normal distribution. As the value of p moves away from 0.5, a larger n will be needed in order for the

FIGURE 6-10
Binomial Distributions

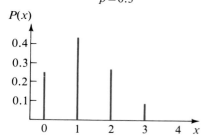
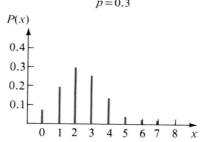
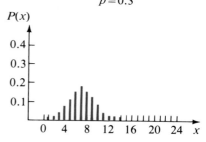

FIGURE 6-11
Binomial Distributions

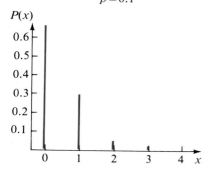
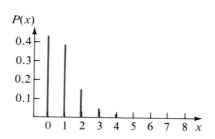
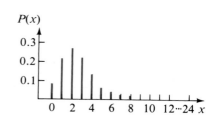

normal approximation to be reasonable. The following rule of thumb is generally used as a guideline.

Rule

The normal distribution provides a reasonable approximation of a binomial probability distribution whenever the values of np and $n(1 - p)$ equal or exceed 5.

Exercises

6-23 Find the normal approximation for the binomial probability $P(x = 5)$ when $n = 12$ and $p = 0.5$. Compare this value with the value of $P(x = 5)$ obtained from table 4, appendix D.

6-24 Find the normal approximation for the binomial probability $P(x = 6, 7, 8)$ when $n = 14$ and $p = 0.5$. Compare this value with the value of $P(x = 6, 7, 8)$ obtained from table 4.

6-25 Find the normal approximation for the binomial probability $P(x \leq 9)$ when $n = 14$ and $p = 0.4$. Compare this value with the value of $P(x \leq 9)$ obtained from table 4.

6-26 Find the normal approximation for the binomial probability $P(x \geq 4)$ when $n = 14$ and $p = 0.4$. Compare this value with the value of $P(x \geq 4)$ obtained from table 4.

6-27 A company asserts that 80% of those individuals who purchase its special lawn mower will have no repairs during the first 2 years of ownership. A study by you has shown that only 70 of the 100 in your sample went the 2 years without repair expenses. Using the normal approximation, what is the probability of a sample outcome of 70% or less if the true expenses-free percentage is 80%?

6-28 One-sixth of the lenses ground by an optical company are correct. Find the probability that more than 8 of the next 36 lenses will be defective.

6-29 Thirty percent of all employees of the Meridian Clock Company voluntarily leave before the end of their first year. Find the probability that less than one-fourth of a random sample of 80 will leave during their first year.

6-30 Sixty percent of all sales made by the Cartlon Furniture Company are credit sales. In a random sample of 200 sales, what is the probability that one-half or more are cash sales?

6-31 The marketing research department of the AMP Machine Company estimated that 50% of all housewives would prefer automatic fabric softener dispensers in their washers. In a poll of 100 housewives, what is the probability that half or more will prefer the dispensers?

6-32 Consider a production process that results in a 10% defective product. A sample of size 50 is taken at random.
  (a) What is the probability that no more than three defective items are found?
  (b) What is the probability that more than five defective items are found?

6-33 It is believed that the stockholders of Estrox are split 50–50 on a merger proposal. Assuming that to be the case, what is the probability that a straw poll of 100 stockholders will show at least 60% in favor?

6-34 A production run of T-shirts is expected to produce 50 seconds in every run of 1,000 T-shirts. If you randomly select 100 from a run of 1,000:
  (a) What is the mean and the standard deviation of the binomial distribution when samples of 100 are taken?
  (b) What is the probability that your sample will contain no T-shirt seconds?
  (c) What is the probability that your sample will contain no more than two T-shirt seconds?

In Retrospect

We now know what a normal distribution is, how to use it, and how it can help us. Let's again consider the news article at the beginning of this chapter, which illustrated a probability distribution for a continuous random variable. The random variable was years of service of an employee, and the curve gives us the probability that he or she will leave the organization. What kind of distribution is this? It certainly does not look like a normal distribution. As we will see in the next chapter, not all probability distributions of continuous variables are normal.

Chapter Exercises

6-35 Do you think incomes of persons 18 to 64 years of age in the United States are normally distributed? Why or why not?

6-36 The middle 50% of a normally distributed population lies between what two standard scores?

6-37 Find the standard score z such that the area above the mean and below z under the normal curve is as follows:
  (a) 0.4099   (b) 0.1331   (c) 0.3133

6-38 Find the standard score z such that the area below the mean and above z under the normal curve is as follows:
  (a) 0.3531   (b) 0.0279   (c) 0.4936

6-39 Find the standard score of a normally distributed variable such that 49% of the distribution falls between the mean and this particular value.

6-40 Find the following values of z:
 (a) $z(0.15)$ (b) $z(0.35)$ (c) $z(0.75)$ (d) $z(0.80)$

6-41 Find the value of a normally distributed random variable x such that 60% of the distribution lies to the left of this specific value; $\mu = 150$ and $\sigma = 20$.

6-42 It is an accepted fact that if customers must wait in line too long for a service, the result is loss of sales. The average length of time a customer must wait in line in a supermarket is normally distributed with a standard deviation of 5 minutes.
 (a) If 10% of the customers must wait at least 12 minutes, what is the mean time the customers wait in line?
 (b) If the addition of a bagger reduces the percentage of customers who must wait at least 12 minutes to 5%, how much does a bagger reduce the mean waiting time?

6-43 Airplanes arriving at a certain airport average 36 arrivals per hour during daylight hours on weekdays. Given that the variance of arrivals per hour is 36 and assuming that the number of arrivals per hour is approximately normally distributed, what is the probability that during a randomly selected 1-hour period,
 (a) fewer than 30 airplanes will arrive
 (b) more than 45 will arrive

6-44 In a large industrial complex, the maintenance department has been instructed to replace light bulbs before they burn out. It is known that the life of light bulbs is normally distributed with a mean life of 600 hours and a standard deviation of 60 hours. When should the light bulbs be replaced so that no more than 5% of them will ever burn out?

6-45 The Federal Reserve gives a bank 2 days clearance on all out-of-district checks deposited with them. That is, the bank may consider a check as cash and use it in its reserve calculations 2 days after it is deposited with the Federal Reserve, regardless of how long it actually takes the Federal Reserve to transfer the money from the bank on which it is written. (Out-of-district checks normally take over 2 days. The time during which both banks have use of the money is called the "float.") The proportion of checks deposited with the Federal Reserve that are out of district is a normally distributed variable with a mean of 0.20 and a standard deviation of 0.05. The Federal Reserve thinks that a bank in Atlanta, Georgia, in cooperation with a bank in Sacramento, California, is abusing the system. If a study shows that the proportion of checks deposited by the Atlanta bank and drawn on the California Federal Reserve District is 0.38, do you think the Federal Reserve is justified in asking for an investigation?

6-46 For a certain brand of TV dinner, the distribution of the weights of meat in the chopped beef dinners has a mean of 3.6 ounces and a standard deviation of 0.2 ounces. The beef portions are measured mechanically, and any deviation from

3.6 ounces is a mechanical error. The result of this system is that the weights of the chopped beef are approximately normally distributed.

    (a) What proportion of the dinners will contain chopped beef weighing less than 3.5 ounces?

    (b) What proportion of the dinners will contain chopped beef weighing over 3.8 ounces?

    (c) What proportion will have weights between 3.5 and 3.8 ounces?

    (d) Eighty-five percent of the dinners will have more than what weight of chopped beef in them?

6-47 Assume that the number of cases sold per week in December by a large group of beverage salesmen is quite close to a normal distribution with a mean of 684 and a standard deviation of 60.

    (a) Given that the top 16% of the salesmen receive bonuses, how many cases must a salesman sell in order to receive a bonus?

    (b) If the break-even number of cases is 600 per salesman, what proportion of the salesmen are above the break-even value?

6-48 For a small airline the probability of arriving at a flight destination more than 20 minutes late is 0.10, based on historical information gathered by the airline's research division. Assuming that the estimate is valid, what is the approximate probability that fewer than 5 of the next 60 flights scheduled will arrive at their destinations more than 20 minutes late?

6-49 The Kinder Kiddie Car Company fails to install horn batteries in 10% of its kiddie cars. Find these probabilities by using the appropriate method:

    (a) that the company fails to install 3 batteries in 5 cars

    (b) that the company fails to install no more than 3 batteries in 5 cars

    (c) that the company fails to install 3 batteries in 15 cars

    (d) that the company fails to install no more than 3 batteries in 100 cars

6-50 Thirty percent of the companies on the New York Stock Exchange have three or more subsidiaries. Use the normal approximation method to compute the probability that a random sample of 500 companies will contain the following:

    (a) exactly 155 companies with three or more subsidiaries

    (b) between 145 and 155 companies with three or more subsidiaries

    (c) more than 145 companies with three or more subsidiaries

6-51 Sixty-eight percent of all the holders of Erie City Bonds sell their bonds before maturity. A random survey of 120 bondholders is taken. Find these probabilities:

    (a) that exactly 75 bondholders will sell before maturity

    (b) that less than 75 will sell before maturity

    (c) that more than 90 will sell before maturity

6-52 To speed payment and hence reduce accounts receivable, American Appliances is considering offering a 1% discount on bills paid within 10 days of the billing date. The company believes that 75% of all customers will take advantage of the discount. Let x be the number of customers out of 80 who pay within 10 days. Use the normal approximation method to find the probability that x is greater than 65.

6-53 Eighty percent of all workers at Estrox favor the proposed union contract. At the union meeting 1,000 of the 10,000 union members are in attendance. The results of the vote were 660 for and 340 against. Would you conclude that the 1,000 who voted constitute a random sample of the total union membership population? (Use probability to justify your answer.)

6-54 To monitor output from machines to make sure that production standards are being met and that the machine does not need to be adjusted, a device called a "control chart" is used in industrial control. The control chart pictured here is commonly used. The product is measured and the value plotted on the chart (the dots). If the value falls between the bounds, production continues. If it falls outside the bounds (e.g., the dot circled), production is considered out of control, and the production is halted and the machine adjusted. Suppose the specification for the width of a steel rod is 3 inches, and the upper band is set at 3.6 inches and the lower band at 2.4 inches.

(a) If the machine is properly adjusted, the width of the steel beams produced is normally distributed with $\mu = 3$ and $\sigma = 0.2$. What proportion of the time will the process be called out of control incorrectly?

(b) If the machine is jolted, it produces beams whose widths are normally distributed with $\mu = 2.6$ and $\sigma = 0.3$. What is the probability that (i) the process will be halted after the first beam has been produced following a jolting of the machine? (ii) five will be produced and all will be within the band and production will continue?

(c) Suppose the bands were changed by lowering the upper band to 3.4 and raising the lower band to 2.6. What would be your answers to parts (a) and (b)?

(d) What is the advantage of narrowing the bands? The disadvantage? What would you want to consider in deciding whether to use narrower or wider bands?

Hands-On Problems

Obtain a sample of size 100 from a population of your choice Choose a population that you expect to be approximately normally distributed.

6-1   Define your population.

6-2   Obtain your sample.

6-3   Classify your data into a grouped frequency distribution and calculate $\bar{x}$ and s.

6-4   Use the $\bar{x}$ and the s found in problem 6-3 to calculate the z scores that correspond to each of the class boundaries in your classes.

6-5   Use the z scores found in problem 6-4 and table 6 of appendix D to calculate the probability associated with each class in your distribution. (These probabilities are those that would have occurred if your distribution was exactly normal with a mean of $\bar{x}$ and a standard deviation of s.)

6-6   Construct an observed probability distribution (a grouped relative frequency distribution) and compare the observed probabilities with the theoretical probabilities.

The calculations for this problem set can most easily be accomplished with the assistance of an electronic calculator or a packaged program on a computer. A list of the available programs can be obtained from your computer center. There are a variety of packaged programs available: Minitab, Biomed (Biomedical Programs), SAS (Statistical Analysis System), IBM Scientific Subroutine Packages, and SPSS (Statistical Package for the Social Sciences) program libraries. Your local computer center will assist you.

7 Sample Variability

Chapter Outline

7-1 Sampling Distributions
A distribution of values for a sample statistic, obtained by repeated sampling.

7-2 The Central Limit Theorem
Describes the sampling distribution of sample means.

7-3 Application of the Central Limit Theorem
To predict the behavior of sample means.

Two Weeks of Trading Days
Market activity from Friday, May 20, through Thursday, June 2

From USA TODAY, June 3, 1983. Reprinted by permission.

Chapter Objectives

In chapters 2 and 3 we discussed how to describe a sample. The description of the sample data is accomplished by using three basic concepts: (1) measures of central tendency (the mean is the most popularly used sample statistic), (2) measures of dispersion (the standard deviation is most commonly used), and (3) kind of distribution (normal, skewed normal, rectangular, etc.). The question that seems to follow is this: What can be deduced about the statistical population from which a sample is taken?

Consider the following quality control problem. We have just taken a sample of 25 rivets made for the construction of airplanes. The rivets were tested for shearing strength, and the force required to break each rivet was the response variable. The various descriptive measures—mean, standard deviation, type of distribution—can be found for this sample. However, the sample itself is not what we are interested in. The rivets that were tested were destroyed during the test, so they can no longer be used in the construction of airplanes. What we are trying to find out is information about the total population, and we certainly cannot test every rivet that is produced. (There would be none left for construction.) Therefore, somehow we must deduce information, or make inferences, about all of the rivets based on the results observed in the sample.

Suppose that we take another sample of 25 rivets and test them by the same procedure. Do you think that we would obtain the same sample mean from the second sample that we obtained from the first? The same standard deviation?

After considering these questions, we might suspect that we would need to investigate the variability in the sample statistics obtained from **repeated sampling**. Thus we need to find (1) measures of central tendency for the sample statistics of importance, (2) measures of dispersion for the sample statistics, and (3) the pattern of variability (distribution) of the sample statistics. Once we have this information, we will be better able to predict the population parameters.

The objective of this chapter is to study the measures and the patterns of variability for the distribution formed by repeatedly observed values of a sample mean.

Section 7-1 Sampling Distributions

To make inferences about a population, we need to discuss a little more about sample results. A sample mean, $\bar{x}$, is obtained from a sample. Do you expect that this value, $\bar{x}$, is exactly equal to the value of the population mean μ? Your answer should be "no." We do not expect that to happen, but we will be satisfied with our sample results if the sample mean is "close" to the value of the population mean. A second question might now be considered: If a second sample is taken, will the second sample have a mean equal to the population mean? Equal to the first sample mean? Again, no, we do not expect it to be equal to the population mean, nor do we expect the sample mean to repeat. We do, however, again expect the values to be "close." (This argument should hold for any other sample statistic and its corresponding population value.)

The next questions should already have come to mind: What is "close"? How do we determine (and measure) this closeness? Just how would repeated sample statistics be distributed? To answer these questions, we must take a look at a *sampling distribution*.

sampling distribution

Sampling Distribution of a Sample Statistic

The distribution of values for that sample statistic obtained from all possible samples of a population. The samples must all be of the same size, and the sample statistic could be any descriptive sample statistic.

Illustration 7-1

To illustrate the concept of a sampling distribution, suppose a firm maintains five separate checking accounts (i.e., one for payroll, one for dividends, one for accounts payable, one for investments, and one for petty cash disbursements). Of the checks written on each account in the first quarter of 1984, the number of checks that were not cashed was 0, 2, 4, 6, and 8, respectively. In an audit by the state, the auditor randomly selects two accounts and fully audits them. The firm contracts with an outside accounting firm to do a similar audit. For both audits, let's consider the mean number of uncashed checks of each possible sample of 2. There are 25 possible samples of size 2:

(0, 0) (2, 0) (4, 0) (6, 0) (8, 0)
(0, 2) (2, 2) (4, 2) (6, 2) (8, 2)
(0, 4) (2, 4) (4, 4) (6, 4) (8, 4)
(0, 6) (2, 6) (4, 6) (6, 6) (8, 6)
(0, 8) (2, 8) (4, 8) (6, 8) (8, 8)

Each of these samples has a mean $\bar{x}$. These means are, respectively,

0 1 2 3 4
1 2 3 4 5
2 3 4 5 6
3 4 5 6 7
4 5 6 7 8

Each of these samples is equally likely, and thus each of the 25 sample means can be assigned a probability of 0.04 ($\frac{1}{25}$). Why? The sampling distribution for the sample mean then becomes:

$\bar{x}$	$P(\bar{x})$
0	0.04
1	0.08
2	0.12
3	0.16
4	0.20
5	0.16
6	0.12
7	0.08
8	0.04

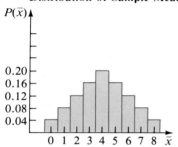

Histogram: Sampling Distribution of Sample Means

This is a probability distribution of $\bar{x}$.

For this same set of all possible samples of size 2, let's find the sampling distribution for sample ranges. Each sample has a range R. These ranges are, respectively,

```
0 2 4 6 8
2 0 2 4 6
4 2 0 2 4
6 4 2 0 2
8 6 4 2 0
```

Again, each possible sample has a probability of 0.04, and, by combining the probabilities of like values, we get the following sampling distribution of sample ranges:

R	$P(R)$
0	0.20
2	0.32
4	0.24
6	0.16
8	0.08

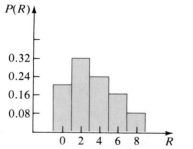

Histogram: Sampling Distribution of Sample Ranges

This is a probability distribution of R. ☐

Most populations that are sampled are much larger than the one used in illustration 7-1, and listing all the possible samples would be a very tedious job. With this in mind, let's investigate a sampling distribution empirically (i.e., by experimentation).

Illustration 7-2

Let's consider the sampling distribution of sample means for samples of size 5 obtained from the rolling of a single die. One sample will consist of five rolls, and we will obtain a sample mean $\bar{x}$ from this sample. We repeat the experiment until 30 sample means have been obtained. Table 7-1 shows 30

TABLE 7-1
Sample Means for Rolling a Single Die Five Times

Trial	Sample	$\bar{x}$	Trial	Sample	$\bar{x}$
1	1, 2, 3, 2, 2	2.0	16	5, 2, 1, 3, 5	3.2
2	4, 5, 5, 4, 5	4.6	17	6, 1, 3, 3, 5	3.6
3	3, 1, 5, 2, 4	3.0	18	6, 5, 5, 2, 6	4.8
4	5, 6, 6, 4, 2	4.6	19	1, 3, 5, 5, 6	4.0
5	5, 4, 1, 6, 4	4.0	20	3, 1, 5, 3, 1	2.6
6	3, 5, 6, 1, 5	4.0	21	5, 1, 1, 4, 3	2.8
7	2, 3, 6, 3, 2	3.2	22	4, 6, 3, 1, 2	3.2
8	5, 3, 4, 6, 2	4.0	23	1, 5, 3, 4, 5	3.6
9	1, 5, 5, 3, 4	3.6	24	3, 4, 1, 3, 3	2.8
10	4, 1, 5, 2, 6	3.6	25	1, 2, 4, 1, 4	2.4
11	5, 1, 3, 3, 2	2.8	26	5, 2, 1, 6, 3	3.4
12	1, 5, 2, 3, 1	2.4	27	4, 2, 5, 6, 3	4.0
13	2, 1, 1, 5, 3	2.4	28	4, 3, 1, 3, 4	3.0
14	5, 1, 4, 4, 6	4.0	29	2, 6, 5, 3, 3	3.8
15	5, 5, 6, 3, 3	4.4	30	6, 3, 5, 1, 1	3.2

such samples and their means. The resulting frequency distribution is shown in figure 7-1. This distribution seems to display characteristics of a normal distribution; it's mounded and nearly symmetric about its mean (approximately 3.5).

FIGURE 7-1
Histogram

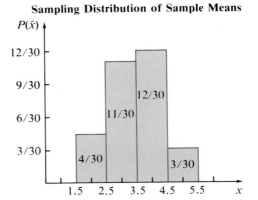

The theory involved with sampling distributions that will be described in the remainder of this chapter requires *random sampling*.

random sample

Random Sample

A sample obtained in such a way that each element of the population has an equal chance for selection and each of the possible samples has an equal chance to be selected.

Exercises

7-1 (a) What is the sampling distribution of sample means?

(b) A sample of size 3 is taken from a population and the sample mean found. Describe how this sample mean is related to the sampling distribution of sample means.

(c) Why is the probability of 0.04 assigned to each of the sample mean values in illustration 7-1?

7-2 Consider the set of interest rates offered by a bank: 4%, 5%, 6%, 7%.

(a) Make a list of all samples of size 2 that can be drawn from this set of interest rates. (Interest rates are not allowed to be repeated.)

(b) Construct the sampling distribution of sample means for samples of size 2 selected from this set.

(c) Construct the sampling distribution of sample ranges for samples of size 2.

7-3 Rework exercise 7-2, except now consider a sample of size 3 rather than 2.

7-4 Suppose that a box contains three identical blocks numbered 2, 4, and 6. A sample of two numbers is drawn with replacement (i.e., the first number is drawn, observed, and returned; then the second number is drawn). The mean of this sample is determined.

(a) Make a list that shows all the possible samples that could result from this sampling. (*Hint*: There should be nine samples.)

(b) Determine the mean of each of these samples and form a sampling distribution of these sample means. (Express this as a probability distribution.)

(c) Find the mean of this sampling distribution, $\mu_{\bar{x}}$.

(d) Find the standard deviation $\sigma_{\bar{x}}$ for the distribution.

7-5 From your local telephone directory, randomly select 20 telephone numbers. Using these 20 numbers as your source, take the fourth, fifth, and sixth digits to create 20 samples of size 3. For example, for 345-8267, you would take the 8, the 2, and the 6 as your sample of size 3.

(a) Calculate the mean of the 20 samples.

(b) Draw a histogram showing the 20 sample means. (Use class boundaries of −0.5 to 0.5, 0.5 to 1.5, 1.5 to 2.5, etc.)

Section 7-2 The Central Limit Theorem

On the preceding pages we discussed two sampling distributions. Many others could be discussed; in fact, these two could themselves be discussed further. However, the only sampling distribution of concern to us here is the **sampling distribution of sample means**. The mean is the most commonly used sample statistic and thus is the most important.

The central limit theorem tells us about the sampling distribution of sample means of random samples of size n. Recall that we basically want three kinds of information about a distribution: (1) where the center is, (2) how widely it is dispersed, and (3) how it is distributed. The central limit theorem tells us all three.

central limit theorem

Central Limit Theorem

If all possible random samples, each of size n, are taken from any population with a mean μ and a standard deviation σ, the sampling distribution of sample means will:

1. have a mean $\mu_{\bar{x}}$ equal to μ
2. have a standard deviation $\sigma_{\bar{x}}$ equal to $\sigma/\sqrt{n}$
3. be approximately normally distributed

The approximation to the normal distribution improves with samples of larger size.

In short, the central limit theorem says the following:

1. $\mu_{\bar{x}} = \mu$; the mean of $\bar{x}$'s equals the mean of x's.
2. $\sigma_{\bar{x}} = \sigma/\sqrt{n}$; the **standard deviation of the sample means** equals the standard deviation of the population divided by the square root of the sample size.
3. The sample means are approximately normally distributed (regardless of the shape of the parent population).

NOTE: The n referred to in the central limit theorem is the size of each sample in the sampling distribution.

standard error of the mean

Standard Error of the Mean

The standard deviation of the sampling distribution of sample means.

COMMENT: When the parent population is normally distributed, the sampling distribution of sample means is normally distributed for all sample sizes. When the parent population is not normally distributed, the

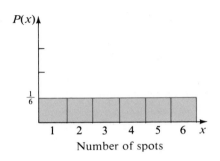

FIGURE 7-2
Probability Distribution for Rolling a Die

approximation improves as n increases. For samples of size 30 or larger, this approximation is adequate.

We are unable to prove this theorem without using advanced mathematics. However, it is possible to check its validity by examining a particular sampling distribution. Let's look at the sampling distribution of sample means for samples of size 5 given in illustration 7-2 and check each of the three conclusions claimed in the theorem.

First let's look at the theoretical probability distribution from which these samples (table 7-1) were taken. A histogram showing the probability distribution of the tossing of a die is shown in figure 7-2. The population mean μ equals **3.5** (see table 7-2). The population standard deviation σ equals $\sqrt{15.17 - (3.5)^2}$, which is $\sqrt{2.92} = \mathbf{1.71}$. (Note that this population has a uniform distribution.)

TABLE 7-2
Probability Distribution and Extensions for Tossing a Die

x	$P(x)$	$x \cdot P(x)$	$x^2 \cdot P(x)$
1	1/6	1/6	1/6
2	1/6	2/6	4/6
3	1/6	3/6	9/6
4	1/6	4/6	16/6
5	1/6	5/6	25/6
6	1/6	6/6	36/6
	6/6 = 1	21/6 = 3.5	91/6 = 15.17

Now let's look at the empirical sampling distribution of the 30 sample means found in illustration 7-2. If we use the 30 values of $\bar{x}$ in table 7-1, the observed mean of the $\bar{x}$'s turns out to be 3.43, and the observed standard deviation $s_{\bar{x}}$ turns out to be 0.73. The histogram appears in figure 7-1.

The central limit theorem (abbreviated as CLT) says that the $\bar{x}$'s should be approximately normally distributed, and the histogram certainly suggests this to be the case. The CLT also says that the mean $\mu_{\bar{x}}$ of the sampling distribution and the mean μ of the population are the same. The mean of the $\bar{x}$'s is 3.43 and $\mu = 3.5$; they seem to be reasonably close. Remember that we have taken only 30 samples, not all possible samples, of size 5.

The theorem says that $\sigma_{\bar{x}}$ should equal $\sigma/\sqrt{n}$. The observed standard deviation of $\bar{x}$'s is $s_{\bar{x}} = 0.73$, and the standard error of the mean is $\sigma/\sqrt{n} = 1.71/\sqrt{5} = 0.76$. These two values are very close.

The evidence seen in this one sampling distribution suggests that the CLT is true, although this one example does not constitute a proof of the theorem, of course.

Suppose that we look at another situation. Let's consider a population in which we can construct the theoretical sampling distribution of all the possible samples. In a situation of this nature, we should be able to observe the exact results claimed by the CLT. For this example let's consider all the possible samples of size 2 that could be drawn from a population that contains the three numbers 2, 4, and 6.

First let's look at the population itself. To calculate the mean μ and the standard deviation σ, we must use the formulas from chapter 5 for discrete probability distributions:

$$\mu = \sum [x \cdot P(x)] \quad \text{and} \quad \sigma = \sqrt{\sum [x^2 \cdot P(x)] - \{\sum [x \cdot P(x)]\}^2}$$

These formulas are necessary because we are not drawing the samples but discussing the theoretical possibilities. See table 7-3.

TABLE 7-3
Probability Distribution and Extensions for x = 2, 4, 6

x	P(x)	x·P(x)	x²·P(x)
2	1/3	2/3	4/3
4	1/3	4/3	16/3
6	1/3	6/3	36/3
	3/3 ck	12/3	56/3

$$\mu = \frac{12}{3} = 4.0$$

$$\sigma = \sqrt{\frac{56}{3} - \left(\frac{12}{3}\right)^2} = \sqrt{18.67 - 16.0} = \sqrt{2.67} = 1.63$$

Table 7-4 gives a list of all the possible samples that could be drawn if samples of size 2 were to be drawn from this population. (One number is drawn, observed, and then returned to the population before the second number is drawn.) Table 7-4 also lists the means of these samples. The

TABLE 7-4
All Possible Samples of Size 2 and Their Means

Possible Samples	$\bar{x}$
2, 2	2
2, 4	3
2, 6	4
4, 2	3
4, 4	4
4, 6	5
6, 2	4
6, 4	5
6, 6	6

probability distribution for these means and the extensions are given in table 7-5. Thus we have

$$\mu_{\bar{x}} = \frac{36}{9} = 4.0$$

$$\sigma_{\bar{x}} = \sqrt{\frac{156}{9} - \left(\frac{36}{9}\right)^2} = \sqrt{17.33 - 16} = \sqrt{1.33} = 1.15$$

TABLE 7-5
Probability Distribution for Means of All Possible Samples of Size 2

$\bar{x}$	$P(\bar{x})$	$\bar{x} \cdot P(\bar{x})$	$\bar{x}^2 \cdot P(\bar{x})$
2	1/9	2/9	4/9
3	2/9	6/9	18/9
4	3/9	12/9	48/9
5	2/9	10/9	50/9
6	1/9	6/9	36/9
	9/9 ✓	36/9	156/9

The histogram for the distribution of possible $\bar{x}$'s is shown in figure 7-3.

FIGURE 7-3
Histogram

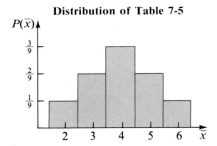

The CLT says that three things will occur in this sampling distribution:

1. It will be approximately normally distributed. The histogram (figure 7-3) suggests this very strongly.
2. The mean $\mu_{\bar{x}}$ of the sampling distribution will equal the mean of the population. They both have the value 4.0.
3. The standard deviation $\sigma_{\bar{x}}$ of the sampling distribution (standard error) will equal the standard deviation of the population divided by the square root of the sample size ($\sigma/\sqrt{n}$):

$$\sigma_{\bar{x}} = 1.15 \quad \text{and} \quad \frac{\sigma}{\sqrt{n}} = \frac{1.63}{\sqrt{2}} = \frac{1.63}{1.41} = 1.15$$

This illustration shows that the CLT is true for a *theoretical* probability distribution. The preceding illustration showed the CLT to be approximately true in a *sampling* situation.

Having taken a look at these two specific illustrations that support the CLT, let's now look at four graphic illustrations that present the same information in slightly different form. In each of these graphic illustrations, there are four distributions. The first is a distribution of the parent population, the distribution of the individual x values. Each of the other three graphs shows a sampling distribution of sample means, using three different sample sizes. In figure 7-4 we have a uniform distribution, much like figure 7-2 for the die illustration, and the resulting distributions of sample means for samples of size 2, 5, and 30. Figure 7-5 shows a U-shaped population

FIGURE 7-4
Uniform Distribution

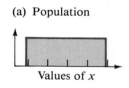
(a) Population
Values of x

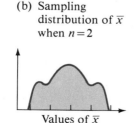
(b) Sampling distribution of $\bar{x}$ when $n=2$
Values of $\bar{x}$

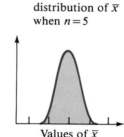
(c) Sampling distribution of $\bar{x}$ when $n=5$
Values of $\bar{x}$

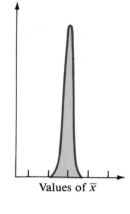
(d) Sampling distribution of $\bar{x}$ when $n=30$
Values of $\bar{x}$

FIGURE 7-5
U-Shaped Distribution

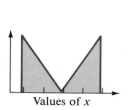
(a) Population
Values of x

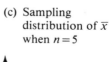
(b) Sampling distribution of $\bar{x}$ when $n=2$
Values of $\bar{x}$

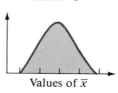
(c) Sampling distribution of $\bar{x}$ when $n=5$
Values of $\bar{x}$

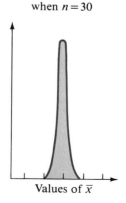
(d) Sampling distribution of $\bar{x}$ when $n=30$
Values of $\bar{x}$

The Central Limit Theorem Section 7-2

and the corresponding sampling distributions. Figure 7-6 shows a J-shaped population and the three corresponding distributions. Figure 7-7 shows a normal distribution population and the three sampling distributions.

All four illustrations seem to verify the CLT. Note that the sampling distributions of three nonnormal distributions produced sample means with an approximately normal distribution for samples of size 30. In the normal population (figure 7-7) the sampling distributions for all sample sizes appear to be normal. Thus you have seen an amazing phenomenon: no matter what the shape of a population, the sampling distribution of the mean becomes approximately normally distributed when n becomes sufficiently large.

FIGURE 7-6
J-Shaped Distribution

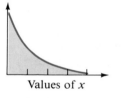
(a) Population
Values of x

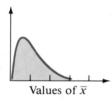
(b) Sampling distribution of $\bar{x}$ when $n=2$
Values of $\bar{x}$

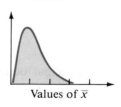
(c) Sampling distribution of $\bar{x}$ when $n=5$
Values of $\bar{x}$

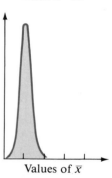
(d) Sampling distribution of $\bar{x}$ when $n=30$
Values of $\bar{x}$

FIGURE 7-7
Normal Distribution

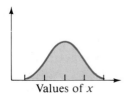
(a) Population
Values of x

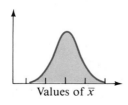
(b) Sampling distribution of $\bar{x}$ when $n=2$
Values of $\bar{x}$

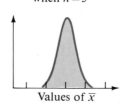
(c) Sampling distribution of $\bar{x}$ when $n=5$
Values of $\bar{x}$

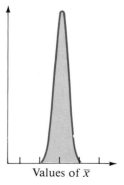
(d) Sampling distribution of $\bar{x}$ when $n=30$
Values of $\bar{x}$

You should notice one other point: the sample mean becomes less variable as the sample size increases. Notice that as n increases from 2 to 30, all the distributions become narrower and taller. Can you explain how this increase in sample size implies less variability? How does the CLT state this decrease in variability? (See exercise 7-6.) This point is discussed further in section 7-3.

Exercises

7-6 (a) What is the measure of the total area for any probability distribution?

(b) How does the CLT state that as n becomes larger, the sample mean becomes less variable?

7-7 The diameter of industrial hose manufactured by the Roberts Rubber Hose Company is normally distributed with a mean of 50 centimeters and a standard deviation of 10 centimeters. Several samples of size 25 are taken and their means, $\bar{x}$, recorded.

(a) What value would you expect to find for the mean of these several $\bar{x}$'s?

(b) If you were to calculate the standard deviation among the several $\bar{x}$'s, approximately what value would you expect to find? Explain.

7-8 If a population has a standard deviation σ of 16 units, what would be the standard error of the mean, $\sigma_{\bar{x}}$, for samples of size 16? What would the standard error be if samples of size 100 were taken?

7-9 If the variance of the prices of municipal bonds is 144 and a sample of 25 municipal bonds is taken, find the standard error of the mean price of municipal bonds, $\sigma_{\bar{x}}$.

7-10 The Golden Jewelry firm buys gold monthly. Records indicate that the distribution of prices has been approximately normal and had a mean price of $412.50 and a standard deviation of $20. A random sample of prices for 10 months is selected, and the sample mean $\bar{x}$ is $416.00. A sampling distribution of means would be formed by the means of all such groups of prices for 10 months.

(a) Determine the mean of this sampling distribution.

(b) Determine the standard error of the mean for this sampling distribution.

Section 7-3 Application of the Central Limit Theorem

The central limit theorem tells us about the sampling distribution of sample means by describing the shape of the distribution of all possible sample means. It also specifies the relationship between the mean μ of the population and the mean $\mu_{\bar{x}}$ of the sampling distribution, and the relationship between the standard deviation σ of the population and the standard error $\sigma_{\bar{x}}$ for the sampling distribution. Since sample means are approximately normally distributed, we will be able to answer probability questions by using table 6 of appendix D.

Illustration 7-3

Consider a population with $\mu = 10.0$ and $\sigma = 3.0$. If a sample of size 36 is selected at random, what is the probability that this sample will have a mean value between 9.0 and 11.0? That is, what is $P(9.0 < \bar{x} < 11.0)$?

Solution The CLT says that the distribution of $\bar{x}$'s is approximately normally distributed. To determine probabilities associated with a normal distribution, we will need to convert the statement

$$P(9.0 < \bar{x} < 11.0)$$

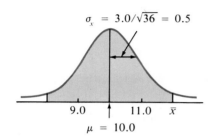

to a probability statement concerning z in order to use table 6, the standard normal distribution table. The sampling distribution is shown in the accompanying figure, with $P(9.0 < \bar{x} < 11.0)$ represented by the shaded area.

The formula for finding z, knowing a value of $\bar{x}$, is

$$z = \frac{\bar{x} - \mu_{\bar{x}}}{\sigma_{\bar{x}}} \qquad (7\text{-}1)$$

However, the CLT tells us that $\mu_{\bar{x}} = \mu$ and $\sigma_{\bar{x}} = \sigma/\sqrt{n}$. Therefore we will rewrite formula (7-1) in terms of μ and σ:

$$z = \frac{\bar{x} - \mu}{\sigma/\sqrt{n}} \qquad (7\text{-}2)$$

Using formula (7-2), we find that $\bar{x} = 9.0$ has a standard score of

$$z = \frac{9.0 - 10.0}{3.0/\sqrt{36}} = \frac{-1.0}{0.5} = -2.0$$

$\bar{x} = 11.0$ has a standard score of

$$z = \frac{11.0 - 10.0}{3.0/\sqrt{36}} = \frac{1.0}{0.5} = 2.0$$

Therefore

$$P(9.0 < \bar{x} < 11.0) = P(-2.0 < z < 2.0)$$
$$= 2(0.4772) = \mathbf{0.9544} \qquad \square$$

Before we look at more illustrations, let's consider for a moment what is implied by saying that $\sigma_{\bar{x}} = \sigma/\sqrt{n}$. For demonstration purposes let's suppose that $\sigma = 2.0$ and let's use a sampling distribution of samples of size 4. Now $\sigma_{\bar{x}}$ would be $2.0/\sqrt{4}$, or 1.0, and approximately 95% (0.9544) of all such sample means should be within the interval from 2.0 below to 2.0 above the population mean (within two standard deviations of the population mean). However, if the sample size were increased to 16, $\sigma_{\bar{x}}$ would become $2.0/\sqrt{16} = 0.5$, and approximately 95% of the sampling distribution would be within 1 unit of the mean; and so on. As the sample size increases, the size of $\sigma_{\bar{x}}$ becomes smaller, so that the distribution of sample means becomes much narrower. Figure 7-8 illustrates what happens to the distribution of $\bar{x}$'s as the size of the individual samples increases.

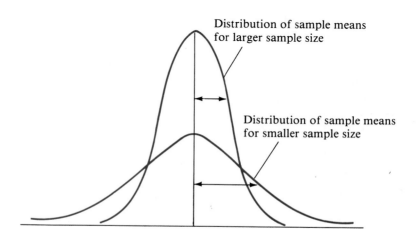

FIGURE 7-8
Distribution of Sample Means

Recall that the area under the normal curve is always exactly 1 unit of area. So as the width of the curve narrows, the height will have to increase to maintain this area.

Illustration 7-4

Application of the preceding discussion to investments would imply that diversification (i.e., purchase of a large number of different stocks and bonds) would lower the standard deviation of the average return $\bar{x}$ and hence minimize risk. That is, if you buy a large number of stocks and bonds, the price fluctuation on each stock and bond (measured by σ) would tend to cancel out, and the fluctuations in the portfolio return ($\sigma/\sqrt{n}$) would be small. This has led many people to suggest that you should "buy the Dow" (invest in every stock in the Dow–Jones average). □

Illustration 7-5

AR Research was interested in studying the attention span of adult listeners to spot television advertisements. To determine how long a commercial should be to be effective, it set up an experiment in which persons were asked to preview a TV program. Unknown to the participants, they were being filmed, and the length of time of eye contact with the TV set during a commercial screening was recorded. If the true mean of attention span is 39 seconds and the standard deviation is 2 seconds, what is the probability that the sample mean of 25 randomly selected participants will be between 38.5 and 40 seconds?

Solution We want to find $P(38.5 < \bar{x} < 40.0) = P(? < z < ?)$, where the z scores are as follows:

$$\text{When } \bar{x} = 38.5: \quad z = \frac{38.5 - 39.0}{2/\sqrt{25}} = \frac{-0.5}{0.4} = -1.25$$

$$\text{When } \bar{x} = 40.0: \quad z = \frac{40.0 - 39.0}{2/\sqrt{25}} = \frac{1.0}{0.4} = 2.5$$

(See the accompanying figure.) Therefore

$$P(38.5 < \bar{x} < 40.0) = P(-1.25 < z < 2.50)$$
$$= 0.3944 + 0.4938 = \mathbf{0.8882}$$

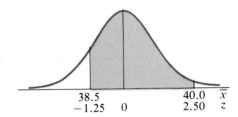

Illustration 7-6

For illustration 7-5, within what limits would the middle 90% of the sampling distribution of sample means of sample size 100 fall?

Solution The basic formula is

$$z = \frac{\bar{x} - \mu}{\sigma/\sqrt{n}}$$

$$\sigma_{\bar{x}} = \frac{\sigma}{\sqrt{n}} = \frac{2}{\sqrt{100}} = \frac{2}{10} = 0.2$$

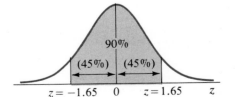

[Recall that the area to the right of the mean, 45% (0.4500), is related to the z score of 1.65, according to table 6.]

If $z = -1.65$:

$$-1.65 = \frac{\bar{x} - 39}{0.2}$$

$$(-1.65)(0.2) = \bar{x} - 39$$

$$\bar{x} = 39 - 0.33 = \mathbf{38.67}$$

If $z = 1.65$:

$$1.65 = \frac{\bar{x} - 39}{0.2}$$

$$(1.65)(0.2) = \bar{x} - 39$$

$$\bar{x} = 39 + 0.33 = \mathbf{39.33}$$

Thus

$$P(38.67 < \bar{x} < 39.33) = \mathbf{0.90}$$

or the middle 90% of the sampling distribution of $\bar{x}$'s lies within the limits of 38.67 and 39.33. □

Exercises

7-11 A random sample of size 25 is to be selected from a normal population that has a mean μ of 50 and a standard deviation σ of 10.

(a) What is the probability that this sample mean will be between 45 and 55?

(b) What is the probability that the sample mean will have a value greater than 48?

(c) What is the probability that the sample mean will be within 3 units of the mean?

7-12 A shipment of steel bars will be accepted if the mean breaking strength of a random sample of 10 is greater than 250 psi. Based on past experience, the breaking strength of such bars has a distribution with a mean of 235 psi and a variance of 400.

(a) What is the probability that one randomly selected steel bar will have a breaking strength in the range from 245 to 255 psi, assuming that the breaking strengths are normally distributed?

(b) What is the probability that the shipment will be accepted?

7-13 The mean weekly earnings of employed males was $243.30 for a random sample of 50 taken from all of the employees of a large firm. Given the current wage structure, it has been estimated in labor negotiations that the standard deviation is $34.10. What is the probability that a sample mean of $243.30 or less would occur if the population mean is really $275, as the management negotiator insists?

7-14 A trucking firm delivers appliances for a large retail operation. The packages (or crates) have a mean weight of 300 pounds and a variance of 2,500.

(a) If a truck has a capacity of 4,000 pounds and 25 appliances are to be picked up, what is the probability that the 25 appliances will have an aggregate weight in excess of the truck's capacity? (Assume the 25 appliances represent a random sample.)

(b) If the truck has a capacity of 8,000 pounds, what is the probability that it will be able to handle the entire lot of 25 appliances?

7-15 Nature's Grain Bakery has several retail outlet stores whose distances from the factory are normally distributed with a mean of 6.9 miles and a standard deviation of 0.4 miles. A random sample of nine stores is selected and their distances from the bakery recorded. The mean of the nine stores belongs to a sampling distribution.

(a) What is the mean of this sampling distribution?

(b) What is the standard error of the mean?

(c) What is the shape of the sampling distribution?

(d) Find $P(\bar{x} > 7.0)$.

(e) Find $P(\bar{x} > 6.7)$.

(f) Why is it unnecessary to use the continuity correction factor in answering (d) and (e)?

In Retrospect

In both chapters 6 and 7 we used the standard normal probability distribution. We have now learned two formulas for calculating a z score:

$$z = \frac{x - \mu}{\sigma} \quad \text{and} \quad z = \frac{\bar{x} - \mu}{\sigma/\sqrt{n}}$$

You must distinguish between these two formulas. The first gives the standard score when dealing with individual values from a normal population (x values). The second uses information provided by the central limit theorem. Sampling distributions of sample means are approximately normally distributed. Therefore the standard scores and the probabilities in table 6 may be used in connection with sample means ($\bar{x}$ values). The key to distinguishing between the formulas is to decide whether the problem deals with individual values of x from the population or deals with sample means from the sampling distribution. If it deals with the individual values of x, we use the first formula, as presented in chapter 6. If, on the other hand, the problem deals with sample means, we use the second formula and proceed as illustrated in this chapter.

As indicated in this chapter, the basic purpose for repeated sampling is to form a sampling distribution. The sampling distribution could then be used to measure the variability that occurs from one sample to the next. Once this pattern of variability is known and understood for a specific sample statistic, we will be able to make accurate predictions about the corresponding population parameters. The central limit theorem describes the distribution for sample means. We will begin to make inferences about population means in chapter 8.

Repeated samples are commonly used in the field of production control, in which samples are taken to determine whether a product is the proper size or quality. When the sample is defective, a mechanical adjustment of the machinery is necessary. The adjustment is then followed by another sampling.

Another use of repeated sampling is demonstrated by the Dow-Jones average. These averages are reported daily with an accompanying graph, as illustrated by the news article at the beginning of the chapter. Plotted over a period of time, the graph depicts trends in the stock market.

Chapter Exercises

7-16 Compare the probability distribution found in exercise 7-2 with the one found in exercise 7-3. Does the distribution of sample means for the larger sample size seem to be "more normal" than the distribution of the sample means for the smaller sample, as the CLT suggests?

7-17 A random sample of 40 part-time employees of fast-food chain outlets in a large metropolitan area showed a mean weekly income of $125.60. The weekly income for all such employees is known to have a mean of $130.60 and a standard deviation of $15.00.
  (a) What is the probability of a sample mean of $125.60 or less?
  (b) If the sample size is set at 40, what is the probability that a sample will have a mean between $128.00 and $135.00?

7-18 The Whirler Blender Company suggests a retail price for each of its models when it signs a contract with a retailer. It recently conducted a study of its retailers' prices and found them to be normally distributed with a mean of $26.30 and a standard deviation of $2.50.
  (a) What percentage of the retailers charge less than $22.50?
  (b) What percentage of the prices are more than $25.60?

A random sample of 100 retailer prices is gathered for a study, and the mean price obtained is $\bar{x} = \$25.60$.

  (c) If another sample of 100 prices is taken, what is the probability that its sample mean will be greater than $25.60?

(d) Why is the z score used in answering questions (a) and (b)?

(e) Why is the formula for z used in (c) different from the formula used in (a) and (b)?

7-19 A pop music record firm seeks to have the cuts on its records average a length of 2 minutes and 15 seconds (135 seconds) and a standard deviation of 10 seconds, so that disc jockeys will have plenty of time for commercials within each 5-minute period. The population of times for cuts is approximately normally distributed with only a negligible skew to the right. You have just timed the cuts on a new release and have found that the 10 cuts average 150 seconds.

(a) What percent of the time would such an occurrence take place if the new release can be considered to be randomly selected?

(b) If the music firm had wanted 10 cuts not to average over 150 seconds more than 5% of the time, what must be the population mean given that the standard deviation remains at 10 seconds?

7-20 Find the value for A such that 98% of the retailer prices in exercise 7-18 are larger than A. That is, find A so that $P(x > A) = 0.98$.

7-21 Find the value for D such that 95% of all retail prices in exercise 7-18 are within D dollars of the mean $26.30. That is, find the value for D such that $P(26.30 - D < x < 26.30 + D) = 0.95$.

7-22 Find a value for E such that 95% of the samples of 100 retailer prices taken in exercise 7-18 will have a mean value within E dollars of the mean $26.30. That is, find E such that $P(26.30 - E < \bar{x} < 26.30 + E) = 0.95$.

7-23 The incomes of nonbusiness customers with checking accounts at the First Federal Bank are distributed about a mean of $12,000 with a standard deviation of $3,000. The shape of this distribution is unknown. Sixty-four customers' incomes are randomly selected, and the mean income for the sample is computed.

(a) Describe the sampling distribution of this sample mean.
(b) What is the probability that $\bar{x}$ is greater than $12,000?
(c) What is the probability that $\bar{x}$ is less than $12,500?
(d) What is the probability that $\bar{x}$ is greater than $11,000?
(e) What is the probability that $\bar{x}$ is greater than $13,500?

7-24 If all the possible samples of size 36 are selected from a population whose mean is 50 and whose standard deviation is 10, between what limits would the middle 80% of the sample means be expected to fall?

7-25 If a random sample of size 36 is drawn from a population whose mean is 72 and whose standard deviation is 12, find the probability that the sample mean $\bar{x}$ is greater than 75.2. (Retain this solution for use in illustration 8-4.)

7-26 If the mean of monthly commissions earned by brokers of Butcher and Wilson, Inc. is $4,750 and the standard deviation is $1,115, what is the probability that a random sample of 36 brokers will earn a total commission in excess of $175,000?

7-27 Many street signs in a small city are defaced around Halloween time every year. The distribution of repair costs per sign has a mean of $68.00 and a standard deviation of $12.40.

(a) If 300 signs are damaged this year, there is a 5% chance that the total repair costs for the 300 signs will exceed what value?

(b) You are about 68% certain that the total repair costs for 300 signs will fall within what interval?

Hands-On Problems

Consider a population that contains the three numbers 0, 3, and 6. We are going to draw, randomly, samples of size 3 from this set of digits by drawing one value, observing it, and then returning it to the container. The three numbers are mixed and a second number drawn; and so on. Three numbers will result from each sample taken.

7-1 (a) Construct the theoretical probability distribution for the drawing of a single number from this population.

(b) Draw a histogram of this probability distribution.

(c) Calculate the mean and standard deviation of this distribution.

7-2 Construct a list showing all the possible samples of size 3 that could be drawn from this population. (There are 27 possibilities.)

7-3 Find the mean, median, and range for each of the 27 possible samples listed in problem 7-2.

7-4 Construct a probability distribution and a histogram for each of these three sample statistics: $\bar{x}$, $\tilde{x}$, and R.

7-5 Calculate the mean and the standard error for the sampling distribution of the sample means.

7-6 Show that the results found in the answers to problems 7-1c, 7-4, and 7-5 support the three facts claimed by the central limit theorem.

7-7 Take 50 samples of size 3 from the population. Record the 50 samples and the mean for each sample. You may take three identical tags numbered 0, 3, and 6, put them in a hat, and draw a tag. Or you may use three dice, letting 1 and 2 represent 0, 3 and 4 represent 3, and 5 and 6 represent 6 (or you may use the random number table to produce your samples). Describe the method used.

7-8 (a) Construct a frequency distribution of the 50 sample means found in problem 7-7.

(b) Calculate the mean and the standard deviation of the 50 sample means found in problem 7-7.

(c) Compare the histogram and the values of $\bar{x}$ and $s_{\bar{x}}$ with the histogram and the values of $\mu_{\bar{x}}$ and $\sigma_{\bar{x}}$ obtained in the answer to problems 7-4 and 7-5. Do they agree?

The calculations for this problem set can most easily be accomplished with the assistance of an electronic calculator or a packaged program on a computer. A list of the available programs can be obtained from your computer center. There are a variety of packaged programs available: Minitab, Biomed (Biomedical Programs), SAS (Statistical Analysis System), IBM Scientific Subroutine Packages, and SPSS (Statistical Package for the Social Sciences) program libraries. Your local computer center will assist you.

8 Introduction to Statistical Inferences

Chapter Outline

8-1 The Nature of Hypothesis Testing
*A hypothesis test studies the **validity of an inference** regarding some aspect of a distribution.*

8-2 The Hypothesis Test (a Classical Approach)
*To test a claim, we must formulate a **null hypothesis** and an **alternative hypothesis**.*

8-3 The Hypothesis Test (a Probability-Value Approach)
An alternative approach to the decision-making process.

8-4 Estimation
*Another type of inference involves estimation, and we learn to make both **point estimates** and **interval estimates**.*

We've matched these tires against some of the toughest roads in the world! In the Baja, East Africa, Greece, Morocco, and the United States, Sears Steel Belted Radials have proven their rugged durability. That's because they're built tough! 2 steel belts and 2 rayon radial plies team-up to give you traction, mileage, and outstanding durability. We even guarantee them for 40,000 miles. So hurry in now and save on the Sears Steel Belted Radial. The tire that proved itself on the tough roads of the world.

Radial Gas-saving Story

Independent tests prove that Sears Steel Belted Radial, when compared to our leading fiber glass belted tire, improved gas mileage by an average of 7.4%, when driven at different constant speeds. That's in both foreign and domestic categories!

Reprinted by permission of Sears, Roebuck & Co.

Chapter Objectives

A random sample of 36 pieces of data yields a mean of 4.64. What can be deduced about the population from which the sample was taken? We will be asked to answer two types of questions:

1. Is the sample mean significantly different in value from a hypothesized mean value of 4.5?
2. Based on the sample, what statement can we make about the value of the population mean?

The first question requires us to make a decision, whereas the second question requires us to make an estimation.

In this chapter and the next two chapters, we will find out how a hypothesis test is used to make a statistical decision about three basic population parameters: the mean μ, the standard deviation σ, and the proportion p. We will also see how an estimation of these three parameters is made. In this chapter we will concentrate our attention on learning about the basic concepts of hypothesis testing and estimation. We will deal mostly with questions about the population mean, using two methods that assume that the population standard deviation is known. This assumption will seldom be realized in real life problems, but it will make our first look at inferences much easier.

Section 8-1 The Nature of Hypothesis Testing

hypothesis test

To illustrate the decision-making process called a **hypothesis test**, let's consider the decision process called acceptance sampling. Burton Products offers for sale a very large lot of irregular sheets. Burton Products claims that only 20% of the sheets are defective enough to be unsalable. The bedding buyer at Tracy's Department Stores is considering purchasing the lot if the claim is true. A random sample of 10 sheets is sent to the buyer, and she finds that 7 are so defective that they would be unsalable. Should the buyer purchase the lot?

This question calls for a decision. A choice must be made between two possibilities: (1) the lot is 20% unsalable and the sample happened to be 70% unsalable or (2) the lot is considerably more than 20% unsalable and the sample shows 70% unsalable. If every sheet in the lot was inspected, we would know the right decision, but that process would cost too much time and money. So the buyer must decide which alternative she thinks is correct on the basis of the sample information. What do you think, should she buy the lot or not? How do you justify the conclusion you have reached?

Let's investigate the statistical decision-making process, the hypothesis test procedure. The statistical hypothesis test is a five-step procedure. Each part of this procedure will be demonstrated and justified as we investigate the acceptance sampling problem outlined above.

The first two steps of the hypothesis test procedure are to formulate two hypotheses.

hypothesis

Hypothesis
A statement that something is true.

STEP 1: Formulate the null hypothesis.

null hypothesis

Null Hypothesis, H_0
The hypothesis that we wish to focus our attention on. Generally, this hypothesis is a statement that a population parameter has a specified value. Often the phrase "there is no difference" is used in its interpretation—thus the name "null" hypothesis.

STEP 2: Formulate the alternative hypothesis.

alternative hypothesis

Alternative Hypothesis, H_a H_1
A statement about the same population parameter that is used in the null hypothesis. Generally, this hypothesis is a statement that specifies that the population parameter has a value different from the value given in the null hypothesis. The rejection of the null hypothesis will imply the acceptance of the alternative hypothesis.

The null hypothesis and the alternative hypothesis are formulated by inspecting the problem or statement to be investigated and then forming two alternative statements. For our illustration we inspect Burton Products' statement and formulate the two alternatives: "Only 20% of the sheets are unsalable in the complete lot" or "more than 20% of the sheets are unsalable in the complete lot." One of these statements becomes the null hypothesis; the other becomes the alternative hypothesis. If Burton's claim is true, then the proportion p in the lot that is unsalable is 0.2. This statement about the population parameter p becomes the null hypothesis:

$$H_0: p = 0.2 \quad \text{(Burton's claim is true)}$$

The alternative hypothesis is

$$H_a: p > 0.2 \quad \text{(Burton's claim is not true)}$$

decision

From this point on in the hypothesis test procedure, we will work under the assumption that the null hypothesis is a true statement. This situation might be compared to a courtroom trial, where the accused is assumed to be innocent until sufficient evidence has been presented to show otherwise. At the conclusion of the hypothesis test, we will make one of two possible **decisions**. We will decide in agreement with the null hypothesis and say that we fail to reject H_0 (this statement corresponds to "fail to convict" or an acquittal of the accused in a trial). Or we will decide in opposition to the null hypothesis and say that we reject H_0 (this statement corresponds to conviction of the accused in a trial).

Four possible outcomes can be reached as a result of the null hypothesis being either true or false and the decision being either "fail to reject" or "reject." Table 8-1 shows these four possible outcomes.

TABLE 8-1
Possible Outcomes in a Hypothesis Test

	Null Hypothesis Is	
Decision	True	False
Fail to reject H_0	Type A correct decision	Type II error
Reject H_0	Type I error	Type B correct decision

type I and type II errors

A **type A correct decision** occurs when the null hypothesis is true and we decide in its favor. A **type B correct decision** occurs when the null hypothesis is false and our decision is in opposition to the null hypothesis. A **type I error** will be committed when a true null hypothesis is rejected, that is, when the null hypothesis is true but we decided against it. A **type II error** is committed when we decide in favor of a null hypothesis that is actually false.

When a decision is to be made, it would be nice if we could always make a correct decision. However, this is statistically impossible, since we will be making our decision on the basis of sample information. The best we can hope for is to control the **risk**, or probability, with which an error occurs. The **probability** assigned to the type I error is called **alpha**, α (α is the Greek letter alpha). The **probability** of the type II error is called **beta**, β (β is the Greek letter beta). See table 8-2. To control these errors, we will assign a small probability to them. The most frequently used probability values are 0.01 or 0.05. The probability assigned to each error will depend on the seriousness of the error. The more serious the error, the less often we will be willing to allow it to occur, and therefore a smaller probability will

risk
alpha, beta

TABLE 8-2
Probability with Which Error Occurs

Outcome	Error	Probability
Rejection of a true null hypothesis	Type I	α
Failure to reject a false null hypothesis	Type II	β

be assigned to it. In this text we are going to devote our attention to α, the P(type I error). The discussion of β, the P(type II error), is beyond the scope of this book.

Let's now return to our illustration and to step 3 of the hypothesis test procedure.

STEP 3: Determine the test criteria.

test criteria

Test Criteria

Consist of (1) determining a test statistic, (2) specifying a level of significance α, and (3) determining the critical region.

test statistic

Test Statistic

A random variable whose value will be used to make the decision "fail to reject H_0" or "reject H_0."

The test statistic will be a numeric value obtained from the sample results. The probability distribution of this test statistic is determined as a result of the assumption that the null hypothesis is true. In our illustration the number of unsalable sheets (7) in the sample of 10 will be used as the test statistic.

Before we discuss the level of significance and the critical region, let's look at the probability distribution of x, the possible number of unsalable sheets in the sample. x is binomially distributed with $n = 10$ and $p = 0.2$. According to the null hypothesis, that is, if, in fact, Burton's claim is true, the probabilities for each x value may be found in table 4 of appendix D. These probabilities are given in table 8-3. Inspection of the distribution suggests that the probability of randomly selecting 10 sheets and finding 5 or more unsalable from a large lot that contains only 20% defective is unlikely. Thus we could say that the occurrence of values for x of 5, 6, 7, 8, 9, or 10 would not support the null hypothesis $p = 0.2$. If the lot were 20% unsalable and we observed only 0, 1, 2, 3, or 4 unsalable in our sample of 10, we would feel that these events were likely and agree that $p = 0.2$. Thus we have specified a critical region—namely, $x = 5, 6, 7, 8, 9, 10$.

critical region

Critical Region

The set of values for the test statistic that will cause us to reject the null hypothesis.

The Nature of Hypothesis Testing Section 8-1

TABLE 8-3
Probability Distribution for x

x	P(x)
0	0.107
1	0.268
2	0.302
3	0.201
4	0.088
5	0.026
6	0.006
7	0.001
8	0.0+
9	0.0+
10	0.0+

critical value

The **critical value** is the "first" value in the critical region. Thus $x = 5$ is the critical value in our illustration. That is, if we take a sample and the observed value of x is 5 or more, we will reject the null hypothesis.

level of significance

Level of Significance
The probability of committing the type I error, α.

Typically, α will be determined first, and then it will dictate the critical value, as we will see in section 8-2; however, for our example, the probability of incorrectly rejecting H_0 if it is true is given by $P(x \geq 5)$ for the binomial random variable x when $n = 10$ and $p = 0.2$.

$$P(x \geq 5) = 0.026 + 0.006 + 0.001 + 3(0+) = \mathbf{0.033}$$

That is, $\alpha = 0.033$ for our critical region.

Now that the test criteria are determined—that is, $\alpha = 0.033$ and the critical region is $x \geq 5$ for the random variable x, the number of unsalable sheets—we are ready to proceed to step 4.

STEP 4: Obtain the observed value of the test statistic.

Having completed the ground rules for the test (steps 1, 2, and 3), we are ready to obtain our sample and either observe or calculate the value of the test statistic. For our illustration $x = 7$; that is, the value of the test statistic is the number of unsalable sheets in our sample. Typically, we will use a formula and calculate the value of the test statistic based on sample information.

STEP 5: Make a decision and interpret it.

The decision is made by comparing the value of the test statistic found in step 4 with the test criteria of step 3.

decision rule

noncritical region or acceptance region

Decision Rule

If the test statistic falls within the critical region, we will reject H_0. If the test statistic does not fall in the critical region, we will fail to reject H_0. *Note*: The set of values that are not in the critical region is called the noncritical region or sometimes the acceptance region.

The test is then completed by interpreting the decision reached. For our illustration $x = 7$ falls in the critical region and we reject H_0. Thus we conclude that the evidence ($x = 7$) does not support Burton's claim that only 20% of the sheets in the lot are unsalable.

Note that we did not prove that Burton's claim is untrue. If only 20% of the lot is really unsalable, Burton encountered a "rare event" in having randomly selected the 10 it did for inspection by the buyer. In retrospect, we have one of two choices to make here: (1) the null hypothesis is correct and we drew a very unrepresentative sample of 10 (a very unlikely event, since α is only 0.033) or (2) the null hypothesis is incorrect and we have sample results that are representative of some other situation. The latter choice seems to be more likely here.

Before we leave this illustration, let's look at the other side. Suppose Burton also offers a lot of irregular towels of which it claims "only 20% are so defective as to be unsalable." The buyer receives a random sample of 10 towels, and she finds that only 3 of the 10 are unsalable. The same test procedure (steps 1, 2, and 3) would be used to test this claim. However, in step 4 we have an observed $x = 3$, and therefore we make the decision "fail to reject H_0" in step 5. What does this imply? $x = 3$ certainly doesn't prove that only 20% of all the towels in the lot are unsalable. **Failing to reject the null hypothesis implies only that we did not find sufficient evidence to reject the statement of the null hypothesis.**

NOTE: Some people prefer to use the phrase "accept H_0" in place of "fail to reject H_0." However, returning to our courtroom analogy, we do not prove the defendant innocent (accept H_0). We acquit him (fail to reject H_0). In that sense, failure to reject the null hypothesis states the decision more accurately, and thus we use the statement "fail to reject H_0."

Exercises

8-1 As described in this section, the hypothesis-testing procedure has many similarities to a courtroom procedure. The null hypothesis "the accused is innocent" is being tested.

(a) Relate each of the four possible outcomes shown in table 8-1 to the courtroom analogy.

(b) If the accused is acquitted, does this "prove" him innocent? (Perhaps, in this respect, the phrase "fail to reject H_0" is more accurate in expressing the situation than is "accept H_0.")

(c) If the accused is found guilty, does this "prove" his guilt? Explain.

8-2 Consider the following nonmathematical situation as a hypothesis test: analyze a certain stock and test the null hypothesis "the stock will go up in price."

(a) Describe how the four possible outcomes indicated in table 8-1 apply to this statement.

(b) Decide on the seriousness of the two possible errors.

(c) If you could choose the values of α and β, which set would you prefer? Explain.

  (i) $\alpha = 0.01$ and $\beta = 0.10$
  (ii) $\alpha = 0.05$ and $\beta = 0.05$
  (iii) $\alpha = 0.10$ and $\beta = 0.001$

8-3 Consider the acceptance sampling hypothesis test discussed in this section (Burton Products' sheets).

(a) Describe how the four possible outcomes indicated in table 8-1 apply to this situation.

(b) Three inspection plans that have the following values as probabilities are considered:

  (i) $\alpha = 0.01$ and $\beta = 0.10$
  (ii) $\alpha = 0.05$ and $\beta = 0.05$
  (iii) $\alpha = 0.10$ and $\beta = 0.001$

Which plan do you think Burton Products would prefer? Which plan do you think Tracy's Department Stores would prefer?

8-4 The director of ADCOM advertising agency is concerned with the effectiveness of a television commercial.

(a) What null hypothesis is he testing if he commits a type I error when he erroneously says that the commercial is effective?

(b) What null hypothesis is he testing if he commits a type II error when he erroneously says that the commercial is effective?

8-5 (a) If the null hypothesis is true, the probability of an error in decision is identified by what name?

(b) If the null hypothesis is false, the probability of an error in decision is identified by what name?

(c) If the test statistic falls in the critical region, what error could be made?

(d) If the test statistic falls in the acceptance region, what error might occur?

8-6 (a) Suppose that a hypothesis test is to be carried out using $\alpha = 0.05$. What is the probability of committing a type I error?
(b) What proportion of the probability distribution is in the noncritical or acceptance region provided the null hypothesis is correct?
(c) What is a critical region?
(d) What is a critical value?

8-7 The chairman of Rapid America is considering making a tender offer for Data Systems Control, Inc. The chairman claims that half the Data Systems stockholders will accept the tender offer. Each stockholder in a random sample of 14 stockholders is asked if he or she would tender the stock. Define x to be the number of stockholders who say "yes."
(a) Construct the probability distribution for x. Let the null hypothesis be "the chairman is correct." That is, x is binomial with $n = 14$ and $p = 0.50$.
(b) For what values of x is the probability of x greater than or equal to 0.05?
(c) For what values of x is the probability of x less than 0.05? (Let these values form the critical region.)
(d) Find α, the P(type I error). (*Hint*: α is the sum of the probabilities for the x values in the critical region.)
(e) What decision with regard to the null hypothesis would you reach if 2 stockholders responded "yes"? If 6 responded "yes"? If 13 responded "yes"?

Section 8-2 The Hypothesis Test (a Classical Approach)

In section 8-1 we very generally surveyed the steps for, and some of the reasoning behind, a hypothesis test. In this section we are going to study the hypothesis test procedure as it applies to statements concerning the mean μ of a population. To simplify the introduction of these techniques, we will impose the restriction that the population standard deviation is known. (This restriction will be removed in chapter 9.) The first three illustrations deal with the procedures for formulating the null and alternative hypotheses.

Illustration 8-1

Carlye Furniture Company currently runs an advertisement every Sunday in the *New York Chronicle*. The advertisement contains a box address to write to for obtaining a current catalog of products. The firm averages 490 requests per week. The *New York News* advertising sales representative would like to show Carlye Company that for the same priced advertisement, it will receive a greater response in the *News* than in the *Chronicle*.

Specifically, he would like to show that the mean number of catalog requests per week from an advertisement in the *New York News* is greater than 490. State the null and alternative hypotheses.

Solution   To state the hypothesis, we first need to identify the population parameter in question and the value with which it is being compared. The "mean weekly number of catalog requests resulting from an advertisement in the *News*" is the parameter μ, and 490 (the rate from the *Chronicle*) is the specific value. The mean rate from the *News* could be related to 490 in any one of three ways: (1) $\mu < 490$, (2) $\mu = 490$, or (3) $\mu > 490$. These three statements must be arranged to form two statements, one that states what the *News* representative is trying to show and the other that states the opposite. $\mu > 490$ represents the statement "the mean number of requests per week is higher than 490," whereas $\mu < 490$ and $\mu = 490$ ($\mu \leq 490$) represent the opposite, "the mean number of requests is not higher than 490." One of these two statements will become the null hypothesis H_0, and the other becomes the alternative hypothesis H_a.

RECALL:

1. The null hypothesis states that the parameter in question has a specified value.
2. A sample mean is going to be used as the basis for an inference about the population mean, and sample means have an approximately normal distribution as described by the central limit theorem.
3. A normal distribution is determined when its mean and standard deviation are specified.

All this information is suggesting that the statement containing the equal sign will become the null hypothesis; the other statement becomes the alternative hypothesis. Thus we have

$$H_0: \mu = 490 \quad \text{and} \quad H_a: \mu > 490$$

Recall that once the null hypothesis is stated, we proceed with the hypothesis test under the assumption that the null hypothesis is true. Thus $\mu = 490$ locates the center of the sampling distribution of sample means. For that reason the null hypothesis will be written with an equal sign only. If $\mu = 490$ and $\mu < 490$ are stated together as the null hypothesis, the null hypothesis will be expressed as

$$H_0: \mu = 490 \ (\leq)$$

where the ($\leq$) serves as a reminder of the grouping that took place when the hypothesis was formed.   □

NOTE: **The equal sign must be in the null hypothesis**, regardless of the statement of the original problem.

Illustration 8-1 is an expression of the viewpoint that the *News* representative might take. Now let's consider how the *Chronicle* representative might view the same basic situation.

Illustration 8-2

The *New York Chronicle* sales representative would naturally like to conclude that "the mean response for catalogs from an advertisement in the *New York News* is less than 490 per week." State the null and alternative hypotheses related to this viewpoint.

Solution   Again the parameter of interest is the mean number of weekly catalog requests, μ, and 490 is the specified value. $\mu < 490$ corresponds to "the mean response is less than 490," whereas $\mu \geq 490$ corresponds to "the mean level is not less than 490." Therefore the hypotheses are

$$H_0: \mu = 490 \ (\geq) \quad \text{and} \quad H_a: \mu < 490$$

A more neutral point of view is suggested by illustration 8-3.

Illustration 8-3

The "mean response for catalogs from an advertisement in the *New York News* is not 490" (i.e., is not the same as that from the *New York Chronicle*). State the null and alternative hypotheses that correspond to this statement.

Solution   The mean response for catalogs from an advertisement in the *News* is equal to 490 ($\mu = 490$), or the mean is not equal to 490 ($\mu \neq 490$). (*Note:* "Less than or greater than" is customarily expressed as "not equal to.") Therefore

$$H_0: \mu = 490 \quad \text{and} \quad H_a: \mu \neq 490$$

The viewpoint of the experimenter has an effect on the way the hypotheses are formed, as we have seen in these three illustrations. Generally, the experimenter is trying to show that the parameter value is different from the value specified. Thus the experimenter is usually hoping to find a rejection of the null hypothesis. Illustrations 8-1, 8-2, and 8-3 represent the three arrangements possible for the <, =, and > relationships between the parameter μ and the specified value 490.

The next illustration demonstrates the complete hypothesis test procedure as it is used in questions dealing with the population mean.

Illustration 8-4

From many years of prior experimentation, the Consumers Union has found that the mean life of a certain brand of 100-watt light bulbs is 72 days of continuous usage and the standard deviation is 12 days. The light bulb company has recently marketed their bulb as "new and improved, yielding longer life." The Consumers Union purchases 36 bulbs and tests them. The average life of the bulbs tested is 75.2 days. Does the mean $\bar{x}$ of

75.2 present sufficient evidence to support the producer's claim that the new bulbs are superior? Use $\alpha = 0.05$ and $\sigma = 12.0$.

Solution To be superior, the new bulb must have a mean life that is greater than that of the old bulb. To be "equal or less than" would not be superior.

STEP 1: $H_0: \mu = 72$ ($\leq$) (bulb is not superior).

STEP 2: $H_a: \mu > 72$ (bulb is superior).

STEP 3: The level of significance $\alpha = 0.05$ is given in the statement of the problem. The test statistic is the standard score z when the null hypothesis is about a population mean and the standard deviation is known. Recall that the central limit theorem tells us that the sampling distribution of sample means is approximately normally distributed. Thus the normal probability distribution will be used to complete the hypothesis test. The critical region—values of the standard score z that will cause a rejection of the null hypothesis—has an area of 0.05 ($\alpha = 0.05$) and is located at the extreme right of the distribution. The critical region is on the right because large values of the sample mean suggest "superior," whereas values near or below 72 support the null hypothesis.

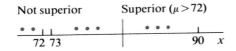

If random samples of size 36 are taken from a population with a mean value equal to 72, many of the sample means would have values near 72, such as 71, 72, 71.8, 72.5, 73, and so on. Only sample means that are considerably larger than 72 would cause us to reject the null hypothesis. The critical value, the cutoff between "not superior" and "superior," is determined by α, the probability of the type I error. $\alpha = 0.05$ was given. Thus the critical region (shaded region of the diagram) has an area of 0.05 and a critical value of $+1.65$. (This value is obtained by using table 6 of appendix D.)

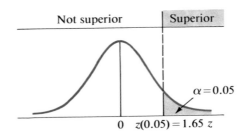

STEP 4: The value of the test statistic z can be found by using formula (7-2) and the sample information.

$$\bar{x} = 75.2 \quad \text{and} \quad n = 36$$

$$z = \frac{\bar{x} - \mu}{\sigma/\sqrt{n}}$$

RECALL: We assumed that μ was equal to 72 in the null hypothesis and that $\sigma = 12.0$ was known.

$$z = \frac{75.2 - 72}{12.0/\sqrt{36}} = \frac{3.2}{2.0} = 1.60$$

$$z^* = 1.60$$

calculated value (*) [We will use an asterisk, *, to identify the **calculated value** of the test statistic; the asterisk will also be used to locate its value relative to the test criterion (next step).]

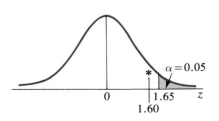

STEP 5: We now compare the calculated test statistic, z^*, with the test criterion set up in step 3 by locating the calculated value on the diagram and placing an asterisk at that value. Since the test statistic (calculated value) falls in the noncritical region (unshaded portion of the diagram), we must reach the following decision:

Decision: Fail to reject H_0.

Recall that the critical region was to be shaded in the diagram, and when the test statistic falls in the critical region, we must reject H_0. Step 5 is then completed by stating a **conclusion**.

conclusion

Conclusion: There is not sufficient evidence to show that the *new bulb* is superior.

Does this conclusion seem realistic? 75.2 is obviously larger than 72. Recall that 75.2 is a sample mean, and if a random sample of size 36 is drawn from a population whose mean is 72 and whose standard deviation is 12, then the probability that the sample mean is 75.2 or larger is greater than the risk, α, with which we are willing to make the type I error. [What is $P(\bar{x} > 75.2)$? See exercise 7-25. □

Illustration 8-5

Teletype Corporation has retained a private consulting firm to determine methods that will reduce its inventory costs. The suggestions of the consulting firm are based on the assumption that mean time between the date of order and date the items are received is 120 days. To test this assumption, the production manager of Teletype randomly selected 100 orders and found that the mean time until delivery was 118.5 days. Is this evidence sufficient to reject the null hypothesis? Past experience indicates that $\sigma = 12$ days, and $\alpha = 0.05$ is to be used.

Solution The statement suggests that the three possible relationships ($<$, $=$, $>$) between the mean time μ and the hypothesized 120 days be split ($=$) or ($<$, $>$). Therefore we have the following steps:

STEP 1: $H_0: \mu = 120$ (mean time is 120 days).

STEP 2: $H_a: \mu \neq 120$ (mean time is not 120 days).

The test statistic will be the standard score z, and the normal distribution is used, since we are using the sampling distribution of sample means. The sample mean is our estimate for the population mean. Since the alternative hypothesis is "not equal to," a sample mean considerably larger than 120, as well as one considerably smaller than 120, will be in opposition to the null hypothesis. The values of the sample mean around 120 will support the null hypothesis. Therefore the critical region will be split into two equal parts, one at each extreme of the normal distribution. The area of each region will be $\alpha/2$. Since $\alpha = 0.05$, each part of the critical region will have a probability of 0.025. Look in table 6 of appendix D for the area 0.4750 (0.4750 = 0.5000 − 0.025). The z score of 1.96 is found and becomes 1.96 on the right of the mean and −1.96 on the left.

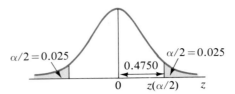

STEP 3: The test criteria are shown in the diagram:

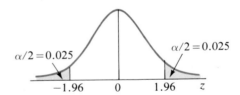

STEP 4: Calculate the test statistic z. The sample information is $\bar{x} = 118.5$ and $n = 100$. σ was given as 12, and the null hypothesis assumed that $\mu = 120$.

$$z = \frac{\bar{x} - \mu}{\sigma/\sqrt{n}} = \frac{118.5 - 120}{12/\sqrt{100}} = \frac{-1.5}{1.2} = -1.25$$

$$z^* = -1.25$$

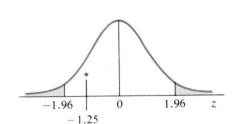

Locate z^* on the diagram constructed in step 3.

Now we are ready to identify the decision.

RECALL: If the test statistic falls in the critical region, we must reject H_0. If the test statistic falls in the noncritical region, we must fail to reject H_0.

STEP 5: The calculated value of z, z^*, falls between the critical values cited in step 3. Therefore our decision is as follows:

Decision: Fail to reject H_0.

The interpretation of our decision is the only thing left to do. This part of the hypothesis test may very well be the most significant part, because the conclusions reached express the results found. We must be very careful to state precisely what is meant by the decision. We are often tempted to overstate a conclusion. In general, a decision to "fail to reject H_0" may be interpreted to mean that the evidence found does not disagree with the null hypothesis. Note that this decision does not "prove" the truth of H_0. A decision to "reject H_0" will mean that the evidence found implies the null hypothesis to be false and thus indicates the alternative hypothesis to be the case.

Conclusion: The evidence (sample mean) found does not contradict the assumption that the population mean time is 120 days. ☐

Before we look at another illustration, let's summarize briefly some of the details we have seen thus far.

1. The null hypothesis specifies a particular value of a population parameter.

2. The alternative hypothesis can take three forms: each form dictates a specific location of the critical regions, as shown here:

	The Sign in the Alternative Hypothesis	$<$	$\neq$	$>$
one-tailed and two-tailed test	Typical Critical Region	One region, left side; one-tailed test	Two regions, one on each side; two-tailed test	One region, right side; one-tailed test

The value assigned to α is called the significance level of the hypothesis test. Alpha cannot be interpreted to be anything other than the risk (or probability) of rejecting the null hypothesis when it is actually true. We will seldom be able to determine whether the null hypothesis is true or false; we will only decide to reject H_0 or to fail to reject H_0. The relative frequency with which we reject a true hypothesis is α, but we will never know the relative frequency with which we make an error in decision. The two ideas are actually quite different; that is, a type I error and an error in decision are two different things altogether.

Let's look at some more illustrations of the hypothesis test.

Illustration 8-6

Argon Advertising Agency has suggested to Brant Specialty Store that it used mail circulars for the area within a 15-mile radius of its store as its major source of advertising. Asked, "Why not extend the area beyond 15 miles?" Argon's representative answered: "Your average customer lives no more than 9 miles from the store, so that 15 miles will cover almost all your potential customers." The owner of Brant was not convinced of the truth of this answer and decided to test the statement. A sample of 50 customers was taken, and a mean traveling distance of 10.22 miles was found. Test the hypothesis stated above at a significance level of $\alpha = 0.05$. Experience indicates that $\sigma = 5$ miles.

Solution Our first two steps are as follows:

STEP 1: $H_0: \mu = 9.0$ ($\leq$) (no more than 9 miles).

STEP 2: $H_a: \mu > 9.0$ (more than 9 miles).

The distance traveled by the average customer would be the same as the mean distance traveled by all customers. Therefore the parameter of concern is μ. The claim "no more than 9.0 miles" implies that the three possible relationships should be grouped ($\leq$) or ($>$).

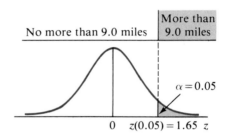

STEP 3: The critical region is on the right; H_0 will be rejected if it appears that the $\bar{x}$ observed is significantly greater than the 9.0 claimed. Notice that the sign in H_a ($>$) points toward the critical region.

STEP 4:

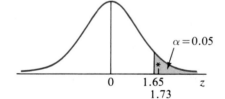

$$z = \frac{\bar{x} - \mu}{\sigma/\sqrt{n}} = \frac{10.22 - 9.0}{5/\sqrt{50}} = \frac{1.22}{0.707} = 1.73$$

$$z^* = 1.73$$

STEP 5: *Decision*: Reject H_0. (z^* fell in the critical region.)

Conclusion: At the 0.05 level of significance, we conclude that the average customer probably lives more than 9.0 miles from the store. □

Illustration 8-7

Draw a sample of 40 single-digit numbers from the random number table and test the null hypothesis $\mu = 4.5$. Use $\alpha = 0.10$ and $\sigma = 2.87$. The standard deviation of the random digits was found earlier in chapter 5 (see exercise 5-12).

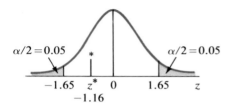

Solution

STEP 1: $H_0: \mu = 4.5$.

STEP 2: $H_a: \mu \neq 4.5$.

STEP 3: $\alpha = 0.10$ (see the diagram).

STEP 4: The following random sample was drawn from table 1 of appendix D.

$$
\begin{array}{cccccccccc}
2 & 8 & 2 & 1 & 5 & 5 & 4 & 0 & 9 & 1 \\
0 & 4 & 6 & 1 & 5 & 1 & 1 & 3 & 8 & 0 \\
3 & 6 & 8 & 4 & 8 & 6 & 8 & 9 & 5 & 0 \\
1 & 4 & 1 & 2 & 1 & 7 & 1 & 7 & 9 & 3
\end{array}
$$

$$\sum x = 159 \qquad n = 40 \qquad \bar{x} = 3.975$$

$$z = \frac{\bar{x} - \mu}{\sigma/\sqrt{n}} = \frac{3.975 - 4.50}{2.87/\sqrt{40}} = \frac{-0.525}{0.454} = -1.156$$

$$z^* = -1.16$$

z^* falls in the acceptance region, as shown in the diagram in step 3.

STEP 5: *Decision:* Fail to reject H_0.

Conclusion: The observed sample mean is not significantly different from 4.5. □

Suppose that we were to take another sample of size 40 from the table of random digits. Would we obtain the same results? Suppose that we took a third sample or a fourth. What results might we expect? What is the level of significance α? Yes, its value is 0.10, but what is it a measure of? Table 8-4 lists the means obtained from 10 different random samples of size 40 that

TABLE 8-4
Random Sample of Size 40 Taken from Table 1, Appendix D

Sample Number	Sample Mean, $\bar{x}$	Calculated z, z^*	Decision Reached
1	4.62	+0.26	Fail to reject H_0
2	4.55	+0.11	Fail to reject H_0
3	4.08	−0.93	Fail to reject H_0
4	5.00	+1.10	Fail to reject H_0
5	4.30	−0.44	Fail to reject H_0
6	3.65	−1.87	Reject H_0
7	4.60	+0.22	Fail to reject H_0
8	4.15	−0.77	Fail to reject H_0
9	5.05	+1.21	Fail to reject H_0
10	4.80	+0.66	Fail to reject H_0

were taken from table 1. The calculated value of z that corresponds to each $\bar{x}$ and the decision each would dictate are listed also. Each of the 10 calculated z scores is shown in figure 8-1, using the sample number to identify it. Note that one of the samples caused us to reject the null hypothesis, although we know that it is true for this situation.

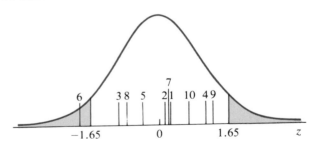

FIGURE 8-1
z Scores from Table 8-4; $\alpha = 0.10$

REMEMBER: α is the probability that we reject H_0 when it is actually a true statement. Therefore we can anticipate that a type I error will occur α of the time when testing a true null hypothesis.

Exercises

8-8 State the null hypothesis H_0 and the alternative hypothesis H_a that would be used to test each of the following statements.

(a) The mean closing price of 90-day futures for pork bellies on the Chicago Mercantile Exchange during May and June is less than $55.

(b) The mean age of Geritol customers is greater than 45.

(c) The mean number of items purchased on a sales receipt at Rimbels Department Store is 6.

(d) The mean rate of return on corporate A-rated bonds is less than 8%.

(e) The average customer at the Sedwick branch of Lincoln Bank lives or works within 2 miles of the branch.

(f) During months when the consumer price index goes up, the mean percentage increase in money supply exceeds 5%.

(g) The mean time between order and delivery of a certain type of computer terminal is 65 days.

8-9 Suppose that we want to test the hypothesis that during the tourist season the mean price charged for a motel room in Florida is more than $51.50. Explain under what conditions we would be committing a type I error. Explain under what conditions we would be committing a type II error.

8-10 All drugs must be approved by the Food and Drug Administration (FDA) before a drug manufacturer can market it. The FDA must weigh the error of marketing an ineffective drug with the usual risks of side effects versus the consequences of not allowing an effective drug to be sold. Suppose, using standard medical treatment, the mortality rate (r) of a certain disease is known to be a. A manufacturer submits for approval a drug that is supposed to treat this disease. The FDA sets up the hypothesis to test the mortality rate for the drug as (i) $H_0: r = a$, $H_a: r < a$, $\alpha = 0.005$; or (ii) $H_0: r = a$, $H_a: r > a$, $\alpha = 0.005$.

(a) If $a = 0.95$, do you think the FDA would use hypothesis (i) or (ii)? Explain.

(b) If $a = 0.05$, do you think the FDA would use hypothesis (i) or (ii)? Explain.

8-11 Determine the test criteria (critical values and critical region for z) that would be used to test the null hypothesis at the given level of significance, as described in each of the following. [Sketch the normal curve and shade the critical region(s).]

(a) $H_0: \mu = 10$ ($\alpha = 0.05$)
$H_a: \mu \neq 10$

(b) $H_0: \mu = 15$ ($\geq$)   ($\alpha = 0.05$)
$H_a: \mu < 15$

(c) $H_0: \mu = 20$ ($\leq$)   ($\alpha = 0.05$)
$H_a: \mu > 20$

(d) $H_0: \mu = 8$ ($\leq$)   ($\alpha = 0.01$)
$H_a: \mu > 8$

(e) $H_0: \mu = 103$ ($\geq$)   ($\alpha = 0.02$)
$H_a: \mu < 103$

(f) $H_0: \mu = 17.5$ ($\alpha = 0.10$)
$H_a: \mu \neq 17.5$

(g) $H_0: \mu = 10.5$ ($\leq$)   ($\alpha = 0.10$)
$H_a: \mu > 10.5$

8-12 A sample of 36 ads was taken to test the hypothesis stated in illustration 8-1. Does the sample mean of 510 provide sufficient evidence to reject the null hypothesis in the illustration if $\sigma = 180$ and $\alpha = 0.05$?

8-13 The Ornamental Brass Foundry has invested a great deal of time and money in occupational safety training for its employees and believes its occupational-related sick days are now below the national average. The population of all brass foundries was found to have a mean of 1.5 occupational-related sick days per 100 employees and a standard deviation of 0.3 days. Ornamental Brass randomly selected 100 employees and analyzed their employment files for the last year. The sample had a mean of 1.3 occupational-related sick days. Is there sufficient evidence to support Ornamental's belief? (Use $\alpha = 0.05$.)

8-14 Insurance company rates for fire insurance depend on the distance a home is from the nearest fire department. A progressive community claims that the average home in its town is within 5.5 miles (i.e., 5.5 miles or less) of the nearest fire department. The insurance company took a sample of 64 homes, which produced a mean of 5.8 miles. Is there sufficient evidence to refute the town's contention that the mean distance is not greater than the claimed 5.5 if $\sigma = 2.4$ miles. Use $\alpha = 0.05$.

8-15 The average age of employees at Blaser, Inc. is 42. In a retrenchment 49 persons were laid-off. Their mean age was 45. The Equal Employment Opportunity Commission (EEOC) claims that Blaser is guilty of age discrimination, since it laid-off older than average workers. Is the information sufficient evidence to support EEOC's claim if $\sigma = 10.8$ years? Use $\alpha = 0.05$.

8-16 The mean length of time between billing date and check received and deposited for nondelinquent accounts of an oil company's monthly credit card customers is 23.5 days; the standard deviation is 3 days. A bank claims that if the firm uses a "lockbox" (i.e., has the payments returned directly to a box number at the bank, where the bank opens the letter, deposits the check, and forwards the returned statement to the company), it can reduce the firm's average time. Forty accounts are randomly selected and tested using the lockbox. The average time for the 40 accounts was 21.5 days. Is this evidence sufficient to support the bank's claim? Use $\alpha = 0.05$.

8-17 An economist claims that when the Dow-Jones average increases, the number of block trades of 10,000 shares or more tends to increase. Over the past 2 years, the average daily number of block trades is 1,510 and the standard deviation is 175. A random sample of 81 days on which the Dow-Jones average increased was selected, and the average daily number of trades was computed. If the average was 1,560, would you conclude that the economist is correct? Use $\alpha = 0.01$.

8-18 The average starting salary for accounting majors last year was $16,500 and the standard deviation was $1,200. If the starting salary for a random sample of 42 accounting majors from Temple University was $16,900, can you conclude that Temple University accounting majors start at a higher than average salary? Use $\alpha = 0.05$.

Section 8-3 The Hypothesis Test (a Probability-Value Approach)

The classical (or traditional) hypothesis-testing procedure was described in section 8-2. An alternative approach to the decision-making process in hypothesis testing has gained popularity in recent years. This alternative process is to determine the area under the curve of the standard normal curve (table 6, appendix D) related to the calculated z^*. The value derived from the table is known as the **probability-value**, the **prob-value**, or simply the **P-value** (P is capital P; not to be confused with "small p," the binomial parameter). The P-value is then compared with the level of significance α, the probability of the type I error. This comparison becomes the basis for the decision to reject H_0 or fail to reject H_0.

prob-value

Prob-Value, P

The prob-value, P, of a hypothesis test is the smallest level of significance for which the observed sample information becomes significant, provided the null hypothesis is true.

Let's return to illustration 8-4 and see how its solution would be different if handled using this prob-value method.

Illustration 8-8

From many years of prior experimentation, the Consumers Union has found that the mean life of a certain brand of 100-watt light bulbs is 72 days of continuous usage and the standard deviation is 12 days. The company that makes this light bulb has recently marketed this same bulb as "new and improved, yielding longer life." The Consumers Union is not convinced and purchases 36 bulbs for testing. Does the resulting sample mean $\bar{x}$ of 75.2 present sufficient evidence to support the producer's claim that the new bulbs have a longer life at the 0.05 level of significance? At the 0.10 level?

Solution To have a longer life, the new bulb must have a mean life that is "greater than" the old bulb's. To be "equal to or less than" would not show longer life.

STEP 1: H_o: $\mu = 72$ ($\leq$) (bulb does not have longer life).

STEP 2: H_a: $\mu > 72$ (bulb does have longer life).

     Notice that steps 1 and 2 are exactly the same when using either the classical or the prob-value approach to hypothesis testing.

STEP 3:   Determine α, the probability of the type I error: (a) $\alpha = 0.05$, (b) $\alpha = 0.10$.

STEP 4:   The calculated value of the test statistic z^* will be obtained using formula (7-2) and the sample information.

$$z = \frac{\bar{x} - \mu}{\sigma/\sqrt{n}} \quad \text{and} \quad \bar{x} = 75.2 \text{ and } n = 36$$

RECALL: We assume that μ was equal to 72 in the null hypothesis and that $\sigma = 12.0$ was known.

$$z = \frac{75.2 - 72.0}{12.0/\sqrt{36}} = \frac{3.2}{2.0} = 1.60$$

$$z^* = 1.60$$

REMINDER: We use an asterisk, *, to identify the calculated value of the test statistic; the asterisk will also be used to locate its value on the probability distribution in the next step.

Notice that step 4 in the prob-value approach is exactly the same as step 4 in the classical approach.

STEP 5: Draw a sketch of the probability distribution curve and locate the value of z^* on the axis.

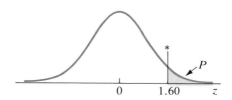

Determine the area under the curve that is defined by z^* and the alternative hypothesis.

There are three possible cases that the alternative hypothesis may dictate:

CASE 1: If H_a contains ">," then $P = P(z > z^*)$, the area under the curve and to the right of z^*.

CASE 2: If H_a contains "<," then $P = P(z < z^*)$, the area under the curve and to the left of z^*.

CASE 3: If H_a contains "$\neq$," then $P = P(z < -|z^*|) + P(z > |z^*|)$, the sum of the area to the left of the negative value of z^* and the area to the right of the positive value of z^*. Since both areas are equal, you will probably find one and double it. Thus $P = 2 \times P(z > |z^*|)$.

Returning to our illustration, the alternative hypothesis indicates that we are interested in that part of the probability distribution that lies to the right of z^*, since the "greater than" sign was used [$P = P(z > z^*) = P(z > 1.60)$]. From table 6 in appendix D we find that the area associated with $z = 1.60$ is 0.4452 and calculate $P = 0.5000 - 0.4452 = 0.0548$.

The results of the statistical analysis are then reported.

STEP 6: The prob-value for this hypothesis test is 0.0548. There is evidence that the new light bulbs do have a longer life, provided you are working at a significance level greater than 0.0548.

(a) If the Consumers Union wishes to make their decision at the 0.05 significance level (or any α value less than or equal to 0.0548), then they will fail to reject the null hypothesis and conclude that there is no evidence to show a longer life.

(b) If, however, they have chosen to use a level of significance of 0.10 (or any α value greater than 0.0548), their decision will be to reject H_0, and their conclusion would be that the new bulbs do have a longer life.

α less than P; z^* in noncritical region

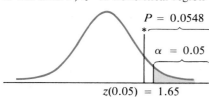

α greater than P; z^* in critical region

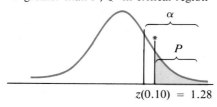

Before looking at another illustration, let's summarize the details for this prob-value approach to the hypothesis testing.

1. The null and alternative hypotheses are formulated in the same manner as before.
2. Determine the level of significance, α, to be used.
3. The value of the test statistic is calculated in step 4 in exactly the same manner as it was calculated before in step 4.
4. Under the curve of the test statistic curve, the prob-value, P, is the area defined by the calculated statistic and the alternative hypothesis. This area is always located in the tail(s) of the distribution and includes all values of the test statistic that are more extreme than the calculated value.
5. The decision will be made by comparing P with the previously established value of α.
   (a) If the **calculated prob-value is less than or equal to the desired α**, then the decision must be **reject H_0**.
   (b) If the **calculated prob-value is greater than the desired α**, then the decision must be **fail to reject H_0**.
6. Conclusions should be worded in the same manner as previously instructed.

Let's look at an illustration involving the two-tailed procedure.

Illustration 8-9

Many of the large companies in our city have for years used the Kelley Employment Agency for the testing of prospective employees. The employment selection test used has historically resulted in scores distributed about a mean of 82 with a standard deviation of 8. The Brown Agency has developed a new test that is quicker and easier to administer and therefore less expensive. Brown claims that their test results are the same as those obtained on the Kelley test. Many of the companies are considering a change from the Kelley Agency to the Brown Agency to cut costs. However, they are unwilling to make the change if the Brown test results have a different mean value.

An independent testing firm tested 36 prospective employees. A sample mean of 80 resulted. Determine the prob-value associated with this hypothesis test.

Solution The Brown Agency's sample test results will be different from the Kelley Agency's if the mean test score is not equal to 82. They will be the same if the mean is 82. Therefore:

STEP 1: H_0: $\mu = 82$ (test results have same mean).

STEP 2: H_a: $\mu \neq 82$ (test results have different mean).

STEP 3: Step 3 is omitted when question asks for the prob-value and not a decision.

STEP 4: The sample information $n = 36$ and $\bar{x} = 80$ and formula (7-2) is used to calculate z^*.

$$z = \frac{\bar{x} - \mu}{\sigma/\sqrt{n}}$$

$$z = \frac{80 - 82}{8/\sqrt{36}} = \frac{-2}{8/6} = -1.50$$

$$z^* = -1.50$$

STEP 5: The value of z^* is located on the axis of the normal distribution curve.

Since the alternative hypothesis indicates a two-tailed test, we must find the probability associated with two areas, namely $P(z < -|z^*|)$ and $P(z > |z^*|)$. And since the left tail $z^* = -1.50$, the value of the right tail $z^* = 1.50$. Thus $P = P(z < -1.50) + P(z > 1.50)$, as shown in the accompanying figure.

$$P = P(z < -1.50) + P(z > 1.50)$$
$$= (0.5000 - 0.4332) + (0.5000 - 0.4332)$$
$$= 0.0668 + 0.0668$$
$$P = 0.1336$$

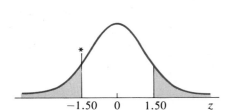

The prob-value for this hypothesis test is 0.1336. Each individual company now will make a decision as to whether to (a) continue to use Kelley's services or (b) change to the Brown Agency. Each will need to establish the level of significance that best fits their own situation and then make their decision following the decision rule described previously. □

Exercises

Calculate the prob-value for each of the following:

8-19   (a) $H_0: \mu = 10$ $z^* = 1.48$
           $H_a: \mu > 10$

       (b) $H_0: \mu = 105$ $z^* = -0.85$
           $H_a: \mu < 105$

       (c) $H_0: \mu = 13.4$ $z^* = 1.17$
           $H_a: \mu \neq 13.4$

       (d) $H_0: \mu = 8.56$ $z^* = -2.11$
           $H_a: \mu < 8.56$

       (e) $H_0: \mu = 110$ $z^* = -0.93$
           $H_a: \mu \neq 110$

       (f) $H_0: \mu = 54.2$ $z^* = 0.46$
           $H_a: \mu > 54.2$

8-20 The calculated prob-value for a hypothesis test is $P = 0.084$. What decision about the null hypothesis would occur:

(a) if the hypothesis test is to be completed at the 0.05 level of significance

(b) if the hypothesis test is to be completed at the 0.10 level of significance

8-21 An economist claims that when the Dow-Jones average increases, the volume of shares traded on the New York Stock Exchange tends to increase. Over the past 2 years, the average daily volume on the exchange has been 21.5 million shares with a standard deviation of 2.5 million. A random sample of 64 days on which the Dow-Jones average increased was selected, and the average daily volume was computed. The sample average was 22 million. Calculate the prob-value for this hypothesis test.

8-22 The marketing director of A & B Cola is worried that their product is not attracting enough young consumers. To test this theory, she randomly surveys 100 A & B Cola consumers. The mean age in the community surveyed is 32 years and the standard deviation is 10 years. The mean age of the surveyed consumers of A & B Cola is 35. At the 0.01 level of significance, is this sufficient evidence to conclude that A & B Cola consumers are, on the average, older than the average person living in the community? Complete this hypothesis test using the prob-value approach.

Section 8-4 Estimation

The second type of statistical inference is that of **estimation**. Estimation is the procedure to use when answering a question that asks for the value of a population parameter. For example, "what is the distance to Brant's Specialty Store from its average customer's home?"

If you needed an answer to this question, you might take a sample from the population and calculate the sample mean $\bar{x}$. Suppose that you did draw a random sample of 100 distances and a sample mean of 10.22 miles resulted. What is your estimate for the mean value of the population? If you report the sample mean $\bar{x}$ as your estimate, you will be making a point estimate.

point estimate

Point Estimate for a Parameter

The value of the corresponding sample statistic.

That is, the sample mean, $\bar{x} = 10.22$ miles, is the best point estimate for the mean distance of this population. In other words, the mean distance for all Brant's customers is estimated to be 10.22 miles. We don't really mean to imply that the population mean μ is exactly 10.22 miles. We intend this point estimate to be interpreted to say, "μ is close to 10.22." From our study so far, you probably realize that when we say μ is 10.22, there is very little chance that the statement is true. In previous chapters we drew samples from probability distributions where μ was known, and seldom did we obtain a sample mean exactly equal in value to μ. Why, then, should we expect a single sample to yield an $\bar{x}$ equal to μ when μ is unknown? We shouldn't.

What does it mean to say, "μ is close to 10.22 miles"? Closeness is a relative term, but perhaps in this case "close" might be arbitrarily defined to be "within 1 mile" of μ. If 1 mile satisfies our intuitive idea of closeness, then to say "μ is close to 10.22" is comparable with saying "μ is between 9.22 (10.22 − 1.0) and 11.22 (10.22 + 1.0)." This suggests that we might make estimations by using intervals. The confidence interval estimate employs this interval concept and assigns a measure to the interval's reliability in estimating the parameter in question.

confidence interval

Confidence Interval
An interval bounded by two values such that the probability that the parameter value is within the interval is known.

level of confidence

Level of Confidence, $1 - \alpha$
The probability that the values bounding the confidence interval will lie on opposite sides of the parameter being estimated. The level of confidence is sometimes called the **confidence coefficient**.

confidence coefficient

The central limit theorem is the source for the information needed to construct confidence interval estimates. Our sample mean ($\bar{x} = 10.22$) is a member of the sampling distribution of sample means. The CLT describes this distribution as being approximately normally distributed with a mean $\mu_{\bar{x}} = \mu$ and a standard deviation $\sigma_{\bar{x}} = \sigma/\sqrt{n}$. The population mean is unknown; however, it exists and is constant. Assume that the standard deviation of the population is $\sigma = 5$.

RECALL: In this chapter we are studying the inferences about μ under the assumption that σ is known. This assumption is for convenience, and the restriction will be removed in chapter 9.

The sampling distribution of sample means to which our sample mean ($\bar{x} = 10.22$ and $n = 100$) belongs is shown in figure 8-2.

When a sample of size 100 is randomly selected from a population whose mean is μ and whose standard deviation is 5, what is the probability that the sample mean is within 1 unit of the population mean μ? In other

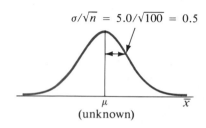

FIGURE 8-2
Sampling Distribution, $n = 100$

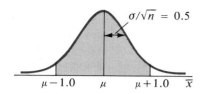

FIGURE 8-3
Probability That x Is Within 1 Unit of μ

words, what is $P(\mu - 1 < \bar{x} < \mu + 1)$? (See figure 8-3.) This probability may be found by using the normal probability distribution (table 6) and formula (7-2):

$$z = \frac{\bar{x} - \mu}{\sigma/\sqrt{n}}$$

If $\bar{x} = \mu - 1$,

$$z = \frac{(\mu - 1) - \mu}{0.5} = \frac{-1.0}{0.5} = -2.00$$

If $\bar{x} = \mu + 1$,

$$z = \frac{(\mu + 1) - \mu}{0.5} = \frac{+1.0}{0.5} = +2.00$$

Therefore

$$P(\mu - 1 < \bar{x} < \mu + 1) = P(-2.00 < z < +2.00)$$
$$= 2 \cdot P(0 < z < 2.00)$$
$$= 2(0.4772) = 0.9544$$

The probability that the mean of a random sample is within 1 unit of this population mean is 0.9544. Therefore the interval 9.22 to 11.22 is a 0.9544 confidence interval estimate for the mean distance that Brant's customers live from the store.

maximum error of estimate

Maximum Error of Estimate, E

One-half the width of the confidence interval. In general, E is a multiple of the standard error.

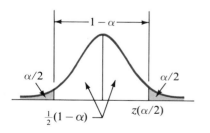

FIGURE 8-4
Each Tail Contains $\alpha/2$

The preceding illustration started with the maximum error being assigned the value of 1 mile. Typically, the maximum error of estimate is determined by the level of confidence that we want our confidence interval to have. That is, $1 - \alpha$ will determine the maximum error. The level of confidence will be split so that half of $1 - \alpha$ is above the mean and half below the mean (figure 8-4). This split will leave $\alpha/2$ as the probability in each of the two tails of the distribution. The z score at the boundary of the confidence interval will be z of $\alpha/2$, $z(\alpha/2)$. [Recall that $\alpha/2$ is the probability (or area) under the curve to the right of this point. See chapter 6.] The

Estimation Section 8-4 **293**

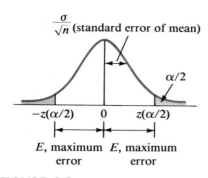

FIGURE 8-5
Maximum Error of Estimate

standard score z is a number of standard deviations. Therefore the maximum error of estimate E is

$$E = z(\alpha/2) \cdot \frac{\sigma}{\sqrt{n}} \tag{8-1}$$

See figure 8-5.
The $1 - \alpha$ confidence interval for μ is

$$\bar{x} - z(\alpha/2) \cdot \frac{\sigma}{\sqrt{n}} \quad \text{to} \quad \bar{x} + z(\alpha/2) \cdot \frac{\sigma}{\sqrt{n}} \tag{8-2}$$

lower confidence limit $\bar{x} - z(\alpha/2) \cdot [\sigma/\sqrt{n}]$ is called the **lower confidence limit (LCL)** for the confidence interval, and $\bar{x} + z(\alpha/2) \cdot [\sigma/\sqrt{n}]$ is called the **upper confidence limit (UCL)** for the confidence interval.

Illustration 8-10

Construct the 0.95 confidence interval for the estimate of the mean distance that Brant's customers live from the store.

Solution

$$1 - \alpha = 0.95$$
$$\alpha = 0.05$$
$$\frac{\alpha}{2} = 0.025$$

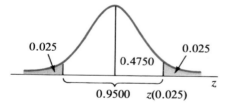

z	$\cdots$	0.06	$\cdots$
$\vdots$			
1.9		0.4750	
$\vdots$			

$$z(0.025) = 1.96$$

$$\bar{x} \pm z(\alpha/2) \cdot \frac{\sigma}{\sqrt{n}}$$

$$10.22 \pm (1.96) \cdot \frac{5}{\sqrt{100}}$$

$$10.22 \pm (1.96)(0.50)$$

$$10.22 \pm 0.98$$

$$10.22 - 0.98 = 9.24 \quad \text{and} \quad 10.22 + 0.98 = 11.20$$

Therefore with 0.95 confidence we can say that the mean distance is between 9.24 and 11.20, and we will write this as **(9.24 to 11.20)**, the **0.95 confidence interval for μ**.

NOTE: Many of the confidence interval formulas that you will be using will have a format similar to formula (8-2), namely, $\bar{x} - E$ to $\bar{x} + E$. However, perhaps the simplest way to handle the arithmetic involved is to think of the formula as $\bar{x} \pm E$. First calculate E. Then calculate the lower bound by subtracting and calculate the upper bound by adding. This format was used in illustration 8-10.

Illustration 8-11

To schedule the production of statistics books, a publisher wanted to know the average number of letters that would have to be reset per 100 lines of page proofs. A random sample of 38 sets of 100 lines of page proofs was examined, and a sample mean of 74.3 was found. Construct the 0.98 confidence interval estimate for the overall mean number of errors per 100 lines of page proofs. Past experience shows that $\sigma = 14$.

Solution

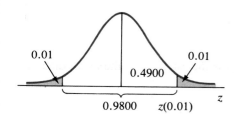

$$1 - \alpha = 0.98$$

$$\alpha = 0.02$$

$$\frac{\alpha}{2} = 0.01$$

z	...	0.03	...
⋮			
2.3		0.4901	(nearest table value to 0.4900)
⋮			

$$z(0.01) = 2.33$$

$$\bar{x} \pm z(\alpha/2) \cdot \frac{\sigma}{\sqrt{n}}$$

$$74.3 \pm (2.33) \cdot \frac{14}{\sqrt{38}}$$

$$74.3 \pm (2.33)(2.27)$$

$$74.3 \pm 5.29$$

$$74.3 - 5.29 = \mathbf{69.01} \qquad 74.3 + 5.29 = \mathbf{79.59}$$

The 0.98 confidence interval for μ is 69.01 to 79.59. That is, with 0.98 confidence we can say that the average number of letters that will have to be reset per 100 lines of page proofs is between 69.01 and 79.59.

Illustration 8-12

Sealan Trucking Company charges for delivery by the weight of the cargo and the distance conveyed. Determine the size of the sample that will be necessary to estimate the weight of full truck delivery between South Bend and Chicago if we desire our estimate to be accurate to within 10 pounds with 95% confidence. Assume that the standard deviation of such weights is 30 pounds.

Solution $1 - \alpha = 0.95$; therefore $z(0.025) = 1.96$, as found by using table 6. The maximum error $E = 10$ and $\sigma = 30.0$. Use formula (8-1).

$$E = z(\alpha/2) \cdot \frac{\sigma}{\sqrt{n}}$$

$$10 = 1.96 \cdot \frac{30}{\sqrt{n}}$$

$$10 = \frac{58.8}{\sqrt{n}}$$

$$\sqrt{n} = 5.88$$

$$n = 34.57$$

Therefore

$$n = 35 \qquad \square$$

NOTE: When solving for the sample size n, all fractional (decimal) values are to be *rounded up* to the next larger integer.

Does a sample size of 35 determine a maximum error of 10 pounds in illustration 8-12?

$$E = z(\alpha/2) \cdot \frac{\sigma}{\sqrt{n}} = 1.96 \cdot \frac{30}{\sqrt{35}}$$

$$= (1.96)(5.07) = 9.939$$

This maximum error is just under 10.

The use of formula (8-1) can be made a little easier by rewriting the formula in a form that expresses n in terms of the other quantities:

$$n = \left[\frac{z(\alpha/2) \cdot \sigma}{E} \right]^2 \qquad (8\text{-}3)$$

Illustration 8-13

A sample of what size would be needed to estimate the population mean to within one-fifth of a standard deviation with 99% confidence?

Solution $1 - \alpha = 0.99$; therefore $\alpha/2 = 0.005$, and $z(0.005) = 2.58$, as found by using table 6. The maximum error E is to be one-fifth of σ; that is, $E = \sigma/5$. Using formula (8-3), we have

$$n = \left[\frac{z(\alpha/2) \cdot \sigma}{E}\right]^2 = \left[\frac{(2.58)(\sigma)}{\sigma/5}\right]^2 = \left[\frac{(2.58\sigma)(5)}{\sigma}\right]^2$$

$$= [(2.58)(5)]^2 = (12.90)^2 = 166.41 = 167 \quad \square$$

Illustration 8-14

Suppose that the sample in illustration 8-7 had been obtained for the purpose of estimating the population mean μ. The sample results ($n = 40$, $\bar{x} = 3.975$) and $\sigma = 2.87$ are used in formula (8-2) to determine the 0.90 confidence interval.

$$\bar{x} \pm z(\alpha/2) \cdot \frac{\sigma}{\sqrt{n}}$$

$$3.975 \pm 1.65 \cdot \frac{2.87}{\sqrt{40}}$$

$$3.975 \pm (1.65)(0.454)$$

$$3.975 \pm 0.749$$

$$3.975 - 0.749 = 3.226 \qquad 3.975 + 0.749 = 4.724$$

With 0.90 confidence, we think that μ is somewhere within this interval.

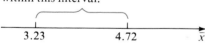

The 0.90 confidence interval estimate for μ is 3.23 to 4.72.
 Since this sample was taken from a known population—namely, the random digits—we can look back at our answer and say that the true value of μ, 4.5, does fall within the confidence interval. $\quad \square$

Suppose that we were to take another sample of size 40; would we obtain the same results? Suppose we took a third and a fourth. What would happen? What is the level of confidence, $1 - \alpha$? Yes, it has the value 0.90, but what does that mean? Table 8-5 lists the means obtained from 10 different random samples of size 40 taken from table 1. The 0.90 confidence interval for the estimate of μ based on each of these samples is also listed in table 8-5. Since all the interval estimates were for the mean of the random digits ($\mu = 4.5$), we can inspect each confidence interval and see that 9 of the 10 contain μ and would therefore be considered accurate estimates. One sample (sample 6) did not yield an interval that contains μ. This sample is shown in figure 8-6.

TABLE 8-5

Samples of Size 40 Taken from Table I, Appendix D

Sample Number	Sample Mean, $\bar{x}$	0.90 Confidence Interval Estimate for μ
1	4.64	3.89 to 5.39
2	4.56	3.81 to 5.31
3	3.96	3.21 to 4.71
4	5.12	4.37 to 5.87
5	4.24	3.49 to 4.99
6	3.44	2.69 to 4.19
7	4.60	3.85 to 5.35
8	4.08	3.33 to 4.83
9	5.20	4.45 to 5.95
10	4.88	4.13 to 5.63

REMEMBER: $1 - \alpha$ is the probability that we will obtain a confidence interval such that μ is contained within it.

FIGURE 8-6

Interval Estimates from Table 8-5

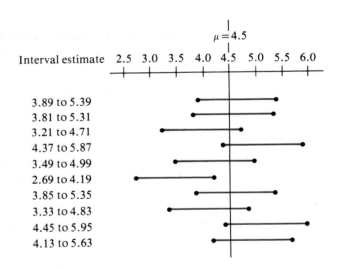

Exercises

8-23 A sample is taken from a population to estimate the mean. $\bar{x} = 25.3$ and $n = 64$.

(a) Give a point estimate for μ.

(b) If the standard deviation of this population is believed to be 16, determine a 95% confidence interval estimate for the population mean.

8-24 A certain population of unskilled laborers' annual incomes has a standard deviation of $800. A random sample of 36 results in $\bar{x} = \$5,650$.
  (a) Give a point estimate for the population mean annual income.
  (b) Estimate μ with a 95% confidence interval.
  (c) Estimate μ with a 99% confidence interval.

8-25 A sample of the ages of 50 corporation presidents is obtained to estimate the mean age of corporation presidents. $\bar{x} = 61$ years; the population standard deviation is 3 years.
  (a) Give a point estimate for μ.
  (b) Find the 95% confidence interval estimate for μ.
  (c) Find the 99% confidence interval estimate for μ.

8-26 The annual private health insurance claims for 200 men between the ages of 26 and 30 were obtained to determine the necessity of a rate increase. The mean claim was $322, and the population standard deviation is $23.
  (a) Find the 90% confidence interval for the population mean claim.
  (b) Find the 98% confidence interval for the population mean claim.

8-27 A random group of 36 American Stock Exchange stocks obtained an annual growth with a mean of 9% and a standard deviation of 0.78%. Find the 99% confidence interval estimate for the population mean.

8-28 A sample of what size would be needed to estimate a population mean to within 4 units with 95% confidence if the population has a standard deviation of 10 units?

8-29 How large a sample should be taken if the population mean is to be estimated with 98% confidence to within $100? The population has a standard deviation of $800.

8-30 How large a sample needs to be taken to estimate the true mean percentage of growth for the stocks in exercise 8-27 to within one-tenth of one point (0.1%) at the 95% level of confidence?

8-31 A medical products firm has developed a new drug that requires very strict control on the amount placed in each capsule, inasmuch as it can become hazardous if too much is used and ineffective if not enough is used. The maximum allowable error of the mean is 2 milligrams. The standard deviation is believed to be 7 milligrams. If the level of significance is set at 0.01, what size sample should be taken to be certain that the mean amount of the drug per capsule is correct?

8-32 In measuring the amount of time it takes a component of a product to move from one work station to the next, an engineer has estimated that the standard deviation is 5 seconds.
  (a) How many measurements should be made to be 99% certain that the maximum error of estimation will not exceed 1 second?
  (b) What sample size is required for a maximum error of 2 seconds?

In Retrospect

Two forms of inference were studied in this chapter: hypothesis testing and estimation. They may be, and often are, used separately. However, it would seem quite natural for the rejection of a null hypothesis to be followed by a confidence interval estimate. (If the value claimed is wrong, we will usually want to know just what the true value is.)

These two forms of inference are quite different, but they are related. There is a certain amount of crossover between the use of the two inferences. For example, suppose that you had sampled and calculated a 90% confidence interval for the mean of a population. The interval was 10.5 to 15.6. Following this calculation, someone claims that the true mean is 15.2. Your confidence interval estimate can be compared with this claim. Since this claimed value falls within your interval estimate, you would fail to reject the null hypothesis that $\mu = 15.2$ at a 10% level of significance in a two-tailed test. If the claimed value (say 16.0) falls outside the interval, you would then reject the null hypothesis that $\mu = 16.0$ at $\alpha = 0.10$ in a two-tailed test. If a one-tailed test is required, or if you prefer a different value of α, a separate hypothesis test must be used.

The news article at the beginning of this chapter suggests that Sears's tests have proved that its steel-belted radial tire will give you 40,000 miles of rugged service. This advertisement, like many others, goes a long way in selling the consumer on a particular product. The sophistication of the presentation creates a lasting and very suggestive impression. Now examine this advertisement in the light of what you have learned in this chapter.

In this chapter we have restricted our discussion of inferences to the mean of a population for which the standard deviation is known. In the next two chapters, we will discuss inferences about other population parameters and eliminate the restriction about the known value for standard deviation.

Chapter Exercises

8-33 The mean of a certain population is believed to be 100, and its standard deviation is 12. A sample of 50 measurements gives a sample mean of 96. Using a level of significance of 0.01, complete a test to decide between the hypothesis "population mean is really 100" or "it is different from 100." State or calculate the desired answer in parts (a) through (l).

(a) H_0

(b) H_a

(c) α

(d) $z(\alpha/2)$

(e) μ (based on H_0)

(f) $\bar{x}$

(g) σ

(h) $\sigma_{\bar{x}}$

(i) $z\ast$

(j) decision

(k) Sketch the z distribution and locate $\alpha/2$ in each tail, sketch in $\pm z(\alpha/2)$, shade the critical regions for testing H_0, and locate $z\ast$.

(l) prob-value

8-34 Over several years a foreman kept records of how long it took assembly line workers to assemble a unit. He claimed that workers required an average time of only 15 minutes with a standard deviation of 1.5 minutes for completion. A group of 60 workers was timed. Does a sample mean of 16.4 minutes show sufficient evidence to support the claim that these 60 represent a group of slow workers? Use $\alpha = 0.02$.

8-35 A company manufactures rope. From a large number of tests over a long period of time, they have found a breaking strength with a mean of 300 pounds and a standard deviation of 24 pounds. Assume that these values are μ and σ. It is believed that by using a newly developed process, the mean breaking strength can be increased.
(a) Design a null and alternative hypothesis such that rejection of the null hypothesis will imply that the mean breaking strength has increased.
(b) If the preceding decision rule is used with $\alpha = 0.01$, what is the critical value for the test statistic and what value of $\bar{x}$ corresponds to it if samples of size 45 are used?
(c) Using the decision rule established in part (a), what is the prob-value associated with rejecting the null hypothesis when a sample mean of 305 results from 45 tests?

8-36 At a very large firm, the clerk-typists were sampled to see if different departments paid different salaries to workers in similar categories. A sample of 50 accounting clerks with the firm averaged $16,010 annual salary. The firm's personnel office asserts that the salaries paid to all clerk-typists with the firm average $15,650 and that the standard deviation is $1,800. At the 0.05 level of significance, can we conclude that the accounting clerks receive a higher average salary than other clerk-typists do?

8-37 At Memorial Hospital the length of time required to clean and sterilize the equipment used for a typical operation has a mean of 60 minutes and a standard deviation of 10 minutes. A team of efficiency experts recently redesigned the sterile supplies department. The employees feel the efficiency team failed to improve the mean time. They collected information from a sample of 40 operations and determined that $\bar{x}$ was 63.7 minutes. Does their sample show sufficient evidence, at the 0.02 level of significance, to support their contention?

8-38 An experiment is set up to test whether a certain sales training program is effective. The weekly sales of salespeople is approximately normally distributed with a mean of $3,000 and a standard deviation of $1,000. Forty-nine salespersons are randomly selected and given the training program. Their average weekly sales following the training is $3,250. Is there sufficient evidence to conclude that the training is effective? Complete this hypothesis test using a 0.02 level of significance.

8-39 A major bank credit card company charges member merchants 5% of the charged amount. To justify this rate, the bank claims it is lending money to the merchant, since it pays the merchant when the charge is received, but the bank does not receive payment from the cardholder until, on the average, 60 days later. A merchant questions if it really takes that long. He conducts a study of 100 randomly drawn charges and finds the average time to be 55 days. Is this sufficient evidence to reject the bank's claim? Complete this hypothesis test at a 0.05 level of significance and assume $\sigma = 20$ days.

8-40 In the news article at the beginning of the chapter, Sears says it has proved that its radial tires improve gas mileage by an average of 7.4%. The word "proved" seems a bit strong. If a statistical test were to be carried out, state the null and alternative hypotheses such that the rejection of H_0 would show significant evidence that the Sears steel-belted radial tire does improve gas mileage.

8-41 A sample of 64 measurements is taken from a continuous population, and the sample mean is found to be 32.0. The standard deviation of the population is known to be 2.4. An interval estimation is to be made of the mean with a level of confidence of 0.90. State or calculate the following items:

(a) $\bar{x}$
(b) σ
(c) n
(d) $1 - \alpha$
(e) $z(\alpha/2)$
(f) $\sigma_{\bar{x}}$
(g) E (maximum error of estimate)
(h) upper confidence limit
(i) lower confidence limit

8-42 Suppose that a confidence interval is assigned a level of confidence of $1 - \alpha = 0.95$. How is 0.95 used in constructing the confidence interval?

8-43 A banker wishes to estimate the average savings account size in a particular branch office. She conducts a preliminary sampling, obtaining a mean of $487 and a standard deviation of $60. Construct a 95% confidence interval for the average account assuming a sample of (a) size 36 and (b) size 81.

8-44 A random sample of the scores of 100 applicants for clerk-typist positions at a large insurance company showed a mean score of 72.6. The preparer of the test asserted that good applicants should average 75.0 and have a standard deviation of 10.5.

(a) Determine the 99% confidence interval estimate for the mean score of all applicants at the insurance company.

(b) Can the insurance company conclude that it is getting good applicants (as measured by this test)?

8-45 A random sample of 50 bank customers opening a new checking account in a particular bank showed that the mean processing time to open the account was 12.6 minutes and the standard deviation was 3.0 minutes. Using a 90% confidence interval, estimate the mean processing time for all such customers.

8-46 A natural gas utility is considering a contract for the purchase of tires for its fleet of service trucks. The decision will be made on the basis of expected mileage. A sample of 100 tires was tested; the mean mileage was 36,000 and the standard deviation was 2,000 miles. Using a 96% level of confidence interval, estimate the mean mileage that the utility should expect from these tires.

8-47 An estimation is to be made of a population mean so that the maximum error of estimate is 2.5. If the population standard deviation is 7.5 and $1 - \alpha = 0.95$, what should the sample size be?

8-48 The following computer output (Minitab) presents (1) a test of the hypothesis that the mean number of days elapsed between sales and return is 10 days and (2) a confidence interval estimate for the mean number of days elapsed based on the data presented in exercise 2-77.

```
-- ZTEST ON MU = 10 SIGMA = 4.9, FOR C2
C2           N =   94        MEAN =      10.362      ST.DEV. =       4.87

   TEST OF MU =     10.0000 VS. MU N.E.      10.0000
   THE ASSUMED SIGMA =     4.9000
   Z = 0.716
   THE TEST IS SIGNIFICANT AT   0.4744
   CANNOT REJECT AT ALPHA = 0.05

-- ZINTERVAL 95 PERCENT CONFIDENCE, ASSUMING SIGMA =4.9, ON C2
C2           N =   94        MEAN =      10.362      ST.DEV. =       4.87

   THE ASSUMED SIGMA =     4.9000

   A 95.00   PERCENT C.I. FOR MU IS (     9.3697,    11.3537)
```

With respect to the test of hypothesis, state or give the value(s) for the following:

(a) H_0 (b) H_a (c) α
(d) prob-value   (e) $\bar{x}$ (f) σ
(g) z^* (h) decision

With respect to the confidence interval estimate, state or give the value(s) for the following:

(i) n (j) $1 - \alpha$ (k) E (maximum error)
(l) upper confidence limit   (m) lower confidence limit

8-49 The data in the accompanying table represent the monthly consumer price index for the Los Angeles–Long Beach area for 2 years. (To simplify the data for our purposes, the first two digits were deleted. Thus 114.6 became 4.6.)

	L.A.	
Month	Year 1	Year 2
Jan.	4.6	2.8
Feb.	3.8	3.5
Mar.	4.0	3.6
Apr.	2.8	3.9
May	3.8	2.7
June	4.2	2.4
July	3.7	3.0
Aug.	4.6	2.8
Sept.	3.5	3.3
Oct.	3.4	3.1
Nov.	3.2	3.7
Dec.	2.7	3.7

The mean consumer price index in the New York–Northeast New Jersey area is 5.1. The following computer output (Minitab) presents (1) a test of the hypothesis that the consumer price index in Los Angeles is the same as that in New York (namely, 5.1) and (2) a confidence interval estimate for the mean consumer price index for Los Angeles. Answer the questions asked in exercise 8-44 (a) through (m).

```
-- NAME C1 = 'LA'

-- ZTEST OF MU = 5.1 ALTERNATIVE = -1 SIGMA =0.86, FOR C1
   LA          N =   24      MEAN =        3.4500      ST.DEV. =        0.593

   TEST OF MU =       5.1000 VS. MU L.T.      5.1000
   THE ASSUMED SIGMA =       0.8600
   Z = -9.399
   THE TEST IS SIGNIFICANT AT     0.0000

-- ZINTERVAL 90 PERCENT CONFIDENCE, ASSUMING SIGMA = .86, ON C1
   LA          N =   24      MEAN =        3.4500      ST.DEV. =        0.593

   THE ASSUMED SIGMA =       .8600

   A 90.00   PERCENT C.I. FOR MU IS (       3.1609,       3.7391)
```

8-50 Rework, Inc. selects applicants for sales jobs on the basis of the Balsa Sales Aptitude Test. All applicants are given the test, and anyone who scores at least 75 is hired. Studies have shown that the distribution of scores of successful salesmen has a mean of 80 and a standard deviation of 10, whereas the distribution for unsuccessful salesmen has a mean of 60 and a standard deviation of 10.

(a) Structure Rework's selection process as a test of the hypothesis stating (i) H_0, (ii) H_a, and (iii) the value of α.

(b) What does a cutoff score of 75 tell you about Rework's assessment of the costs of hiring someone who will be an unsuccessful salesman compared with the cost of rejecting someone who will be a successful salesman.

8-51 (a) "The level of confidence for an interval estimate is like a probability." To what extent is this statement true? Explain.

(b) "The level of confidence is not a probability when we look at the interval estimate after it has been obtained." Explain why this statement is the case.

Hands-On Problems

Calco Calculator Company produces a wide range of pocket calculators that cover the full price range. They are interested in mounting a direct mail advertising campaign to college students. The marketing manager feels that the average price college students are willing to pay for a calculator is $25 with a standard deviation of $5. The production manager thinks that the average college student wants to spend less and feels that Calco should advertise its least expensive model (it sells for $10). Calco hires you as a market research consultant to test the marketing manager's claim. To test the validity of this claim, you are to take a sample of size 36 and test it by using a 0.05 level of significance. Define your population as the full-time students at your college who own or wish to own a pocket calculator.

8-1 (a) State the null and alternative hypotheses to be used to test the marketing manager's claim about average price.

(b) Describe what it would mean to "fail to reject H_0."

(c) Describe what it would mean to "reject H_0."

8-2 Obtain your sample, include a list of the original data, and calculate the sample mean $\bar{x}$.

8-3 Complete the test of the marketing manager's hypothesis. Be sure to state your decision and your conclusion.

8-4 In the event that you found evidence to reject the null hypothesis, construct the 95% confidence interval estimate for the mean amount students want to spend for a pocket calculator.

8-5 Determine the size of sample that will be required to estimate the true amount students wish to spend for a pocket calculator to within $1.00 at the 0.90 level of confidence.

9 Inferences Involving One Population

Chapter Outline

9-1 Inferences About the Population Mean
When the standard deviation of the population is unknown, we use the ideas of *Student's t distribution* and *degrees of freedom*.

9-2 Inferences About Proportions
The observed sample proportion p' is *approximately normally distributed*.

9-3 Inferences About Variance and Standard Deviation
The *chi-square distribution* is employed when testing hypotheses concerning variance or standard deviation.

The Rating Game

With much self-congratulation, the nightly newscasts of ABC and NBC set out to rate TV's own ratings system last week. Both shows broadcast three-part series on the all-powerful A. C. Nielsen Co., but perhaps predictably, the No. 1 and No. 2 networks barely nibbled at the hand that feeds them.

The question they focused on was whether a ratings service that monitors a mere 1,170 of the nation's 73 million TV households can accurately gauge what America is watching—and each answered in the affirmative. That view is shared by most network and advertising executives, who say a larger sample would not add enough accuracy to be worth the extra cost. But the two series studiously avoided discussing some major deficiencies in the Nielsen ratings, which are the prime determinants of network profits and programs.

To produce its ratings, Nielsen attaches a black box called an Audimeter to TV sets in 1,170 homes—each selected as a demographic representative of part of the nationwide population. Every 30 seconds, the Audimeters record which channels the families are tuned to. Each meter is connected by phone to a central computer in Florida and, at least twice a day, the computer electronically collects the data. After being tabulated, the returns are received by Teletype at the networks. There they are regarded as the most important factor in deciding not only the fate of shows but what rates networks can charge their advertisers.

Viewers of the ABC and NBC reports never would have known it, but the Nielsen findings, like all such samplings, have a margin of statistical error. For a show accorded a rating of 18,* which is considered to be borderline for survival, the error margin is 2.2 points. Thus the show could actually have a rating as low as 15.8 (approximately the average of "Ball Four," one of last season's early casualties) or as high as 20.2 (that of such hits as "Baretta" and "The Love Boat"). And such small differences can be crucial to a network's income. It can charge sponsors $150,000 for a one-minute spot on a 20.2-rated half-hour show, versus only $85,000 on a 15.8-rated show. Over the course of a season, that difference in revenue can total as much as $3 million.

---

* According to Nielsen, an 18 rating means that 18 percent of its sample—and by extrapolation, the same percentage of the nation's 73 million television homes—is watching a particular show.

Condensed from "The Rating Game," *Newsweek*, 21 November 1977, p. 142. Copyright 1977 by Newsweek, Inc. All rights reserved. Reprinted by permission.

Chapter Objectives

In chapter 8 we discussed two forms of statistical inference: (1) the hypothesis test and (2) the confidence interval estimation. The study of these two inferences was restricted to inferences about the population parameter μ under the restriction that σ was known. However, the population standard deviation is generally not known. Thus the first section of this chapter deals with the inferences about μ when σ is unknown. In addition, we will learn how to perform hypothesis tests and confidence interval estimations about the population parameters σ (standard deviation) and p (proportion).

Section 9-1 Inferences About the Population Mean

Until now we have made inferences about the mean in situations where the standard deviation of the population is known. Since such situations seldom occur, it is important to know what to do when σ is not given. If the sample size is large enough (generally speaking, samples of more than $n = 30$ pieces of data are considered large enough), the sample standard deviation s is a good estimate of the population standard deviation σ, and we may simply substitute s for σ in the procedures discussed in chapter 8. [Recall that the sample standard deviation s is the value calculated by using formula (2-8) or one of its equivalent forms.] If the population that we are sampling is approximately normal and $n \leq 30$, we will base our procedures on **Student's t distribution**. Student's t distribution (or just t distribution) is the distribution of the t **statistic**, which is defined as

$$t = \frac{\bar{x} - \mu}{s/\sqrt{n}} \tag{9-1}$$

In 1908 W. S. Gosset, an Irish brewery employee, published a paper about this t distribution under the pseudonym "Student." In deriving the t distribution, Gosset assumed that the samples were taken from normal populations. Although this concept might seem to be quite restrictive, satisfactory results are also obtained when sampling from many nonnormal populations.

The t distribution has the following properties (see also figure 9-1):

Properties of the t Distribution

1. t is distributed with a mean of 0.
2. t is distributed symmetrically about its mean.
3. t is distributed with a variance greater than 1, but as the sample size n increases, the variance approaches 1.

4. t is distributed so as to be less peaked at the mean and thicker at the tails than the normal distribution is.

5. t is distributed so as to form a family of distributions, a separate distribution for each sample size. The t distribution approaches the normal distribution as the sample size increases.

FIGURE 9-1
Normal and Student's t Distribution

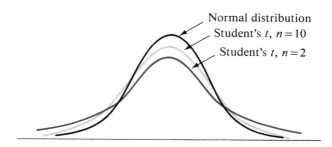

Although there is a separate t distribution for each sample size, $n = 2$, $n = 3$, $n = 4$, and so on, only certain key critical values of t will be necessary for our work. Consequently, the table for Student's t distribution (table 7 of appendix D) is a table of critical values rather than a complete table, such as that for the standard normal distribution for z. As you look at table 7, you will note that the left side of the table is identified by df, which means *degrees of freedom*. This left-hand column starts at 1 at the top and ranges to 29, then jumps to z at the bottom. As stated previously, as the sample size increases, the distribution approaches that of the normal. By inspecting table 7, you will note that as you read down any one of the columns, the entry value approaches a familiar z value at the bottom. By now you should realize that as the sample size increases, so does the number of degrees of freedom, df.

degrees of freedom

Degrees of Freedom (df)

A parameter in statistics that is very difficult to define completely. Perhaps it is best thought of as an "index number" used for the purpose of identifying the correct t distribution to be used. In the methods presented in this chapter, the value of df for a given situation will be $n - 1$, the sample size minus 1.

The critical value of t to be used either in a hypothesis test or in the construction of a confidence interval will be obtained from table 7. To obtain the value of t, you will need to know two values: (1) df, the number of degrees of freedom, and (2) α, the area under the curve to the right of the right-hand critical value. A notation much like that used with z will be used to identify a critical value. $t(df, \alpha)$, read "t of df, α," is the symbol for the value of t described in the previous sentence and shown in figure 9-2.

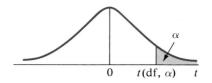

FIGURE 9-2
t Distribution Showing t(df, α)

Illustration 9-1

Find the value of $t(10, 0.05)$. See the accompanying diagram.

Solution There are 10 degrees of freedom. So in table 7, appendix D, we look for the column marked $\alpha = 0.05$ and come down to df = 10.

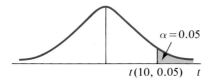

	Amount of α in One Tail		
df	...	0.05	...
⋮			
10		1.81	
⋮			

From the table we see that $t(10, 0.05) = \mathbf{1.81}$. □

For values of t on the left-hand side of the mean, we can use one of two notations. The t value shown in figure 9-3 could be $t(df, 0.95)$, since the area to the right of it is 0.95. Or it could be identified by $-t(df, 0.05)$, since the t distribution is symmetric about its mean 0.

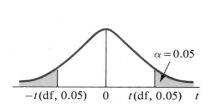

FIGURE 9-3
t Value on Left Side of Mean

Illustration 9-2

Find the value of $t(15, 0.95)$.

Solution There are 15 degrees of freedom. In table 7 we look for the column marked $\alpha = 0.05$ and come down to df = 15. The table value is 1.75. Thus $t(15, 0.95) = \mathbf{-1.75}$ (the value is negative because it is to the left of the mean).

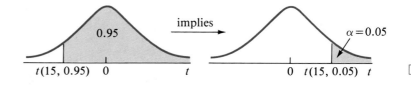

The t statistic is used in problems concerned with μ in much the same manner as z was used in chapter 8. In hypothesis-testing situations we will use formula (9-1) to calculate the test statistic value in step 4 of our procedure.

Illustration 9-3

Let's return to the hypothesis of illustration 8-1 (p. 276) concerning the response rate from an advertisement in the *New York News*. Does a random sample of the response to 25 advertisements in the *News* (sample results: $\bar{x} = 510$ and $s = 210$) present sufficient evidence to cause us to reject this claim? Use $\alpha = 0.05$.

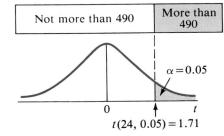

Solution

STEP 1: $H_0: \mu = 490 \ (\leq)$.

STEP 2: $H_a: \mu > 490$.

STEP 3: $\alpha = 0.05$, df $= 25 - 1 = 24$, and $t(24, 0.05) = 1.71$, from table 7, appendix D.

STEP 4:

$$t = \frac{\bar{x} - \mu}{s/\sqrt{n}} = \frac{510 - 490}{210/\sqrt{25}} = \frac{20}{210/5} = \frac{20}{42} = 0.476$$

$$t^* = 0.48$$

Comparing this value with the test criteria, we have the situation shown in the diagram on the left.

STEP 5: *Decision*: Fail to reject H_0 (t^* is in the noncritical region).

Conclusion: We cannot reject the claim that the mean number of catalog requests resulting from an advertisement in the *New York News* is no more than 490 (the mean rate from the *New York Chronicle*).

NOTE: **If the value of df (df $= n - 1$) in an exercise similar to illustration 9-3 is larger than 29**, then the critical value for $t(\text{df}, \alpha)$ actually becomes nearly the same as $z(\alpha)$, the z score listed at the bottom of table 7. Therefore $t(\text{df}, \alpha)$ is not given when df > 29; $z(\alpha)$ is used instead.

Because table 7 has only critical values for the Student's t distribution, the prob-value cannot be calculated for a hypothesis test that involves the use of t. However, the prob-value can be estimated.

Illustration 9-4

Let's return to illustration 9-3. Note that $t^* = 0.48$, df $= 24$, and $H_a: \mu > 490$. Thus for step 5 of the prob-value solution, we have

$$P = P(t > 0.48, \text{ knowing df} = 24)$$

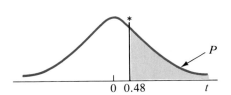

Inferences About the Population Mean   Section 9-1

By inspecting the df = 24 row of table 7, you can determine that the prob-value is greater than 0.25.

Portion of Table 7

df	0.25
⋮	
24	0.685

The 0.685 entry tells us that the $P(t > 0.685) = 0.25$, as shown in the accompanying figure.

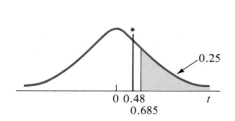

By comparing $t = 0.685$ with $t^* = 0.48$, we see that the **prob-value P is greater than 0.25**. ☐

Illustration 9-5

Determine the P-value for the following hypothesis test:

$$H_0: \mu = 55$$
$$H_a: \mu \neq 55$$
$$df = 15 \quad \text{and} \quad t^* = -1.84$$

Solution

$$P = P(t < -1.84) + P(t > 1.84) = 2P(t > 1.84)$$

By inspecting the df = 15 row of table 7 in the appendix,

Portion of Table 7

df	0.05	0.025
15	1.75	2.13

we find $P(t > 1.84)$ is between 0.025 and 0.05. Since this is a two-tailed test, the values from the table are doubled and the prob-value is between 0.05 and 0.10, $0.05 < P < 0.10$. ☐

confidence interval

The population mean may be estimated when σ is unknown in a manner similar to that used when σ is known. The difference is the use of Student's t in place of z and the use of s, the sample standard deviation, as an estimate of σ. The formula for the $1 - \alpha$ **confidence interval** of estimation then becomes

$$\bar{x} - t(df, \alpha/2) \cdot \frac{s}{\sqrt{n}} \quad \text{to} \quad \bar{x} + t(df, \alpha/2) \cdot \frac{s}{\sqrt{n}} \qquad (9\text{-}2)$$

where $df = n - 1$.

Illustration 9-6

A random sample of 20 municipal bonds due in 1990 was selected. A mean rate of return of 6.87% and a standard deviation of 1.76% were found for the sample. Estimate the mean rate of return of all municipal bonds due in 1990 with a 95% confidence interval.

Solution The information we are given is $\bar{x} = 6.87$, $s = 1.76$, and $n = 20$. $1 - \alpha = 0.95$ implies that $\alpha = 0.05$ and $\alpha/2 = 0.025$; $n = 20$ implies that $df = 19$. From table 7 we get $t(19, 0.025) = 2.09$. See the diagram.

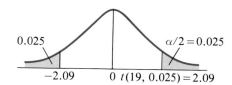

The confidence interval can now be found:

$$\bar{x} \pm t(19, 0.025) \cdot \frac{s}{\sqrt{n}}$$

$$6.87 \pm 2.09 \cdot \frac{1.76}{\sqrt{20}}$$

$$6.87 \pm \frac{(2.09)(1.76)}{4.472}$$

$$6.87 \pm 0.82$$

$$6.87 - 0.82 = \mathbf{6.05} \quad \text{and} \quad 6.87 + 0.82 = \mathbf{7.69}$$

Thus the 95% confidence interval for μ is 6.05 to 7.69. That is, with 95% confidence we estimate the mean rate of return to be between 6.05% and 7.69%.

Exercises

9-1 Find these critical values by using table 7, appendix D.
(a) $t(15, 0.05)$ (b) $t(20, 0.10)$
(c) $t(5, 0.01)$ (d) $t(11, 0.025)$
(e) $t(11, 0.95)$ (f) $t(16, 0.975)$
(g) $t(29, 0.99)$ (h) $t(50, 0.025)$

9-2 Name (use the notation of exercise 9-1) and find the following critical values of t:

(a) 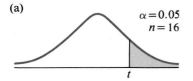 (b) 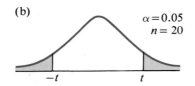 (c)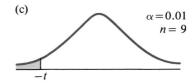

(d) 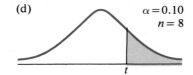 (e)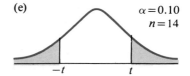

9-3 Ninety percent of Student's t distribution lies between $t = -1.89$ and $t = 1.89$ for what number of degrees of freedom?

9-4 (a) Find the first percentile of Student's t distribution with 24 degrees of freedom.
(b) Find the 95th percentile of Student's t distribution with 24 degrees of freedom.
(c) Find the first quartile of Student's t distribution with 24 degrees of freedom.

9-5 (a) State two ways in which the normal distribution and Student's t distribution are alike.
(b) State two ways in which they are different.

9-6 The mean valuation of homes in a nearby college town is $58,950. Homes in the college area are thought to be of higher value. To test this hypothesis, a random sample of 12 homes is chosen from the college area. Their mean valuation is found to be $62,460 and the standard deviation is $5,200. Complete a hypothesis test using $\alpha = 0.05$.
(a) Solve using the classical approach.
(b) Solve using the prob-value approach.

9-7 A union representing the employees of a small industrial plant claims that it takes an employee, on the average, 25 minutes to reach the closest local health clinic by cab; therefore an industrial nurse should be hired. Management obtained a random sample of 50 one-way travel times to the clinic. The sample had a mean of 19.4 minutes and a standard deviation of 9.6 minutes. Does management have sufficient evidence to show that the average time is less than 25 minutes? Use $\alpha = 0.01$.

9-8 Merit increases are given at the discretion of the division director. The average increase at the company is $100.00. The personnel manager is concerned that a certain division supervisor is too lenient in awarding merit increases. She takes a sample of 20 randomly selected employees from that division and finds a mean increase of $107.50 and a standard deviation of $10.50. Assuming the increases are normally distributed, is there enough evidence to support the personnel manager's concern at a 0.01 level of significance?

(a) Solve using the classical approach.

(b) Solve using the prob-value approach.

9-9 The *bad debt ratio*—the dollars of loans defaulted divided by the total amount loaned—on new-car loans by commercial banks in Philadelphia is 4.36%. It has been hypothesized that credit unions, since they generally grant loans to all members, screen clients less and hence will have a higher bad debt ratio. Twelve credit unions in Philadelphia were randomly selected and their new-car-loan bad debt ratio obtained. These ratios (expressed as percentages) were as follows:

3.56 5.00 4.88 4.93 4.25 5.12

5.13 4.79 5.35 4.81 3.48 4.45

At the 0.05 level of significance, do we have sufficient evidence to claim that the new-car-loan bad debt ratio of credit unions is greater than that of commercial banks?

9-10 The weights of the drained fruit found in 15 randomly selected cans of peaches packed by Sunny Fruit Cannery were (in ounces) as follows:

11.0 11.6 10.9 12.0 11.5

10.5 12.2 11.8 12.1 11.6

11.2 12.0 11.4 10.8 11.8

(a) Calculate the sample mean and the standard deviation.

(b) Construct the 0.95 confidence interval for the estimate of the mean weight of drained peaches per can.

9-11 A television station was interested in stressing to the Federal Communications Commission its public service effort. Consequently, it took a random sample of 51 prime-time television hours and recorded, for each hour, the number of seconds of free public service announcements. The random variable x is the number of seconds of free time aired in a sampled hour. The sample data can be summarized by $n = 51$, $\sum x = 647$, and $\sum (x - \bar{x})^2 = 2{,}636.5996$.

(a) Find the sample mean $\bar{x}$.
(b) Find the sample standard deviation s.
(c) Find the 0.95 confidence interval to estimate the true mean seconds per hour devoted to public service announcements.

Section 9-2 Inferences About Proportions

Proportion of or percentage of a population and the probability associated with the occurrence of a particular event all involve the **binomial parameter** p. Recall that p was defined to be the theoretical, or population, probability of success on a single trial in a binomial experiment. Also, the random variable x is the number of successes that occur in a set of n trials. By combining the definition of empirical probability, $P'(A) = \text{No.}(A)/n$ [formula (4-1)], with the notation of the binomial experiment, we define p', **the observed, or sample, binomial probability**, to be $p' = x/n$. Also recall that the mean and standard deviation of the binomial random variable x are found by use of formulas (5-7) and (5-8): $\mu = np$ and $\sigma = \sqrt{npq}$, where $q = 1 - p$. This distribution of x is considered to be approximately normal if n is larger than 20 and if np and nq are both larger than 5. This commonly accepted rule of thumb allows us to use the normal distribution when making inferences concerning a binomial parameter p.

Generally, it is easier to work with the distribution of p' rather than the distribution of x. Consequently, we will convert formulas (5-7) and (5-8) from the units of x to units of proportions. If we divide formulas (5-7) and (5-8) by n, we should change the units from those of x to those of proportion. The mean of x is np; thus the mean of p', $\mu_{p'}$, should be np divided by n (np/n), or just p. (Does it seem reasonable that the mean of the distribution of observed values of p' should be p, the true proportion?) Furthermore, the standard error of p' in this sampling distribution is

$$\sigma_{p'} = \sqrt{npq}/n = \sqrt{npq/n^2} = \sqrt{pq/n}$$

We summarize this information as follows:

An observed value of p' belongs to a sampling distribution that:

1. is approximately normal
2. has a mean $\mu_{p'}$ equal to p
3. has a standard error $\sigma_{p'}$ equal to $\sqrt{pq/n}$

This approximation to the normal distribution is considered reasonable whenever n is greater than 20 and both np and nq are greater than 5.

RECALL: The standard deviation of a sampling distribution is called the standard error.

As a result of these new definitions for μ and σ, the calculated **value of z** in step 4 of a hypothesis test concerning p is obtained by using the following formula:

$$z = \frac{p' - p}{\sqrt{pq/n}}, \quad \text{where } p' = \frac{x}{n} \quad (9\text{-}3)$$

The value of p to be used in formula (9-3) will be the value stated in the null hypothesis.

Illustration 9-7

While discussing the possibility of marketing a new nonprescription cold capsule, the director of research and development made the claim that it would capture at least 15% of the nonprescription cold capsule market. The marketing manager decided to check the validity of this claim; so he set up a market test study. Two hundred potential customers were randomly selected and asked to try the new drug. Only 17 said they preferred the new cold capsule. At a level of significance of 0.10, does the market manager have sufficient evidence to reject the research and development director's claim?

Solution p represents the proportion of market captured.

STEP 1: $H_0: p = 0.15$ ($\geq$) (at least 15%).

STEP 2: $H_a: p < 0.15$ (less than 15%).

STEP 3: $\alpha = 0.10$. The critical value of z is found by using table 6. See the diagram.

$$p' = \frac{17}{200} = 0.085$$

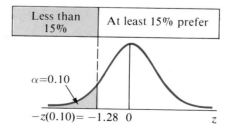

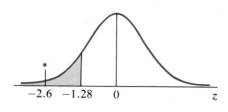

STEP 4: Calculate z^* using formula (9-3).

$$z = \frac{p' - p}{\sqrt{pq/n}} = \frac{0.085 - 0.150}{\sqrt{(0.15)(0.85)/200}} = \frac{-0.065}{\sqrt{0.00064}}$$

$$= \frac{-0.065}{0.025} = -2.6$$

$$z^* = -2.6$$

Comparing this value with the test criteria, we have the situation shown in the accompanying diagram.

STEP 5: *Decision*: Reject H_0 (z^* is in critical region).

Conclusion: The evidence found contradicts the claim. It appears that the new drug will capture less than 15% of the market. □

The solution to illustration 9-7 could have been carried out using the prob-value procedure. This alternative solution is shown below.

Solution

STEP 1: H_0: $p = 0.15$ ($\geq$) (at least 15%).

STEP 2: H_a: $p < 0.15$ (less than 15%).

STEP 3: $\alpha = 0.10$.

STEP 4: $z = \dfrac{p' - p}{\sqrt{pq/n}} = \dfrac{0.085 - 0.150}{\sqrt{(0.15)(0.85)/200}}$

$z^* = -2.60$

STEP 5:

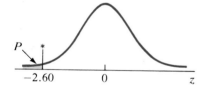

$$P = P(z < z^*) = P(z < -2.60) = 0.5000 - 0.4953$$
$$P = 0.0047$$

STEP 6: At the 0.10 level of significance, the sample information is significant. That is, it appears that the new drug will capture less than 15% of the market. □

When the true population proportion p is to be estimated, we will base our estimations on the observed value p'. The **confidence interval formula** is similar to the previous confidence interval formula.

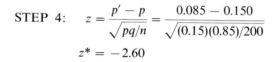

$$p' - z(\alpha/2) \cdot \sqrt{\frac{p'q'}{n}} \quad \text{to} \quad p' + z(\alpha/2) \cdot \sqrt{\frac{p'q'}{n}} \quad (9\text{-}4)$$

where $p' = x/n$ and $q' = 1 - p'$.

Notice that the standard error, $\sqrt{pq/n}$, has been replaced by $\sqrt{p'q'/n}$. Since we do not know the value of p, we must use the best replacement available. That replacement is p', the observed value or the point estimate for p. This replacement will cause little change in the width of our confidence interval.

Illustration 9-8

Suppose the marketing manager (illustration 9-7) had taken his sample with the intention of estimating the value of p, the proportion of cold capsule users who would use the new drug (i.e., the market share of customers).

(a) Find the best point estimate for p that he could use.
(b) Determine the 90% confidence interval estimate for the true value of p by using formula (9-4).

Solution

(a) The best point estimate of p is 0.085, the observed value of p'.
(b) The confidence interval is $p' \pm z(\alpha/2) \cdot \sqrt{p'q'/n}$, $q' = 1 - p'$, and $1 - \alpha = 0.90$; therefore $z(\alpha/2) = z(0.05) = 1.65$. The confidence interval is

$$0.085 \pm (1.65) \cdot \sqrt{\frac{(0.085)(0.915)}{200}}$$

$$0.085 \pm (1.65)\sqrt{0.000389}$$

$$0.085 \pm (1.65)(0.020)$$

$$0.085 \pm 0.033$$

$$0.085 - 0.033 = \mathbf{0.052} \qquad 0.085 + 0.033 = \mathbf{0.118}$$

$$0.052 \quad \text{to} \quad 0.118$$

That is, the true proportion of new drug users is between 0.052 and 0.118, with 90% confidence. □

maximum error of estimate

By using the maximum error part of the confidence interval formula, we can determine the size of the sample that must be taken to estimate p with a desired accuracy. The **maximum error of estimate for a proportion** is

$$E = z(\alpha/2) \cdot \sqrt{\frac{pq}{n}} \tag{9-5}$$

When using this formula, we must decide how accurate we desire our answer to be. (Remember that we are estimating p. Therefore E will be expressed in hundredths.) We need to establish the level of confidence with which we wish to work. If you have any indication of the value of p, use this value for $p(q = 1 - p)$. If there is no indication of an approximate value for p, then by assigning p the value 0.5, you will obtain the largest possible sample size that may be required.

For ease of use, formula (9-5) can be expressed as

$$n = \frac{[z(\alpha/2)]^2 \cdot p \cdot q}{(E)^2} \tag{9-6}$$

Illustration 9-9

Determine the sample size that is required to estimate the true proportion of married couples in which both are employed, if you want your estimate to be within 0.02 with 90% confidence.

Solution

STEP 1: $1 - \alpha = 0.90$; therefore $z(\alpha/2) = z(0.05) = 1.65$.

STEP 2: $E = 0.02$.

STEP 3:   Use $p = 0.5$; therefore $q = 1 - p = 0.5$.

STEP 4:   Use formula (9-6) to find n:

$$n = \frac{(1.65)^2 \cdot (0.5) \cdot (0.5)}{(0.02)^2} = \frac{0.680625}{0.0004} = 1{,}701.56$$

$$= 1{,}702 \qquad \square$$

Illustration 9-10

A manufacturer of automobiles purchases bolts from a supplier who claims his bolts to be approximately 5% defective. Determine the sample size that will be required to estimate the true proportion of defective bolts if we want our estimate to be within 0.02 with 90% confidence.

Solution

STEP 1: $1 - \alpha = 0.90$; therefore $z(\alpha/2) = z(0.05) = 1.65$.

STEP 2: $E = 0.02$.

STEP 3:   The supplier's claim is "5% defective"; thus $p = 0.05$. Therefore $q = 1 - p = 0.95$.

STEP 4:   Use formula (9-6) to find n:

$$n = \frac{(1.65)^2 (0.05)(0.95)}{(0.02)^2} = \frac{0.12931875}{0.0004} = 323.3$$

$$= 324 \qquad \square$$

Notice the difference in the sample size required in illustrations 9-9 and 9-10. The only real difference between the problems is the value that was used for p. In illustration 9-9 we used $p = 0.5$, and in illustration 9-10 we used $p = 0.05$. Recall that $p = 0.5$ gives a sample of maximum size. Thus it will be of great advantage to have an indication of the value expected for p if p is much different from 0.5.

Exercises

9-12 We are testing $H_0: p = 0.2$. We decide to reject H_0 if in 15 trials we observe more than five successes.
  (a) State an appropriate alternative hypothesis.
  (b) What is the level of significance of this test?
  (c) If we observe five successes, do we reject H_0?
  (d) If we observe six successes, do we reject H_0?
  (e) Suppose that H_0 is $p = 0.1$ and we use the same decision rule. Then what happens to the level of significance?

9-13 You are testing the hypothesis $p = \frac{1}{3}$ and have decided to reject this hypothesis if in 25 trials you observe either 3 successes or fewer or 14 successes or more.
  (a) If the null hypothesis is true and you observe 13 successes, then you will: (1) correctly fail to reject H_0, (2) correctly reject H_0, (3) commit a type I error, (4) commit a type II error.
  (b) Find the significance level of your test.
  (c) If the probability of success is $\frac{1}{2}$ and you observe 13 successes, then you will: (1) correctly fail to reject H_0, (2) correctly reject H_0, (3) commit a type I error, (4) commit a type II error.
  (d) Calculate the prob-value for your hypothesis test having observed 13 successes.

9-14 A poll of 1,200 randomly selected union members was taken to estimate the proportion of the union that was in favor of a strike vote. If only 562 responded in favor of a strike, do we have sufficient evidence to reject the union president's claim that the rank-and-file union members are willing to strike? Use a 0.05 level of significance.

9-15 Penn Milk Company has decided to discontinue its line of yogurt if it has not captured at least 15% of the yogurt market in the stores carrying its brand. A random sample of 1,200 yogurt shoppers in its retail outlets included 198 who bought Penn Yogurt. Does this sample suggest that Penn Yogurt should be discontinued? Carry out this hypothesis test using a 0.10 level of significance.
  (a) Solve using the classical approach.
  (b) Solve using the prob-value approach.

9-16 Fifteen percent of all salesroom demonstration models of a major car manufacturer are shades of red. The Dealers Association feels that more cars should be red, since it claims that more than 15% of all consumers prefer a shade of red. The vice-president of marketing for the manufacturer, however, claims that no more than 15% of all customers prefer a shade of red. If a survey of 400 randomly selected customers yields 64 who preferred a shade of red, is there enough evidence to dispute the vice-president's claim? Use $\alpha = 0.05$.

9-17 Yearly orders for Penn Brand Milk are comprised of 58% whole milk and 42% skim milk. Does a random sample of orders received for the week of 1 October for 30 cases of whole milk and 20 cases of skim show sufficient evidence to support the hypothesis that the mix of milk types ordered for any given week is the same as for the entire year? Test by using $\alpha = 0.05$.

9-18 An opinion poll based on 400 randomly selected employees yields a sample proportion $p' = 0.37$ of the voters favoring an increase in fringe benefits rather than wages. Find a 99% confidence interval estimate for p, the true population proportion favoring fringes over wages. Interpret this interval.

9-19 Construct 90% confidence intervals for the binomial parameter p for each of the following pairs of values.

	Observed Proportion $p' = x/n$	Sample Size
(a)	$p' = 0.3$	$n = 30$
(b)	$p' = 0.7$	$n = 30$
(c)	$p' = 0.5$	$n = 10$
(d)	$p' = 0.5$	$n = 100$
(e)	$p' = 0.5$	$n = 1{,}000$

(f) Compare answers (a) and (b).
(g) Compare answers (c), (d), and (e).

9-20 A bank randomly selected 150 checking account customers and found that 68 of them also had savings accounts at this same bank. Construct a 0.90 confidence interval estimate for the true proportion of checking account customers that also have savings accounts.

9-21 Refer to exercise 9-17. Construct the 0.95 confidence interval for the estimate of the true proportion of orders for skim milk received on 1 October (30 cases of whole milk and 20 cases of skim milk).

9-22 A bank believes that approximately one-third of its checking account customers have used at least one other service provided by the bank within the last 6 months. How large a sample will be needed to estimate the true proportion to within 4% at the 0.95 level of confidence?

9-23 A corporation is considering changing the method by which sales commissions are computed. It feels that if 60% of the salespeople favor the new method, it should be adopted. How large a sample needs to be taken to enable the corporation to estimate the proportion of salespeople who favor the proposal to within 5% with 95% confidence?

Section 9-3 Inferences About Variance and Standard Deviation

Often problems arise that require us to make inferences about variability. For example, a soft drink bottling company has a machine that fills 32-ounce bottles. They need to control the variance σ^2 (or standard deviation σ) among the amount x of soft drink put in each bottle. The mean amount placed in each bottle is important, but the mean amount being correct does not ensure that the filling machine is working correctly. If the variance is too large, there could be many bottles that are overfilled and many that are underfilled. Thus this bottling company will want to maintain as small a variance (or standard deviation) as possible.

Two kinds of inferences will be studied in this section: (1) the hypothesis test concerning the variance (or standard deviation) of one population and (2) the estimation of the variance or standard deviation of one population. Often in these two inferences we talk about the variance instead of the standard deviation. This is because the techniques employ the sample variance rather than the standard deviation. However, remember that the standard deviation is the square root of the variance; thus to talk about the variance of a population is comparable with talking about the standard deviation.

Returning to our problem, suppose the soft drink bottling company wishes to detect when the variability in the amount of soft drink placed in each bottle gets out of control. A variance of 0.0004 is considered acceptable, and the company will want to adjust the bottle-filling machine when the variance becomes larger. This decision will be made by use of the hypothesis test procedure. The null hypothesis is that the variance is no larger than the specified value 0.0004; the alternative hypothesis is that the variance is larger than 0.0004.

$$H_0: \sigma^2 = 0.0004 \ (\leq) \qquad \text{(variance not out of control)}$$

$$H_a: \sigma^2 > 0.0004 \qquad \text{(variance out of control)}$$

chi-square The test statistic that will be used in making a decision about the null hypothesis is **chi-square**, χ^2 (χ is the Greek lowercase letter chi, pronounced "kī" as in "sky"). The calculated value of chi-square will be obtained by using the following formula:

$$\chi^2 = \frac{(n-1)s^2}{\sigma^2} \qquad (9\text{-}7)$$

where s^2 is the sample variance, n is the sample size, and σ^2 is the value specified in the null hypothesis.

When random samples are drawn from a normal population of a known variance σ^2, the quantity $(n-1)s^2/\sigma^2$ possesses a probability distribution that is known as the **chi-square distribution**. The equations that define the chi-square distribution are not given here, as they are beyond the level of this book. However, to use the chi-square distribution, we must be aware of the following properties (see figure 9-4):

FIGURE 9-4
Chi-square Distribution

Properties of the Chi-square Distribution

1. χ^2 is nonnegative in value; it is 0, or positively valued.
2. χ^2 is not symmetrical; it is skewed to the right.
3. There are many χ^2 distributions. Like the t distribution, a different χ^2 distribution exists for each degree-of-freedom value.

NOTE: The mean value of the chi-square distribution is $n - 1$. The mean is located to the right of the mode (the value where the curve reaches its high point). See the figure.

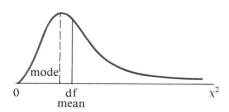

For the inferences discussed in this section, the number of degrees of freedom df is equal to $n - 1$.

The critical values for chi-square will be obtained from table 8 in appendix D. The critical values will be identified by two values: degrees of freedom df and the area under the curve to the right of the critical value being sought. Thus χ^2 (df, α) is the symbol used to identify the critical value of chi-square with df degrees of freedom and with α being the area to the right of χ^2 (df, α), as shown in figure 9-5. Since the chi-square distribution is not symmetrical, the critical values associated with both tails are given in table 8.

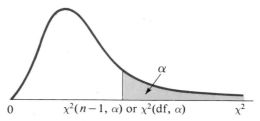

FIGURE 9-5
Chi-square Distribution
Showing χ^2(df, α)

df = $(n-1)$, degrees of freedom;
α is the area under curve to the right of a particular value

Illustration 9-11

Find χ^2 (20, 0.05).

Solution In table 8 you will find the value shown in the table here. Therefore χ^2 (20, 0.05) = 31.4.

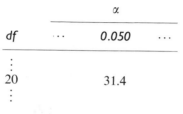

df	...	α 0.050	...
⋮			
20		31.4	
⋮			

Illustration 9-12
Find $\chi^2(14, 0.90)$.

Solution df = 14, and the area to the right of the critical value is 0.90, as shown in the figure. Therefore $\chi^2(14, 0.90) = \textbf{7.79}$.

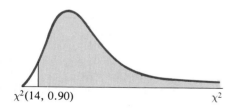

$\chi^2(14, 0.90)$

df	...	Area Under Curve to the Right 0.90	...
⋮			
14		7.79	
⋮			

Illustration 9-13
Recall that the soft drink bottling company wanted to control the variance by not allowing the variance to exceed 0.0004. Does a sample of size 28 with a variance of 0.0010 indicate that the bottling process is out of control (with regard to variance) at the 0.05 level?

Solution

STEP 1: $H_0: \sigma^2 = 0.0004$ ($\leq$) (not out of control).

STEP 2: $H_a: \sigma^2 > 0.0004$ (out of control).

STEP 3: $\alpha = 0.05$ and $n = 28$; therefore df = 27. The test statistic is χ^2 and the critical region is the right tail, with an area of 0.05. $\chi^2(27, 0.05)$ is found in table 8. See the diagram.

STEP 4:
$$\chi^2 = \frac{(n-1)s^2}{\sigma^2} = \frac{(28-1)(0.0010)}{0.0004}$$
$$= \frac{(27)(0.001)}{0.0004} = \frac{0.0270}{0.0004} = 67.5$$
$$\chi^{2*} = \textbf{67.5}$$

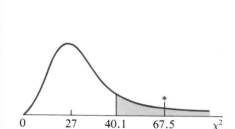

STEP 5: *Decision*: Reject H_0 (χ^{2*} is in the critical region).

Conclusion: The bottling process is out of control with regard to the variance. □

The prob-value can be estimated for hypothesis tests using the chi-square test statistic in much the same manner as when Student's t was used.

Illustration 9-14

Find the prob-value for the following hypothesis test:

$$H_0: \sigma^2 = 150$$
$$H_a: \sigma^2 > 150$$

$$\text{df} = 18 \quad \text{and} \quad \chi^{2*} = 32.7$$

Solution

$$P = P(\chi^2 > 32.7)$$

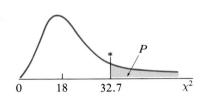

By inspecting the df $= 18$ row of table 8, we find that 32.7 is between 31.5 and 34.8. Therefore the prob-value is between 0.010 and 0.025. □

Illustration 9-15

In many jobs where an employee turns out individual pieces of work (e.g., a lathe operator may turn out bolts or a clerk may record payments received), the employee receives a bonus for producing items above a certain fixed quota amount. This bonus is known as "piecework incentive rates." In setting the quota rate, we must consider the normal "spread" in output that occurs in a shift production. The quota is normally set somewhat above the mean rate of production. If the spread is too small, too few employees will exceed the quota and the incentive aspect is lost. If the spread is too large, some employees will receive so large an incentive pay as to create large wage imbalances and personnel problems.

A quota has been suggested on a certain job based on the belief that the amount of production per employee per shift has a standard deviation of 12 items. Currently, the job is not a piecework incentive job. To help to determine whether to implement this suggestion, production records of 28 employees on a given shift showed the standard deviation of production per employee to be 10.5. Does this evidence, at the 0.05 level of significance, dispute the claim that the overall standard deviation is 12?

Solution The information given is $n = 28$, $s = 10.5$, and $\alpha = 0.05$.

STEP 1: $H_0: \sigma = 12$.

STEP 2: $H_a: \sigma \neq 12$.

STEP 3: $\alpha = 0.05$; the critical values are $\chi^2(27, 0.975) = 14.6$ and $\chi^2(27, 0.025) = 43.2$. See the diagram.

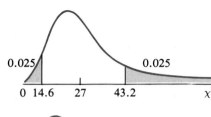

STEP 4:

$$\chi^2 = \frac{(n-1)s^2}{\sigma^2} = \frac{(27)(10.5)^2}{(12)^2} = \frac{2{,}976.75}{144} = 20.6719$$

$$\chi^{2*} = 20.67$$

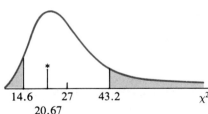

The diagram on the left shows this value compared with the test criteria.

STEP 5: *Decision*: Fail to reject H_0.

Conclusion: There is not sufficient evidence to claim that the standard deviation is different from 12. □

These tests for variance may be one-tailed or two-tailed tests, in accordance with the statement of the claim being tested.

The formula for chi-square may be reworked to give the values at the extremities of the confidence interval:

$$\chi^2 = \frac{(n-1) \cdot s^2}{\sigma^2}$$

or if solved for σ^2,

$$\sigma^2 = \frac{(n-1) \cdot s^2}{\chi^2} \qquad (9\text{-}8)$$

When constructing a $1 - \alpha$ confidence interval, the critical values of chi-square are separately substituted into formula (9-8) to obtain the two endpoints of the confidence interval of estimation. Note that $\chi^2(\text{df}, 1 - \alpha/2)$ is less than $\chi^2(\text{df}, \alpha/2)$. Therefore, after dividing, the numbers will be in the opposite order, yielding the following **confidence interval for variance**:

$$\frac{(n-1)s^2}{\chi^2(\text{df}, \alpha/2)} \quad \text{to} \quad \frac{(n-1)s^2}{\chi^2(\text{df}, 1 - \alpha/2)} \qquad (9\text{-}9)$$

If the confidence interval for the standard deviation is desired, we need only take the square root of each of the numbers in formula (9-9).

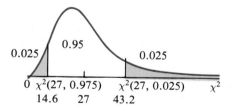

Illustration 9-16

Using the sample results from illustration 9-15 ($n = 28$, $s = 10.5$), calculate the 95% confidence interval for the population variance and standard deviation.

Solution The given information is $n = 28$ and $s = 10.5$. For a 95% confidence interval, $\alpha = 0.05$ and hence $\alpha/2 = 0.025$. The critical values for χ^2 are shown in the diagram. The confidence interval, using formula (9-9), is

$$\frac{(27)(10.5)^2}{43.2} \quad \text{to} \quad \frac{(27)(10.5)^2}{14.6}$$

$$\frac{2{,}976.75}{43.2} \quad \text{to} \quad \frac{2{,}976.75}{14.6}$$

$$68.9 \quad \text{to} \quad 203.9$$

That is, with 95% confidence we estimate the population variance to be between 68.9 and 203.9.

The confidence interval for the standard deviation can be found by taking the square root of 68.9 and of 203.9. The 95% confidence interval estimate for the standard deviation is **8.3 to 14.3**. □

Exercises

9-24 Find these critical values by using table 8.
(a) χ^2 (10, 0.01) (b) χ^2 (8, 0.025)
(c) χ^2 (18, 0.10) (d) χ^2 (25, 0.01)
(e) χ^2 (12, 0.95) (f) χ^2 (20, 0.975)
(g) χ^2 (40, 0.90) (h) χ^2 (4, 0.99)

9-25 Name (use the notation of exercise 9-24) and find these critical values of χ^2.

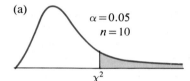
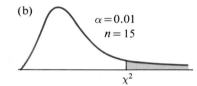
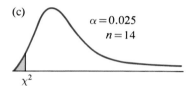
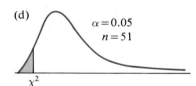
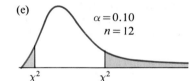
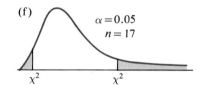
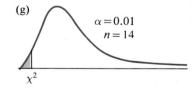
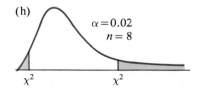

9-26 (a) What value of chi-square for five degrees of freedom subdivides the area under the distribution curve such that 5% is to the right and 95% is to the left?

(b) What is the value of the 95th percentile for the chi-square distribution with five degrees of freedom?

(c) What is the value of the 90th percentile for the chi-square distribution with five degrees of freedom?

9-27 (a) The central 90% of the chi-square distribution with 11 degrees of freedom lies between what values?

(b) The central 95% of the chi-square distribution with 11 degrees of freedom lies between what values?

(c) The central 99% of the chi-square distribution with 11 degrees of freedom lies between what values?

9-28 The standard deviation of monthly sales, adjusted for normal growth, over the past 60 months is $16,000. The standard deviation of adjusted sales for the past five Decembers is only $10,000. Does it appear that the variation in adjusted sales between Decembers is less than the normal monthly variation? Test by using $\alpha = 0.05$. (If that is the case, it would imply that the month has an effect on sales. That is called a seasonal effect.) Within what range does the prob-value fall?

9-29 Management Service, Incorporated has recommended to its bank clients that they use a single line feeding to the various tellers to serve their clients. It claims that the average time a customer will have to wait in line will not be lowered, but the standard deviation will. Hence the chances of a client waiting a very long time and becoming angry enough to change banks will be reduced. Currently, Lincoln Bank does not use the suggested system, and the waiting time averages 5 minutes with a standard deviation of 2.1 minutes. The bank tries the system, and a random sample of 30 customers yields an average waiting time of 5.1 minutes and a standard deviation of 1.65 minutes. Does this sample show a significant lowering in the standard deviation to support Management Service's claim? Use a 0.05 level of significance.

9-30 In the past the standard deviation of weights of certain 32.0-ounce packages filled by a machine was 0.25 ounce. A random sample of 20 packages showed a standard deviation of 0.40 ounce. Is the apparent increase in variability significant at the 0.10 level of significance?

9-31 (a) Determine the best point estimate for σ^2, the variance of the seconds per hour devoted to public service in exercise 9-11.

(b) Construct the 0.95 confidence interval for σ^2 using the sample information in exercise 9-11.

9-32 Suppose a sample of size 12 had a sum of squared deviations about the mean, $\sum (x - \bar{x})^2$, of 3.57.

(a) What would be the point estimate for the population variance?

(b) What would be the point estimate for the population standard deviation?

(c) What would be the 0.95 confidence interval estimate for σ^2 (variance)?

(d) What would be the 0.95 confidence interval estimate for σ (standard deviation)?

9-33 Suppose a sample of size 22 had a sum of x equal to 397.3 and a sum of x^2 equal to 7,374.09.

(a) What would be the point estimate for the population variance?

(b) What would be the 0.90 confidence interval estimate for the population variance?

(c) What would be the 0.90 confidence interval estimate for the population standard deviation?

In Retrospect

We have studied inferences, both hypothesis testing and confidence interval estimation, for three basic population parameters—mean μ, proportion p, and standard deviation σ. When we make inferences about a single population, we are usually concerned with one of these three values. Table 9-1 identifies the formula that is used in each of the inferences for problems involving a single population.

In this chapter we also used the maximum error of estimate term of formula (9-4) to determine the size of sample required to make estimates about the population proportion with the desired accuracy.

The news article at the beginning of the chapter discussed the confidence one can place in TV ratings, which estimate the population proportion of viewers. The inferences drawn here have great impact, and the margin of error is important in properly interpreting television ratings. Do you think after reading the article that the confidence intervals are tight enough? How could they be narrowed?

In the next chapter we will discuss inferences about two populations, whose respective means, proportions, and standard deviations are to be compared.

TABLE 9-1
Formulas to Use for Inferences Involving a Single Population

	Test Statistic	Formula to Be Used	
		Hypothesis Tests	Interval Estimate
One mean			
σ known	z	Formula (7-2)	(8-2)
σ unknown	t	Formula (9-1)	(9-2)
One proportion	z	Formula (9-3)	(9-4)
One standard deviation	χ^2	Formula (9-7)	(9-9)
One variance	χ^2	Formula (9-7)	(9-9)

Chapter Exercises

9-34 A consumer panel of 10 persons was asked to rate a new product on the basis of three factors: (1) taste, (2) price, and (3) appearance. The following table gives the results:

Factor	Panel Member's Rating									
	1	2	3	4	5	6	7	8	9	10
Taste	2	2	3	2	4	2	4	4	2	2
Price	4	3	3	4	3	5	3	4	3	3
Appearance	6	5	8	5	3	3	3	4	3	6

(a) Calculate the mean and standard deviation for each of the three factors.

(b) "Taste" is rated and interpreted according to the following scale:

```
  Great    Good    Fair    Poor
L___.___.___|___.___|___.___|___.___J
  1        3        5        7        9
```

Does the sample show sufficient evidence to reject the claim that the taste is not "great," that is, $\mu \geq 3$ at $\alpha = 0.05$?

(c) "Price" is measured and interpreted according to the following scale:

```
   High         Average         Low
L___.___.___|___.___.___|___.___.___J
  1             4             6             9
```

Does the sample show sufficient evidence to allow us to conclude that the price level is high, that is, $\mu < 4$ at $\alpha = 0.05$?

(d) "Appearance" is measured and interpreted according to this scale:

```
  Pleasant     Average     Poor
L___.___.___|___.___.___|___.___J
  1             4             7         9
```

Does the sample present sufficient evidence to reject the null hypothesis that the appearance is pleasant, that is, $\mu \leq 4$ at $\alpha = 0.05$?

Construct the following confidence intervals:

(e) 90% confidence interval for estimating the mean rating of taste
(f) 99% confidence interval for estimating the mean rating of price.
(g) 95% confidence interval for estimating the mean rating of appearance

9-35 A manufacturer of television sets claims that the maintenance expenditures for his product will average less than $50 during the first year following the expiration of the warranty. A consumer group has asked you to substantiate or discredit the claim. A random sample of 50 owners of such television sets was taken; the resulting sample had a mean expenditure of $61.60 and a standard deviation of $32.46. At the 0.01 level of significance, should you conclude that the producer's claim is true or not likely to be true?

(a) Solve using the classical approach.

(b) Solve using the prob-value approach.

9-36 A random sample of size 72 is taken from very extensive records on cash awards made to employees of a company for helpful suggestions. If the mean of this sample is $140 and its standard deviation is $25, construct a 99% confidence interval for the true mean of all cash awards.

9-37 A large retailer of television sets in Chicago claims that at least 75% of all service calls for color TV sets are due to the malfunctioning of a single type of tube. A random sample of 150 service calls showed that 102 were due to this type of tube. Does this sample present sufficient evidence to reject the retailer's claim? Use $\alpha = 0.01$.

(a) Solve using the classical approach.

(b) Solve using the prob-value approach.

9-38 Reliance Appliances includes free installation with the purchase of any window air conditioner. In attempting to study the markup it needs on a unit to assure proper profit margins, the company makes a study of actual costs of installation. Construct a 95% confidence interval for an estimate of the average cost of installing a window air conditioner based on the following information: $n = 25$, $\bar{x} = \$10.20$, $s = \$5.70$.

9-39 The manager of a pension fund feels a certain investment counselor is incompetent. She claims that by picking stocks out of a hat you could get the same proportion of "winners" (stocks that advance) as the counselor does. Forty percent of all stocks advanced this month, and the investment counselor's recommendations moved as shown in the accompanying table. Is there sufficient evidence to reject the pension fund manager's claim? Use $\alpha = 0.10$.

Stock Recommended	1	2	3	4	5	6	7	8	9	10	11	12	13
Price Movement	U†	U	D‡	U	D	D	U	D	U	U	U	D	U

Stock Recommended	14	15	16	17	18	19	20	21	22	23	24	25
Price Movement	U	D	D	U	D	U	U	D	U	D	U	D

† Up
‡ Down

9-40 You are interested in testing the hypothesis $p = 0.8$ against the alternative $p < 0.8$. In 100 trials you observe 73 successes. Calculate the prob-value associated with this result.

9-41 A large industrial firm wishes to estimate the proportion of its employees who have adequate knowledge of the new safety regulations adopted by the firm. A random sample of 400 employees is selected and given a test on the new rules. Of those tested, 80 made a passing score. Construct a 90% confidence interval for the true proportion of employees having adequate knowledge of the rules.

9-42 A sample of 250 walkie-talkies contains 31 that do not function properly. Construct a 98% confidence interval estimate for the true proportion that do not function properly.

9-43 M & R Welding Company is testing the shearing strength of a particular weld. They would like the strength of such welds to have little variability. The sample of weld strengths are as follows:

$$2190 \quad 2280 \quad 2283 \quad 2275 \quad 2340$$
$$2305 \quad 2250 \quad 2235 \quad 2270 \quad 2280$$

(a) Does the sample of weld strengths show sufficient evidence to allow M & R to claim that the variance of shear strength is no more than 950 at the 0.05 level of significance?

(b) How large a sample will be needed for M & R Welding Company to estimate the mean shearing strength of a particular weld if the maximum allowable error is 6 pounds at 98% confidence? Assume that the variance in the weld strengths is approximately 950.

9-44 A survey of brand loyalty (percentage of customers who repeatedly buy the same brand) is taken to estimate the percentage of Pres users who have purchased Pres at least six consecutive times to within two percentage points at a 98% level of confidence. How large a sample should be taken? (It is expected that the percentage will be approximately one-fourth of all Pres purchasers.)

9-45 According to the norms established by the publisher of a certain employment test, all prospective employees should average 82.6. The test is given to 40 minority applicants. The mean score was 79.9 and the standard deviation was 10.6. The corporation's equal employment opportunity (EEO) officer informs the personnel department that "it should not use the test because it has a statistically significant disparate effect upon minorities and hence the firm is exposed to a Title VII lawsuit." Is the EEO officer correct? (The law requires that $\alpha = 0.05$.)

9-46 The marketing research department of an instant-coffee firm conducted a survey of married males to determine the proportion of married males who prefer their brand. Twenty of the one hundred in the random sample preferred the company's brand. Use a 95% confidence interval to estimate the proportion of all married males that prefer this company's brand of instant coffee. Interpret your answer.

9-47 A radio station on the West Coast is promoting a popular music group named Warren Peace and his Atom Bombs. In the past, 60% of the listeners of the station have liked music groups promoted by the station. You have randomly selected a sample of 200 listeners, and 102 of them like the group. At the 0.02 level of significance, test the hypothesis that there is no difference between the attitude of the current listeners and listeners in the past.

(a) Solve using the classical approach.

(b) Solve using the prob-value approach.

9-48 A company is drafting an advertising campaign that will involve endorsements by noted athletes. For this campaign to work, the endorser must have a high level of recognizability and respect. Photos of various athletes are shown to a random sample of 100 prospective customers, and if the respondent recognizes the athlete, then he or she is asked if they respect the athlete. In the case of a top women golfer, 16 of the 100 recognized her picture and indicated that they also respected her. At the 95% level of confidence, what is the true proportion of prospective customers who both recognize and respect this golfer?

9-49 A local auto dealership advertises that 90% of those having their autos serviced by the dealership's service department are pleased with the results. As a researcher you take exception to this statement, since you are aware that many people are reluctant to express dissatisfaction even when they are not pleased. A research experiment was set up in which those being sampled had received service by this dealer within the last 2 weeks. During the interview, the individuals were led to believe that the person inquiring was new in town and was considering taking his or her car to this dealer's service department. Of the 60 sampled, 14 said that they were dissatisfied and would not recommend the department.

(a) Estimate the proportion of dissatisfied customers using a 95% confidence interval.

(b) What can be concluded about the dealer's claim in light of your answer in part (a)?

9-50 An industrial firm has just installed a new furnace. To be able to install a furnace, the law requires a firm to obtain a satisfactory government environmental impact study. Such a study was obtained. The report stated that the furnace would not significantly raise the average sulfur dioxide pollutant index nor would the waste dumped in Lake Michigan raise significantly the mineral pollutant index. The Public Interest Law Corporation (PILCOP) has sued the firm and the U.S. Department of Interior, claiming that the environmental impact study is wrong. To defend against the suit, the firm has collected the sulfur dioxide pollutant index for the past 20 days, and it was

$$3.1 \quad 2.1 \quad 2.9 \quad 2.8 \quad 3.4 \quad 4.5 \quad 2.1 \quad 4.1 \quad 2.7 \quad 2.8$$
$$3.1 \quad 4.3 \quad 2.1 \quad 3.9 \quad 3.8 \quad 3.2 \quad 3.7 \quad 2.8 \quad 1.6 \quad 3.9$$

During the 3 months prior to the installation of the furnace, the average index was 2.8.

(a) Calculate the mean and standard deviation of the sample data.

(b) Does the sample information support the firm or PILCOP? (The burden of proof is on the plaintiff PILCOP, and $\alpha = 0.05$.)

The average mineral pollutant index before the furnace was installed was 31, and for the past 20 days since the furnace was installed, it was as follows:

$$32 \quad 48 \quad 33 \quad 22 \quad 29 \quad 30 \quad 45 \quad 25 \quad 26 \quad 43$$
$$36 \quad 27 \quad 34 \quad 20 \quad 35 \quad 55 \quad 52 \quad 38 \quad 34 \quad 37$$

(c) Calculate the mean and standard deviation of the sample data.

(d) Does the sample information support the firm or PILCOP? (The burden of proof is on the plaintiff PILCOP, and $\alpha = 0.05$.)

9-51 The investment manager of a mutual fund is making a study of the option of treasury notes as opposed to grade Aaa bonds. He collects the data shown in the following table on a random sample of the fund's investments in bonds. Treasury notes for the period paid 8%.

(a) Estimate the proportion of bonds held that yield less than treasury notes, and construct a 95% confidence interval for the true proportion.

(b) Do the data support the claim that the rate of return for less than half the grade Aaa bond investments is less than the rate of return for the no-risk treasury notes? Use $\alpha = 0.05$.

Bond	Rate of Return†	Bond	Rate of Return†
1	7.8	9	18.3
2	7.9	10	7.4
3	7.9	11	13.8
4	7.6	12	7.4
5	6.5	13	7.9
6	10.3	14	7.6
7	7.9	15	7.6
8	7.7	16	7.8

† Interest plus price appreciation or minus price depreciation

9-52 The following computer output (Minitab) presents an analysis similar to that in exercise 8-48 of the data in exercise 8-49.

```
-- TINTERVAL 90 PERCENT CONFIDENCE FOR C1
   LA         N =  24       MEAN =       3.4500      ST.DEV. =      0.593

   A 90.00   PERCENT C.I. FOR MU IS (     3.2426,      3.6574)

-- TTEST OF MU = 0.86 ALTERNATIVE = -1 FOR C1
   LA         N =  24       MEAN =       3.4500      ST.DEV. =      0.593

   TEST OF MU =          0.8600 VS. MU L.T.       0.8600
   T = 21.407
   CANNOT REJECT SINCE T IS G.T. 0.
```

With respect to the test of hypothesis, state or give the value(s) for the following:

(a) H_0 (b) H_a (c) α
(d) prob-value (e) $\bar{x}$ (f) s
(g) t^* (h) the decision

With respect to the confidence interval estimate, state or give the value(s) for the following:

(i) n (j) $1 - \alpha$ (k) E (maximum error)
(l) upper confidence limit (m) lower confidence limit

9-53 Compare the results presented in exercise 9-52 with those presented in exercise 8-49. Why was t used in exercise 9-52 and z used in exercise 8-49.

9-54 The following computer output (Minitab) presents a confidence interval estimate for the mean number of days elapsed based on the data presented in exercise 2-77. Compare these results with those presented in exercise 8-48. Be sure to include an explanation of why the t statistic was used here and the z statistic was used in exercise 8-48.

```
-- TINTERVAL 99 PERCENT CONFIDENCE FOR C2
   C2         N =  94       MEAN =       10.362      ST.DEV. =       4.87

   A 99.00   PERCENT C.I. FOR MU IS (     9.0394,     11.6840)
```

Challenging Problem

9-55 As a measure of a stock's performance during a given year, Valworth Investment Service computes the difference between the percentage change in the stock price during the year and the percentage change in the Dow–Jones index during the year. It calls this number the Valworth Performance Index (VPI). If VPI is greater than 0, then the stock has outperformed the general market. If VPI is less than 0, the stock has performed worse than the general market. A value of 0 means the stock has performed the same as the general market. According to Valworth's studies, the distribution of the VPI for *all* stocks follows the normal distribution with a mean of 0 and standard deviation of 0.20. The VPI for a random sample of 36 stocks Valworth has recommended over the past 5 years yielded a mean $\bar{x}$ of 0.04 and a standard deviation s of 0.05.

(a) Test the following hypotheses (use $\alpha = 0.05$):
   (i) The mean VPI for all stocks recommended by Valworth is the same as that of all stocks.
   (ii) The standard deviation of VPI for all stocks recommended by Valworth is the same as that of all stocks.

(b) Based on your results in part (a), how would you evaluate Valworth's investment advice for the past 5 years.

(c) Assume that the VPI for all stocks recommended by Valworth is normally distributed with $\bar{x} = 0.04$ and $\sigma = 0.05$. If you were to randomly select 50 stocks recommended by Valworth, build a 95% confidence interval on the proportion of stocks that would outperform the general market.

Hands-On Problems

For the following problems you are to collect three samples, as indicated.

9-1 From the full-time male student population, select a random sample of 12 who have part-time jobs and record the number of hours that each worked last week.

9-2 From the full-time female student population, select a random sample of 10 who have a part-time job and record the number of hours that each worked last week.

9-3 From a daily newpaper observe, but do not list, a random sample of 100 stocks on the New York Stock Exchange and record the number that reported a positive (+) net change for the day.

9-4 Calculate the mean and standard deviation for each of the samples in problems 9-1 and 9-2.

9-5 For the sample in problem 9-3, $n = 100$, the number of positive net changes; $x = $ _____; and $p' = x/100 = $ _____(fill in the blanks).

For the following problems complete the hypothesis tests using the samples and their statistics.

9-6 It has been claimed that working students work an average of 12 or more hours per week with a standard deviation of 5 hours.

  (a) Does the sample evidence allow you to reject the null hypothesis that the male working students work a mean of 12 or more hours per week? Use $\alpha = 0.05$.

  (b) Does the sample evidence allow you to reject the null hypothesis that the standard deviation of the number of hours worked by the working female students is 5 hours? Use $\alpha = 0.10$.

9-7 Does the sample evidence allow you to conclude that the percentage of stocks that made gains was different from 50%? Use $\alpha = 0.02$.

For the following problems provide estimations using the samples and their statistics.

9-8 Construct the 99% confidence interval estimate for the standard deviation of the number of hours worked by the male students.

9-9 Suppose that we wanted to take a survey of the full-time students with the idea of estimating the mean number of hours worked by the full-time female students. If a maximum error of 1 hour with 95% confidence is required, a sample of what size should be taken? Assume that $\sigma = 5$.

9-10 How large a sample is needed to estimate the true percentage of stocks reporting a (+) net change for the day observed to within 4% at the 0.95 level of confidence?

The calculations for this problem set can most easily be accomplished with the assistance of an electronic calculator or a packaged program on a computer. A list of the available programs can be obtained from your computer center. There are a variety of packaged programs available: Minitab, Biomed (Biomedical Programs), SAS (Statistical Analysis System), IBM Scientific Subroutine Packages, and SPSS (Statistical Package for the Social Sciences) program libraries. Your local computer center will assist you.

10 Inferences Involving Two Populations

Chapter Outline

10-1 Independent and Dependent Samples
Independent samples are obtained by using unrelated sets of subjects; dependent samples result from using paired subjects.

10.2 Inferences Concerning the Difference Between Two Independent Means (Variances Known or Large Samples)
The comparison of the mean values of two populations is a common objective.

10-3 Inferences Concerning Two Variances
To investigate the relationship between the variances of two populations, we need the concept of the F distribution.

10-4 Inferences Concerning the Difference Between Two Independent Means (Variances Unknown and Small Samples)
We distinguish between cases in which the variances are equal and cases in which they are not equal.

10-5 Inferences Concerning Two Dependent Means
The use of dependent samples helps control otherwise untested factors.

10-6 Inferences Concerning Two Proportions
Questions about the percentages or proportions of two populations are answered by means of hypothesis testing or a confidence interval estimate.

An Assessment of the Personal and Consumer Finance Literacy Among Undergraduate Students at Brigham Young University

One of the fundamental competencies needed by all people in today's society is that of handling and managing personal finances. Everyone makes choices regarding employment, money management, consumer spending, the use of credit, savings and investments, and other personal financial matters that ultimately affect one's financial security and standard of living. Since a person's life may be affected by the pattern of financial decisions he makes, the question of whether or not competent and sensible decisions will be made becomes a critical one....

Research Design

To acquire the data for this study, undergraduate students at Brigham Young University completed the Oregon Personal Finance Test during the fall semester of 1974....

The mean scores of students in each of the five sections of the test as well as the composite scores were compared against the competency level scores established by a faculty jury....

Findings

...The mean score achieved was 58.14, which was lower, but not significantly lower, than the established competency score of 60.25.

Table 1 provides detailed data concerning achievement in each student category....

One significant difference was observed in comparing the performance of students tested according to the main student categories. In the test section which dealt with credit concepts, students who had completed a course in personal and consumer finance scored significantly higher than students who had not completed such a course. No significant differences in test scores were observed in the categories of sex, marital status, class standing, college major, or whether or not the student had served as a missionary.

As a result, the hypothesis which stated that there is no significant difference at the 0.05 level in the personal and consumer finance literacy of students who have and have not taken a course in personal and consumer finance was accepted....

TABLE I
Mean Scores Achieved per Student Category on the Oregon Personal Finance Test

Category	Employment and Income	Credit	Purchase of Goods and Services	Rights and Responsibilities in the Marketplace	Money Management	Composite	Number
Male	5.14	18.39	12.02	8.49	14.57	58.61	182
Female	5.04	18.23	12.24	8.41	14.00	57.93	231
Not returned missionary	5.07	18.20	12.22	8.35	14.09	57.92	301
Returned missionary	5.14	18.57	11.95	8.71	14.70	59.06	112
Single	5.09	18.20	12.14	8.41	14.20	58.02	364
Married	5.10	19.10	12.16	8.71	14.67	59.76	49
Personal finance class	5.03	19.08	12.36	8.59	14.29	59.35	99
No personal finance class	5.11	18.06	12.07	8.40	14.24	57.88	314
Business and family living	5.13	18.94	12.43	8.82	14.61	59.92	115
Other colleges	5.07	18.06	12.03	8.30	14.11	57.58	298
Freshman	5.09	17.70	11.88	8.08	13.87	56.63	144
Sophomore	5.11	18.23	12.13	8.44	14.23	58.13	111
Junior	5.09	18.75	12.48	8.75	14.53	59.60	104
Senior	5.04	19.20	12.22	8.83	14.80	60.09	54
Mean score achieved	5.08	18.30	12.14	8.45	14.25	58.14	413
Minimum competency score*	4.25	19.25	10.75	9.50	16.50	60.25	

* Mean of competency level scores suggested by faculty panel.

From Boyd G. Worthington, "An Assessment of the Personal and Consumer Finance Literacy Among Undergraduate Students at Brigham Young University." Reprinted from *Business Education Forum* (vol. 31, no. 5, February 1977), pp. 35–37, by permission of the National Business Education Association.

Chapter Objectives

In chapters 8 and 9 we introduced the basic concepts of hypothesis testing and confidence interval estimation in connection with inferences about one population and the following parameters: mean, standard deviation, and proportion. In this chapter we continue to investigate inferences about those same three parameters; however, here we will use them to compare two populations.

Section 10-1 Independent and Dependent Samples

In this chapter we are going to study the procedures for making inferences about two populations. When comparing two populations, we need two samples, one from each population. Two basic kinds of samples can be used: independent and dependent. The dependence or independence of a sample is determined by the sources used for the data. A **source** can be a person, an object, or anything that yields a piece of data. If the same set of sources is used, or if elements from different populations are matched on some criteria, to obtain the data representing both populations, we have **dependent sampling**. If two unrelated sets of sources are used, one set from each population, we have **independent sampling**. The following illustrations should amplify these ideas further.

Illustration 10-1

Broston Publishing Company is planning to sell a bound set of 40 great classics. The set is to be sold by direct home sales, following leads from newspaper and magazine advertisements. Salespeople are trained to give a standard sales delivery, following a basic "script" and using supporting materials, which they show to the customer. The firm has two potential scripts to use, and the sales manager would like to determine which one is better. To decide which script to use, the market research department proposes the following two sampling procedures:

> Plan A: Randomly select 100 salespeople and 500 leads. Train 50 in one script and 50 in the other. Give each salesperson five leads to attempt to sell.

> Plan B: Randomly select 50 salespeople and 500 leads. Train each salesperson in both scripts, and give each salesperson five leads to attempt to sell by using one script and five leads to attempt to sell by using the other.

Plan A illustrates independent sampling—the sources (salespeople) used for each sample (the type of script) were selected separately. Plan B illustrates dependent sampling—the sources used for both samples are the same. ☐

Illustration 10-2

A test is being designed to compare the wearing quality of two brands of tires. The automobiles will be selected and equipped with the new tires and then driven under "normal" conditions for 1 month. A measurement then will be taken to determine how much wear took place. Two plans will be used:

Plan C: n cars will be selected randomly and equipped with brand A and driven for the month, and n other cars will be selected and equipped with brand B and driven for the month.

Plan D: n cars will be selected randomly, equipped with one tire of brand A and one tire of brand B (the other two tires are not part of the test) and driven for the month.

In this illustration we might suspect that many other factors must be taken into account when testing automobile tires—such as age, weight, and mechanical condition of the car; driving habits of drivers; location of the tire on the car; and where the car is driven. However, at this time we are only trying to illustrate dependent and independent samples. Plan C is independent (unrelated sources), and plan D is dependent (common sources). □

Illustration 10-3

A multidivisional hospital wishes to compare length of stay for patients receiving a proposed new therapy with those receiving an existing one. Patients with specific symptoms from division A are paired with patients having the same symptoms from division B. Division A continues the existing therapy, and division B institutes the new one. The length of stay is compared.

This example illustrates dependent samples from unrelated sources. Although the subjects (patients) are unrelated, these samples are dependent because the elements (symptoms) have been paired. □

Independent and dependent cases each have their advantages, which will be emphasized later. Either method of sampling is often used.

Exercises

10-1 In trying to estimate the amount of economic growth that took place in a wealthy suburb, a marketing agency randomly selected 36 families from the population of 4,000 and recorded their incomes. One year later another set of 42 families and their incomes were randomly selected from the same population and recorded. Do the two sets of data (36 incomes, 42 incomes) represent dependent or independent samples? Explain.

10-2 Twenty people were selected to participate in a promotion experiment. The 20 answered a short multiple-choice quiz about their attitudes toward several products. They then viewed a 45-minute film presentation, which included several commercials. The following day the same 20 people were asked to answer a follow-up questionnaire about their attitudes toward the same products. At the completion of this experiment, the experimenter will have two sets of scores. Do these two samples represent dependent or independent samples? Explain.

10-3 Merit Stockbrokers was interested in studying the investment needs and preferences of young, single adults. The research department conducts an in-depth interview with a sample of 20 young, single men and 15 young, single women. After the interviewing is complete, there will be two sets of data: one from the 20 men and another from the 15 women. Do these samples represent dependent or independent samples? Explain.

10-4 Four hundred students are in the Philadelphia Summer Youth Employment Training Program. Describe how you would obtain two independent samples of size 25 from these 400 students to test a pretraining program skill against the same skill after completing the training program.

10-5 Describe how you would obtain your samples in exercise 10-4 if you were to use dependent samples.

Section 10-2 Inferences Concerning the Difference Between Two Independent Means (Variances Known or Large Samples)

When comparing the means of two populations, we typically consider the difference between their means, $\mu_1 - \mu_2$. The inferences to be made about $\mu_1 - \mu_2$ will be based on the difference between the observed sample means, $\bar{x}_1 - \bar{x}_2$. This observed difference belongs to a sampling distribution, the characteristics of which are described in the following statement:

independent means

If independent samples of sizes n_1 and n_2 are drawn randomly from large populations with means μ_1 and μ_2 and variances σ_1^2 and σ_2^2, respectively, **the sampling distribution of $\bar{x}_1 - \bar{x}_2$, the difference between the means,**

1. is approximately normally distributed
2. has a mean of $\mu_{\bar{x}_1 - \bar{x}_2} = \mu_1 - \mu_2$
3. has a standard error of $\sigma_{\bar{x}_1 - \bar{x}_2} = \sqrt{(\sigma_1^2/n_1) + (\sigma_2^2/n_2)}$

This normal approximation is good for all sample sizes if the populations involved are approximately normal. When nonnormal populations are involved, the approximation is good if the sample sizes (n_1 and n_2) are both larger than 30.

Since the sampling distribution is approximately normal, we will use the z **statistic** in our inferences. In the **hypothesis test** z will be determined by

$$z = \frac{(\bar{x}_1 - \bar{x}_2) - (\mu_1 - \mu_2)}{\sqrt{(\sigma_1^2/n_1) + (\sigma_2^2/n_2)}} \qquad (10\text{-}1)$$

if both σ_1 and σ_2 are known quantities.

Illustration 10-4

Distributors of products buy shelf space in supermarkets. Eye-level shelf space costs more than any other shelf level, since it is customarily believed that items at eye level are most visible and will lead to highest sales. The sales manager of Spott Paper, however, thinks the average sales of Spott products would be the same no matter whether he purchased the shelf below eye level or if he purchased eye-level shelf space. To test his claim, the research group at Spott selects two similar supermarkets and sets Spott towels on the eye-level shelf in one and on the next lower shelf in the other. A random sample of 40 customers passing the Spott shelf in each store is observed, and their purchases (if any) are recorded. The mean sales obtained were 2.03 for the 40 customers at the lower shelf (l) and 2.20 for the 40 customers at the eye-level shelf (e). Assume that the standard deviation of both populations is $\sigma = 0.6$. Complete a hypothesis test of the Spott manager's claim using $\alpha = 0.05$.

Solution

STEP 1: $H_0: \mu_l = \mu_e$ or $\mu_l - \mu_e = 0$ ($\geq$) (sales on lower shelf are not less).

STEP 2: $H_a: \mu_l < \mu_e$ or $u_l - \mu_e < 0$ (sales on lower shelf are less).

The null hypothesis is usually interpreted as being "there is no difference between the means," and therefore it is customarily expressed by $\mu_l - \mu_e = 0$.

STEP 3: The test statistic used will be z. The test criteria for $\alpha = 0.05$ will be as shown in the diagram.

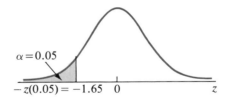

STEP 4: The formula for the test statistic is formula (10-1):

$$z = \frac{(\bar{x}_l - \bar{x}_e) - (\mu_l - \mu_e)}{\sqrt{(\sigma_l^2/n_l) + (\sigma_e^2/n_e)}}$$

$$= \frac{(2.03 - 2.20) - 0}{\sqrt{[(0.6)^2/40] + [(0.6)^2/40]}} = \frac{-0.17}{\sqrt{(0.36/40) + (0.36/40)}}$$

$$= \frac{-0.17}{\sqrt{0.009 + 0.009}} = \frac{-0.17}{\sqrt{0.018}} = \frac{-0.17}{0.134} = -1.269$$

$$z^* = -1.27$$

This value is compared with the test criteria in the next diagram.

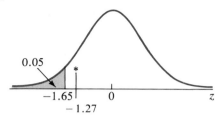

STEP 5: *Decision*: Fail to reject H_0 (z^* is in the noncritical region).

Conclusion: The claim that there is no difference in mean sales between the two shelf levels cannot be rejected. □

We often wish to estimate the difference between the means of two different populations. When independent samples are involved, we will use the information about the sampling distribution of $\bar{x}_1 - \bar{x}_2$ and the z statistic to construct our **confidence interval estimate for $\mu_1 - \mu_2$**.

$$(\bar{x}_1 - \bar{x}_2) - z(\alpha/2) \cdot \sqrt{\frac{\sigma_1^2}{n_1} + \frac{\sigma_2^2}{n_2}}$$

to (10-2)

$$(\bar{x}_1 - \bar{x}_2) + z(\alpha/2) \cdot \sqrt{\frac{\sigma_1^2}{n_1} + \frac{\sigma_2^2}{n_2}}$$

Illustration 10-5

Construct the 95% confidence interval estimate for the difference in the two independent means of illustration 10-4, mean sales from the eye-level shelf and mean sales from the lower shelf level. Use the sample values found in illustration 10-4.

Solution The given information is $\bar{x}_l = 2.03$, $\sigma_l = 0.6$, $n_l = 40$, $\bar{x}_e = 2.20$, $\sigma_e = 0.6$, $n_e = 40$, and $1 - \alpha = 0.95$. (See illustration 10-4.)

$$(\bar{x}_l - \bar{x}_e) \pm z(0.025) \cdot \sqrt{\frac{\sigma_l^2}{n_l} + \frac{\sigma_e^2}{n_e}}$$

$$(2.03 - 2.20) \pm (1.96) \cdot \sqrt{\frac{(0.6)^2}{40} + \frac{(0.6)^2}{40}}$$

$$(-0.17) \pm (1.96)(0.134)$$

$$-0.17 \pm 0.26$$

$$-0.43 \quad \text{to} \quad 0.09$$

This interval is the 0.95 confidence interval for $\mu_l - \mu_e$. That is, with 95% confidence we estimate the difference between the means to be between -0.43 and $+0.09$, or we are 95% confident that the true average sales for the lower shelf will be between 43¢ less and 9¢ more than sales from the eye-level shelf. □

As noted in previous chapters, the variance of a population is generally unknown. Therefore when we wish to make an inference about the mean, it is necessary to replace σ_1 and σ_2 with the best estimates available, namely s_1 and s_2. If both samples have sizes that exceed 30, we may replace σ_1 and σ_2 in formulas (10-1) and (10-2) with s_1 and s_2, respectively, without appreciably affecting our level of significance or confidence. Thus for inferences about the difference between two population means, based on independent samples where the σ's are **unknown** and both $n_1 > 30$ and $n_2 > 30$, we will use formula (10-3) for the calculation of the **test statistic in the hypothesis test**. For calculating the endpoints of the $1 - \alpha$ **confidence interval estimate**, we will use formula (10-4).

$$z = \frac{(\bar{x}_1 - \bar{x}_2) - (\mu_1 - \mu_2)}{\sqrt{(s_1^2/n_1) + (s_2^2/n_2)}} \qquad (10\text{-}3)$$

$$(\bar{x}_1 - \bar{x}_2) - z(\alpha/2) \cdot \sqrt{\frac{s_1^2}{n_1} + \frac{s_2^2}{n_2}}$$

to (10-4)

$$(\bar{x}_1 - \bar{x}_2) + z(\alpha/2) \cdot \sqrt{\frac{s_1^2}{n_1} + \frac{s_2^2}{n_2}}$$

Illustration 10-6

Two independent samples are taken to compare the means of two populations. The sample statistics are given in the accompanying table. Can we conclude that the mean of population A is greater than the mean of population B at the 0.02 level of significance?

	n	$\bar{x}$	s
Sample A	50	57.5	6.2
Sample B	60	54.4	10.6

Solution This problem calls for a hypothesis test for the difference of two independent means. Both n's are larger than 30; therefore formula (10-3) will be used to calculate z.

STEP 1: $H_0: \mu_A - \mu_B = 0 \ (\leq)$.

STEP 2: $H_a: \mu_A - \mu_B > 0$.

STEP 3: $\alpha = 0.02$. The test criteria are shown in the diagram.

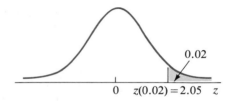

STEP 4: Using formula (10-3),

$$z = \frac{(57.5 - 54.4) - 0}{\sqrt{[(6.2)^2/50] + [(10.6)^2/60]}}$$

$$= \frac{3.1}{\sqrt{0.7688 + 1.8727}} = \frac{3.1}{1.625} = 1.908$$

$z^* = 1.91$

STEP 5: *Decision*: Fail to reject H_0 (z^* is in the noncritical region).

Conclusion: We do not have sufficient evidence to conclude that $\mu_A > \mu_B$.

NOTE: If the null hypothesis states a difference between the two means such as $\mu_A - \mu_B = 10$, the difference is used in formula (10-3). ☐

Illustration 10-7

Suppose that the samples given in illustration 10-6 were taken for the purpose of estimating the difference between the two population means. Construct the 0.99 confidence interval for the estimation of this difference.

Solution This estimation is based on the difference between the means of two independent samples whose sizes are both greater than 30. Therefore we will use formula (10-4).

$$(57.5 - 54.4) \pm 2.58 \cdot \sqrt{\frac{(6.2)^2}{50} + \frac{(10.6)^2}{60}}$$

$$3.1 \pm (2.58)(1.625)$$

$$3.1 \pm 4.19$$

$$-1.09 \quad \text{to} \quad 7.29$$

This interval is the 0.99 confidence interval for $\mu_A - \mu_B$. That is, our 99% confidence interval estimate for the difference between the two population means is -1.09 to 7.29. □

Exercises

10-6 Independent samples are taken from two normal populations, each of which has a variance of 900. Do the sample means shown in the following table provide sufficient evidence to reject the hypothesis that the means of these two populations are equal? Use $\alpha = 0.05$.

Sample	n	$\bar{x}$
A	30	35.6
B	35	38.3

(a) Solve using the classical approach.
(b) Solve using the prob-value approach.

10-7 The purchasing department for a regional supermarket chain is considering two sources of 10-pound bags of potatoes. A random sample is taken from each source, with the following results:

	Idaho Supers	Idaho Best
Number of bags weighed	100	100
Mean weight	10.2 lb	10.4 lb
Sample variance	0.36	0.25

At the 0.05 level of significance, is there a difference between the mean weights of the 10-pound bags of potatoes?

(a) Solve using the classical approach.
(b) Solve using the prob-value approach.

10-8 Use the sample information in exercise 10-6 to construct the 0.99 confidence interval estimate for the difference between the two population means.

10-9 An investment counselor is interested in estimating the difference between the average rate of return of load mutual funds (funds that have a sales fee) and no-load funds (funds that do not have a sales fee). She randomly selects 36 mutual funds of each type. The mean rate of returns were $\bar{x}_{NL} = 7.9$ (no load) and $\bar{x}_L = 8.5$ (load). Assume a variance for each type of mutual fund of 0.10. Construct a 98% confidence interval for the difference in the mean rates of return.

10-10 Does the sample information given in the following table provide sufficient evidence to support the contention that μ_A is not equal to μ_B? Use $\alpha = 0.10$.

Sample	n	$\bar{x}$	s^2
A	100	152.3	120
B	80	149.6	144

10-11 Small metal clips used in a furnace mechanism are known to have a mean weight of 0.6 ounce and a variance of 0.0004. Two random samples, one of 100 observations and the other of 80, are taken on two consecutive days. Assuming that the production process has not changed, what is the probability that the two sample means differ by:
(a) more than 0.002 ounce
(b) less than 0.0015 ounce

10-12 The purchasing agent for Star Tool Company has narrowed the purchase of a new automated lathe to two brands. Brand A is less costly than brand B, but brand B's manufacturer claims its output per hour is higher. A cost analysis has determined that if the average output per hour of brand B is no more than 2.5 greater than A, then A is more profitable; otherwise, B is more profitable. To determine which machine to buy, Star leased both machines for 1 week. The results of using the machines are given in the following table. Do these data provide sufficient evidence to warrant the purchase of brand B at the 0.01 level of significance?

Machine	Hours Run	Total Output, $\sum x$	$\sum (x - \bar{x})^2$
A	30	1,952	74.2
B	40	2,827	284.3

10-13 The sample statistics shown in the following table were obtained to estimate the difference in mean management fees of load and no-load mutual funds. Construct the 0.98 confidence interval estimate for the difference in average management fees of load and no-load mutual funds.

Type of Fund	n	$\bar{x}$ (× $1,000)	s (× $1,000)
Load	40	119.5	11.9
No load	50	173.8	18.9

10-14 Monarch Record Company markets its records by direct sales to the customer through television advertisement. It is planning to advertise in a new market, a major midwestern area. The market area is covered by two television stations. To determine which station to use the majority of the time, Monarch sets up the following experiment. It purchases various spot times on each station. It presents the same 1-minute commercial on each station, but gives different post office boxes to respond to each time so that it can trace the source of the sale to each commercial time slot and station. The results are shown in the following table. Use these sample data to construct the 95% confidence interval estimate for the difference in the average units sold from commercials on the two stations.

Station	Minutes of Commercial	Total Sales Units	$\sum (x - \bar{x})^2$
1	32	2,253	32,426
2	35	5,157	138,600

Section 10-3 Inferences Concerning Two Variances

When comparing two populations, it is quite natural that we compare their variances or standard deviations. The inferences concerning two population variances (or standard deviations) are much like those comparing means. We will study two kinds of inferences about the comparison of the variances of two populations: (1) the **hypothesis test** for the **equality of the two variances** and (2) the **estimation** of the **ratio of the two population variances**, σ_1^2/σ_2^2.

The soft drink bottling company discussed in section 9-3 is trying to decide whether to install a modern high-speed bottling machine. There are, of course, many concerns in making this decision. The variance in the amount of fill per bottle is one of them. In this respect the manufacturer of the new system contends that the variance in fills is no larger with the new machine than it was with the old. A hypothesis test for the equality of the two variances can be used to make a decision in this situation. The null hypothesis will be that the variance of the modern high-speed machine (m) is no larger than the variance of the present machine (p); that is, $\sigma_m^2 \leq \sigma_p^2$. The alternative hypothesis will then be $\sigma_m^2 > \sigma_p^2$.

$$H_0: \sigma_m^2 = \sigma_p^2 \ (\leq) \quad \text{or} \quad \frac{\sigma_m^2}{\sigma_p^2} = 1 \ (\leq)$$

$$H_a: \sigma_m^2 > \sigma_p^2, \quad \text{or} \quad \frac{\sigma_m^2}{\sigma_p^2} > 1$$

The test statistic that will be used in making a decision about the null hypothesis is F. The calculated value of F will be obtained by the following formula:

$$F = \frac{s_1^2}{s_2^2} \tag{10-5}$$

where s_1^2 and s_2^2 are the variances of two independent samples of sizes n_1 and n_2, respectively.

F distribution

When independent random samples are drawn from normal populations with equal variances, the ratio of the sample variances, s_1^2/s_2^2, will possess a probability distribution known as the *F distribution* (see figure 10-1).

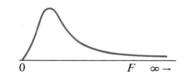

FIGURE 10-1
F Distribution

Properties of the F Distribution

1. F is nonnegative in value; it is 0 or positively valued.
2. F is nonsymmetrical; it is skewed to the right.
3. There are many F distributions, much like the t and χ^2 distributions. Each pair of degree-of-freedom values has a distribution.
4. The upper limit is infinity (the curve never touches the axis).

For the inferences discussed in this section, the degrees of freedom for each of the samples are $df_1 = n_1 - 1$ and $df_2 = n_2 - 1$.

The critical values for the F distribution may be obtained from tables 9a, 9b, and 9c in appendix D. Each critical value will be determined by three identification values: (1) df_n, the degrees of freedom associated with the sample whose variance is in the numerator of the calculated F, (2) df_d, the degrees of freedom associated with the sample whose variance is in the denominator, and (3) the area under the curve to the right of the critical value being sought. Therefore the symbolic name for a critical value of F will be $F(df_n, df_d, \alpha)$, as shown in figure 10-2.

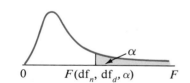

FIGURE 10-2
F Distribution Showing $F(df_n, df_d, \alpha)$

Table 9a in appendix D shows the critical values for $F(df_n, df_d, \alpha)$, where α is equal to 0.05; table 9b gives the critical values when $\alpha = 0.025$; table 9c gives values when $\alpha = 0.01$.

Illustration 10-8

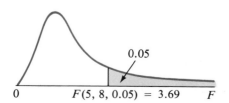

Find $F(5, 8, 0.05)$, the critical F value for samples of size 6 and 9 with 5% of the area in the right tail.

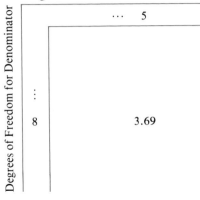

Solution From table 9a ($\alpha = 0.05$) we obtain the value shown in the table here. Therefore $F(5, 8, 0.05) = \mathbf{3.69}$.

Notice that $F(8, 5, 0.05)$ is 4.82. The degrees of freedom associated with the numerator and with the denominator must be kept in the correct order. (3.69 is quite different from 4.82; check some other pairs to verify this fact.)

NOTE: When the correct degrees of freedom are not found in the table, the critical value is found by using the linear interpolation technique (a technique that is not discussed in this book). For example, find $F(21, 31, 0.05)$. From table 9a we see that the actual value is somewhere between 1.79 and 1.93. Seldom will we need to interpolate on the hypothesis test problem. Record the interval in which the value lies and interpolate only when necessary.

Illustration 10-9

Recall that our soft drink bottling company was to make a decision about the equality of the variance of amounts of fill between its present machine and a modern high-speed outfit. Does the sample information given in the accompanying table present sufficient evidence to reject the manufacturer's claim that the modern high-speed bottle-filling machine fills bottles with no more variance than the company's present machine? Use $\alpha = 0.01$.

	n	s^2
Present machine	22	0.0008
Modern high-speed machine	25	0.0018

Solution

STEP 1: $H_0: \sigma_m^2 = \sigma_p^2$, or $\sigma_m^2/\sigma_p^2 = 1$ ($\leq$) (no more variance).

STEP 2: $H_a: \sigma_m^2 > \sigma_p^2$, or $\sigma_m^2/\sigma_p^2 > 1$ (more variance).

STEP 3: $\alpha = 0.01$. The test statistic to be used is F, since the null hypothesis is about the equality of the variances of two populations. The critical region is one-tailed and on the right because the alternative hypothesis says "greater than." $F(24, 21, 0.01)$ is the critical value. The number of degrees of freedom for the numerator is 24 ($25 - 1$) because the sample from the modern high-speed machine is associated with the numerator, as specified by the null hypothesis. $df_d = 21$ because the sample associated with the denominator has size 22. The critical value is found in table 9c and is 2.80. See the diagram.

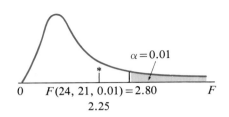

STEP 4:
$$F = \frac{s_m^2}{s_p^2} = \frac{0.0018}{0.00008}$$
$$= F^* = \mathbf{2.25}$$

This value is shown in the diagram.

STEP 5: *Decision*: Fail to reject H_0 (F^* is in the noncritical region).

Conclusion: The samples do not present sufficient evidence to reject the manufacturer's claim. □

Illustration 10-10

Does the sample information shown in the accompanying table provide sufficient evidence to reject the claim that the variance among stock prices on the New York Exchange is the same as the variance among stock prices on the American Exchange? Use $\alpha = 0.05$.

Exchange	n	s^2
American, A	16	105.4
New York, B	25	136.3

Solution

STEP 1: $H_0: \sigma_A^2 = \sigma_B^2$, or $\sigma_B^2/\sigma_A^2 = 1$.

STEP 2: $H_a: \sigma_A^2 \neq \sigma_B^2$, or $\sigma_B^2/\sigma_A^2 \neq 1$.

NOTE: The order in which the two variances are mentioned in the hypotheses is your choice. However, when a ratio is used in stating the hypotheses, then the same ordering should be used when calculating F. That is, if H_0 mentions σ_A^2/σ_B^2, then $F^* = s_A^2/s_B^2$.

STEP 3: This test is two-tailed with $\alpha = 0.05$.

NOTE: Tables showing the critical values for the F distribution give only the right-hand critical value. Since, however, F is nonnegative and nonsymmetrical (it is 0 or positive), we cannot obtain the left-hand critical value as we did with the t and z distributions. If the critical value for the left-hand tail is needed, we will obtain it by calculating the reciprocal of the related critical value obtained from the table. Expressed in formula form this value is

$$F(df_1, df_2, 1 - \alpha) = \frac{1}{F(df_2, df_1, \alpha)} \tag{10-6}$$

See figure 10-3. Notice that when the reciprocal is taken, the degrees of freedom for the numerator and the denominator are switched also. (Why do you suppose this switch is necessary?)

FIGURE 10-3
Finding the Critical Value for the Left Tail of the F Distribution

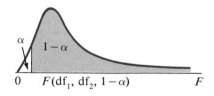
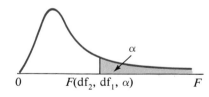

Returning to step 3, $F(24, 15, 0.025)$ is read directly from table 9b and is 2.70. $F(24, 15, 0.975)$ must be found by using formula (10-6). See the figure.

$$F(24, 15, 0.975) = \frac{1}{F(15, 24, 0.025)} = \frac{1}{2.44}$$

$$= 0.4098 = 0.410$$

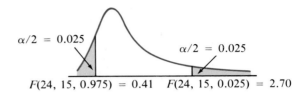

STEP 4:

$$F = \frac{s_B^2}{s_A^2} = \frac{136.3}{105.4} = 1.293$$

$$F^* = 1.29$$

See the diagram.

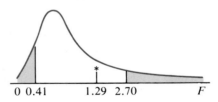

STEP 5: *Decision*: Fail to reject H_0 (F^* is in the noncritical region).

Conclusion: It appears that no real difference in the variance among prices on the New York and American Exchanges has been demonstrated. □

Optional Technique

When we complete a hypothesis test about the equality of two population variances, as discussed previously, it would be convenient if we could always use a right-hand critical value for F without the need to calculate the left-hand critical value. This can be accomplished by minor adjustments in the null hypothesis and in the calculation of F in step 4. The two cases we would like to change are (1) the two-tailed test and (2) the one-tailed test where the critical region is on the left. The one-tailed test with the critical region on the right already meets our criterion.

CASE 1: When a two-tailed test is to be completed, we will state the hypotheses in the normal way. The calculated value of F, F^*, will be the larger of s_1^2/s_2^2 or s_2^2/s_1^2. (One of these values will be between 0 and 1; the other will be larger than 1.) The critical value of F will be $F(df_n, df_d, \alpha/2)$, where df_n is the number of degrees of freedom for the sample whose variance is used in the numerator: df_d represents the degrees of freedom used in the denominator. Only the right-tail critical value will be needed. The test is completed in the usual fashion.

CASE 2: When a one-tailed test where the critical region is on the left is to be completed, we will interchange the position of the two variances in the statement of the hypotheses. This interchange will reverse the direction of the alternative hypothesis and put the critical region on the right. From this point on the procedure is the same as that for the one-tailed test with a critical region on the right.

ratio of variances

Even though the hypothesis test is the most commonly used inference about two variances, occasionally you may be asked to estimate the **ratio of two variances**, σ_A^2/σ_B^2. The best point estimate is s_A^2/s_B^2. The $1 - \alpha$ **confidence interval** for the estimate is constructed by using the following formula:

$$\frac{s_A^2/s_B^2}{F(df_A, df_B, \alpha/2)} \quad \text{to} \quad \frac{s_A^2/s_B^2}{F(df_A, df_B, 1 - \alpha/2)} \tag{10-7}$$

Recall that left-tail values of F are not in the tables. These left-tail values are found by using formula (10-6). Using formula (10-6), we can rewrite formula (10-7) as

$$\frac{s_A^2/s_B^2}{F(df_A, df_B, \alpha/2)} \quad \text{to} \quad \frac{s_A^2}{s_B^2} F(df_B, df_A, \alpha/2) \tag{10-8}$$

(Notice that when the reciprocal is used, the degrees of freedom for the numerator and denominator are also switched.)

ratio of standard deviation

If the **ratio of standard deviations** is desired, you need only take the positive square root of each of the bounds of the interval found by using formula (10-8).

Illustration 10-11

Suppose that the samples in illustration 10-10 had been taken for the purpose of estimating the ratio of the variance among stocks on the American Exchange to that on the New York Exchange. Construct the 90% confidence interval estimate for the ratio of the variances and the standard deviations.

Solution Use formula (10-8):

$$\frac{105.4/136.3}{F(15, 24, 0.05)} \quad \text{to} \quad \frac{105.4}{136.3} F(24, 15, 0.05)$$

$$\frac{0.77329}{2.11} \quad \text{to} \quad (0.77329)(2.29)$$

The 90% confidence interval for the ratio of variances is

$$0.37 \quad \text{to} \quad 1.77$$

The 90% confidence interval for the ratio of standard deviations is

$$0.61 \quad \text{to} \quad 1.33 \qquad \square$$

NOTE: The formula for estimating the ratio of population variances or population standard deviations requires the use of the ratio of sample variances. If the standard deviation of the sample is given, it must be squared to obtain the variance.

Exercises

10-15 What is the formula used to calculate the value of F? Explain it.

10-16 Using the $F(df_1, df_2, \alpha)$ notation, name each of the critical values shown in the following figures. (For two-tail cases, use $\alpha/2$ in each tail.)

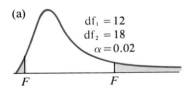

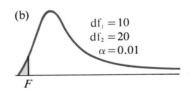

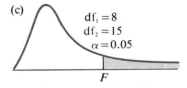

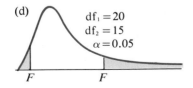

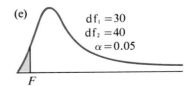

10-17 Find the following critical values for the F distribution from tables 9a, 9b, or 9c in appendix D.
(a) $F(20, 12, 0.05)$ (b) $F(24, 40, 0.01)$
(c) $F(12, 20, 0.05)$ (d) $F(5, 20, 0.025)$
(e) $F(15, 18, 0.01)$ (f) $F(5, 9, 0.05)$
(g) $F(40, 30, 0.01)$ (h) $F(8, 40, 0.025)$

10-18 Find the following critical values for F. (*Hint*: Use formula 10-6.)
(a) $F(20, 12, 0.95)$ (b) $F(5, 24, 0.975)$
(c) $F(18, 15, 0.99)$ (d) $F(7, 10, 0.95)$

10-19 An investment counselor feels one has to be more careful in selecting a growth-oriented mutual fund than an income fund. To prove his point, he samples 25 growth funds and 16 income funds and computes the variance among their rates of return for last year.

$$\text{Growth:} \quad n = 25, \quad s^2 = 82.3$$
$$\text{Income:} \quad n = 16, \quad s^2 = 34.7$$

Do the data provide sufficient evidence to support the investment counselor? Use $\alpha = 0.05$.
(a) Solve using the classical approach.
(b) Solve using the prob-value approach.

10-20 A bakery is considering buying one of two ovens. One of its requirements is that the temperature remains very constant during a baking operation. The variance in temperature before the thermostat restarted the flame for the Monarch oven in the study was 2.4 for 16 measurements. The variance for the Kraft oven was 3.2 for 12 measurements. Does this information provide sufficient reason to conclude that there is a difference in the variances for the two ovens? Use a 0.02 level of significance.
(a) Solve using the classical approach.
(b) Solve using the prob-value approach.

10-21 An economist claims that price competition is greater among discount department stores than other department stores. A sample of prices of a brand of shaving lotion in discount department stores and other department stores yielded these data:

$$\text{Discount:} \quad s = 9.0¢, \quad n = 36$$
$$\text{Others:} \quad s = 6.5¢, \quad n = 45$$

Do these statistics provide significant evidence to justify the economist's claim? Use $\alpha = 0.05$.

10-22 Assuming that the data in exercise 10-14 were collected to test the ratio of variances, do they provide sufficient reason to reject the null hypothesis that the variance in unit sales resulting from the commercials is the same for both stations? Use $\alpha = 0.05$.

10-23 Assuming that the sample statistics given in exercise 10-13 were gathered for the purpose of testing the populations' standard deviations, does the sample provide sufficient evidence to conclude that the variance of the management fees among load funds is less than the variance among no-load funds? Use $\alpha = 0.05$.

10-24 Use the sample information given in exercise 10-19 to complete the following:
(a) Construct the 0.95 confidence interval for the ratio of the variance of growth funds to the variance of income funds.
(b) Construct the 0.95 confidence interval for the ratio of the standard deviations.

10-25 Using the sample information about the management fees (exercise 10-13), construct the 0.98 confidence interval for estimating the ratio of the standard deviations, no-load to load funds.

10-26 Assuming that the data in exercise 10-14 were collected for the purpose of estimation, answer the following:
(a) What would be your point estimate for the ratio of the two variances?
(b) Construct the 95% confidence interval for the estimation of the ratio of the two variances.

Section 10-4 Inferences Concerning the Difference Between Two Independent Means (Variances Unknown and Small Samples)

In section 10-2 we treated the cases for inferences about the difference between the population means based on two independent samples where the samples were both large or the population variances were known. We will now investigate the inference procedures to be used in situations where two independent small samples (i.e., one or both samples are of size less than or equal to 30) are taken from approximately normal populations for the purpose of comparing their means. For these inferences we must use Student's t distribution. However, we must distinguish between two possible cases: (1) the variances of the two populations are equal, $\sigma_1^2 = \sigma_2^2$, or (2) the variances of the two populations are unequal, $\sigma_1^2 \neq \sigma_2^2$.

The two population variances are unknown; therefore we will employ the F test studied in the preceding section to determine whether we have case 1 or case 2. The two sample variances will be used in a two-tailed test of the null hypothesis $H_0: \sigma_1^2 = \sigma_2^2$ or $\sigma_1^2/\sigma_2^2 = 1$. If we fail to reject H_0, we will proceed with the methods for case 1. If we reject H_0, we will proceed with case 2.

CASE 1: The procedures here are very similar to those used when Student's t distribution was employed. The standard error of estimate must be estimated by

$$s_p \cdot \sqrt{\frac{1}{n_1} + \frac{1}{n_2}}$$

pooled estimate where s_p symbolizes the **pooled estimate for the standard deviation**. ("Pooled" means that the information from both samples is combined so as to give the best possible estimate.) The formula for s_p is

$$s_p = \sqrt{\frac{(n_1 - 1)s_1^2 + (n_2 - 1)s_2^2}{n_1 + n_2 - 2}} \qquad (10\text{-}9)$$

The number of degrees of freedom, df, is the sum of the number of degrees of freedom for the two samples, $(n_1 - 1) + (n_2 - 1)$; that is,

$$\text{df} = n_1 + n_2 - 2 \qquad (10\text{-}10)$$

With this information we can now write the formula for the **test statistic** t that will be used in a **hypothesis test**.

$$t = \frac{(\bar{x}_1 - \bar{x}_2) - (\mu_1 - \mu_2)}{s_p \sqrt{(1/n_1) + (1/n_2)}} \qquad (10\text{-}11)$$

with $n_1 + n_2 - 2$ degrees of freedom.

Illustration 10-12

Peton Learning markets executive development seminars primarily through direct mail advertisements. The sales department is interested in the following question: Is the sales response the same for a one-color advertisement brochure as for a two-color advertisement brochure? To answer this question, 25 mailings of 1,000 each were made with each of the different types of brochures. The results were as shown in the accompanying table. Does this evidence contradict the null hypothesis that the brochure coloring (one or two colors) makes no difference on the average response at $\alpha = 0.10$?

Brochure	Mailing, n	Response per Mailing, $\bar{x}$	s^2
1 color	25	10.22	33.95
2 color	25	10.55	24.47

Solution Since the σ's are unknown, we must first test the variances to determine whether we have case 1 or case 2.

STEP 1: $H_0: \sigma_1^2 = \sigma_2^2$, or $\sigma_1^2/\sigma_1^2 = 1$.

STEP 2: $H_a: \sigma_1^2 \neq \sigma_2^2$.

STEP 3: $\alpha = 0.10$. See the diagram for the test criteria.

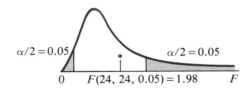

STEP 4: $F^* = 33.95/24.47 = 1.387$ (F^* is in the noncritical region).

STEP 5: *Decision*: Fail to reject H_0.

Conclusion: The test for the difference between means will be completed according to case 1 procedures, using t and formula (10-11).

STEP 1: $H_0: \mu_1 = \mu_2$, or $\mu_1 - \mu_2 = 0$.

STEP 2: $H_a: \mu_1 \neq \mu_2$.

STEP 3: $\alpha = 0.10$; df $= 25 + 25 - 2 = 48$; $t(48, 0.05) = 1.65$. See the diagram.

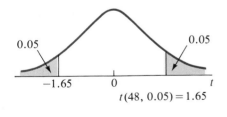

$$t = \frac{(\bar{x}_1 - \bar{x}_2) - (\mu_1 - \mu_2)}{s_p \cdot \sqrt{(1/n_1) + (1/n_2)}}$$

$$= \frac{(10.22 - 10.55) - (0)}{\sqrt{[(24)(33.95) + (24)(24.47)]/48} \cdot \sqrt{(1/25) + (1/25)}}$$

$$= \frac{-0.33}{\sqrt{29.21} \cdot \sqrt{0.08}} = \frac{-0.33}{\sqrt{2.3368}}$$

$$= \frac{-0.33}{1.52866} = -0.2158$$

$t^* = -0.22$

STEP 5: *Decision*: Fail to reject H_0 (t^* is in the noncritical region).

Conclusion: There appears to be no significant difference between the mean response per 1,000 mailed pieces for one-color and two-color brochures. □

Having completed our discussion of case 1, let's go on to case 2.

CASE 2: If we must assume that the two populations have **unequal variances**, then the hypothesis test for the difference between two independent means is completed by using the test statistic t. The **calculated value of t** is obtained by using

$$t = \frac{(\bar{x}_1 - \bar{x}_2) - (\mu_1 - \mu_2)}{\sqrt{(s_1^2/n_1) + (s_2^2/n_2)}} \tag{10-12}$$

with the number of degrees of freedom for the critical value being given by the smaller of $n_1 - 1$ or $n_2 - 1$.

Illustration 10-13

An investment banker feels that the commission (measured as a percentage of the total underwriting) for underwriting new stock issues is less if the corporation's headquarters are located in the middle Atlantic states than if it is located elsewhere. (*Note*: Commissions are negotiated between the company and the broker selling the stock.) To test this belief, samples of offerings were taken and the commissions recorded, with the results as shown in the following table:

Location of Company	n	$\bar{x}$	s
Middle Atlantic, A	10	5.38	1.59
Elsewhere, B	12	5.92	0.83

Does this evidence contradict the hypothesis that the mean commission on underwritings of middle Atlantic corporations is less than the mean commission for corporations located elsewhere? Use $\alpha = 0.05$.

Solution First we test for the equality of the variances.

STEP 1: $H_0: \sigma_A^2 = \sigma_B^2$, or $\sigma_A^2/\sigma_B^2 = 1$.

STEP 2: $H_a: \sigma_A^2 \neq \sigma_B^2$.

STEP 3: $\alpha = 0.05$. See the diagram.

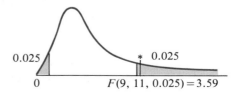

STEP 4: $F^* = (1.59)^2/(0.83)^2 = 3.67$.

STEP 5: *Decision*: Reject H_0.

Conclusion: Assume that $\sigma_A^2 \neq \sigma_B^2$ with $\alpha = 0.05$.

Now we are ready to test the difference between the means by use of case 2 procedures.

STEP 1: $H_0: \mu_A = \mu_B$, or $\mu_A - \mu_B = 0$ ($\geq$).

STEP 2: $H_a: \mu_A < \mu_B$, or $\mu_A - \mu_B < 0$.

STEP 3: df $= 10 - 1 = 9$ (smaller sample is of size 10); $\alpha = 0.05$. See the diagram.

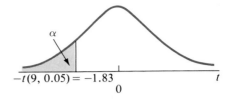

STEP 4:

$$t = \frac{(5.38 - 5.92) - (0)}{\sqrt{[(1.59)^2/10] + [(0.83)^2/12]}}$$

$$= \frac{-0.54}{\sqrt{0.2528 + 0.0574}} = \frac{-0.54}{0.557} = -0.969$$

$t^* = -0.97$.

STEP 5: *Decision*: Fail to reject H_0 (t^* is in the noncritical region).

Conclusion: There is not sufficient evidence shown by these samples to conclude that the mean commission is less for companies located in middle Atlantic states than for companies located elsewhere.

NOTE: However, we could conclude (see step 5) that the variance of spread of commission for companies located in the middle Atlantic region was greater than that for companies located elsewhere. Perhaps the investment broker had heard of some of the commissions on the very low side for companies in the middle Atlantic region and had too quickly generalized. □

If confidence interval estimates are to be constructed to estimate the difference between the means of two populations using the means of two independent small samples, the F test will again be used to determine which of the following two formulas is to be used:

CASE 1: σ's unknown but assumed equal. The $1 - \alpha$ confidence interval is given by

$$(\bar{x}_1 - \bar{x}_2) - t(\text{df}, \alpha/2) \cdot s_p \cdot \sqrt{\frac{1}{n_1} + \frac{1}{n_2}}$$

to (10-13)

$$(\bar{x}_1 - \bar{x}_2) + t(\text{df}, \alpha/2) \cdot s_p \cdot \sqrt{\frac{1}{n_1} + \frac{1}{n_2}}$$

where df $= n_1 + n_2 - 2$ and s_p is the pooled estimate for the standard deviation found by using formula (10-9).

CASE 2: σ's unknown but assumed unequal. The $1 - \alpha$ confidence interval is given by

$$(\bar{x}_1 - \bar{x}_2) - t(\text{df}, \alpha/2) \cdot \sqrt{\frac{s_1^2}{n_1} + \frac{s_2^2}{n_2}}$$

to (10-14)

$$(\bar{x}_1 - \bar{x}_2) + t(\text{df}, \alpha/2) \cdot \sqrt{\frac{s_1^2}{n_1} + \frac{s_2^2}{n_2}}$$

where df is the smaller of $n_1 - 1$ or $n_2 - 1$.

Illustration 10-14

The personnel department at Eton Corporation is worried that the employment test used for choosing management trainees may have a different (and damaging) effect on females than on males and possibly may be in violation of the equal employment laws. To study the problem, a sample of 20 female applicants and 30 male applicants was randomly selected. Find (a) a point estimate and (b) a 95% confidence interval for the difference between the means. The following table gives the results of the samples:

	Applicants	Employment Mean Test Score	s
Female	20	63.8	2.18
Male	30	69.8	1.92

Solution

(a) The point estimate for $\mu_m - \mu_f$ is 6.0, the difference between the two observed sample means (69.8 − 63.8).

(b) First we must decide whether we should assume that $\sigma_f^2 = \sigma_m^2$ or $\sigma_f^2 \neq \sigma_m^2$.

$$H_0: \sigma_f^2 = \sigma_m^2 \quad \text{or} \quad \frac{\sigma_f^2}{\sigma_m^2} = 1$$

$$H_a: \sigma_f^2 \neq \sigma_m^2$$

$\alpha = 0.05$; see the diagram for the test criteria.

$$F^* = \frac{(2.18)^2}{(1.92)^2} = 1.289$$

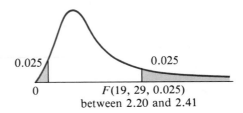

F(19, 29, 0.025) between 2.20 and 2.41

So we fail to reject H_0. Therefore we assume that $\sigma_f^2 = \sigma_m^2$. The 0.95 confidence interval is then found by using formula (10-13).

$$6.0 \pm t(48, 0.025) \cdot \sqrt{\frac{19(2.18)^2 + 29(1.92)^2}{48}} \cdot \sqrt{\frac{1}{20} + \frac{1}{30}}$$

$6.0 \pm 1.96\sqrt{4.1084}\sqrt{0.0833}$

$6.0 \pm (1.96)(2.0269)(0.2886)$

6.0 ± 1.1466

6.0 ± 1.15

4.85 to 7.15

This interval is the 0.95 confidence interval for $\mu_m - \mu_f$. (No specific illustration is shown for estimation where $\sigma_1^2 \neq \sigma_2^2$. All parts of such a solution can be found in various illustrations in this chapter.) □

Exercises

10-27 Do the sample data shown in the following table support the hypothesis that the two population means are significantly different? Use $\alpha = 0.01$. Assume that the population variances are equal.

Sample	n	$\bar{x}$	s^2
1	12	74.6	8.2
2	15	78.1	7.1

(a) Solve using the classical approach.
(b) Solve using the prob-value approach.

10-28 If a random sample of 18 homes south of Center Street in Provo showed a mean selling price of $15,000 and a variance of 2,400, and a random sample of 18 homes north of Center Street had a mean selling price of $16,000 and a variance of 4,800, can you conclude that there is a significant difference between the mean selling prices of homes in these two areas of Provo at the 0.05 level?

(a) Solve using the classical approach.
(b) Solve using the prob-value approach.

10-29 The results from two independent samples were as shown in the following table. Does this information provide sufficient reason to reject the null hypothesis in favor of the claim that the mean of population R is significantly larger than the mean of population S? Use $\alpha = 0.05$. (Remember to perform the F test on the variances first.)

Sample	n	$\sum x$	$\sum (x - \bar{x})^2$
R	8	295	70
S	6	195	80

10-30 Two independent samples of 10 housewives were asked to rate, on a scale from 0 to 10 (10 being highest), two brands of dishwasher detergent, with the hope of showing that the population mean rating for brand A is larger than the mean of brand B. Do the samples (see the accompanying table) provide significant evidence to justify that hope? Use $\alpha = 0.05$. (Assume the populations are normally distributed.)

Brand A	6	7	7	6	6	5	6	8	5	4
Brand B	8	3	5	3	4	5	7	6	2	5

(a) Solve using the classical approach.
(b) Solve using the prob-value approach.

10-31   Independent samples were taken from each of two normal populations to compare the two population means. The sample data are summarized in the accompanying table. Assume that $\sigma_1 \neq \sigma_2$.

Sample	n	$\bar{x}$	s^2
1	10	12.5	6.8
2	15	13.9	23.5

(a)   What effect does the assumption $\sigma_1 \neq \sigma_2$ have on the situation?

(b)   Determine the number of degrees of freedom to be used in comparing μ_1 and μ_2.

(c)   Is there sufficient reason to reject the hypothesis that $\mu_1 = \mu_2$ at $\alpha = 0.05$?

10-32   Certain "adjustments" are made between the test for the equality of two population means when the variances are unknown and judged to be equal and the test for equality of two population means when variances are unknown and judged to be unequal. (The t distribution is used in both cases.) Explain how each of the following adjustments is accomplished:

(a)   The test statistic is modified.

(b)   The critical region is modified.

(c)   The number of degrees of freedom is modified.

Section 10-5 Inferences Concerning Two Dependent Means

The procedure for comparing the means of two populations based on dependent samples is quite different from that of comparing the means of two populations based on two independent samples. When comparing two independent means, we make inferences about the two population means by using the difference between the two observed sample means. However, because of the relationship between two dependent samples, we will actually compare each piece of data of the first sample with the data in the second sample that came from the same source. The two data, one from each set, that come from the same source are often referred to as being **paired**. These pairs of data are compared by using the difference in their numerical values. This difference is called a **paired difference**. The two dependent means are then compared by using the observed mean of the resulting paired differences. The concept of using paired data for this purpose has a built-in ability to remove many otherwise uncontrollable factors. The tire wear problem (illustration 10-2) is an excellent example of such additional factors. The wearing ability of a tire is greatly affected by a multitude of factors: the size and weight of the car, the age and condition of the car, the driving habits of the driver, the number of miles driven, the condition of and types of roads driven on, the quality of the material used to make the tire, and so on.

paired difference

Illustration 10-15

If we were to test the wearing quality of two tire brands by plan D, as described in illustration 10-2, all the factors above will have an equal effect on both tire brands. The accompanying chart shows the amount of wear (in thousandths of an inch) that took place in such a test. One tire of each brand was placed on each of the six test cars. The position was determined with the aid of a random number table.

Car	1	2	3	4	5	6
Brand A	125	64	94	38	90	106
Brand B	133	65	103	37	102	115

Since the various cars, drivers, and conditions are the same for each paired set of data, it would make sense to introduce a new measure, the paired difference d, that was observed in each pair of related data. Therefore we add a third row to the chart, $d = $ brand $B - $ brand A.

Car	1	2	3	4	5	6
$d = B - A$	8	1	9	-1	12	9

Do the sample data provide sufficient evidence for us to conclude that the two brands show unequal wear at the 0.05 level of significance? ☐

Before we can solve the problem posed in this illustration, we need some new formulas. Our two sets of sample data have been combined into one set, a set of n values of d, $d = x_1 - x_2$. The sample statistics that we will use are $\bar{d}$, the observed **mean** value of d,

$$\bar{d} = \frac{\sum d}{n} \qquad (10\text{-}15)$$

and s_d, the observed **standard deviation** of the d's,

$$s_d = \sqrt{\frac{n(\sum d^2) - (\sum d)^2}{n(n-1)}} \qquad (10\text{-}16)$$

The observed values of d in our sample are from a population of **paired differences**, whose distribution is assumed to be approximately normal with a mean value of μ_d and a standard deviation of σ_d. Since σ_d is unknown, it will be estimated by s_d. The inferences about μ_d are completed in the same manner as the inferences about μ in chapter 8 and in section

9-1, by using $\bar{d}$ as the point estimate of μ_d. The inferences are completed by using the t distribution and $s_d/\sqrt{n}$ as the approximate measure for the standard error.

$$t = \frac{\bar{d} - \mu_d}{s_d/\sqrt{n}} \qquad (10\text{-}17)$$

with $n - 1$ degrees of freedom.

Now let's answer the question posed in illustration 10-15 (p. 368).

Solution

mean difference STEP 1: H_0: μ_d (**mean difference**) $= 0$.

STEP 2: H_a: $\mu_d \neq 0$.

STEP 3: $\alpha = 0.05$. See the diagram for the test criteria.

NOTE: n is the number of paired differences (d). df $= n - 1$, and $t(5, 0.025)$ is obtained from table 7, appendix D.

STEP 4:

$$t = \frac{\bar{d} - \mu_d}{s_d/\sqrt{n}}$$

First we must find $\bar{d}$ and s_d; some preliminary calculations are shown in the following table:

d	d^2
8	64
1	1
9	81
-1	1
12	144
9	81
38	372

$$\bar{d} = \frac{\sum d}{n} = \frac{38}{6} = 6.333 = \mathbf{6.3}$$

$$s_d = \sqrt{\frac{(6)(372) - (38)(38)}{(6)(5)}} = \sqrt{26.27} = 5.13 = 5.1$$

Therefore

$$t = \frac{6.3 - 0}{5.1/\sqrt{6}} = \frac{(6.3) \cdot \sqrt{6}}{5.1} = \frac{(6.3)(2.45)}{5.1} = 3.026$$

$$t^* = \mathbf{3.03}$$

See the diagram.

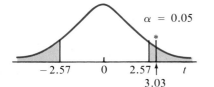

STEP 5: *Decision*: Reject H_0 (t^* is in the critical region).

Conclusion: There is a significant difference in the mean amount of wear.

Since the calculated value of t fell in the critical region on the right, we might want to conclude that one brand of these tires was better than the other. This conclusion is possible, but we must be extremely careful to interpret it correctly. Recall that the recorded data concerned the amount of wear. Therefore the tire with the lesser amount of wear should be the better one. Our hypothesis test showed that tire brand B had significantly more wear, thereby implying that brand A is the longer-wearing tire. □

It is also possible to estimate the mean difference in paired data. The formula used for this **confidence interval** estimate is

$$\bar{d} - t(dt, \alpha/2) \cdot \frac{s_d}{\sqrt{n}} \quad \text{to} \quad \bar{d} + t(df, \alpha/2) \cdot \frac{s_d}{\sqrt{n}} \quad (10\text{-}18)$$

Illustration 10-16

Construct the 95% confidence interval for the estimation of the mean difference in the paired data on tire wear, as found in illustration 10-16. The given information is $n = 6$ pieces of paired data, $\bar{d} = 6.3$, and $s_d = 5.1$.

Solution

$$\bar{d} \pm t(df, \alpha/2) \cdot \frac{s_d}{\sqrt{n}}$$

$$6.3 \pm 2.57 \cdot \frac{5.1}{\sqrt{6}}$$

$$6.3 \pm 5.4$$

$$0.9 \quad \text{to} \quad 11.7$$

This interval is the 0.95 confidence interval for $\mu_{d=B-A}$. That is, with 95% confidence we can say that the mean difference in the amount of wear is between 0.9 and 11.7.

NOTE: This confidence interval is quite wide, which is due, in part, to the small sample size. Recall from the central limit theorem that as the sample size increases, the standard error (estimated by $s_d/\sqrt{n}$) decreases. □

Exercises

10-33 Dependent samples were taken to compare the means of two populations. The sample statistics found were

$$n = 25, \quad \bar{d} = 3.3, \quad s_d = 11.8$$

Do we have sufficient evidence to reject the null hypothesis that the mean difference is 0 at the $\alpha = 0.05$ level of significance?

(a) Solve using the classical approach.
(b) Solve using the prob-value approach.

10-34 Dependent samples were obtained to test the claim that the mean difference is greater than 0. The data are summarized by

$$n = 14, \quad \sum d = 36, \quad \sum d^2 = 152$$

Do the data support the claim? Use $\alpha = 0.01$.

(a) Solve using the classical approach.
(b) Solve using the prob-value approach.

10-35 We want to know which of two types of filters should be used over an oscilloscope to help the operator pick out the image of the cathode ray tube. A test was designed in which the strength of a signal could be varied from 0 up to the point where the operator first detects the image. At this point the intensity setting is read. The lower the setting, the better the filter. Twenty operators were asked to make one reading for each filter. Does the following data show a significant difference in the filters at the 0.10 level?

Operator	Filter 1	Filter 2	Operator	Filter 1	Filter 2	Operator	Filter 1	Filter 2
1	96	92	8	91	90	15	90	89
2	83	84	9	100	93	16	92	90
3	97	92	10	92	90	17	91	90
4	93	90	11	88	88	18	78	80
5	99	93	12	89	89	19	77	80
6	95	91	13	85	86	20	93	90
7	97	92	14	94	91			

(a) Solve using the classical approach.
(b) Solve using the prob-value approach.

10-36 Eston Corporation's legal department is concerned that female employees are not advancing as fast as males are and that the firm might be cited by the OFCC (Office of Federal Contract Compliance). To see if a potential liability exists, 36 pairs of males and females were selected such that each pair, male and female, was hired at the same time at the same rate of pay. The data collected are summarized by

$$n = 36, \quad \sum d = 270, \quad \sum d^2 = 6{,}220$$

where d is the difference, in hundreds of dollars, between the current salary of the male and female employees (male salary minus female salary). Does this sample provide sufficient reason to conclude that female employees are not advancing as fast as males are? Use $\alpha = 0.01$.

10-37   Construct the 95% confidence interval estimate for the mean difference based on the sample information given in exercise 10-34.

10-38   A shoe manufacturer has developed a new artificial leather heel for its shoes and would like to test the new material against leather in hopes of showing that the new material is longer wearing than leather. Forty-eight people are each outfitted with a pair of shoes that have one leather heel and one heel of the new material. After 3 months of normal wear, the following summary was reported for the data collected:

$$n = 48, \quad \sum d = 88, \quad \sum (d - \bar{d})^2 = 373.65$$

where d is defined as the amount of wear on the leather heel minus the amount of wear on the new material. Can you conclude that this sample evidence indicates that the new material does wear longer? Use $\alpha = 0.05$.

10-39   To evaluate the effectiveness of a "zero-defective" campaign, 12 employees were randomly selected. One hundred items produced by each employee were randomly selected and inspected for defects. Each employee was then given a lecture, shown a film, and given reading matter, all stressing the importance of making no mistakes in production (i.e., produce zero defectives). Another 100 items produced by each employee after the training were randomly selected and the number of defective items recorded. The results are shown in the following table. Construct a 90% confidence interval estimate for the average reduction in the number of defective items produced per 100 items.

	Employee											
	1	2	3	4	5	6	7	8	9	10	11	12
Before	10	13	18	12	9	8	14	12	17	20	7	11
After	5	9	13	17	4	5	11	14	13	18	7	12

Section 10-6 Inferences Concerning Two Proportions

We often are interested in making statistical comparisons between the proportions, percentages, or probabilities associated with two populations. Such questions as the following are frequently asked: Is the proportion of mortgages that will default higher in one area of the city than in another? Is the percentage of items not meeting production specification higher for company A than for company B? Do employees have a different opinion about unionization under a participative management style than under a paternalistic management style? You can probably see the many types of questions involved. In this section we will compare two population proportions by using the difference between the observed proportions, $p'_1 - p'_2$, of two independent samples.

RECALL:

1. The observed probability is $p' = x/n$, where x is the number of observed successes in n trials.
2. $q' = 1 - p'$.
3. p is the probability of success for an individual trial in a binomial probability experiment of n repeated independent trials. (See p. 192.)

The sampling distribution of $p'_1 - p'_2$ is approximately normally distributed with a mean $\mu_{p'_1 - p'_2} = p_1 - p_2$ and with a standard error of

$$\sqrt{\frac{p_1 q_1}{n_1} + \frac{p_2 q_2}{n_2}}$$

if $(n_1 + n_2) > 20$, $(n_1 + n_2)p > 5$, and $(n_1 + n_2)q > 5$. Therefore when the null hypothesis that there is no difference between two proportions is to be tested, the **test statistic** will be z. The calculated value of z will be determined by formula (10-19):

$$z = \frac{p'_1 - p'_2}{\sqrt{pq[(1/n_1) + (1/n_2)]}} \tag{10-19}$$

NOTES:

1. The null hypothesis is $p_1 = p_2$, or $p_1 - p_2 = 0$.
2. The numerator of formula (10-19) written in the usual manner is $(p'_1 - p'_2) - (p_1 - p_2)$; however, this simplifies to $p'_1 - p'_2$, since the null hypothesis is $p_1 - p_2 = 0$.
3. Since the null hypothesis is $p_1 = p_2$, the standard error of $p_1 - p_2$ can be written as $\sqrt{pq[(1/n_1) + (1/n_2)]}$.

Illustration 10-17

A salesman for a new manufacturer of walkie-talkies claims that the percentage of defective walkie-talkies found among his products will be no higher than the percentage of defectives found in a competitor's line. To test the salesman's statement, random samples were taken of each manufacturer's product. The sample summaries are shown in the following table. Can we reject the salesman's claim at the 0.05 level of significance?

Sample	No. Defective	No. Checked
Salesman's, 1	8	100
Competitor's, 2	2	100

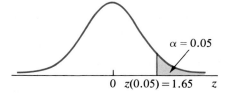

Solution

STEP 1: $H_0: p_1 - p_2 = 0 \;(\leq)$ (no higher than).

STEP 2: $H_a: p_1 - p_2 > 0$ (higher than).

STEP 3: $\alpha = 0.05$. The test criteria are shown in the diagram.

STEP 4: $p'_1 = 8/100 = 0.08$; $p'_2 = 2/100 = 0.02$.

Since the values of p_1 and p_2 are unknown but assumed equal (H_0), the best estimate we have for p ($p = p_1 = p_2$) is obtained by pooling the two samples:

$$p^* = \frac{x_1 + x_2}{n_1 + n_2} \qquad (10\text{-}20)$$

Our pooled estimate for p, p^*, then becomes

$$p^* = \frac{8 + 2}{100 + 100} = \frac{10}{200} = 0.05$$

$$q^* = 1 - p^* = 1 - 0.05 = 0.95$$

The value of the test statistic z is calculated by using formulas (10-19) and (10-20) together. p^* and q^* are used as estimates for p and q.

$$z = \frac{p'_1 - p'_2}{\sqrt{p^* q^* [(1/n_1) + (1/n_2)]}}$$

$$= \frac{0.08 - 0.02}{\sqrt{(0.05)(0.95)[(1/100) + (1/100)]}}$$

$$= \frac{0.06}{\sqrt{(0.05)(0.95)(0.02)}} = \frac{0.06}{0.031} = 1.9354$$

$$z^* = 1.94$$

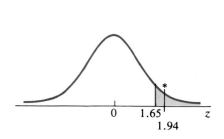

See the diagram.

STEP 5: *Decision*: Reject H_0 (z^* is in the critical region).

Conclusion: There is sufficient evidence to reject the salesman's claim. □

Illustration 10-18

The Equal Opportunity Commission's guidelines (equivalent to law) state that if an employment test has a disparate impact upon a protected class (i.e., any racial minority group, any ethnic group, females, or older workers), its use is in violation of the law unless it can be demonstrated that the test is a valid predictor of job performance. A test is said to have a disparate impact if the passing rate for a protected class (or, equivalently, the failure rate) is statistically significantly different from that of a nonprotected class at the 0.05 level. Elon Corporation is considering the use of a certain test to screen applicants for sales positions. It is worried, however, that the test may have a disparate impact upon blacks and thus be in possible violation of the law. Before instituting the employment test, it gives the test to a random sample of 500 blacks and 600 whites. The resulting sample statistics are shown in the accompanying table. Should the company use the test? Use $\alpha = 0.05$.

Race of Applicant	n	Number Failing Test
Blacks, b	500	30
Whites, w	600	28

Solution

STEP 1: $H_0: p_b = p_w$, or $p_b - p_w = 0$ (proportion of failures is same for both races).

STEP 2: $H_a: p_b \neq p_w$, or $p_b - p_w \neq 0$ (test has disparate effect).

STEP 3: $\alpha = 0.05$. The test criteria are shown in the diagram.

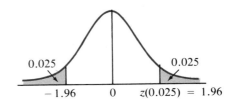

STEP 4:

$$z = \frac{p'_b - p'_w}{\sqrt{p^* q^* [(1/n_b) + (1/n_w)]}}$$

$p'_b = \dfrac{30}{500} = 0.060$ $p'_w = \dfrac{28}{600} = 0.04667 = 0.047$

The pooled estimate for p is

$$p^* = \frac{30 + 28}{500 + 600} = \frac{58}{1{,}100} = 0.0527 = \mathbf{0.053}$$

and thus $q^* = 1 - 0.053 = \mathbf{0.947}$. Therefore

$$z = \frac{0.060 - 0.047}{\sqrt{(0.053)(0.947)(1/500 + 1/600)}}$$

$$= \frac{0.013}{\sqrt{0.05019(0.002 + 0.0017)}}$$

$$= \frac{0.013}{\sqrt{0.000186}} = \frac{0.013}{0.0136} = 0.956$$

$$z^* = \mathbf{0.96}$$

STEP 5: *Decision*: Fail to reject H_0 (z^* is in the noncritical region).

Conclusion: There is no evidence to indicate that the probability of a black applicant failing is different from that of a white (i.e., no evidence of a disparate effect) at $\alpha = 0.05$. ☐

To estimate the difference between the proportions of two populations, you should use the observed difference $(p'_1 - p'_2)$ as the point estimate. The $1 - \alpha$ **confidence interval** is given by

$$(p'_1 - p'_2) - z(\alpha/2) \cdot \sqrt{\frac{p'_1 q'_1}{n_1} + \frac{p'_2 q'_2}{n_2}}$$

to (10-21)

$$(p'_1 - p'_2) + z(\alpha/2) \cdot \sqrt{\frac{p'_1 q'_1}{n_1} + \frac{p'_2 q'_2}{n_2}}$$

Illustration 10-19

In studying his campaign plans for union president of the United Metal Workers, Mr. Stanski wishes to estimate the difference in his voter appeal among craft and noncraft union employees. He asks his campaign manager to take two samples and find the 99% confidence interval estimate of the difference. A sample of 1,000 voters was taken from each population, with 388 craft and 459 noncraft employees favoring Stanski.

Solution The campaign manager calculated the confidence interval estimate by using formula (10-21), as follows:

$$p'_c = \frac{388}{1{,}000} = 0.388 \qquad p'_n = \frac{459}{1{,}000} = 0.459$$

$$(0.459 - 0.388) \pm 2.58 \sqrt{\frac{(0.459)(0.541)}{1{,}000} + \frac{(0.388)(0.612)}{1{,}000}}$$

$$0.071 \pm 2.58 \sqrt{0.000248 + 0.000237}$$

$$0.071 \pm 2.58 \sqrt{0.000485}$$

$$0.071 \pm (2.58)(0.022)$$

$$0.071 \pm 0.057$$

$$\mathbf{0.014 \quad to \quad 0.128}$$

This interval is the 0.99 confidence interval for $p_n - p_c$. That is, with 99% confidence we can say that the proportion of noncraftworkers who favor Mr. Stanski will exceed the proportion of craft workers who favor Mr. Stanski by between 1.4 and 12.8 percent. □

Exercises

10-40 Two manufacturing firms, producing equivalent products, claim the same rate of defectives for their products. A random sample of each product shows 14 of 300 and 25 of 400 to be defective for firms A and B, respectively. Does this evidence indicate a significant difference in the proportion of defectives? Use $\alpha = 0.05$.

(a) Solve using the classical approach.

(b) Solve using the prob-value approach.

10-41 Two randomly selected groups of citizens were exposed to different media campaigns dealing with the image of a political candidate. One week later they were surveyed to see if they would vote for the candidate. The results were:

	Exposed to Conservative Image	Exposed to Moderate Image
Number in sample	100	100
Proportion saying they would vote for the candidate	0.40	0.50

Is there sufficient evidence to show a difference in the effectiveness of the two image campaigns at the 0.05 level of significance?

(a) Solve using the classical approach.

(b) Solve using the prob-value approach.

10-42 Two areas are sampled to determine the interest in a new product. In area I a sample of 50 people indicated that 40% were interested in the product. In area II a sample of 100 people indicated that 30% were interested in the new product. Is there a significant difference in the percentage of people interested in the product in each area? Use $\alpha = 0.10$.

(a) Solve using the classical approach.

(b) Solve using the prob-value approach.

10-43 In a random sample of 40 brown-haired individuals, 22 indicated that they used hair coloring. In another random sample of 40 blonde-haired individuals, 26 indicated they used hair coloring. Use a 0.92 confidence interval to estimate the difference in the proportion of these groups that use hair coloring.

10-44 In a survey exploring the relationship between successful managers and management orientation, 22 of 71 successful managers were familiar with the managerial grid, whereas 23 of 91 unsuccessful managers were familiar with the managerial grid. Can the null hypothesis, "there is no difference in the proportions of these groups who were familiar with the managerial grid," be rejected at the $\alpha = 0.05$ level of significance?

10-45 In a survey of 300 people who watch the ABC news, 128 were in the age group at which the marketing effort of a product line is directed. In a survey of the NBC news, 149 of 400 people were in the target age group. Find the 98% confidence interval estimate for the difference in the two proportions.

10-46 A random sample of 100 stocks on the New York Exchange showed that 31 made a gain today. A random sample of 100 stocks on the American Stock Exchange showed 29 stocks making a gain.

(a) Construct the 99% confidence interval estimating the difference in the proportion of stocks making a gain.

(b) Does the answer to part (a) suggest that there is a significant difference between the proportions of stocks making gains on the two different stock exchanges?

In Retrospect

In this chapter we began the comparisons of two populations by first distinguishing between independent and dependent samples, which are statistically very important and useful sampling procedures. We then proceeded to examine the inferences concerning the comparison of means, variances, and proportions for two populations.

The use of hypothesis testing and interval estimates can sometimes be interchanged—that is, the calculation of a confidence interval can often be used in place of a hypothesis test. For example, in illustration 10–19 a confidence interval estimate was called for. Now suppose that Mr. Stanski asked: "Is there a difference in my voter appeal between craft and noncraft employees?" To answer his question you would not need to calculate a test statistic if you chose to test at $\alpha = 0.01$ with a two-tailed test. "No difference" would mean a difference of 0, which is not included in the interval from 0.014 to 0.128 (the interval determined in illustration 10-19). Therefore a null hypothesis of "no difference" would be rejected, thereby substantiating the conclusion that there is a significant difference in voter appeal between the two groups.

The comparison of two populations is a common occurrence. We often compare two means or two proportions and probably don't even recognize it. Look back at the Sears advertisement at the beginning of chapter 8. That gas-saving story is actually a comparison of two population means: the difference between the mean number of miles per gallon obtained by using two different types of tires. The article at the beginning of this chapter concluded, "there is no significant difference at the 0.05 level...."

For convenience table 10-1 identifies the formulas to use when making inferences about comparisons between two populations.

In chapters 8 through 10, we have introduced and completed hypothesis testing and confidence interval estimation for questions dealing

TABLE 10-1
Formulas to Use for Inferences Involving Two Populations

	Test Statistic	Formula to Be Used	
		Hypothesis Test	Interval Estimate
Two independent means			
σ known	z	Formula (10-1)	(10-2)
σ unknown (large samples)	z	Formula (10-3)	(10-4)
σ unknown (equal and small samples)	t	Formulas (10-9), (10-10), (10-11)	(10-13)
σ unknown (unequal and small samples)	t	Formula (10-12)	(10-14)
Two dependent means	t	Formulas (10-15), (10-16), (10-17)	(10-15), (10-16), (10-18)
Two proportions	z	Formula (10-19)	(10-21)
Two variances	F	Formula (10-5)	(10-8)

with one or two means, proportions, and variances. There is much more to be done. However, this amount of inferential statistics is sufficient to answer many questions. So we now are ready in succeeding chapters to look at some other useful tests for other types of questions that may arise.

Chapter Exercises

10-47 Two summers ago T. R. Brewery hired college students to operate three additional production lines to meet increased summer demand. The average hourly production for 150 hours was 419 cases. Last year, instead of college students, the firm used its existing work force on an overtime basis to operate the extra production lines. The average hourly production for 150 hours was 430 cases. Assume that $\sigma = 90$ for both years. Whereas the use of existing employees was more costly because of overtime rates, the personnel manager felt that their productivity would be greater than that of the college student summer help. Do the data support the manager's contention? Use $\alpha = 0.05$.

(a) Solve using the classical approach.

(b) Solve using the prob-value approach.

10-48 To comply with the Office of Federal Contract Compliance, a firm collected the data shown in the accompanying table on the test results from their management trainee selection examination. Construct a 98% confidence interval estimate for the difference in the mean scores of male and female test takers.

	n	$\bar{x}$	s
Males	86	73.2	6.3
Females	100	70.5	5.9

10-49 The score on a certain psychological test is used as an index of status frustration. The scale ranges from 0 (low frustration) to 10 (high frustration). The test was administered to independent random samples of seven managers and eight hourly workers.

Managers: 6 10 3 8 8 7 9

Hourly workers: 3 5 2 0 3 1 0 4

At the 10% level of significance, test the hypothesis that the mean score for both groups is the same against the alternative that they are not the same.

(a) Solve using the classical approach.

(b) Solve using the prob-value approach.

10-50 A survey was taken to appraise the level of sales per bookstore for a recently published technical book. In the Northeast the mean sales were

17.3 books and the variance was 41.2. Stores in the South had a mean of 20.3 books and a variance of 68.4. Forty-two stores were sampled in the Northeast, whereas 36 were sampled in the South. At the 0.05 level of significance, is there a difference in mean number of books sold in the two regions?

(a) Solve using the classical approach.

(b) Solve using the prob-value approach.

(c) Estimate the mean number of books sold per bookstore in the South using a 0.90 confidence interval.

10-51 Two sales approaches were to be compared. Eighty individuals were randomly selected from a population of individuals who responded to an advertisement that they would like additional information about a certain product. Of this group 45 were offered the product at a reduced price, and the other 35 were offered the item at full price but with a bonus gift that was equivalent, in cost to the company, to the price reduction. The following table shows the results of the experiment:

Offer	Sample Size	Number Purchasing
Price reduction, P	45	9
Gift, G	35	14

(a) Do the data substantiate the conclusion that the expected sales are greater with a gift than with a price reduction? Test at the 0.10 level. Draw the appropriate conclusion.

(b) Construct a 90% confidence interval for $p_G - p_P$.

10-52 A person might use the variance or standard deviation of the daily change in the stock market price as a measure of stability. Suppose you wish to compare the stability of a company's stock this year with its stability last year. You are given the following results of taking random samples from the daily gains and losses (changes) over the last 2 years:

$$\text{This year:} \quad n = 25, \quad s = 1.57$$
$$\text{Last year:} \quad n = 25, \quad s = 0.96$$

(a) Construct the 95% confidence interval estimate for the ratio of last year's standard deviation to this year's.

(b) Is the daily gain more, less, or about the same, as far as stability is concerned, for this year when compared with last year? Explain.

10-53 A soft drink distributor is considering two new models of dispensing machines. The Harvard Company machine and the Fizzit machine can both be adjusted to a certain mean amount to fill; however, the variation in the amount dispensed from cup to cup is of major concern. Ten cups were dispensed from the Harvard Company machine, and the variance was

0.065. Fifteen cups from the Fizzit machine showed a variance of 0.033. The factory representative from Harvard Company maintains that his machine had no more variability than the Fizzit machine had.

(a) At the 0.05 level of significance, does the sample refute the salesman's assertion?

(b) Estimate the variance for the amount of fill dispensed by the Harvard Company machine using a 90% confidence interval.

10-54 In planning the construction of a new manufacturing facility, the purchasing department has narrowed its choice of paints to that of two brands. Brand A is less expensive than brand B, but the department is concerned that brand A's quality is inferior. To aid in a decision, several chips are painted and exposed to weather conditions for a period of 6 months. Each chip is then judged as to several qualities, and a score is determined. The paint samples scored as shown in the accompanying table (higher scores are better). Do these results provide sufficient evidence to reject the null hypothesis that brand A is at least as good as brand B? Use $\alpha = 0.01$.

(a) Solve using the classical approach.
(b) Solve using the prob-value approach.

Paint A	84	86	91	93	84	88	
Paint B	90	88	92	94	84	85	92

10-55 The advertisement used at the beginning of chapter 8 states that Sear's radial tires improve gas mileage by 7.4% over its leading fiber-belted tires. Can you give any reasons why Sears might report 7.4% rather than the value of the mean mileage increase?

10-56 Altan Steel Company's personnel department is interested in knowing whether formal education increases the productivity of its production and maintenance workers. It selects two employees who were hired at the same time (to eliminate the effect of length of service on productivity), one who has a high school diploma and one who does not, and records their average incentive rate for the week. (The higher the rate, the more productive the employee.) It does this for 10 such pairs of employees. Based on the results shown in the following table, construct a 95% confidence interval for the difference in productivity of the two groups.

High School	145	133	116	128	85	100	105	150	97	110
Less Than High School	131	119	103	93	108	100	111	130	135	113

10-57 Do the data in exercise 10-56 support the personnel department's decision to require a high school diploma for its production and maintenance personnel? Use $\alpha = 0.05$.

10-58 Independent samples from two normal populations yielded the statistics shown in the following table. Can you conclude that the mean of population Q is less than the mean of population R at the 0.05 level of significance?

	$\bar{x}$	s^2	n
Q	21.3	2.35	25
R	23.9	8.52	21

10-59 In compliance with the Occupational Safety and Health Law, a firm installed certain expensive safety procedures. The director of safety has complained to the OSHA commission that the new procedures are a waste of the company's money. To support her claim, she presents the data shown in the following table. Do the data support the introduction of the new procedures? Use $\alpha = 0.10$.

Average Number of Accident Days Lost per Person-Year Worked

	Production Divisions									
	1	2	3	4	5	6	7	8	9	10
Before new procedures	7.9	5.6	9.2	6.7	8.1	7.3	8.1	5.4	6.9	6.1
After new procedures	7.7	6.1	8.9	7.1	7.9	6.7	8.2	5.0	6.2	5.7

(a) Solve using the classical approach.

(b) Solve using the prob-value approach.

10-60 The publisher of Follege Press feels that the physical appearance of an elementary text greatly affects its adoption, but that for advanced texts physical appearance has little effect. (Content and reputation of the author in the field dominate for advanced texts, she believes.) In a survey of 60 professors teaching elementary courses, 35 responded that appearance of the text was important in the decision to adopt it. Of 60 professors teaching advanced courses, only 27 responded that appearance was important. Do these data support the publisher's belief? Use $\alpha = 0.05$.

10-61 Two randomly selected groups, of 60 employees each, of a very large firm are taught an assembly operation by two different methods and then tested for performance. The first group averaged 150 units and a standard deviation of 10 units, whereas the second group averaged 146 units and a standard deviation of 12 units. Test, at the 0.05 level of significance, whether the difference between the means is significant (i.e., different from 0).

10-62 To cut down on paperwork, 10 randomly selected executives of the firm were given a course in "Effective Written Communication." To evaluate the program in order to determine if it should be given to all executives, the data shown in the following table were collected on the average number of minutes per day the executives' secretaries spent typing memos. Can you conclude that the program resulted in a significant improvement? Use $\alpha = 0.05$.

Before	30	26	25	35	33	31	32	54	50	43
After	29	22	25	29	26	24	31	46	34	28

10-63 To speed the flow of tax revenues to the city, the finance director suggests that the city offer a percentage discount on real estate taxes paid 30 days in advance of the due date. The assistant finance director feels that the discount is too small to have an effect. On a trial basis eight tax statements are sent out offering the discount and ten are sent without the offer. Those with a discount are paid in an average of 36 days with a standard deviation of 15 days, whereas those without a discount average 52 days with a standard deviation of 19 days. Do these data support the director's suggestion? Use $\alpha = 0.01$.

10-64 Amat Corporation currently subcontracts the keypunching aspects of its data-processing operation. It is considering hiring and supervising its own keypunch staff, but is worried about whether it can get the same quality internally as it can from a firm that specializes in that operation. To assist in making a decision, it sends 12 orders to the subcontracting firm and processes 12 identical orders internally. The cards are then checked, and the average number of keystroke errors per 100 is recorded (see the following table).

Internally	1.21	1.25	1.24	1.20	1.19	1.21
Subcontractor	1.22	1.16	1.12	1.13	1.18	1.21

Internally	1.21	1.22	1.22	1.25	1.23	1.24
Subcontractor	1.15	1.16	1.13	1.22	1.15	1.18

(a) Test the hypothesis that $\sigma_I = \sigma_S$ at $\alpha = 0.01$.

(b) Is there sufficient evidence to conclude that the subcontractor is more accurate? Use $\alpha = 0.05$.

10-65 United Bank has two branch offices located in Chester, Pennsylvania. The vice-president in charge of consumer loans notices that fewer "bill payer" loans are approved at branch A than at branch B. Each loan is first "scored" from 0 to 100 by a junior loan officer, 100 being the best risk and 0 the worst. A senior loan officer then gives final approval or disapproval. The vice-president suspects that the junior loan officials at branch A are more stringent than those at branch B; that is, the difference is not due to the quality of applicants at the different branches. To test this suspicion, he randomly selects 20 loan applications from both branches and randomly assigns 10 to each branch. The resultant loan ratings are shown in the following table. Can the vice-president conclude that branch B scores more leniently than branch A? Use $\alpha = 0.05$.

Branch A	72	29	62	60	68	59	61	73	38	48
Branch B	75	43	63	63	61	72	73	82	47	43

10-66 Consider the data and the problem in exercise 10-65. Suppose that, instead of selecting 20 loan applications, the vice-president had selected 10 and given the same 10 to each branch to evaluate. Consider the data in exercise 10-65 as being paired; that is, each score in the same column represents a branch A and a branch B scoring of the same application.

(a) Does the set of 10 pairs of data show that branch B scores more leniently? Use $\alpha = 0.05$.

(b) Explain why the dependent sample procedure used in part (a) might be considered a better statistical procedure than the independent samples used in exercise 10-65.

10-67 A group of 17 management trainees participated in an evaluation of a special training that claimed to improve memory. The employees were randomly assigned to two groups: group A, the test group, and group B, the control group. All 17 were tested for the ability to remember certain material. Group A was given the special training, and group B was not. After 1 month both groups were tested again, with the results as shown in the following table. Do these data support the alternative hypothesis that the special training is effective at the $\alpha = 0.01$ level of significance?

Management Trainees

	Group A									Group B						
	1	2	3	4	5	6	7	8	9	10	11	12	13	14	15	16	17
Before	23	22	20	21	23	18	17	20	23	22	20	23	17	21	19	20	20
After	28	29	26	23	31	25	22	26	26	23	25	26	18	21	17	18	20

10-68 The data in the accompanying table represent the monthly consumer price index for the Los Angeles–Long Beach and New York–Northeastern New

Jersey areas for 2 years. (The first two digits, which were always 11, were deleted to simplify the data; thus 114.6 became 4.6.)

Month	LA Year 1	LA Year 2	NY Year 1	NY Year 2
Jan.	4.6	2.8	6.4	5.0
Feb.	3.8	3.5	6.4	4.9
Mar.	4.0	3.6	6.3	4.7
Apr.	2.8	3.9	5.9	4.7
May	3.8	2.7	5.9	4.4
June	4.2	2.4	6.0	3.9
July	3.7	3.0	6.2	3.6
Aug.	4.6	2.8	5.9	3.8
Sept.	3.5	3.3	5.6	4.4
Oct.	3.4	3.1	5.2	4.5
Nov.	3.2	3.7	5.0	4.5
Dec.	2.7	3.7	4.6	4.5

The following SPSS computer printout presents a t test analysis of the data:

VARIABLE	NUMBER OF CASES	MEAN	STANDARD DEVIATION	STANDARD ERROR	(DIFFERENCE) MEAN	STANDARD DEVIATION	STANDARD ERROR
A LA		3.4500	0.593	0.121			
	24				-1.6458	0.645	0.132
		5.0958	0.862	0.176			
B NY							

CORR.	2-TAIL PROB.	F VALUE	DEGREES OF FREEDOM	2-TAIL PROB.
0.663	0.000	-12.49	23	0.000

Answer the following questions based on the computer printout:
(a) Is the data considered paired or independent samples?
(b) What is the mean price index for NY?
(c) What is the mean price index for LA?
(d) What is the standard deviation of the price index in LA?

(e) What is the standard error of the mean of the price index in LA?

(f) What is the standard deviation of d, the difference between the price index in LA and NY?

(g) What is the standard error of d?

(h) Construct a 95% confidence interval estimate for the difference in the means of the price indexes.

(i) State the null hypothesis being tested.

(j) How many degrees of freedom does the test statistic t have?

(k) What is the value of t^*?

(l) What is the prob-value?

(m) At the 0.01 level of significance, what decision is reached?

Challenging Problem

10-69 Meyer Drug Company is testing the effectiveness of a new weight control drug compared with its current drug. It conducts two experiments. In experiment 1 it matches 15 pairs of patients based on their weight, then administers the new drug to one member of each pair and the old drug to the other and records the weight loss after 1 week. In experiment 2 it randomly selects 15 patients and gives them the new drug and randomly selects another 15 patients and gives them the old drug. Again it records the weight loss of the patients after 1 week. The results are given in the following table:

Patient's Initial Weight	Experiment 1 Weight Loss (pounds)		Experiment 2 Weight Loss (pounds)	
	New Drug	Old Drug	New Drug	Old Drug
110	3	2	5	7
120	4	3	8	2
130	5	2	7	4
140	4	3	4	3
150	7	4	4	3
160	6	4	6	4
170	8	7	10	13
180	10	8	15	15
190	13	10	20	8
200	15	13	3	21
210	14	10	14	2
220	18	13	25	10
230	17	15	18	10
240	20	17	17	13
250	25	21	13	17

(a) Test the hypothesis that the new drug is no better than the old drug (use the average weight loss after 1 week of medication as the measure of effectiveness of a drug). Use (i) the data from experiment 1 and (ii) the data from experiment 2. Use $\alpha = 0.05$.

(b) Based on the results in part (a), would you conclude that the new drug is better? Compare the two test results. What can you say about weight loss and initial weight? Matching versus nonmatching?

Hands-On Problems

10-1 Use the student body of your college as the population of source objects to obtain a sample of 50 pieces of data that will enable you to answer any one of the questions in problem 10-2.
  (a) Define the response variable for which you are going to obtain a sample of 50 values.
  (b) Exactly how will you obtain the sample?

10-2 (a) Estimate (with a point estimate and with a 90% confidence interval) the difference in the proportion of male and female college students who favor a specific position on some topic of interest, such as favorite class, student governance problem, or a question about drugs, sex, or marriage.
  (b) Test the hypothesis that male college students obtain lower grades than female students do. (Use a specific required or elective course, specific semester grade point averages, or cumulative grade point averages.)
  (c) Estimate (with a point estimate and a 95% confidence interval) the standard deviation of last semester's grade point averages. (Use just male, just female, or a mixed population; use last semester's grades; or possibly restrict your variable to first-semester or second-semester averages.)
  (d) Test the hypothesis that there is no difference in the variances of the male and the female grades. (These grades could be from specific courses, last semester's grade point, the first semester's grade point, and so on.)

The calculations for this problem set can most easily be accomplished with the assistance of an electronic calculator or a packaged program on a computer. A list of the available programs can be obtained from your computer center. There are a variety of packaged programs available: Minitab, Biomed (Biomedical Programs), SAS (Statistical Analysis System), IBM Scientific Subroutine Packages, and SPSS (Statistical Package for the Social Sciences) program libraries. Your local computer center will assist you.

11 Additional Applications of Chi-Square

Chapter Outline

11-1 Chi-Square Statistic
The chi-square distribution will be used to test hypotheses concerning **enumerative data**.

11-2 Inferences Concerning Multinomial Experiments
A multinomial experiment differs from a binomial experiment in that **each trial has many outcomes** rather than two outcomes.

11-3 Inferences Concerning Contingency Tables
The chi-square distribution will be used to test hypotheses about the **independence of the two variables**.

Consumer Awareness of Truth in Lending

On July 1, 1969, consumers were given the right to know all the direct and indirect costs, terms and conditions of a credit arrangement without having to ask for them. They were given this right with the passage by the Ninetieth Congress of the Consumer Credit Protection Act of which Truth in Lending is a major part.

The purpose of Truth in Lending was and is to foster the informed use of credit by consumers. In the words of the Law itself: "It is the purpose of this title to assure a meaningful disclosure of credit terms so that the consumer will be able to compare more readily the various credit terms available to him and avoid the uninformed use of credit."

Purpose/Methodology

The purpose of the research was (1) to determine if consumers were aware of Truth in Lending and (2) to determine if there was a relationship between consumer awareness of Truth in Lending and the demographic characteristics of consumers. In other words, were there certain characteristics which were found to be related to knowledge of the Law?

...In examining the demographic characteristics of the two groups, significant facts were revealed ... and ... supported by the use of chi-square statistical procedures.

TABLE I
Selected Demographic Characteristics of Consumers Who Had Knowledge of Truth in Lending as Compared to Consumers Who Did Not Have Knowledge of the Law, by Number and Percent of Respondents

Characteristic	Knew the Law[1]		Did Not Know Law[2]	
	Number	Percent	Number	Percent
Marital status				
Single	12	11.3	31	10.5
Married	87	82.1	244	83.0
Divorced	7	6.6	19	6.5
Chi-square 0/05402 (Not Significant) D.F. -2				
Income				
Under $5M	2	1.9	70	23.8
$5M/Less $10M	13	12.3	122	41.5
$10M/Less $15M	41	38.7	75	25.5
$15M/Less $20M	30	28.3	16	5.4
$20M and over	20	18.9	11	3.7
Chi-square 103.59236 (Significant) D.F. -4				

[1] N = 106
[2] N = 294

Note: Percentages may be less than or greater than 100 percent due to rounding.

From William H. Bolen, "Consumer Awareness of Truth in Lending," *Business Ideas and Facts* (Eastern Michigan University), vol. 7, no. 2 (Fall 1974): 38–39. Reprinted by permission.

Chapter Objectives

Previously we discussed and used the chi-square distribution both to test and to estimate the value of the variance (or standard deviation) of a single population. The chi-square distribution may also be used for tests in other types of situations. In this chapter we are going to investigate some of these uses. Specifically, we will look at two tests: a multinomial experiment and the contingency table. These two types of tests will be used to compare experimental results with expected results to determine (1) preferences, (2) independence, and (3) homogeneity.

The data that we will be using in these techniques will be enumerative; that is, the data will result from counts of occurrences.

Section 11-1   Chi-Square Statistic

There are many problems for which information is categorized and the results are shown by way of counts. For example, loans can be categorized with respect to payment as "good," "slow," or "default," and a bank may tabulate the frequency with which loans fall into these categories. The result of this tabulation is a frequency table, an idea that was first presented in chapter 2. Our discussion in this chapter differs from what was done in chapter 2, however, in that now we will work *only* with the frequencies.

Suppose that we have a number of **cells** into which n observations have been sorted. [The terms *cell* (or class) and *frequency* were defined and first used in earlier chapters. Before you continue, a brief review of sections 2-2 and 3-1 might be beneficial.] The **observed frequencies** in each cell are denoted by $O_1, O_2, O_3, \ldots, O_k$ (see table 11-1). Note that the sum of all observed frequencies is equal to n $(O_1 + O_2 + \cdots + O_k = n)$. What we would like to do is compare the observed frequencies with some **expected, or theoretical, frequencies,** denoted by $E_1, E_2, E_3, \ldots, E_k$, for each of these cells. Again, the sum of these expected frequencies must be exactly n $(E_1 + E_2 + \cdots + E_k = n)$. We will decide whether the observed frequencies seem to agree or seem to disagree with the expected frequencies. This decision will be accomplished by a hypothesis test using the **chi-square distribution,** χ^2.

expected frequency

The calculated value of the test statistic will be

$$\chi^2 = \sum_{\text{all cells}} \frac{(O - E)^2}{E} \qquad (11\text{-}1)$$

This calculated value for chi-square will be the sum of several positive numbers, one from each cell (or category). The numerator of each term in the formula for χ^2 is the square of the difference between the values of the

TABLE 11-1
Observed Frequencies

	k Categories					Total
	1st	2d	3d	...	kth	
Observed frequency	O_1	O_2	O_3	...	O_k	n

observed and the expected frequencies. The closer together these values are, the smaller the value of $(O - E)^2$; the farther apart, the larger the value of $(O - E)^2$. The denominator for each cell puts the size of the numerator into perspective. That is, a difference $(O - E)$ of 10 resulting from frequencies of 110 (O) and 100 (E) seems quite different from a difference of 10 resulting from 15 (O) and 5 (E).

The preceding ideas suggest that small values of chi-square indicate agreement, whereas larger values indicate disagreement, between the two sets of frequencies. Therefore these tests are customarily one-tailed, with the critical region on the right.

In repeated sampling the calculated value of χ^2 [formula (11-1)] will have a sampling distribution that can be approximated by the chi-square probability distribution when n is large. This approximation is generally considered adequate when all the expected frequencies are equal to or greater than 5. Recall that there is a separate chi-square distribution for each possible value of degrees of freedom, df. The appropriate value of df will be described with each specific test. The critical value of chi-square, χ^2 (df, α), will be found in table 8 of appendix D.

In this chapter we will permit a certain amount of "liberalization" with respect to the null hypothesis and its testing. Prior to this chapter the null hypothesis has always been a statement about a population parameter (μ, σ, or p). However, other types of hypotheses can be tested, such as "the type of loan and its repayment categorization (as discussed previously) are independent" or "a person's age and brand preference of cigarettes are independent." Notice that these hypotheses are not claims about a parameter, although sometimes they could be stated with parameter values specified.

Suppose I claimed that any fractional part of a stock price is equally likely to occur, and you wanted to test my statement. What would you do? Was your answer something like this? "Take a random sample of stocks from the stock pages of your newspaper and record the frequency of each fractional price." Suppose you randomly select 80 stocks. If fractional prices *are* random, what would you expect to happen? Each fractional price, none, $\frac{1}{8}, \frac{1}{4}, \frac{3}{8}, \ldots, \frac{7}{8}$, should appear approximately $\frac{1}{8}$ of the time (i.e., 10 times). If it happens that approximately 10 of each fractional price occurs, you will certainly accept the claim of randomness. If it happens that a particular fractional price or prices appear disproportionately (i.e., suppose none or $\frac{1}{2}$ occurs in 40 of the 80 cases), you will reject the claim. (The test statistic χ^2 will have a large value in that case, as we will see shortly.)

Section 11-2 Inferences Concerning Multinomial Experiments

multinomial experiment

The preceding problem is a good illustration of a **multinomial experiment**. Let's consider this problem again. Suppose we want to test the claim that every fractional price is equally likely to occur (at $\alpha = 0.05$). We randomly select 80 stocks and record the number of times each fractional price occurs, as shown in the following table:

Fractional Price	0	1/8	1/4	3/8	1/2	5/8	3/4	7/8
Occurrences	12	10	9	11	13	7	10	8

The null hypothesis that each fractional price is equally likely is assumed to be true. This assumption allows us to calculate the expected frequencies. If the null hypothesis claim is true, we certainly would expect 10 occurrences for each fractional price.

Let's now calculate an observed value of χ^2. These calculations are shown in table 11-2. The calculated value is $\chi^{2*} = 2.8$.

NOTE: $\sum (O - E)$ must equal 0, since $\sum O = \sum E = n$. You can use this fact as a check, as shown in table 11-2.

Before continuing, let's set up the hypothesis-testing format.

TABLE 11-2
Computations for Calculating χ^2

Fractional Price	Observed, O	Expected, E	O − E	$(O - E)^2$	$\dfrac{(O - E)^2}{E}$
0	12	10	+2	4	0.4
1/8	10	10	0	0	0
1/4	9	10	−1	1	0.1
3/8	11	10	+1	1	0.1
1/2	13	10	+3	9	0.9
5/8	7	10	−3	9	0.9
3/4	10	10	0	0	0
7/8	8	10	−2	4	0.4
Total	80	80	0 ck		2.8

STEP 1: $H_0: p_0 = p_{1/8} = p_{1/4} = p_{3/8} = \cdots = p_{7/8} = \frac{1}{8}$ (each fractional price is equally likely to occur).

STEP 2: H_a: More than one $p_i \neq \frac{1}{8}$. Some fractional prices are more likely to occur than others.

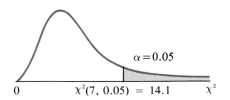

STEP 3: $\alpha = 0.05$. See the figure for the test criteria. In the multinomial test, df $= k - 1$, where k is the number of cells.

STEP 4: The test statistic was calculated in table 11-2 and was found to be $\chi^{2*} = 2.8$.

STEP 5: *Decision*: Fail to reject H_0 (χ^{2*} is not in the critical region).

Conclusion: The observed frequencies are not significantly different from those expected if all prices are equally likely to occur.

Before we look at other illustrations, we must define the term *multinomial experiment* and we must state the guidelines for completing the chi-square test for it.

Multinomial Experiment

An experiment with the following characteristics:

1. It consists of n identical independent trials.
2. The outcome of each trial fits into exactly one of k possible cells.
3. There is a probability associated with each particular cell, and these individual probabilities remain constant during the experiment. (It must be true that $p_1 + p_2 + \cdots + p_k = 1$.)
4. The experiment will result in a set of observed frequencies, $O_1, O_2, \ldots, O_k$, where each O_i is the number of times a trial outcome falls into that particular cell. (It must be the case that $O_1 + O_2 + \cdots + O_k = n$.)

The preceding illustration with the stock prices meets the definition of a multinomial experiment, since it has all four of the characteristics described in the definition.

1. Each stock represents a trial and, since the stocks were randomly selected, the trials are independent.
2. Each stock's fractional price could be only one of eight possible values, and each value forms a cell.
3. The probability associated with each cell is $\frac{1}{8}$, which was constant from stock to stock (trial to trial).
4. When the experiment was complete, we had a list of frequencies (12, 10, 9, 11, 13, 7, 10, and 8) that summed to 80, indicating that each outcome was accounted for.

The **testing procedure** for multinomial experiments is very much as it was in previous chapters. The biggest change comes with the statement of the null hypothesis. It may be a verbal statement, such as in the example of stock prices: "Each fractional price is equally likely." Often the alternative to the null hypothesis is not stated. However, in this book the alternative hypothesis will be shown, since it seems to aid in the organization and understanding of the problem. However, it will not be used to determine the location of the critical region, as was the case previously. **For multinomial experiments we will always use a one-tailed critical region, and it will be the right-hand tail of the χ^2 distribution.**

The critical value will be determined by the level of significance that is assigned (α) and the number of degrees of freedom. The number of degrees of freedom (df) will be 1 less than the number of cells (k) into which the data are divided:

$$\text{df} = k - 1 \tag{11-2}$$

Each expected frequency, E_i, will be determined by multiplying the corresponding probability (p_i) for the cell by the total number of trials n. That is,

$$E_i = n \cdot p_i \tag{11-3}$$

One guideline should be met to ensure a good approximation to the chi-square distribution: each expected frequency should be at least 5 (i.e., each $E_i \geq 5$). If this guideline cannot be met, corrective measures to ensure a good approximation should be used. These corrective measures are not covered in this book, but are discussed in other statistics books.

Illustration 11-1

Convenience Food Market is open 7 days a week, 24 hours a day. Its employees are scheduled 5 days a week for the same shift (three shifts a day). The manager currently schedules the employees so that the same proportion of the total workforce works each day, based on his assumption that the proportion of shoppers on any given day is approximately the same. The accompanying table shows the number of shoppers (in hundreds) who were in the store each day last week. Do the data justify the manager's personnel assignment? Or do the data indicate certain days attract more customers? Use $\alpha = 0.05$.

Day	Sun.	Mon.	Tues.	Wed.	Thurs.	Fri.	Sat.	Total
Shoppers	18	12	25	23	8	19	14	119

Solution If there were no preference shown in the days, we would expect the 119 (hundred) shoppers to be equally distributed among the 7 days.

Thus if no preference is the case, we would expect 17 (hundred) shoppers each day. The test is completed as shown next, at the 5% level of significance.

STEP 1: H_0: $p_{Su} = p_M = p_T = p_W = \cdots = \frac{1}{7}$.

No preference was shown (equally distributed).

STEP 2: H_a: More than one p_i is not equal to the others. A preference was shown (not equally distributed).

STEP 3: $\alpha = 0.05$. See the diagram for the test criteria.

$\chi^2(6, 0.05) = 12.6$, $\alpha = 0.05$

STEP 4: The test statistic is calculated in the following table:

	Sun.	Mon.	Tues.	Wed.	Thurs.	Fri.	Sat.	Total
O	1,800	1,200	2,500	2.300	800	1,900	1,400	11,900
E	1,700	1,700	1,700	1,700	1,700	1,700	1,700	11,900
$(O - E)^2/E$	5.88	147.06	376.47	211.76	476.47	23.53	52.94	1,294.11

$$\chi^{2*} = 1{,}294.11$$

STEP 5: *Decision*: Reject H_0 (χ^{2*} falls in the critical region).

Conclusion: A preference seems to be shown.

We cannot determine, from the given information, *why* there is a preference. Thus conclusions must be worded carefully to avoid suggesting conclusions that cannot be supported. □

Not all multinomial experiments result in equal expected frequencies, as we will see in the next illustration.

Illustration 11-2

Geico Insurance Company classifies automobile insurance applications into four categories, "high risk," "risk," "regular," and "merit," on the basis of the age and prior driving experience of the applicants. Its actuarial department estimates that a claim is six times more likely on a high-risk policy than a merit policy, three times more likely on a risk policy, and two times more likely on a regular policy. The premiums for the different types of policies are based on this actuarial estimate. A randomly drawn sample of 1,326 policies yielded the information shown in the accompanying table. Do these sample data provide sufficient evidence to question the actuarial estimates and, consequently, the rate structure of policies? Use $\alpha = 0.05$.

Policy Type	High risk, HR	Risk, R	Regular, O	Merit, M	Total
Number of Claims	674	325	224	103	1,326

Solution

STEP 1: $H_0: P(HR) = \frac{1}{2}, P(R) = \frac{1}{4}, P(O) = \frac{1}{6}, P(M) = \frac{1}{12}$; or 6:3:2:1 is the ratio of claims by policy type.

STEP 2: H_a: 6:3:2:1 is not the ratio of claims by policy type.

STEP 3: $\alpha = 0.05$, $k = 4$, and df $= 3$. See the diagram.

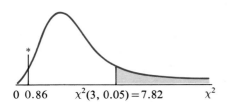

STEP 4: The actuarial estimates mean that of every n policies we would expect that $\frac{6}{12}$ of the claims will be from high-risk policies, $\frac{3}{12}$ from risk, $\frac{2}{12}$ from regular, and $\frac{1}{12}$ from merit. Therefore the expected frequencies are $6n/12$, $3n/12$, $2n/12$, and $n/12$, where n equals the number of claims. The computations for calculating χ^2 are given in the following table:

O	E	O − E	(O − E)²/E
674	663.0	11.0	0.1825
325	331.5	−6.5	0.1275
224	221.0	3.0	0.0407
103	110.5	−7.5	0.5090
1,326	1,326.0	0	0.8597

Thus $\chi^{2}{*} = \mathbf{0.86}$.

STEP 5: *Decision: Fail to reject H_0.*

Conclusion: There is not sufficient evidence to reject the actuarial estimates. □

Exercises

11-1 A random sample of stocks listed on the American Exchange resulted in the data concerning fractional prices shown in the following table. Test the claim that all fractional prices are equally likely to occur. Use $\alpha = 0.05$.

Price	Even dollar	1/8	1/4	3/8	1/2	5/8	3/4	7/8
Number	25	17	15	22	24	14	22	21

(a) Solve using the classical approach.
(b) Solve using the prob-value approach.

11-2 Use the data in exercise 11-1 to test the claim that $\frac{1}{2}$ and even dollar amounts combined occur as frequently as do all other fractional prices combined. Use $\alpha = 0.05$.

11-3 A manufacturer of floor polish conducted a consumer-preference experiment to determine which of five different floor polishes was the most appealing. A sample of 100 housewives viewed five patches of flooring that had received the five polishes. Each housewife indicated which patch she preferred for appearance reasons. The lighting, background, and so forth were approximately the same for all patches. The results were as follows:

Polish	A	B	C	D	E	Total
Frequency	27	17	15	22	19	100

(a) State the hypothesis for "no preference" in statistical terminology.
(b) What test statistic will be used in testing this null hypothesis?
(c) Complete the hypothesis test using $\alpha = 0.10$.
  (i) Solve using the classical approach.
  (ii) Solve using the prob-value approach.

11-4 Previous sales records show that Inboard Custom Boats sold 20% of their boats in their Northeast sales district, and 28%, 8%, 12%, and 32%, respectively, in their Southeast, North Central, South Central, and West Coast sales districts. Of the first 500 boats sold this year, 120, 128, 43, 66, and 143 were sold in each of the five districts, respectively. Does the sales distribution so far this year seem to be the same as in previous years? Use $\alpha = 0.05$.

11-5 Historically, a manufacturer of electronic calculators has found that 30% of its sales are in the basic four-function models, 30% in models with one memory, 25% in scientific models, and 15% in programmable models. To schedule current production, the firm takes a random sample of 100 calculators sold last week and finds the data shown in the accompanying table. Test the hypothesis, at the 0.01 level, that the mix of sales has remained the same.

Type	Basic	Memory	Scientific	Programmable
Number Sold	39	26	17	18

(a) Solve using the classical approach.
(b) Solve using the prob-value approach.

Section 11-3 Inferences Concerning Contingency Tables

In chapter 3 we discussed the use of contingency tables to present data in a two-way classification. The usual question concerning such tables is whether the data indicate that the two variables are independent or dependent.

Test of Independence

To illustrate the analysis of a contingency table, let's consider the sex classification of cigarette smokers and their preference in length of cigarette.

Illustration 11-3

To assist in targeting advertising, the marketing department of Meric Tobacco surveyed 300 purchases of cigarettes and recorded the sex of the person and the length of the brand of cigarettes purchased. The following table shows the frequencies found for these classifications. Does this sample present sufficient evidence to reject the null hypothesis "preference for length of cigarette is independent of the sex of the consumer" at the 0.05 level of significance?

Sex of Purchaser	Length of Cigarette Purchased			Total
	120 mm, MM	King, K	Regular, R	
Male	37	41	44	122
Female	35	72	71	178
Total	72	113	115	300

Solution The steps of the hypothesis test for this problem are as follows:

STEP 1: H_0: Preference for cigarette length is independent of the sex of the consumer.

STEP 2: H_a: Length preference is not independent of the sex of the consumer.

STEP 3: To determine the critical value of chi-square, we need to know df, the number of degrees of freedom involved. In the case of contingency tables, the number of degrees of freedom is exactly the same number as the number of cells in the table that may be filled in freely when you are given the row and column totals. The totals in this problem are shown in the following table:

			122
			178
72	113	115	300

Given these totals, you can fill in only two cells before the others are all determined. (The totals must, of course, be the same.) For example, once we pick two arbitrary values (say 50 and 60) for the first two cell of the first row (see the following table), the other four cell values are fixed.

50	60	C	122
D	E	F	178
72	113	115	300

400 Chapter 11 Additional Applications of Chi-Square

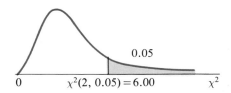

They have to be $C = 12$, $D = 22$, $E = 53$, and $F = 103$. Otherwise, the totals will not be correct. Therefore for this problem there are two free choices. Each free choice corresponds to one degree of freedom. Hence the number of degrees of freedom for our example is 2 (df = 2). Thus if $\alpha = 0.05$ is used, the critical value is $\chi^2(2, 0.05) = 6.00$. See the diagram. (After the conclusion of this illustration, a formula for finding df will be discussed.)

STEP 4: Before the calculated value of chi-square can be found, we need to determine the expected values E for each cell. To do this, we must recall the null hypothesis, which asserts that these factors are independent. Therefore we would expect the values to be distributed in proportion to the marginal totals. There are 122 males; we would expect them to be distributed between MM, K, and R in proportion to the 72, 113, and 115 totals. Thus the expected cell counts for males are

$$\frac{72}{300} \cdot 122, \quad \frac{113}{300} \cdot 122, \quad \frac{115}{300} \cdot 122$$

Similarly, since there are 178 females, we would expect

$$\frac{72}{300} \cdot 178, \quad \frac{113}{300} \cdot 178, \quad \frac{115}{300} \cdot 178$$

Thus the expected values are as shown in the following table. (Always check the new totals with the old totals.)

29.28	45.95	46.77	122.00
42.72	67.05	68.23	178.00
72.00	113.00	115.00	300.00

NOTE: Recall that we assume the null hypothesis to be true until there is evidence to reject it. Having made this assumption, in our example we are saying, in effect, that the event of a smoker picked at random being male and the event of a smoker picked at random preferring a 120-mm length are independent. Our point estimate for the probability that a smoker is male is 122/300, and the point estimate for the probability that the consumer prefers the 120-mm length is 72/300. Therefore the probability that both events occur is the product of the probabilities. [Refer to formula (4-8a).] (122/300)·(72/300) is the probability of a selected smoker being male and preferring a 120-mm length. Therefore the number of smokers out of 300 that are expected to be male and prefer 120 mm is found by multiplying the probability (or proportion) by the total number of smokers, 300. Thus the expected number of males who prefer 120 mm is

$$\left(\frac{122}{300}\right) \cdot \left(\frac{72}{300}\right) \cdot (300) = \left(\frac{122}{300}\right) \cdot (72) = 29.28$$

The other expected values can be determined in the same manner.

Typically, the contingency table is written with all this information in it, as shown in the accompanying table.

Sex of Consumer	Length of Cigarette			Total
	120 mm	King	Regular	
Male	37 (29.28)	41 (45.95)	44 (46.77)	122
Female	35 (42.72)	72 (67.05)	71 (68.23)	178
Total	72	113	115	300

The calculated chi-square is

$$\chi^2 = \sum \frac{(O-E)^2}{E}$$

$$= \frac{(37-29.28)^2}{29.28} + \frac{(41-45.95)^2}{45.95} + \frac{(44-46.77)^2}{46.77}$$

$$+ \frac{(35-42.72)^2}{42.72} + \frac{(72-67.05)^2}{67.05} + \frac{(71-68.23)^2}{68.23}$$

$$= 2.035 + 0.533 + 0.164 + 1.395 + 0.365 + 0.112 = 4.604$$

$$\chi^{2*} = 4.604$$

STEP 5: *Decision*: Fail to reject H_0 (χ^{2*} did not fall in the critical region).

Conclusion: The evidence does not allow us to reject the idea of independence between the sex of a consumer and the preferred cigarette length. □

In general, the $r \times c$ **contingency table** (r is the number of rows; c is the number of columns) will be used to test the independence of the row factor and the column factor. The number of degrees of freedom will be determined by

$$\text{df} = (r-1) \cdot (c-1) \quad (11\text{-}4)$$

where r and c are both greater than 1.

(This value for df should agree with the number of cells counted, according to the general description on pages 400 and 401.)

NOTE: If either r or c is 1, then df $= k - 1$, where k represents the number of cells [formula (11-2)].

The expected values for an $r \times c$ contingency table will be found by means of the formulas found in each cell in table 11-3, where n = grand total.

TABLE 11-3
Expected Values for an $r \times c$ Contingency Table

		Columns				
Rows	1	2	...	j	...	c	Total
1	$\dfrac{R_1 \times C_1}{n}$	$\dfrac{R_1 \times C_2}{n}$	...	$\dfrac{R_1 \times C_j}{n}$	...	$\dfrac{R_1 \times C_c}{n}$	R_1
2	$\dfrac{R_2 \times C_1}{n}$		...	$\vdots$	...		R_2
$\vdots$	$\vdots$		...		...		$\vdots$
i	$\dfrac{R_i \times C_1}{n}$	$\vdots$	...	$\dfrac{R_i \times C_j}{n}$	...	$\vdots$	R_i
$\vdots$	$\vdots$		...	$\vdots$	...		$\vdots$
r	$\dfrac{R_r \times C_1}{n}$		...		...	$\dfrac{R_r \times C_c}{n}$	R_r
Total	C_1	C_2	...	C_j	...	C_c	n

In general, the **expected value** at the intersection of the **ith row and the jth column** is given by

$$E_{i,j} = \frac{R_i \times C_j}{n} \qquad (11\text{-}5)$$

We should again observe the previously mentioned guideline: each $E_{i,j}$ should be at least 5.

NOTE: The notation used in table 11-3 and formula (11-5) may be unfamiliar to you. For convenience in referring to cells or entries in a table, $E_{i,j}$ (or E_{ij}) can be used to denote the entry in the ith row and the jth column. That is, the first letter in the subscript corresponds to the row number, and the second letter corresponds to the column number. Thus $E_{1,2}$ is the entry in the first row, second column, whereas $E_{2,1}$ is the entry in the second row, first column. Referring to the table on page 401, $E_{1,2}$ for that table is 45.95 and $E_{2,1}$ is 42.72. The notation used in table 11-3 is interpreted in a similar manner: R_1 corresponds to the total from row 1, whereas C_1 corresponds to the total from column 1.

Test of Homogeneity

Another type of contingency table problem is called a test of homogeneity. This test arises when one of the two variables is controlled by the experimenter so that the row (or column) totals are predetermined.

For example, suppose that we were to poll employees about a contract proposal: 200 laborers, 200 operatives, and 100 craftspeople will be randomly selected and asked if they favor or oppose the contract. A total of 500 employees are to be polled. But notice that it has been predetermined (before the sample is taken) just how many are to fall within each row category, as shown in table 11-4.

TABLE 11-4
Employee Poll with Predetermined Row Totals

Type of Employee	Contract Proposal		Total
	Favor	Oppose	
Laborer			200
Operative			200
Craft			100
Total			500

In a test of this nature, we are actually testing the hypotheses "the distribution of proportions within the rows is the same for all rows." That is, the distribution of proportions in row 1 is the same as in row 2, is the same as in row 3, and so on. The alternative to this hypothesis is that the distribution of proportions within the rows is not the same for all rows. This type of example may be thought of as a comparison of several multinomial experiments.

Beyond this conceptual difference, the actual testing for independence and homogeneity with contingency tables is the same. Let's demonstrate this by completing the polling illustration.

Illustration 11-4

Each person in a random sample of 500 employees (200 laborers, 200 operatives, and 100 craftspeople) was asked his or her opinion about the contract proposed. Does the sample evidence shown in the accompanying table support the hypothesis that employees within the different groups have different opinions about the contract proposal? Use $\alpha = 0.05$.

| | Contract Proposal | | |
Type of Employee	Favor	Oppose	Total
Laborer	143	57	200
Operative	98	102	200
Craft	13	87	100
Total	254	246	500

Solution

STEP 1: H_0: The proportion of employees favoring the proposed contract is the same in all three groups.

STEP 2: H_a: The proportion of employees favoring the proposed contract is not the same in all three groups. (That is, in at least one group the proportions are different from the others.)

STEP 3: $\alpha = 0.05$ and df $= (3 - 1)(2 - 1) = 2$. See the diagram.

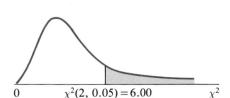
$\chi^2(2, 0.05) = 6.00$

STEP 4: The expected values are found by using formula (11-5) and are as shown in the next table.

NOTE: Each expected value is used twice in the calculation of χ^{2*}, therefore it is a good idea to keep at least two extra decimal places while doing the calculations.

| | Contract Proposal | | |
Type of Employee	Favor	Oppose	Total
Laborer	143 (101.6)	57 (98.4)	200
Operative	98 (101.6)	102 (98.4)	200
Craft	13 (50.8)	87 (49.2)	100
Total	254	246	500

$$\chi^2 = \frac{(143 - 101.6)^2}{101.6} + \frac{(57 - 98.4)^2}{98.4} + \frac{(98 - 101.6)^2}{101.6}$$

$$+ \frac{(102 - 98.4)^2}{98.4} + \frac{(13 - 50.8)^2}{50.8} + \frac{(87 - 49.2)^2}{49.2}$$

$$= 16.87 + 17.42 + 0.13 + 0.13 + 28.13 + 29.04 = 91.72$$

$$\chi^{2*} = \mathbf{91.72}$$

STEP 5: *Decision*: Reject H_0.

Conclusion: The three groups of employees do not all have the same proportions favoring the proposed contract. ☐

Exercises

11-6 A bank conducted a survey to see if the employment status of the wife at the time of a loan to a married couple had any effect on the chances of default. Given the results shown in the following table, test the claim that the wife's employment status does not affect repayment. Use $\alpha = 0.05$.

Present Status of Loan	Employment Status of Wife at Time of Loan	
	Employed	Not Employed
In default	8	15
No default	75	60

11-7 A rating by customers of the quality of service at the service stations for one of the very large petroleum distributors in the United States showed the following results:

Sex of Respondent	Quality of Service			Total
	Above Average	Average	Below Average	
Female	7	24	28	59
Male	8	26	7	41
Total	15	50	35	100

Using $\alpha = 0.05$, test to see if the null hypothesis, the rating of quality of service is independent of the sex of the respondent, can be rejected.

(a) Solve using the classical approach.

(b) Solve using the prob-value approach.

11-8 An industrial engineer claims that the reaction to stress by executives who have exercised for $\frac{1}{2}$ hour each morning is different from those who have not. He thus argues that the firm should install an executive gym and require participation of its key executives. To "prove" his point, he presents the following data based on an experiment where executive stress was measured among executives who exercised regularly and those who did not.

| | Stress Reaction | | | |
Exercised Regularly	Mild	Medium	Strong	Total
Yes	170	100	30	300
No	70	100	30	200
Total	240	200	60	500

Does the data support the industrial engineer's claim? Use $\alpha = 0.05$.

(a) Solve using the classical approach.

(b) Solve using the prob-value approach.

11-9 An insurance company, to project the pension investment choices of employees, surveyed 1,000 employees about their preference for the allocation of the investment of their pension funds. The following table summarizes the findings. At the 0.01 level of significance, do these data contradict the claim of independence of age and investment preference?

| | Type of Investments | | |
Age	All Common Stocks	All Fixed Annuities	Mixed
Over 45	196	131	158
Under 45	223	116	176

11-10 A survey of corporate personnel directors revealed the results shown in the accompanying table. Each director was asked to rank the item that he or she most preferred in a management trainee applicant. Do the data collected show sufficient evidence to reject the null hypothesis that the most preferred characteristic of management trainee applicants is independent of the size of the firm? Use $\alpha = 0.05$.

| | Preference | | | |
Size of Firm	MBA	Quantitative Training	Pleasing Personality	Ambition
Over 1 million sales	17	14	4	9
Under 1 million sales	8	10	7	11

11-11 To evaluate income characteristics of viewers of a certain television show, A & C Advertising Agency conducted a survey of 100 randomly selected individuals in each income category. The results are shown in the accompanying table. At $\alpha = 0.05$, do these data provide sufficient reason to reject the hypothesis that the same proportion of all income levels watched the television show?

	Household Income Level				
	Low	Low Middle	Middle	Upper Middle	Upper
Watching	37	28	25	27	21
Not watching	63	72	75	73	79

11-12 To study the projected impact of a new management stock option offering, a large firm surveyed 200 management personnel in each of the various divisions of the firm. The results are shown in the following table. At $\alpha = 0.01$, do we have sufficient evidence to reject the hypothesis that the same proportion of managers in each division will select the stock option? (*Hint*: Be sure to include the "do not plan to use option" data also.)

Division	Marketing	Accounting	Maintenance	Production	Finance
Plan to Use Option	83	72	49	36	112

In Retrospect

In this chapter we have been concerned with tests of hypotheses using chi-square, with the cell probabilities associated with the multinomial experiment, and with the contingency table. In each case the basic assumptions are that a large number of observations have been made and that the resulting test statistic, $\sum [(O - E)^2/E]$, is approximately distributed as chi-square. In general, if n is large and the minimum allowable expected cell size is 5, this assumption is satisfied.

The news article at the beginning of this chapter is an illustration of the use of the chi-square test on contingency tables to decide if two variables are independent. Be aware that the conclusions from the test of hypothesis do not explain the reasons for the dependence, but simply indicate that a relationship exists between the variables (i.e., the variables are not independent).

The contingency table can also be used to test homogeneity. The test for homogeneity and the test for independence look very similar and, in fact, are carried out in exactly the same way. The concepts being tested, equal distributions and independence, are, however, quite different. The two tests are easily distinguished from one another, for the test of homogeneity possesses predetermined marginal totals in one direction in the table. That is, before the data are collected, the experimenter determines how many subjects will be observed in each category. The only predetermined number in the test of independence is the grand total.

A few words of caution: The correct number of degrees of freedom is critical if the test results are to be meaningful. The degrees of freedom determine, in part, the critical region, and the size of this region is important. Like other tests of hypothesis, failure to reject H_0 does not mean outright acceptance of the null hypothesis.

percent of the males responded yes, whereas seventy-five percent of the females responded yes.

(a) Structure the data as a contingency table and test the hypothesis that sex and preference for a job with a large CPA firm are independent. Use $\alpha = 0.05$.

(b) Use the procedure in section 10-6 to test the hypothesis that the proportion of males preferring to work for a large CPA firm is the same as the proportion of females preferring to work for a large CPA firm.

(c) Compare your results in (a) and (b).

(*Note*: The square of the z statistic is equivalent to the chi-square statistic with one degree of freedom.)

Hands-On Problem

11-1 Use the student body at your school as your source population, and partition the student body according to two classifications. Poll 100 members concerning a question of your choice and summarize your data in a contingency table. Test for independence of the two factors (use $\alpha = 0.05$). (For example, your contingency table might look something like the two tables shown here.)

Favorite Academic Subjects

Male				
Female				

	Work	Not Work
Male		
Female		

The calculations for this problem set can most easily be accomplished with the assistance of an electronic calculator or a packaged program on a computer. A list of the available programs can be obtained from your computer center. There are a variety of packaged programs available: Minitab, Biomed (Biomedical Programs), SAS (Statistical Analysis System), IBM Scientific Subroutine Packages, and SPSS (Statistical Package for the Social Sciences) program libraries. Your local computer center will assist you.

12 Analysis of Variance

Chapter Outline

12-1 Introduction to the Analysis of Variance Technique
Analysis of variance (ANOVA) is used to test a hypothesis about several population means.

12-2 The Logic Behind ANOVA
Between-sample variation and within-sample variation are compared in an ANOVA test.

12-3 Applications of Single-Factor ANOVA
Notational considerations and a mathematical model explaining the composition of each piece of data are presented.

12-4 Two-Factor ANOVA (Without Replication)
*ANOVA techniques are expanded to consider an experiment involving **two independent factors**.*

12-5 Two-Factor ANOVA (with Replication)
*Further expansion to study the **interaction** between the two factors.*

A Durability Test

The Durability Test

This durability test was supervised by an independent laboratory. It consisted of firing an equal number of randomly purchased golf balls of each brand against a flat steel block under air pressure that would yield an impact against the block comparable to that produced by a driver striking a ball on a tee, swung by a professional golfer. Each ball was taken to 300 hits or failure, whichever came first. Failure was defined as the number of hits at which the windings or core could first be seen through a crack in the cover, or at which the ball became distorted in shape. Full test results are available from: Blue Max Durability Test, P.O. Box 1286, Buffalo, N.Y. 14240.

Results of the Durability Test. An equal number of balls from each brand were hit 300 times, or to failure, whichever came first.

Ball No.	Dunlop Blue Max	Golden Ram MK IV-90	Hogan Leader 90	Rawlings Toney Penna DB	Spalding Top Flite	Wilson Pro Staff
1	300	190	228	276	162	264
2	300	164	300	296	175	168
3	300	238	268	62	157	254
4	260	200	280	300	262	216
5	300	221	300	230	200	257
6	261	132	300	175	256	183
7	300	156	300	211	92	93

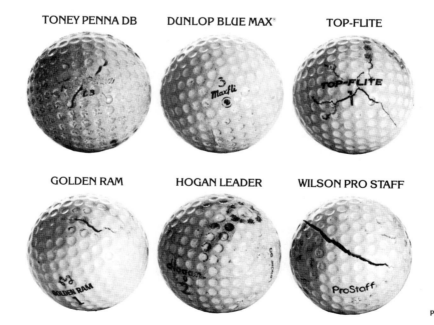

Photograph of balls is unretouched.

From *Golf Magazine*, June 1977, p. 9. Reproduced by permission of Dunlop Sports Company, Division of Dunlop Tire & Rubber Corp., Buffalo, N.Y.

Chapter Objectives

Previously, we tested hypothesis about two means. The analysis of variance techniques (ANOVA), which we are about to explore, is concerned with testing a hypothesis about several means: for example,

$$H_0: \mu_1 = \mu_2 = \mu_3 = \mu_4 = \mu_5$$

By using our former technique for two means, we could test several hypotheses if each states a comparison of two means. For example, we could test

1. $H_0: \mu_1 = \mu_2$
2. $H_0: \mu_1 = \mu_3$
3. $H_0: \mu_1 = \mu_4$
4. $H_0: \mu_1 = \mu_5$
5. $H_0: \mu_2 = \mu_3$
6. $H_0: \mu_2 = \mu_4$
7. $H_0: \mu_2 = \mu_5$
8. $H_0: \mu_3 = \mu_4$
9. $H_0: \mu_3 = \mu_5$
10. $H_0: \mu_4 = \mu_5$

To test the null hypothesis H_0 that all five means are equal, we would have to test these 10 hypotheses using our former technique for two means. Rejection of any one of the 10 hypotheses about two means would cause us to reject the null hypothesis that all five means are equal. If we were to fail to reject all 10 hypotheses about the means, we would fail to reject the main null hypothesis. However, if we were to test a null hypothesis dealing with several means by testing all the possible pairs of two means, we would lose track of the specific value of the probability of the type I error, α. The ANOVA techniques allow us to test the hypothesis that all means are equal against the alternative hypothesis that at least one mean value is different, with a specified value of α.

In this chapter we will present only an introduction to ANOVA. ANOVA experiments can be very complex, depending on the situation.

Section 12-1 The Logic Behind ANOVA

Many experiments are conducted to determine the effect that some test factor has on a response variable. The test factor may be temperature, the manufacturer of a product, the day of the week, or any number of other things. In this section we are investigating the **single-factor analysis of variance.** The design for the single-factor ANOVA is basically to obtain independent random samples at each of the several levels of the factor being tested. We then make a statistical decision concerning the effect that the levels of the test factor have on the response (observed) variable.

Briefly, the reasoning behind the technique proceeds as follows: To compare the means of the levels of the test factor, a measure of the **variation between the levels** (between columns on the data table) is compared with a measure of the **variation within the levels** (within the columns on the data table). If the variation between levels is significantly larger than the variation within levels, we then conclude that the means for each of the factor

variation between levels

variation within levels

levels being tested are not all the same, thus implying that the factor being tested does have a significant effect on the response variable. If, however, the variation between levels is not significantly larger than the variation within levels, we are not able to reject the null hypothesis that all means are equal.

Section 12-2 Introduction to the Analysis of Variance Technique

Let's continue our discussion of the analysis of variance technique by looking at an illustration.

Illustration 12-1

The temperature at which a plant is maintained is believed to affect the rate of production. The data in table 12-1 show the number x of units produced in 1 hour for randomly selected 1-hour periods when the process was operating at each of three temperature **levels**.

level

TABLE 12-1
Sample Results for
Illustration 12-1

	Temperature Levels		
	Sample from 68°F	Sample from 72°F	Sample from 76°F
	10	7	3
	12	6	3
	10	7	5
	9	8	4
		7	
Column totals	$C_1 = 41$	$C_2 = 35$	$C_3 = 15$
	$\bar{x}_1 = 10.25$	$\bar{x}_2 = 7.0$	$\bar{x}_3 = 3.75$

replicate

The data values from repeated sampling are called **replicates**. Four replicates, or data values, were obtained at two of the temperatures, and five were obtained from the third temperature. Does this data suggest that temperature has a significant effect on the production level at the 0.05 level?

The level of production is measured by the mean value, and $\bar{x}_i$ indicates the observed production mean at level i, where $i = 1$, 2, and 3 corresponds to temperatures 68°, 72°, and 76°F, respectively. There is a certain amount of variation among these means. Since sample means do not necessarily repeat when repeated samples are taken from a population, some variation can be expected. The question is, is this variation among the $\bar{x}$'s due to chance or is it due to the effect that temperature has on the production rate?

Solution The null hypothesis that we will test is

$$H_0: \mu_1 = \mu_2 = \mu_3$$

That is, the true production mean is the same at each temperature level tested. In other words, the temperature does not have a significant effect on the production rate.

The alternative to this hypothesis is

$$H_a: \text{Not all temperature level means are equal}$$

Thus we will want to reject the null hypothesis if the data show that one or more of the means are significantly different from the others. □

The decision to reject H_0 or fail to reject H_0 is made by using the F distribution and the F test statistic. Recall from chapter 10 that the calculated value of F is the ratio of two variances. The analysis of variance procedure will separate the variation among the entire set of data into two categories: **sum of squares for factor** (which measures variation between columns) and **sum of squares for error** (which measures variation within columns). Both are components of the **sum of squares for total**, or simply, sum of squares. Sum of squares is a name often used for the numerator of the fraction used to define sample variance [formula (2-7)]; thus:

sum of squares

$$\text{Sum of squares} = \sum (x - \bar{x})^2 \quad (12\text{-}1)$$

The **total sum of squares, SS(total)**, for the total set of data is calculated by using a formula that is equivalent to formula (12-1) but does not require the use of $\bar{x}$. This equivalent formula is

$$\text{SS(total)} = \sum (x^2) - \frac{(\sum x)^2}{n} \quad (12\text{-}2)$$

The SS(total) for our illustration can now be found by using formula (12-2):

$$\sum (x^2) = 10^2 + 12^2 + 10^2 + 9^2 + 7^2 + 6^2 + 7^2$$
$$+ 8^2 + 7^2 + 3^2 + 3^2 + 5^2 + 4^2$$
$$= 731$$
$$\sum x = 10 + 12 + 10 + 9 + 7 + 6 + 7$$
$$+ 8 + 7 + 3 + 3 + 5 + 4$$
$$= 91$$
$$\text{SS(total)} = 731 - \frac{(91)^2}{13}$$
$$= 731 - 637 = \mathbf{94}$$

The sum of squares, SS(total), must now be separated into two parts: (1) the sum of squares factor, SS(factor), and (2) the sum of squares error, SS(error).

$$SS(factor) + SS(error) = SS(total)$$

partitioning This splitting is often referred to as **partitioning**.

The **SS(factor)**, which measures the **variation between columns**, is found by using formula (12-3):

$$SS(factor) = \left(\frac{C_1^2}{k_1} + \frac{C_2^2}{k_2} + \frac{C_3^2}{k_3} + \cdots\right) - \frac{(\sum x)^2}{n} \qquad (12\text{-}3)$$

where C_i represents the ith column total, k_i represents the number of replicates at the ith level of the factor, and n represents the total sample size ($n = \sum k_i$).

SS(factor), the SS(temperature) for our illustration, can now be found by use of formula (12-3):

$$SS(\text{temperature}) = \left(\frac{41^2}{4} + \frac{35^2}{5} + \frac{15^2}{4}\right) - \frac{(91)^2}{13}$$

$$= (420.25 + 245.00 + 56.25) - 637.0$$

$$= 721.5 - 637.0 = \mathbf{84.5}$$

The sum of squares **SS(error)**, which measures the **variation within the rows**, is found by using formula (12-4):

$$SS(error) = \sum x^2 - \left(\frac{C_1^2}{k_1} + \frac{C_2^2}{k_2} + \frac{C_3^2}{k_3} + \cdots\right) \qquad (12\text{-}4)$$

The SS(error) for our illustration can now be found by use of formula (12-4):

$$\sum (x^2) = 731 \text{ (found previously)}$$

$$\left(\frac{C_1^2}{k_1} + \frac{C_2^2}{k_2} + \frac{C_3^2}{k_3}\right) = 721.5 \text{ (found previously)}$$

$$SS(error) = 731.0 - 721.5 = \mathbf{9.5}$$

NOTE: Inspection of the three formulas (12-2), (12-3), and (12-4) will verify that (SS(total) = SS(factor) + SS(error).

For convenience we will use an **ANOVA table** to record the sum of squares and to organize the rest of the calculations. The format of an ANOVA table is shown in table 12-2.

TABLE 12-2
Format for ANOVA Table

Source	SS	df	MS
Factor			
Error			
Total			

For our illustration the three sums of squares have been calculated. The degrees of freedom df associated with each of the three sources are determined as follows:

1. df(factor) is 1 less than the number of levels (columns) at which the factor is tested:

$$\text{df(factor)} = c - 1 \tag{12-5}$$

2. df(total) is 1 less than the total number of data:

$$\text{df(total)} = n - 1 \tag{12-6}$$

where n represents the number of data in the total sample (i.e., $n = k_1 + k_2 + k_3 + \cdots + k_c$, where k_i is the number of replicates at each level tested).

3. df(error) is the sum of the degrees of freedom for all the levels tested (columns in the data table). Each column has $k_i - 1$ degrees of freedom; therefore

$$\text{df(error)} = (k_1 - 1) + (k_2 - 1) + (k_3 - 1) + \cdots + (k_c - 1)$$

or

$$\text{df(error)} = n - c \tag{12-7}$$

The degrees of freedom for our illustration are

$$\text{df(temperature)} = c - 1 = 3 - 1 = \mathbf{2}$$
$$\text{df(total)} = n - 1 = 13 - 1 = \mathbf{12}$$
$$\text{df(error)} = n - c = 13 - 3 = \mathbf{10}$$

The sums of squares and the degrees of freedom are additive. That is,

$$\text{SS(factor)} + \text{SS(error)} = \text{SS(total)} \tag{12-8}$$
$$\text{df(factor)} + \text{df(error)} = \text{df(total)} \tag{12-9}$$

mean square The **mean square** for the factor being tested, **MS(factor)**, and for error, **MS(error)**, is obtained by dividing the sum-of-squares value by the corresponding number of degrees of freedom. That is,

$$\text{MS(factor)} = \frac{\text{SS(factor)}}{\text{df(factor)}} \tag{12-10}$$

$$\text{MS(error)} = \frac{\text{SS(error)}}{\text{df(error)}} \tag{12-11}$$

The mean squares for our illustration are

$$\text{MS(temperature)} = \frac{\text{SS(temperature)}}{\text{df(temperature)}} = \frac{84.5}{2}$$
$$= 42.25$$

$$\text{MS(error)} = \frac{\text{SS(error)}}{\text{df(error)}} = \frac{9.5}{10}$$
$$= 0.95$$

The completed ANOVA table then appears as shown in table 12-3.

TABLE 12-3
ANOVA Table for Illustration 12-1

Source	SS	df	MS
Temperature	84.5	2	42.25
Error	9.5	10	0.95
Total	94.0	12	

The hypothesis test is now completed using the two mean squares as measures of variance. The calculated value of the **test statistic, F^***, is found by dividing MS(factor) by MS(error):

$$F = \frac{\text{MS(factor)}}{\text{MS(factor)}} \qquad (12\text{-}12)$$

The calculated value of F for our illustration is then found by using formula (12-12):

$$F = \frac{\text{MS(temperature)}}{\text{MS(temperature)}} = \frac{42.25}{0.95}$$

$$F^* = \mathbf{44.47}$$

The decision to reject H_0 or fail to reject H_0 is made by comparing the calculated value of F, $F^* = 44.47$, with a one-tailed critical value of F obtained from table 9a of appendix D. See the accompanying figure.

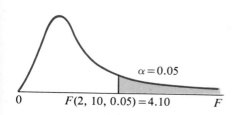

NOTE: Since the calculated value of F, F^*, is found by dividing MS(factor) by MS(error), the number of degrees of freedom for the numerator, df(factor), and the denominator, df(error), are used in that order in obtaining the critical value of F from the table.

For our illustration, df(temperature) = 2 and df(error) = 10.

We reject H_0, since the value F^* falls in the critical region. We can therefore conclude that room temperature does have a significant effect on the production rate. The differences in the mean production rates at the tested temperature levels were found to be significant.

In this section we used the ANOVA technique to separate the variance among the sample data into two measures of variance: (1) MS(factor), the measure of variance between the levels tested, and (2) MS(error), the measure of variance within the levels being tested. These measures of variance were then compared. Since the between-level variance was found to be significantly larger than the within-level variance (experimental error), we concluded that the factor had a significant effect on the variable x. For our illustration this led us to the conclusion that temperature did have a significant effect on the number of units of production completed per hour.

Illustration 12-2

Do the data in table 12-4 show sufficient evidence to conclude that there is a difference in the three population means, μ_1, μ_2, and μ_3?

TABLE 12-4
Sample Results for Illustration 12-2

	Factor Levels		
	Sample from Level 1	Sample from Level 2	Sample from Level 3
	3	5	8
	2	6	7
	3	5	7
	4	5	8
Column totals	$C_1 = 12$ $\bar{x}_1 = 3.00$ $k_1 = 4$	$C_2 = 21$ $\bar{x}_2 = 5.25$ $k_2 = 4$	$C_3 = 30$ $\bar{x}_3 = 7.50$ $k_3 = 4$

Figure 12-1 shows the relative relationship among the three samples. They demonstrate relatively little **within-sample variation**, although there is a relatively large amount of **between-sample variation**, thus implying that the populations being sampled have different mean values.

FIGURE 12-1
Relative Relationship Among the Three Samples in Illustration 12-2

Let's look at another illustration.

Illustration 12-3

Do the data in table 12-5 show sufficient evidence to conclude that there is a difference in the three population means, μ_J, μ_K, and μ_L?

TABLE 12-5 Sample Results for Illustration 12-3

	Factor Levels		
	Sample from Level J	Sample from Level K	Sample from Level L
	3	5	6
	8	4	2
	6	3	7
	4	7	5
Column totals	$C_J = 21$ $\bar{x}_J = 5.25$	$C_K = 19$ $\bar{x}_K = 4.75$	$C_L = 20$ $\bar{x}_L = 5.00$

Figure 12-2 shows the relative relationship among the three samples. For these three samples there is little between-sample variation (i.e., the sample means are relatively close in value), whereas the within-sample variation is relatively large (i.e., the data values within each sample cover a relatively wide range of values). A quick look at the diagram *does not suggest* that the three population means are different from each other.

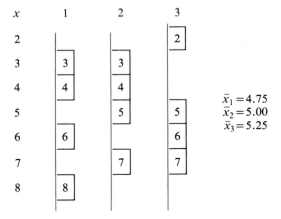

FIGURE 12-2 Relative Relationship Among the Three Samples in Illustration 12-3

To complete a hypothesis test for analysis of variance, we have to make the following three basic assumptions:

1. Our goal is to investigate the effect that various levels of the factor under test have on the response variable. Typically, we will be looking to find the level that yields the most advantageous values of the response variable. This procedure, of course, means that we will probably want to reject the null hypothesis in favor of the alternative. Then a follow-up study could determine the "best" level of the factor.

2. We must assume that the effects due to chance and to untested factors are normally distributed and that the variance caused by these effects is constant throughout the experiment.
3. We must assume independence among all observations of the experiment. (Recall that independence basically means that the results of any one observation of the experiment do not affect the results of any other observation.) We will usually run the tests in a randomly assigned order to ensure independence. This technique will also help to avoid data contamination.

Section 12-3 Applications of Single-Factor ANOVA

Before continuing with the ANOVA discussion, let's identify the notation, particularly the subscripts that are used (see table 12-6). Notice that each piece of data can be double subscripted: the first subscript indicates the column number (test factor level), and the second identifies the replicate (row) number. The column totals, C_j, are listed across the bottom of the table, with C_c being the last column total. The grand total, T, is equal to the sum of all x and is generally found by adding the column totals. Row totals can be used as a cross-check, but serve no other purpose.

TABLE 12-6
Notation Used in Anova

	Factor Levels				
Replication	Sample from Level 1	Sample from Level 2	Sample from Level 3	...	Sample from Level c
$k = 1$	$x_{1,1}$	$x_{2,1}$	$x_{3,1}$		$x_{c,1}$
$k = 2$	$x_{1,2}$	$x_{2,2}$	$x_{3,2}$	...	$x_{c,2}$
$k = 3$	$x_{1,3}$	$x_{2,3}$	$x_{3,3}$		$x_{c,3}$
⋮					
Column totals	C_1	C_2	C_3	...	C_c
	T = grand total = sum of all x's = $\sum x = \sum C_j$				

(*Note*: It is *not* necessary for the number of replications (k) to be the same in each column.)

Table 12-6 presents the notation for a single-factor ANOVA problem in tabular form. As an alternative, a single-factor ANOVA problem can be expressed in more general terms, that is, in the form of a single equation. This single-equation form, known as a mathematical model, is often used to express a particular situation. In chapter 3 we used a mathematical model to help explain the relationship between the values of bivariate data. The equation $\hat{y} = b_0 + b_1 x$ was used as the model when we believed that a straight-line relationship existed. The probability functions that were studied in chapter 5 are also examples of mathematical models. For the single-factor ANOVA, the **mathematical model** is an expression of the composition of each data entry in our data table.

$$x_{c,k} = \mu + F_c + \varepsilon_{k(c)} \tag{12-13}$$

The terms of this model may each be interpreted as follows:

1. μ is the population mean value for all the data without respect to the test factor.
2. F_c is the effect that the factor being tested has on the response variable at each different level c.
3. $\varepsilon_{k(c)}$ (ε is the Greek lowercase letter epsilon) is the **experimental error** that occurs among the k replicates in each of the c columns.

Let's look at another hypothesis test concerning an analysis of variance.

Illustration 12-4

Continent Insurance's major advertising effort is to send advertisements with the monthly bills of three major credit card firms. The advertisements include a return postcard if more information is desired. These "leads" are then followed up in person by Continent's sales force. The number of leads per 1,000 mailings is approximately the same from each credit card firm's mailing. However, the quality of the leads from each source—that is, the proportion of sales Continent can expect to close from each source's leads—is in question. Since budget constraints require Continent to reduce the mailings, an experiment was set up to determine if the quality of leads was different by source. Fifteen salespeople were randomly selected and each given 200 leads from a particular credit card's mailing. The resulting number of sales is shown in table 12-7.

TABLE 12-7
Sample Results for
Illustration 12-4

National	Business	Super Charge
12	10	16
10	17	14
18	16	16
12	13	11
14		20
		21

At the 0.05 level of significance, is there sufficient evidence to reject the claim that the credit cards used for the leads are equally effective?

Solution In this experiment the factor is "credit card company" and the levels are the three different credit card companies (National, Business, Super Charge). The replicates are the number of sales made by each of the different salespeople using leads from each of the different credit card companies. The null hypothesis to be tested is "the leads from the three credit card companies are equally effective" or "the mean number of sales attained using each of the three credit card companies is the same."

STEP 1: H_0: $\mu_N = \mu_B = \mu_{SC}$.

STEP 2: H_a: The means are not all equal (i.e., at least one mean is different).

STEP 3: $\alpha = 0.05$. The test criteria are shown in the diagram.

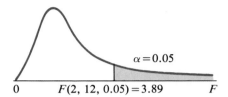

$$df(\text{numerator}) = df(\text{company}) = 3 - 1 = 2 \quad [\text{using formula (12-5)}]$$

$$df(\text{denominator}) = df(\text{error}) = 15 - 3 = 12 \quad [\text{using formula (12-7)}]$$

STEP 4: Calculate the test statistic F^*. Table 12-8 is used to find column totals.

TABLE 12-8 Calculations for Illustration 12-4

Replicates	National	Business	Super Charge
$k = 1$	12	10	16
$k = 2$	10	17	14
$k = 3$	18	16	16
$k = 4$	12	13	11
$k = 5$	14		20
$k = 6$			21
Totals	$C_N = 66$	$C_B = 56$	$C_{SC} = 98$
	$k_N = 5$	$k_B = 4$	$k_{SC} = 6$
	$\bar{x}_N = 13.2$	$\bar{x}_B = 14.0$	$\bar{x}_{SC} = 16.3$

Next, calculate the summations $\sum x$ and $\sum x^2$:

$$\sum x = 12 + 10 + 18 + 12 + 14 + 10 + 17 + \cdots + 21$$
$$= 220$$
$$\sum x^2 = 12^2 + 10^2 + 18^2 + 12^2 + 14^2 + 10^2 + \cdots + 21^2$$
$$= 3{,}392$$

Using formula (12-2), we find

$$SS(\text{total}) = 3{,}392 - \frac{(220)^2}{15}$$
$$= 3{,}392 - 3{,}226.67$$
$$= 165.33$$

Using formula (12-3), we find

$$SS(\text{company}) = \left(\frac{66^2}{5} + \frac{56^2}{4} + \frac{98^2}{6}\right) - 3{,}226.67$$
$$= 3{,}255.87 - 3{,}226.67$$
$$= \mathbf{29.20}$$

Using formula (12-4), we find

$$SS(\text{error}) = 3{,}392 - 3{,}255.87 = \mathbf{136.13}$$

Check the sums of squares by using formula (12-8):

$$SS(\text{company}) + SS(\text{error}) = 29.20 + 136.13 = 165.33$$

and

$$SS(\text{total}) = 165.33$$

The numbers of degrees of freedom are found using formulas (12-5), (12-6), and (12-7):

$$df(\text{company}) = 3 - 1 = \mathbf{2}$$
$$df(\text{total}) = 15 - 1 = \mathbf{14}$$
$$df(\text{error}) = 15 - 3 = \mathbf{12}$$

Using formulas (12-10) and (12-11), we find

$$MS(\text{company}) = \frac{29.20}{2} = \mathbf{14.60}$$

$$MS(\text{error}) = \frac{136.13}{12} = \mathbf{11.34}$$

TABLE 12-9
ANOVA Table for Illustration 12-4

Source	SS	df	MS
Company	29.20	2	14.60
Error	136.13	12	11.34
Total	165.33	14	

The results of these computations are combined in the ANOVA table shown in table 12-9. The calculated value of the test statistic is then found using formula (12-12):

$$F^* = \frac{14.60}{11.34} = \mathbf{1.287}$$

STEP 5: The decision is made by comparing F^* with the critical value shown in the test criteria (step 3).

Decision: Fail to reject H_0. (F^* is in the noncritical region.)

Conclusion: The data show no evidence that would give reason to reject the null hypothesis that the quality of leads from the three credit card companies are equally effective. □

Recall that the null hypothesis is "there is no difference between the means of the response values for the levels of the factor being tested." A "fail to reject H_0" decision must be interpreted as the conclusion that there is no evidence of a difference in response due to the levels of the tested factor, whereas the rejection of H_0 implies that there is a difference between the means of response values for the various levels. If there is a difference, the next problem is to locate the level (or levels) that is different. Locating this difference may be the main object of the analysis. To do this, the only method that is appropriate at this stage of our development is to inspect the data. It may be somewhat obvious which level(s) caused the rejection of H_0. In illustration 12-1 it seems quite obvious that each level is different from the other two. If the higher values are more desirable for finding the "best" level to use, we would choose that corresponding level of the factor.

To this point we have dealt with analysis of variance for data dealing with one factor. It is not unusual for problems to have several factors of concern. The ANOVA techniques presented in this chapter can and will be further developed and applied to these more complex cases.

Exercises

12-1 Suppose that an F test (as described in this chapter) has a critical value of 2.2, as shown in the diagram.

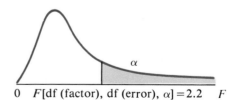

(a) What is the interpretation of a calculated value of F larger than 2.2?

(b) What is the interpretation of a calculated value of F smaller than 2.2?

(c) What is the interpretation if the calculated F is 0.1? 0.01?

12-2 Why does df(factor), the number of degrees of freedom, always appear first in the critical value notation F [df(factor), df(error), α]?

12-3 Consider the accompanying table for a single-factor ANOVA. Find the following:
(a) $x_{1,2}$
(b) $x_{2,4}$
(c) C_1
(d) $\sum x$
(e) $\sum (C_j)^2$

Replicates	Level of Factor		
	1	2	3
1	3	2	7
2	0	5	4
3	1	4	5
4	5	3	6
5	7	6	2

12-4 (a) State the null and alternative hypotheses that would be under test in exercise 12-3.
(b) How would a "reject the null hypothesis" decision be interpreted?
(c) How would a "fail to reject the null hypothesis" decision be interpreted?
(d) How is such a decision reached?

12-5 A certain vending company's soft-drink-dispensing machines are supposed to serve 6 ounces of beverage. Various machines were sampled and the resulting amounts of dispensed drink were recorded in the accompanying table. Does this sample evidence provide sufficient reason to reject the null hypothesis that all four machines dispense the same average amount of soft drink? Use $\alpha = 0.01$.

Machines			
A	B	C	D
3.8	6.8	4.4	6.5
4.2	7.1	4.1	6.4
4.1	6.7	3.9	6.2
4.4			6.9

12-6 Test the hypothesis that "the day of the week does have an effect upon the amount of grocery store sales at Fast Market." The sales in thousands of dollars are shown in the following table. Complete the test at $\alpha = 0.05$.

Replicate	Days						
	Sun.	Mon.	Tues.	Wed.	Thurs.	Fri.	Sat.
1	8	3	18	17	16	8	3
2	4	12	15	11	8	17	17
3	2	10	11	12	14	11	5
4	7	14	9	15	12	13	9

Section 12-4 Two-Factor ANOVA (Without Replication)

Frequently we will wish to consider two factors and their simultaneous effects upon the response variable. To illustrate, let's do another experiment similar to the one in illustration 12-1. This time, instead of **repeated observations** (replicates) of results at each level of a single factor (temperature), we will introduce a second factor, humidity, and conduct a series of observations in which only *one* observation is made each time the level of either temperature and/or humidity is varied. The results of such an experiment are displayed in table 12-10. Note that the rows now indicate the various temperature (factor FI) levels, and the columns indicate the various humidity (factor FII) levels. There is only one observation at each combination of levels, and thus "no replication."

TABLE 12-10
Temperature/Humidity
Experiment Data

Temperature Levels	Humidity Levels				Total
	1	2	3	4	
1	7	6	7	8	28
2	3	3	4	5	15
3	10	12	10	9	41
Total	20	21	21	22	84

Our mathematical model for a two-factor experiment without replication is

$$x_{rc} = \mu + \text{FI}_r + \text{FII}_c + \varepsilon_{rc} \tag{12-14}$$

where x_{rc} is the piece of data in the rth row and the cth column. The model tells us that the value of x_{rc} is determined by the overall average of the population, μ, modified by an effect caused by the row factor (FI), an effect caused by the column factor (FII), and a random experimental error.

Because there are no replications at each factor level, we cannot calculate SS(error) directly. Recall that SS(error) measures the variation that takes place within each treatment level. Instead, SS(error) is estimated by the SS(residual). The residual is that part of the sum of squares that is left over after the SS(row factor) and SS(column factor) have been subtracted out of SS(total). You will frequently find the terms SS(error) and SS(residual) used interchangeably.

To estimate SS(error), we must assume that there is no interaction between factors. When two factors are said *not* to interact, we mean that the effect of one factor upon the dependent variable is the same regardless of the presence, absence, or level of the other factor. However, the effect of one factor often *is* different because of the presence, absence, or level of the other factor, and in such a case there is said to be interaction between the factors.

For our illustration in this section, we must assume that there is no interaction effect between the two factors. Since the physical properties of temperature and humidity are related, in reality they might have a combined effect upon the response variable. It is also possible that some factors that interact with each other can have separate effects on the response variable. In other words, direct interaction between factors is not the same concept as interaction effect upon the response variable.

For certain types of experiments, it is not unusual to assume that there is no interaction between factors. For our illustration we proceed with that assumption.

There are now two null hypotheses to be tested. One is "the mean value of the response variable is the same at all levels of factor I," and the other is "the mean value of the response variable is the same at all levels of factor II." Let's complete the hypothesis test for each of the null hypotheses at $\alpha = 0.01$.

STEP 1: Null hypotheses:

$$H_0(\text{temperature}): \mu_1 = \mu_2 = \mu_3$$

$$H_0(\text{humidity}): \mu_1 = \mu_2 = \mu_3 = \mu_4$$

STEP 2: Alternative hypotheses:

$$H_a(\text{temperature}): \text{At least one } \mu \text{ is different}$$

$$H_a(\text{humidity}): \text{At least one } \mu \text{ is different}$$

STEP 3: $\alpha = 0.01$. The critical values will be determined after the number of degrees of freedom have been determined.

Before we can begin our calculations, we need to agree upon some notation (see table 12-11) and we will need formulas.

TABLE 12-11
ANOVA Notation for a Two-Factor Experiment, No Replication

Factor I Levels	Factor II Levels					Row Totals
	1	2	3	$\cdots j \cdots$	c	
1	$x_{1,1}$	$x_{1,2}$	$x_{1,3}$		$x_{1,c}$	R_1
2	$x_{2,1}$				$x_{2,c}$	R_2
$\vdots$						
i				$x_{i,j}$		R_i
$\vdots$						
r	$x_{r,1}$	$x_{r,2}$			$x_{r,c}$	R_r
Column totals	C_1	C_2	C_3	C_j	C_c	T

NOTE: R_i represents the ith row total; C_j represents the jth column total.

The formulas we will need are

$$SS(\text{total}) = \sum_{\text{all cells}} (x_{i,j})^2 - \frac{(T)^2}{n} \quad (12\text{-}15)$$

$$SS(\text{factor I}) = \frac{\sum (R_i)^2}{c} - \frac{(T)^2}{n} \quad (12\text{-}16)$$

$$SS(\text{factor II}) = \frac{\sum (C_j)^2}{r} - \frac{(T)^2}{n} \quad (12\text{-}17)$$

$$SS(\text{residual}) = SS(\text{total}) - [SS(\text{factor I}) + SS(\text{factor II})] \quad (12\text{-}18)$$

NOTE: Formula (12-15) is the same as the previous formula for SS(total). Formulas (12-16) and (12-17) are basically the same as formula (12-3).

$$df(\text{total}) = n - 1 \quad (n \text{ is total number of pieces of data}) \quad (12\text{-}19)$$

$$df(\text{factor I}) = r - 1 \quad (r \text{ is number of levels of factor I}) \quad (12\text{-}20)$$

$$df(\text{factor II}) = c - 1 \quad (c \text{ is number of levels of factor II}) \quad (12\text{-}21)$$

$$df(\text{residual}) = df(\text{total}) - [df(\text{factor I}) + df(\text{factor II})] \quad (12\text{-}22)$$

STEP 4: Data:

Temperature Levels	Humidity Levels				Row Totals
	1	2	3	4	
1	7	6	7	8	28
2	3	3	4	5	15
3	10	12	10	9	41
Column totals	20	21	21	22	84

ANOVA table:

Source	SS	df(e)	MS(f)
Temperature	84.50 (a)	2	42.25
Humidity	0.66 (b)	3	0.22
Residual	8.84 (d)	6	1.4733
Total	94.00 (c)	11	

(a) Formula (12-16):

$$SS(\text{temperature}) = \frac{(28)^2 + (15)^2 + (41)^2}{4} - \frac{(84)^2}{12} = \mathbf{84.50}$$

(b) Formula (12-17):

$$SS(\text{humidity}) = \frac{(20)^2 + (21)^2 + (21)^2 + (22)^2}{3} - \frac{(84)^2}{12} = \mathbf{0.66}$$

(c) Formula (12-15):

$$SS(\text{total}) = (7)^2 + (6)^2 + \cdots + (9)^2 - \frac{(84)^2}{12} = \mathbf{94.00}$$

(d) Formula (12-18):

$$SS(\text{residual}) = 94.00 - (84.50 + 0.66) = \mathbf{8.84}$$

(e) The numbers of degrees of freedom are found according to formulas (12-19) through (12-22).

(f) The three values for mean square are found by dividing the sum of squares by the number of degrees of freedom.

Now we can test each of the two separate null hypotheses that were stated earlier. The calculated value of F is obtained by dividing the mean square for the factor being tested by the mean square for the experimental error (residual in this case). The critical value for H_0(temperature) is shown in the accompanying figure.

$F(2, 6, 0.01) = 10.9$ F

$$F^* = \frac{42.25}{1.4733} = 28.67$$

STEP 5:

(a) *Decision*: Reject H_0.

(b) *Conclusion*: Temperature does have an effect on the productivity.

The critical value for H_0(humidity) is shown in the accompanying diagram.

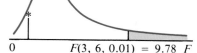
$F(3, 6, 0.01) = 9.78$ F

$$F^* = \frac{0.22}{1.4733} = 0.149$$

STEP 6:

(a) *Decision*: Fail to reject H_0.

(b) *Conclusion*: There is no evidence to indicate an effect due to humidity.

Notice that in each of the F tests the value of the mean square for that factor is tested against the MS(residual). Essentially, we are asking if the measured variation due to that factor is sufficiently bigger than the measure of variation within the whole experiment to enable us to reject the null hypothesis and thereby claim that the variation is due to more than chance error. When this rejection occurs, we say that the variation is due to the various levels of the factor.

Exercises

12-7 A consumer group ran a comparison test to help determine the "best" gasoline for a Toyota. Three Toyotas were used for each year (nine in all). Each car was run for one tank of one of the three brands of gasoline. The following table shows the mileage obtained.

Brand of Gasoline	Distance Traveled (miles)		
	1982 Cars	1983 Cars	1984 Cars
I	21.2	18.7	17.0
II	21.5	19.4	19.2
III	18.7	18.7	18.6

(a) Test the hypothesis that the car used did not have a significant effect on the responses.

(b) Test the hypothesis that the brand of gasoline had no significant effect on the mean mileage.

12-8 Five solutions containing a certain chemical were prepared for industrial use. Three analysts were asked to take a sample from each solution and to measure the content of this certain chemical. The results are shown as follows:

Analyst	Solution				
	1	2	3	4	5
A	5.9	3.9	3.4	8.2	5.6
B	6.4	4.2	3.7	7.5	6.1
C	7.6	3.7	3.1	7.7	5.4

(a) What is the mathematical model for this experiment?

(b) Is there a variation in the mean level of the five different solutions?

(c) Is there a significant difference in the results obtained by the analysts?

Section 12-5 Two-Factor ANOVA (with Replication)

Previously, we chose to ignore the possibility of interaction between the factors being tested. This policy is often inappropriate in real applications of the ANOVA. The following two illustrations demonstrate the idea of interaction:

1. Starting salaries are greatly dependent upon a person's major field of study and his or her grades in school. If a major is not in high demand, then grades may be very critical. On the other hand, if a major is very much in demand, then grades may be much less critical.

2. In certain chemical reactions the effect of temperature and type of catalyst may be greatly dependent upon each other.

Many other illustrations could be cited; however, these examples should demonstrate the idea of interacting factors.

We are going to look at a two-factor experiment in which we suspect interaction; therefore we must replicate. By replicating, we will be able to partition the sum of squares of residual into two cells—experimental error and interaction.

The mathematical model for this experiment is

$$x_{r,c,k} = \mu + \text{FI}_r + \text{FII}_c + (\text{FI} \times \text{FII})_{r,c} + \varepsilon_{k(r,c)} \qquad (12\text{-}23)$$

$x_{r,c,k}$ is the piece of data in the rth row, cth column, and the kth replicate. This piece of data would be the result of the overall average μ, modified by the kth experiment at the rth level of factor I and the cth level of factor II. FI_r represents the factor I effect (rows $1, 2, \ldots, i, \ldots, r$). FII_c represents the factor II effect (columns $1, 2, \ldots, j, \ldots, c$). $(\text{FI} \times \text{FII})_{r,c}$ represents the joint effect of the two factors in the rth row and the cth column. $\varepsilon_{k(r,c)}$ represents the experimental error. This error occurs among the k replicates in each of the r, c cells. [Notice the "nested notation" $k(r, c)$ that is used to represent this concept.]

crossed experiment An experiment that is designed as shown here is often referred to as a **crossed experiment**. Data are obtained from each of the rc cells. Each level of factor I is used in combination with each level of factor II—thus a crossed experiment. When obtaining the data for an experiment of this type, we should randomize the sequence to avoid any systematic errors that might be introduced.

Illustration 12-5

The study example is numerical only; factors I and II are to be investigated with r levels (3) of factor I and c levels (4) of factor II, with k replicates (2).

The data are as follows:

TABLE 12-12
Data for Illustration 12-5

Factor I	Factor II				R_i
	1	2	3	4	
1	0, 2 / 2	4, 4 / 8	0, 0 / 0	10, 10 / 20	30
2	2, 2 / 4	4, 6 / 10	2, 2 / 4	8, 10 / 18	36
3	2, 4 / 6	6, 6 / 12	2, 0 / 2	8, 8 / 16	36
C_j	12	30	6	54	102

NOTE: The number in the corner of each cell is the cell total.

Tables 12-13 and 12-14 show the notation that will be used to identify the individual data and the various totals. The formulas to be used follow the table. Note carefully the expanded notation from the two previous sections.

The accompanying formulas are

$$\text{SS(FI)} = \frac{(R_1)^2 + (R_2)^2 + \cdots + (R_r)^2}{ck} - \frac{(T)^2}{n} \quad (12\text{-}24)$$

$$\text{SS(FII)} = \frac{(C_1)^2 + (C_2)^2 + \cdots + (C_c)^2}{rk} - \frac{(T)^2}{n} \quad (12\text{-}25)$$

$$\text{SS(FI} \times \text{FII)} = \frac{\sum_{\text{all cells}} (T_{i,j})^2}{k} - \frac{(T)^2}{n} - [\text{SS(FI)} + \text{SS(FII)}] \quad (12\text{-}26)$$

$$\text{SS(error)} = \sum_{\text{all cells}} \left[\sum_1^k x^2 - \frac{\left(\sum_1^k x\right)^2}{k} \right] \quad (12\text{-}27)$$

NOTE: The summation for SS(error) is done one cell at a time.

$$\text{SS(total)} = \sum (x_{i,j,k})^2 - \frac{(T)^2}{n} \quad (12\text{-}28)$$

TABLE 12-13
Data Table

Factor I Levels	Factor II Levels				Row Totals
	1	2	$\cdots j \cdots$	c	
1	$x_{1,1,1}$ $x_{1,1,2}$ $\vdots$ $x_{1,1,k}$ $T_{1,1}$	$T_{1,2}$		$T_{1,c}$	R_1
$\vdots$					$\vdots$
i			$x_{i,j,1}$ $x_{i,j,2}$ $\vdots$ $x_{i,j,k}$ Cell totals $T_{i,j}$		R_i
$\vdots$					$\vdots$
r				$x_{r,c,1}$ $x_{r,c,2}$ $\vdots$ $x_{r,c,k}$ $T_{r,c}$	R_r
Column totals	C_1	C_2	$\cdots C_j \cdots$	C_c	T Grand total

TABLE 12-14
ANOVA Table

Source	SS	df	MS
Factor I	(12-24)	$r - 1$	
Factor II	(12-25)	$c - 1$	
FI × FII	(12-26)	$(r - 1)(c - 1)$	
Error	(12-27)	$rc(k - 1)$	
Total	(12-28)	$n - 1$	

The formula for SS(error) is a little tricky to use. We must be very careful to obtain a separate total for each of the rc cells; then the sum of all cell totals becomes SS(error). The other sum-of-squares formulas are extensions of the previous formulas. The formula for number of degrees of freedom associated with each of these sums of squares is shown in table 12-14.

CAUTION: When working with these formulas, do not round off while calculating the sums of squares. If rounding off is necessary, keep as many decimal places as possible.

The three null hypotheses to test are:
1. There is no interaction.
2. There is no effect due to factor II.
3. There is no effect due to factor I.

Each of these hypotheses are tested by comparing the mean square of each factor with the mean square of error.

Illustration 12-5 is finished as follows at the 5% level of significance.

ANOVA table:

Source	SS	df	MS
Factor I	3.00	2	1.500
Factor II	232.50	3	77.500
FI × FII	13.00	6	2.167
Error	10.00	12	0.833
Total	258.50	23	

Calculations (data from table 12-12):

$$SS(I) = \frac{(30)^2 + (36)^2 + (36)^2}{8} - \frac{(102)^2}{24} = 436.5 - 433.5 = 3.0$$

$$SS(II) = \frac{(12)^2 + (30)^2 + (6)^2 + (54)^2}{6} - \frac{(102)^2}{24} = 666 - 433.5 = 232.5$$

$$SS(\text{total}) = (0^2 + 2^2 + 2^2 + 2^2 + 2^2 + 4^2 + \cdots + 8^2) - \frac{(102)^2}{24}$$
$$= 692 - 433.5 = 258.5$$

$$SS(I \times II) = \frac{(2^2 + 4^2 + 6^2 + 8^2 + \cdots + 16^2)}{2} - \frac{(102)^2}{24} - (3.0 + 232.5)$$
$$= 682 - 433.5 - 235.5 = 13.0$$

$$SS(\text{error}) = \left[(0^2 + 2^2) - \frac{(2)^2}{2}\right] + \left[(2^2 + 2^2) - \frac{(4)^2}{2}\right]$$
$$+ \cdots + \left[(8^2 + 8^2) - \frac{(16)^2}{2}\right]$$
$$= 10.00$$

NOTES:
1. SS's are *always* positive!
2. Totals must check.
3. SS(error) is the sum over all cells.

Test of hypotheses ($\alpha = 0.05$):

$H_0(I \times II)$: There is no interaction between factors I and II

$H_a(I \times II)$: There is interaction between factors I and II

$\alpha = 0.05$

$$F^* = \frac{MS(I \times II)}{MS(\text{error})} = \frac{2.167}{0.833} = 2.60$$

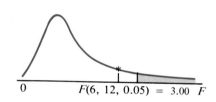

$F(6, 12, 0.05) = 3.00$

Fail to reject $H_0(I)$; that is, there is no evidence of an effect due to the factors I and II.

$H_0(II)$: $\mu_1 = \mu_2 = \mu_3 = \mu_4$

$H_a(II)$: Mean values not all equal

$\alpha = 0.05$

$$F^* = \frac{MS(II)}{MS(\text{error})} = \frac{77.5}{0.833} = 93.04$$

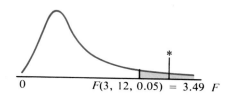

$F(3, 12, 0.05) = 3.49$

Reject $H_0(II)$; that is, it appears that factor II does have an effect on the variable x.

$H_0(I)$: $\mu_1 = \mu_2 = \mu_3$ or $\sigma^2_{\mu_r} = 0$

$H_a(I)$: Mean values not all equal

$\alpha = 0.05$

$$F^* = \frac{MS(I)}{MS(\text{error})} = \frac{1.50}{0.833} = 1.8$$

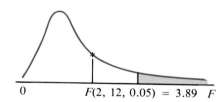

$F(2, 12, 0.05) = 3.89$

Fail to reject $H_0(I)$; that is, there is no evidence of an effect due to the different levels of factor I.

The interaction (or lack of interaction) may also be shown graphically by means of an interaction graph. On such a graph the average value of each cell is graphed at c horizontal positions, and a separate point is plotted on the graph for each level r of factor I.

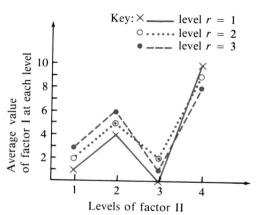

Two-Factor ANOVA (With Replication)   Section 12-5

The lack of interaction is shown by parallel lines. The presence of interaction is suggested by nonparallel (intersecting) line segments. The greater the angle formed by the line segments, the greater the indication of interaction.

Exercises

12-9 The local soft drink vending company has some machines that will dispense ice or no ice, whichever the customer desires. The following results are weights of the amount of soft drink dispensed. In the cases where ice was dispensed, it was immediately removed and only the soft drink weighed.

	Vending Machine			
	I	II	III	IV
Ice	6.5	3.1	7.6	6.0
	6.4	3.0	7.6	5.9
	6.8	3.3	7.5	6.0
No ice				
	6.2	3.3	7.4	6.1
	6.1	3.4	7.0	6.0
	6.2	3.4	7.4	6.0

(a) State the mathematical model.
(b) Test the hypothesis dealing with differences in machine, difference with ice or no ice, and for interaction. ($\alpha = 0.01$.)
(c) Draw an interaction graph and comment on what it shows you and how it seems to support the results of part (b).

12-10 Many industrial psychologists feel that the employee's shift plays an important role in determining his or her productivity. The following data records the time required (in minutes) to complete a production task by five

	Worker				
Shift	1	2	3	4	5
I	7.7	7.0	6.6	7.2	5.8
	6.8	5.5	7.4	7.4	5.1
II	7.0	5.3	6.1	6.7	4.8
	6.7	6.0	5.3	7.1	7.1
III	7.9	7.5	7.5	7.0	7.6
	8.2	7.5	7.7	7.9	7.9

different employees rotated randomly to each of the three shifts. Two measurements were made on each employee for each shift (20 has been subtracted from each entry).

(a) What is the mathematical model for this experiment?
(b) Is there a significant difference due to the shift?
(c) Is there a difference in the workers?
(d) Does the data suggest that some workers are more productive on certain shifts? (i.e., is there an interaction?)
(e) Draw an interaction graph. Comment on what it seems to tell you.

12-11 The data in the accompanying table, showing sales of product in 10,000 units, were obtained for the following tests: (1) Does shelf location affect the sales of the product? (2) Does the color of the package affect the sales? (3) Is there an interaction between the shelf location and the color of the package?

Shelf Location	Color		
	Black	Orange	Tan
Bottom	0.01	4.5	0.4
	0.02	5.3	0.5
	0.01	5.2	0.4
Waist level	0.04	3.1	0.2
	0.05	2.5	0.4
	0.03	3.6	0.4
Eye level	0.01	2.4	0.5
	0.02	1.3	0.4
	0.02	1.5	0.6
Top	0.01	0.8	0.3
	0.00	1.1	0.2
	0.01	1.0	0.4

(a) State the mathematical model for this experiment and complete the ANOVA test at $\alpha = 0.05$.
(b) Sketch an interaction graph.
(c) Does the interaction graph seem to agree with the results of the tests in part (a)?

In Retrospect

In this chapter we have presented an introduction to the statistical technique known as analysis of variance. This technique is used in conjunction with an experiment or situation in which variations in the value of one or more factors are expected to produce a change in the value of a single variable. Several observations of the changed value of the variable are recorded at each variation or level of the influencing factor(s). The mean of the observed values of the variable is computed for each level of each factor, resulting in multiple means. Analysis of variance allows us to test the statistical significance of the differences between the multiple means simultaneously, rather than two at a time as we had done previously.

The methods we have learned require that we restrict ourselves to normally distributed populations and to populations with homogeneous (equal) variances at the various levels. To test multiple means, we divide (partition) a measure of variance, the sum of squares, into segments: one part is assignable to the difference between the levels and the other part is assignable to the difference within the levels of variation of the factor(s).

The chapter started with the simplest experiment, where we had one factor being tested at each of several levels. We then introduced a second factor and considered not only the individual effects of two factors but also the concept of interaction between the two test factors. The concept of interaction was also studied with the aid of the interaction graph, a graphic display of the data that helps to visually explain the experimental findings. As noted previously, this chapter is only an introduction to analysis of variance. The application of this technique can become quite complex as additional concepts are included.

Refer back to the news article at the beginning of the chapter and you will see the results from a durability test. The factor being tested is the brand of golf ball, whereas each ball tested is a replicate. The advertisement wants you to conclude that "the Blue Max is the most durable." Does this conclusion seem reasonable based on what you have learned in this chapter?

Chapter Exercises

12-12 Four types of advertising displays were set up in 20 randomly selected retail stores for the purpose of evaluating the point-of-sale impact of the different displays. The type of display was randomly assigned to each store, and the number of units of the sale item sold in each store was recorded. The sales in hundreds of dollars are shown in the following table. At $\alpha = 0.01$, do the data provide sufficient evidence to reject the null hypothesis that the average effect at the point of sale is the same for each display?

Stores (replicates)	Type of Display			
	1	2	3	4
1	65	64	85	86
2	81	88	74	62
3	71	75	70	70
4	92	84	64	64
5	74		73	68
6	85			

12-13 The industrial relations division of United Works is considering three alternative incentive plans. Plan A is an individual incentive plan, plan B is a group incentive plan (the productivity of a complete production line is shared by all members of the line), and plan C is a plan that has aspects of both individual and group incentives. Thirteen production lines were selected and each was randomly assigned to a plan. The total production for the week for each line was then recorded. The production results ($\times 1{,}000$ units) are shown in the following table. At $\alpha = 0.01$, should we conclude that the type of incentive plan has a significant effect upon productivity?

Incentive Plan		
A	B	C
4	2	9
6	1	12
7	3	14
5	2	10
		15

Production results ($\times 1{,}000$ units)

12-14 In attempting to locate a new plant facility, the personnel department raised the problem of availability of skilled craft labor. The personnel manager maintains that the availability of such labor varies significantly according to size of community. The accompanying table shows the number of skilled craftspersons, according to the U.S. Census, in 18 communities of four different size categories. Entries are rates per 10,000 inhabitants. At $\alpha = 0.05$, is there sufficient evidence to support the personnel manager's claim?

Large Cities (over 250,000)	Cities (100,000 to 250,000)	Cities (under 100,000)	Nonurban
45	23	25	8
34	18	17	16
41	27	19	14
42	21	28	17
37	26		

12-15 Pendon Seminars Incorporated believes that the demand for executive-training seminars is price inelastic (i.e., price levels do not significantly affect sales). To determine if the hypothesis appears to be true, the next 32 presentations of the seminar "Finance for the Nonfinancial Executive" were randomly separated into four groups of eight seminars each, and each group was given a different price. The number of persons who enrolled for each seminar presentation is shown in the accompanying table. At the 0.01 level of significance, do the data contradict the hypothesis that the demand for the seminar is price inelastic?

	Price			
325	375	425	475	
12	9	11	7	
15	7	13	8	
16	10	9	9	
12	6	9	10	Number of participants
10	8	12	13	
13	11	11	9	
10	9	17	6	
8	6	8	7	

12-16 Beck Industries wants to increase the speed of collecting its accounts receivable. A consulting firm suggests three options: A, use a lockbox (e.g., have payment mailed by the customer directly to a box at the bank the firm uses); B, offer a 1% discount for payment within 30 days; and C, include a postage-paid response envelope. To test the consultant's suggestion, the firm chooses 16 accounts and randomly uses each plan on 4 accounts and mails invoices as it normally does (D). The results (time, in days) are shown in the following table:

Method of Collection			
A	B	C	D
37	37	33	41
34	40	34	36
38	37	38	40
36	42	40	39

Time (in days)

(a) At the 0.05 level of significance, do the data indicate that any one of the consultant's suggestions will work?

(b) Assuming you did not reject the null hypothesis, explain why the inclusion of the last treatment group (D), known as a *control group*, would allow you to conclude that none of the consultant's suggestions work rather than that they all work equally well.

12-17 A large department store has six entrances. The store plans to add a new department, and the buyer for the department requests that it be located next to the busiest entrance to attract customers. The store manager maintains, however, that traffic through all entrances is equal. A head count of customers entering the store between 9 A.M. and 10 A.M. on five different mornings from each entrance yields the results shown in the accompanying table. Do the data contradict the store manager's claim? Use a 0.05 level of significance.

Chapter 12 Analysis of Variance

	Entrances					
Mornings	1	2	3	4	5	6
1	245	225	260	239	252	241
2	263	253	237	263	251	253
3	238	230	263	210	250	236
4	264	253	246	253	265	278
5	241	252	236	274	255	248

12-18 Using the data in the article in the beginning of the chapter, test the hypothesis that the mean durability of the different balls is equal. Use $\alpha = 0.01$.

12-19 A sample of quarterly earnings (in hundreds of dollars) of four salespersons is shown in the following table:

	Salesperson			
Quarter	A	B	C	D
1	820	886	908	746
2	866	902	894	776
3	812	794	776	812
4	874	886	856	786

(a) What is the mathematical model for this experiment?

(b) Is there a variation in the mean commissions of the salespeople?

(c) Is there a variation in the mean commissions by quarter?

12-20 In exercise 12-17 consider each morning to represent a particular day of the week. That is, morning 1 is Monday, morning 2 is Tuesday, and so on.

(a) Test the hypothesis that the mean traffic on each day is the same. Consider each entrance as a replicate.

(b) Consider both day of week and entrance as factors and redo your analysis. Compare your results with your results in part (a) and exercise 12-17.

12-21 An automobile insurance company employs three claims adjusters. Usually, only one adjuster estimates damages to a car, but it is advantageous to the company if all three adjusters are consistent enough so that it doesn't matter which of the three estimators is assigned to a particular car. To check on the consistency of the estimators, several damaged cars are selected and all three adjusters are asked to make estimates. The estimates in dollars for each adjuster are shown in the following table:

	Car				
Estimator	1	2	3	4	5
1	273	667	1,048	876	545
2	265	673	1,021	856	556
3	282	659	1,008	865	559

(a) Does this data provide sufficient evidence that the means of the estimators differ? Use $\alpha = 0.05$ and treat the cars as replicates.

(b) Does the data provide sufficient evidence that the means of the estimators differ? Use $\alpha = 0.05$ and treat the cars as the second factor.

(c) Does the data provide sufficient evidence that the means of the cars differ? Use $\alpha = 0.05$ and treat estimators as a factor.

(d) Compare the results in parts (a), (b), and (c).

12-22 A time-and-motion study was conducted to determine the best process (work design) for assembling computer terminals. Four possible work designs were tried on each of the three shifts. The average number of assembled units produced per hour were:

	Design			
Shift	1	2	3	4
Day	10	8	14	10
Swing	14	12	15	14
Night	8	7	10	7

(a) What is the mathematical model for this experiment?

(b) Is there a variation in the average number of terminals assembled by the work designs?

(c) Is there a variation in the average number of terminals assembled per shift?

12-23 A firm has a large volume of phone orders. It wishes to test two systems of handling phone orders and see if there is any difference in how male and female operators handle each system. Five male operators and five female operators are randomly assigned to use each method, and the number of phone orders taken in a day is recorded.

	Method	
Sex	A	B
Male	110, 108, 113 120, 105,	123, 121, 126 126, 118
Female	125, 120, 119 123, 117	140, 127, 131 126, 123

(a) State the mathematical model for this experiment and complete the ANOVA test at $\alpha = 0.10$.

(b) Sketch an interaction graph.

(c) Does the interaction graph seem to agree with the results of the tests in part (a)?

12-24 Renlon Tobacco, Inc. is considering two different advertising campaigns for its new product. One campaign would stress the image of the cigarette, whereas the other would stress the flavor. To test the campaigns, it exposes 100 smokers to one campaign and 100 smokers to the other. It then records the percent of smokers that would purchase the new brand after viewing the ad. Each trial is repeated four times for four different age groups of smokers. The results were:

Advertising	Percent Willing to Try Brand							
	21–29		30–39		40–49		50–59	
Image	24.3	18.0	24.0	27.1	9.3	12.1	18.4	8.6
	21.1	16.2	22.6	23.1	15.6	12.4	15.1	9.9
Flavor	18.9	16.1	8.8	16.5	23.7	15.9	23.6	22.1
	14.7	16.9	11.7	13.0	11.2	18.0	16.7	18.9

Answer the same questions as in exercise 12-23, except use $\alpha = 0.05$.

12-25 A mail-order record company wishes to spend a fixed amount of dollars on television advertising. It is considering four options: (A) spend it all on prime time; (B) spend it all on off-hour cheaper time, hence buying more air time; (C) spend it mostly on prime time, but buy some off-hour time; or (D) spend it mostly on off-hour time, but buy some prime time. To decide what option to use, the company does an experiment. It runs each option on eight different television stations. Each option asks customers to respond to a different phone number, and the number of responses (in hundreds) from each option is recorded as:

Advertising Options	Station							
	1	2	3	4	5	6	7	8
A	150	138	159	180	77	180	84	103
B	130	135	171	170	89	160	85	89
C	125	116	149	129	78	129	77	94
D	116	128	139	151	73	147	75	84

(a) Considering the stations as replicates, state the model used and test for a significant variation in advertising option. Use $\alpha = 0.05$.

(b) Considering the stations as a factor, state the model used and test for a significant variation in advertising options. Use $\alpha = 0.05$.

(c) Do you think there is a significant difference in advertising options? Explain.

(d) Which option would you use and why?

12-26 A firm was interested in evaluating four brands of calculators. All calculators perform the same functions, but each operates differently owing to differences in design. To decide which calculator brand to purchase, various employees were randomly assigned to each brand and timed doing the same calculations. The results were:

Brand 1	Brand 2	Brand 3	Brand 4
3.2	1.5	2.9	3.4
3.0	1.9	3.2	3.4
1.5	2.6	2.9	2.3
2.2	1.3	2.2	2.0
1.6	2.0	2.4	3.2
2.7	2.1	2.3	2.3
3.0	1.9	3.8	2.9
			2.6

The data were analyzed on a computer using SPSS with the following results:

SOURCE OF VARIATION	SUM OF SQUARES	DF	MEAN SQUARE	F	SIGNIF OF F
MAIN EFFECTS	3.774	3	1.258	3.939	0.828
B	3.774	3	1.258	3.939	0.020
EXPLAINED	3.774	3	1.258	3.939	0.020
RESIDUAL	7.984	25	0.319		
TOTAL	11.759	28	0.420		

State or give the value of:

(a) the mathematical model used
(b) H_0
(c) H_a
(d) SS(error)
(e) df of factor
(f) F^*, the calculated value of F
(g) prob-value
(h) decision using $\alpha = 0.05$

NOTE: "Main effects" appears on this particular computer output. SS(main effects) = SS(factor B).

12-27 An industrial engineer was interested in comparing three different processes for production. To determine which process to use, she set up an experiment where each process was tried on two shifts, day and night. The results of the computer analysis of the data using SPSS were:

SOURCE OF VARIATION	SUM OF SQUARES	DF	MEAN SQUARE	F	SIGNIF OF F
MAIN EFFECTS	2661.079	3	887.026	7.223	0.001
B	317.218	1	317.218	2.583	0.115
C	2242.725	2	1121.363	9.131	0.001
2-WAY INTERACTIONS	836.728	2	418.364	3.407	0.042
B C	836.728	2	418.364	3.407	0.042
EXPLAINED	3497.807	5	699.561	5.696	0.001
RESIDUAL	5649.270	46	122.810		
TOTAL	9147.077	51	179.354		

NOTE: "Main effects" appears on this particular computer output. SS(main effects) = SS(B) + SS(C).

State or give the value of:

(a) the mathematical model
(b) H_0 concerning the process
(c) H_a concerning the process
(d) prob-value concerning the process
(e) decision concerning the process ($\alpha = 0.01$)
(f) H_0 concerning the shift
(g) H_a concerning the shift
(h) prob-value concerning the shift
(i) decision concerning the shift ($\alpha = 0.01$)
(j) H_0 concerning interaction
(k) H_a concerning interaction
(l) prob-value concerning interaction
(m) decision concerning interaction ($\alpha = 0.01$)

Challenging Problem

12-28 In some cases where we have two factors and replicates, we may not believe that an interaction between the two factors exists. In such a case the sum of squares assigned to the source interaction would simply be combined with the sum of squares of the source normal error and labeled error.

(a) Write how table 12-14 would appear if no interaction were assumed. (You need not write out how the SS would be calculated directly. Assume that you would calculate it by subtracting the factors SS from the total.)

(b) Define SS of FI × FII as A and its degrees of freedom as df_A, and SS of error as B and its degrees of freedom as df_B. Show that the calculated MS(error) when no interaction is assumed is less than the calculated MS(error) when interaction is assumed, when F^* of the interaction effect is less than 1.

(c) Show that when F^* of interaction is less than 1 and no interaction is assumed, the F associated with factor I and the F associated with factor II are greater than if interaction were assumed.

Hands-On Problems

12-1 Does the mode of transportation by which a college student travels affect the length of time that it takes him or her to travel, one way, to class each day? The one-factor ANOVA is mode of transportation (possible levels are walking, own car, car pool, bus, bicycle). Use five replicates at each level and test $\alpha = 0.01$. (Be sure to state how you collect the data, the hypothesis under test, and all other pertinent factors.)

12-2 For this problem any timely or interesting topic may be used as your subject. Use four replicates and two factors of your choice and test at $\alpha = 0.05$.

13 Linear Correlation and Simple Regression Analysis

Chapter Outline

13-1 Linear Correlation Analysis
*Analysis of linear dependency uses the concepts of **covariance** and the **coefficient of linear correlation**.*

13-2 Inferences About the Linear Correlation Coefficient
*Ways that we **question and interpret** the correlation coefficient once we have obtained it.*

13-3 Linear Regression: Fitting the Equation
*Finding the **line of best fit**, a mathematical expression for the relationship between two variables.*

13-4 Coefficient of Determination: R Square
*The measure of **how good** a regression line **fits** the data.*

Academic and Leadership Performance of Graduate Business Students

The primary objective of the research reported in this article was to study selected factors which may be of value as valid and objective predictors of various dimensions of a graduate business student's performance. The research reported here deals with two of these dimensions: academic performance in a graduate school of business, the leadership behavior in a simulated decision making contest. ...

The academic performance evaluation form consists of five scales of ten steps each ranging from "less" to "more" on a designated dimension of academic performance. The five scales utilized were as follows: amount of work done, quality of work done, attendance and punctuality, conscientiousness, and over-all academic performance. ...

The Academic Predictors

The following predictors were significantly correlated (at the .05 level) with academic performance. ...

2. Business and economics unadjusted, undergraduate G.P.A. ($r = +0.28$)

3. Cumulative, adjusted, undergraduate G.P.A. ($r = +0.41$)

4. ATGSB (quantitative) ($r = +0.19$)

5. ATGSB (total) ($r = +0.28$)

. . .

13. LHAI, Academic Background. This item relates to what percentage of his fellow students the subject surpassed academically upon graduation from high school ($r = +0.24$).

14. LHAI, Academic Background. This item relates to how many times the subject changed his major in college ($r = -0.32$). The fewer the number of changes, the higher the rated academic performance.

From Larry L. Cummings and William E. Scott, Jr., "Academic and Leadership Performance of Graduate Business Students," *Business Perspectives*, vol. 1, no. 3 (Spring 1965): 11–20. Reprinted by permission.

Chapter Objectives

In chapter 3 we introduced the basic ideas of regression and linear correlation analysis. (If these concepts are not fresh in your mind, review chapter 3 before beginning this chapter.) Chapter 3 was only a first look—a presentation of the basic graphic and descriptive statistical aspects of linear correlation and regression analysis. In this chapter we will take a second, more detailed look at linear correlation and regression analysis.

Previously, we analyzed the linear correlation coefficient to determine whether there was a linear relationship between two variables. Now we will determine if there is a linear relationship by use of a hypothesis test, where the probability of a type I error is fixed by the value assigned to α. In chapter 3 we discussed the concept of fitting a line to the data. Now we will become familiar with a set of formulas for actually finding the equation of the straight line of best fit. After determining the equation of this line, we will ask the question, "How well does the line fit the data?" and we will use a statistic to measure the goodness of fit.

Recall that **bivariate data** are ordered pairs of response variables. They are paired as a result of a common bond (see p. 103). To perform correlation and regression analysis, both variables will need to produce numerical data.

Section 13-1 Linear Correlation Analysis

In chapter 3 we presented the linear correlation coefficient as a quantity that measures the strength of a linear relationship (dependency). Now let's take a second look at this concept and see how r, the coefficient of linear correlation, works. Intuitively, we want to think about how to measure the mathematical linear dependency of one variable on another. As x increases, does y tend to increase (decrease)? How strong (consistent) is this tendency? We are going to use two measures of the dependence, covariance and the coefficient of linear correlation, to measure the relationship between two variables. We'll begin our discussion by examining a set of bivariate data and identifying some related facts as we prepare to define covariance.

Illustration 13-1

Let's consider the following sample of six sets of bivariate data: (2, 1), (3, 5), (6, 3), (8, 2), (11, 6), (12, 1); see the accompanying figure on the top of the next page. The mean of the six x values (2, 3, 6, 8, 11, 12) is $\bar{x} = 7$. The mean of the six y values (1, 5, 3, 2, 6, 1) is $\bar{y} = 3$.

centroid

The point $(\bar{x}, \bar{y})$, which is (7, 3), is located as shown on the graph of the sample points in figure 13-1. The point $(\bar{x}, \bar{y})$ is called the **centroid** of the data. If a vertical and a horizontal line are drawn through the centroid, the graph is divided into four sections, as shown in figure 13-1. Each point (x, y) lies a certain distance from each of these two lines. $(x - \bar{x})$ is the horizontal distance from (x, y) to the vertical line passing through the centroid. $(y - \bar{y})$ is the vertical distance from (x, y) to the horizontal line passing through the centroid. Both the horizontal and the vertical distances of each data point

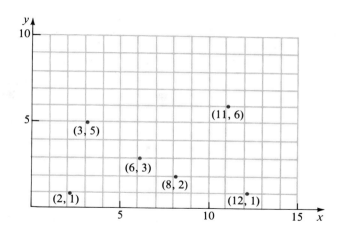

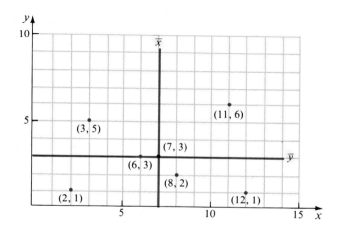

FIGURE 13-1
The Point (7, 3) Is the Centroid

from the centroid can be measured, as shown in figure 13-2. The distances may be positive, negative, or 0, depending on the position of the point (x, y) in reference to $(\bar{x}, \bar{y})$. The distances $(x - \bar{x})$ and $(y - \bar{y})$ are represented by means of braces, with positive or negative signs, as shown in figure 13-2. □

covariance One measure of linear dependency is the covariance. **The covariance of x and y** is defined as the sum of the products of the distances of all values of x and y from the centroid, $\sum[(x - \bar{x})(y - \bar{y})]$, divided by $n - 1$:

$$\operatorname{covar}(x, y) = \frac{\sum[(x - \bar{x})(y - \bar{y})]}{n - 1} \tag{13-1}$$

The covariance for the data given in illustration 13-1 is calculated in table 13-1. The covariance, written as covar(x, y), of the data is $\frac{3}{5} = \mathbf{0.6}$.

Linear Correlation Analysis   Section 13-1   **457**

FIGURE 13-2
Measuring the Distance of Each Data Point from the Centroid

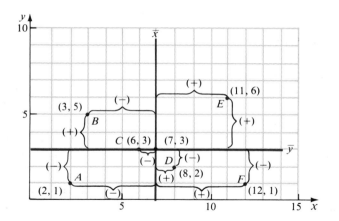

NOTE: $\sum (x - \bar{x}) = 0$ and $\sum (y - \bar{y}) = 0$. This will always happen. Why? (See p. 61.)

The covariance is positive if the graph is dominated by points to the upper right and to the lower left of the centroid. The products of $(x - \bar{x})$ and $(y - \bar{y})$ are positive in these two sections. If the majority of the points are in the upper left and lower right sections relative to the centroid, the sum of the products is negative. Figure 13-3 shows data representing a positive dependency (a), a negative dependency (b), and little or no dependency (c). The covariances for these three situations would definitely be positive in part (a), negative in (b), and near 0 in (c).

TABLE 13-1
Calculations Needed for Finding covar(x, y) for Data of Illustration 13-1

Points	$(x - \bar{x})$	$(y - \bar{y})$	$(x - \bar{x})(y - \bar{y})$
A (2, 1)	−5	−2	10
B (3, 5)	−4	2	−8
C (6, 3)	−1	0	0
D (8, 2)	1	−1	−1
E (11, 6)	4	3	12
F (12, 1)	5	−2	−10
	0	0	3

The biggest disadvantage of covariance as a measure of linear dependency is that it does not have a standardized unit of measure. One reason for this is that the spread of the data is a strong factor in the size of the covariance. For example, if we were to multiply each data point in illustration 13-1 by 10, we would have (20, 10), (30, 50), (60, 30), (80, 20), (110, 60), and (120, 10). The relationship of the points to each other would be changed only in that they would be much more spread out. However,

FIGURE 13-3
Data and Covariance

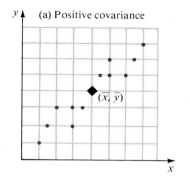
(a) Positive covariance

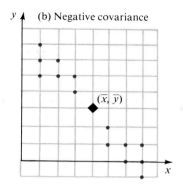
(b) Negative covariance

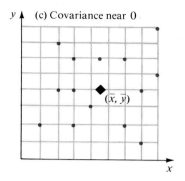
(c) Covariance near 0

the covariance of the new set of data is 60. Does this larger value mean that the amount of dependency between the x and y variables is stronger than in the original case? No, it does not; the relationship is the same, even though each set of points was multiplied by 10. This is the trouble with covariances as a measure. We must find a way to eliminate the effect of the spread of the data when we measure dependency.

If we standardize x and y by dividing the distance of each from the respective mean by the respective standard deviation,

$$x' = \frac{x - \bar{x}}{s_x} \quad \text{and} \quad y' = \frac{y - \bar{y}}{s_y}$$

and then compute the covariance of x' and y', we will have a covariance that is *not* affected by the spread of the data. This is exactly what is accomplished by the linear correlation coefficient. It divides the covariance of x and y by a measure of the spread of x and by a measure of the spread of y (the standard deviations of x and y are used as measures of spread). Therefore, by definition, the **coefficient of linear correlation** is

$$r = \text{covar}(x', y') = \frac{\text{covar}(x, y)}{s_x \cdot s_y} \tag{13-2}$$

The coefficient of linear correlation standardizes the measure of dependency and permits us to compare the relative strength of the dependency of different sets of data. Formula (13-2) for linear correlation is commonly referred to as **Pearson's product moment, r**.

The value of r, the coefficient of linear correlation, of the data in illustration 13-1 can be found by calculating the two standard deviations and then dividing:

$$s_x = 4.099 \quad \text{and} \quad s_y = 2.098$$

$$r = \frac{0.6}{(4.099)(2.098)} = 0.07$$

Finding the correlation coefficient by using formula (13-2) can be a very tedious arithmetic process. The formula can be written in a more workable form, as it was given in chapter 3:

$$r = \frac{\text{covar}(x, y)}{s_x \cdot s_y} = \frac{\sum [(x - \bar{x}) \cdot (y - \bar{y})]/(n - 1)}{s_x \cdot s_y}$$

$$= \frac{\text{SS}(xy)}{\sqrt{\text{SS}(x) \cdot \text{SS}(y)}}$$

(13-3)

Formula (13-3) avoids the separate calculations of $\bar{x}$, $\bar{y}$, s_x, and s_y and, more importantly, the calculations of the deviations from the means. Therefore formula (13-3) is much easier to use. (Refer to chapter 3 for an illustration of the use of this formula.)

The techniques shown in this section for the calculation of the linear correlation coefficient can be modified to allow for the use of coding and frequency distributions, when these additional techniques are needed. Such procedures may be found in other textbooks.

Exercises

13-1 Explain why $\sum (x - \bar{x}) = 0$ and $\sum (y - \bar{y}) = 0$.

13-2 The set of data in the accompanying table was compiled by a coffee bean importing company. It compares the amount of coffee beans imported (in tons) by both its East and West Coast headquarters over 8 years.

	Year							
	1976	1977	1978	1979	1980	1981	1982	1983
East Coast	2	2	4	4	6	6	8	8
West Coast	2	3	3	4	4	5	5	6

(a) Graph this set of data.
(b) Calculate the covariance.

(Retain these solutions for use in answering exercises 13-5 and 14-1.)

13-3 Use this set of data: (20, 10), (30, 50), (60, 30), (80, 20), (110, 60), (120, 10).
  (a) Calculate the covariance.
  (b) Calculate the standard deviation of the six x values and the standard deviation of the six y values.
  (c) Calculate r, the coefficient of linear correlation, by using formula (13-2).
  (d) Compare these results with those found in the text for illustration 13-1.

13-4 The management of a firm wanted to analyze the number of days of sick leave that its employees had taken in the first 6 months of this year (y) as compared with the first 6 months of the last year (x). The data are shown in the following table:

	\multicolumn{8}{c}{Employee}							
	A	B	C	D	E	F	G	H
x	0	1	2	3	4	5	6	7
y	6	7	4	5	2	3	0	1

  (a) Draw a scatter diagram for the data.
  (b) Calculate the covariance.
  (Retain these solutions for use in answering exercises 13-6 and 14-2.)

13-5 Use the data from exercise 13-2.
  (a) Calculate s_x and s_y.
  (b) Calculate r by using formula (13-2).
  (c) Calculate r by using formula (13-3).

13-6 Use the data from exercise 13-4.
  (a) Calculate s_x and s_y.
  (b) Calculate r by using formula (13-2).
  (c) Calculate r by using formula (13-3).

Section 13-2 Inferences About the Linear Correlation Coefficient

After the linear correlation coefficient r has been calculated for the sample data, it seems necessary to ask this question: Does r indicate that there is a dependency between the two variables in the population from which the sample was drawn? To answer this question, we can perform a hypothesis test. The null hypothesis is "the two variables are linearly unrelated" **rho (ρ)** ($\rho = 0$), where ρ (the Greek lowercase letter rho) is the **linear correlation coefficient for the population**. The alternative hypothesis may be either one tailed or two tailed. Frequently, it is two tailed. However, when we suspect that there is only a positive or only a negative correlation, we should use a one-tailed test. The alternative hypothesis of a one-tailed test is $\rho > 0$ or $\rho < 0$.

The critical region for the test is on the right when a positive correlation is expected and on the left when a negative correlation is expected. The test statistic used to test the null hypothesis is the calculated value of r from the sample. Critical values for r are found in table 14 of appendix D at the intersection of the column identified by the appropriate value of α and the row identified by the degrees of freedom. The number of degrees of freedom for the r statistic is 2 less than the sample size, $df = n - 2$.

The rejection of the null hypothesis means that there is evidence of a linear dependency between the two variables in the population.

Caution: The sample evidence may only say that the pattern of behavior of the two variables is related in that one can be used effectively to predict the other. **This does not mean that you have established a cause-and-effect relationship.**

Failure to reject the null hypothesis is interpreted as meaning that linear dependency between the two variables in the population has not been shown.

Now let's look at such a hypothesis test.

Illustration 13-2

In illustration 13-1, when $n = 6$, we found $r = 0.07$. Is this value significantly different from 0 at the 0.02 level of significance?

Solution

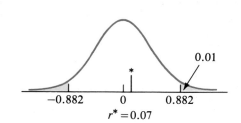

STEP 1: $H_0: \rho = 0$.

STEP 2: $H_a: \rho \neq 0$.

STEP 3: $\alpha = 0.02$, $df = n - 2 = 6 - 2 = 4$. See the figure.

The critical values (-0.882 and 0.882) were obtained from table 14 of appendix D.

STEP 4: The calculated value of r, r^*, was found earlier. It is $r^* = 0.07$.

STEP 5: *Decision*: We have failed to show that x and y are correlated.

Conclusion: We have failed to show that x and y are correlated. □

As in other problems, sometimes a confidence interval estimate of the population correlation coefficient is required. It is possible to estimate the value of ρ, the linear correlation coefficient of the population. Usually, this **confidence belts** estimation is accomplished with the aid of a table showing **confidence belts**. Table 15 in appendix D gives confidence belts for 95% confidence interval estimates. This table is a bit tricky to read, so be extra careful when you use it. The next illustration demonstrates the procedure for estimating ρ.

Illustration 13-3

A sample of 15 accounts of a major credit card company is selected, and the average monthly charges and the number of years the customers have had their credit cards are recorded. The calculated r for the paired data is 0.35. Find the 95% confidence interval estimate for ρ, the population linear correlation coefficient of a customer's average monthly charge and the length of time a customer has had the credit card.

Solution Find $r = 0.35$ at the bottom of table 15—see the arrow on figure 13-4. Visualize a vertical line through that point. Find the two points where the belts marked for the correct sample size cross the vertical line. The sample size is 15. These two points are circled in figure 13-4. Now look horizontally from the two circled points to the vertical scale on the left and read the confidence interval. The values are 0.72 and -0.20. Thus the 95% confidence interval estimate for ρ, the population coefficient of linear correlation, is **-0.20 to 0.72**.

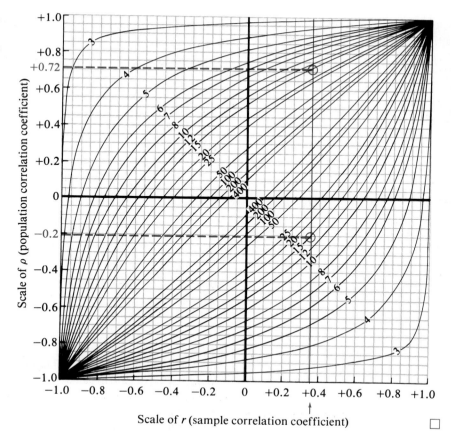

FIGURE 13-4
Using Table 15 of Appendix D, Confidence Belts for the Correlation Coefficient

Inferences About the Linear Correlation Coefficient   Section 13-2

Exercises

13-7 A sample of 15 pieces of bivariate data has a linear correlation coefficient of $r = 0.67$. Does this coefficient provide sufficient evidence to reject the null hypothesis that $\rho = 0$ in favor of a two-sided alternative? Use $\alpha = 0.05$.

13-8 If a sample of size 32 has a linear correlation coefficient of -0.29, do we have significant reason to conclude that the linear correlation coefficient of the population is negative? Use $\alpha = 0.05$.

13-9 A sample of size 10 produced $r = 0.67$. Is this value sufficient evidence to conclude that ρ is different from 0 at the 0.01 level of significance?

13-10 Is a value of $r = +0.30$ significant in trying to show that ρ is greater than 0 for a sample of 60 data?

13-11 Use table 14, appendix D, to determine a 0.95 confidence interval estimate for the true population linear correlation coefficient based on the following sample statistics:
 (a) $n = 10, r = 0.20$
 (b) $n = 50, r = -0.30$
 (c) $n = 8, r = +0.65$
 (d) $n = 100, r = -0.43$

Section 13-3 Linear Regression: Fitting the Equation

When two variables are studied jointly, we often would like to control one variable by means of the other—for example, we might want to change the amount of sales by changing advertising expenditures. Or we might want to predict the value of a variable based on knowledge of another variable—for example, we might want to predict the price of a company's stock on the basis of the company's earnings per share. In either case we want to find the **line of best fit** that will best predict the value of the dependent variable for a given value of the independent variable. The variable that we know or control is called the **input**, or **independent**, **variable**. The value of the dependent variable, determined by the equation for the line of best fit, is called the **predicted value**. It is customary to label the dependent variable y and the independent variable x.

Given a set of bivariate (x, y) data, how do we find the line of best fit? If a straight-line relationship seems appropriate, the best-fitting straight line is found by using the **method of least squares**. Suppose that $\hat{y} = b_0 + b_1 x$ is the equation of a straight line, where $\hat{y}$ (read "y hat") represents the **predicted value** of y corresponding to a particular value of x. The **least squares criterion** requires that we find the constants b_0 and b_1 such that the sum, $\sum (y - \hat{y})^2$, is as small as possible.

Figure 13-5 shows the distance of an observed value of y from a predicted value of y. The length of this distance represents the value $(y - \hat{y})$. Figure 13-6 shows a scatter diagram with what might appear to be the line of best fit, along with all the individual $(y - \hat{y})$'s. The sum of the squares of these differences is **minimized** (made as small as possible) if the line is indeed the line of best fit. Figure 13-7 shows a line that is definitely *not* the line of best fit.

FIGURE 13-5
Observed and Predicted Values of y

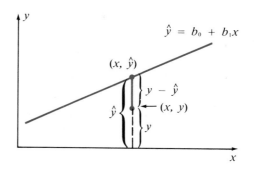

FIGURE 13-6
The Line of Best Fit

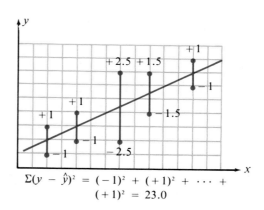

$$\Sigma(y - \hat{y})^2 = (-1)^2 + (+1)^2 + \cdots + (+1)^2 = 23.0$$

FIGURE 13-7
Not the Line of Best Fit

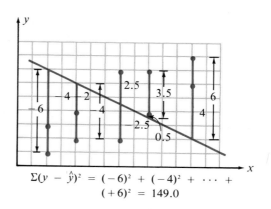

$$\Sigma(y - \hat{y})^2 = (-6)^2 + (-4)^2 + \cdots + (+6)^2 = 149.0$$

The **equation of the line of best fit** is determined by its slope (b_1) and its y-intercept (b_0). (See the accompanying *Study Guide* for a review of the concepts of slope and intercept of a straight line.) The values of these constants that satisfy the least squares criterion are found by using the following formulas:

Slope

$$b_1 = \frac{\sum (x - \bar{x})(y - \bar{y})}{\sum (x - \bar{x})^2} \qquad (13\text{-}4)$$

Intercept

$$b_0 = \frac{1}{n}\left(\sum y - b_1 \cdot \sum x\right) \qquad (13\text{-}5)$$

(The derivation of these formulas is beyond the scope of this text.)

We will use a mathematical equivalent of formula (13-4) that is easier to apply. We will calculate the slope b_1 by use of the formula

$$b_1 = \frac{\text{SS}(xy)}{\text{SS}(x)} \qquad (13\text{-}6)$$

Notice that the numerator of this fraction is the same as the numerator of formula (13-3) for the linear correlation coefficient. Notice also that the denominator is the first radicand (expression under a radical sign) in the denominator of formula (13-3). Thus if you calculate the linear correlation coefficient by using formula (13-3), you can easily find the slope of the line of best fit. If you do not use formula (13-3), set up a table similar to table 3-5 (p. 113) and find the necessary values.

Now let's consider the problem of predicting a salesperson's sales on the basis of his or her entertainment expenses. We want to find the line of best fit, $\hat{y} = b_0 + b_1 x$. The necessary calculations and totals for determining b_1 have already been obtained in table 3-5. We repeat these calculations here in table 13-2. Using formula (13-6), we calculate the slope to be

$$b_1 = \frac{12{,}978 - [(339)(365)/10]}{12{,}481 - [(339)^2/10]} = \frac{604.5}{988.9} = 0.611 = \mathbf{0.61}$$

Formula (13-5) can now be used to calculate the value of the y-intercept, b_0. (Throughout, keep the extra decimal places in the calculations to ensure an accurate answer.)

$$b_0 = \frac{1}{10}[365 - (0.611)(339) = 15.79 = \mathbf{15.8}$$

TABLE 13-2
Computations Needed to Calculate r for the Sales-Entertainment Expenses Data

Salesperson	Entertainment Expenses, x	x^2	Sales, y	y^2	xy
1	27	729	30	900	810
2	22	484	26	676	572
3	15	225	25	625	375
4	35	1,225	36	1,296	1,260
5	33	1,089	33	1,089	1,089
6	52	2,704	36	1,296	1,872
7	35	1,225	32	1,024	1,120
8	40	1,600	54	2,916	2,160
9	40	1,600	50	2,500	2,000
10	40	1,600	43	1,849	1,720
Total	339	12,481	365	14,171	12,978

Thus the equation of the line of best fit is $\hat{y} = 15.8 + 0.61x$. Figure 13-8 shows a scatter diagram of the data and the line of best fit.

FIGURE 13-8
Entertainment Expenses Versus Sales

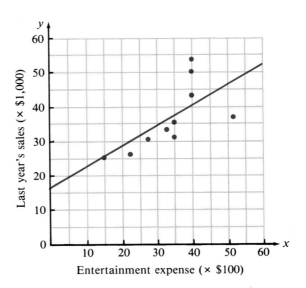

We need to discuss some additional facts about the least squares method:

1. The slope b_1 represents the predicted change in y per unit change in x. In our example $b_1 = 0.61$; thus if a salesperson had spent an additional $100 last year on entertainment (x), we would predict he or she would have sold an additional $610 (0.61 of $1,000) of merchandise ($y$). ($x$ was measured in hundreds of dollars and y in thousands of dollars.)

Linear Regression: Fitting the Equation   Section 13-3   **467**

2. The y-intercept is the value of y where the line of best fit intersects the y-axis, that is, where $x = 0$. However, in interpreting b_0, you first must consider whether $x = 0$ is a reasonable value before you conclude that you would predict $y = b_0$ if $x = 0$. To predict that if a salesperson spent nothing on entertainment, he or she would sell \$15,800 (15.8 × \$1,000) a year in merchandise is probably incorrect. An x value of 0 is outside the domain of the data on which the regression line is based. **In predicting y based on an x value, you should check to be sure that the x value is within the domain of the x values observed.**

3. The line of best fit will always pass through the point $(\bar{x}, \bar{y})$. When drawing the regression line on your scatter diagram, you should use this point as a check. It must lie on the regression line.

Illustration 13-4

A marketing research firm randomly selected eight radio stations operating in the Tampa, Florida, area and recorded their CUME ratings for Monday through Friday from 3 to 7 P.M. and the amount they charged per minute for advertising time. A CUME rating is an estimate of the percentage of the total listening audience that listens to a radio station for at least 5 minutes during a given time period. The data are shown in table 13-3. The sample was taken to find an equation that could be used to predict the advertising rate per minute of a station based on its CUME rating. Thus the pricing policy of the firm's client could be evaluated. Find the equation for the line of best fit and draw it on the scatter diagram.

TABLE 13-3
Data for Illustration 13-4

	Station							
	1	2	3	4	5	6	7	8
CUME, x	6.5	6.5	6.2	6.7	6.9	6.5	6.1	6.7
Cost per minute, y	105	125	110	120	140	135	95	130

Solution Before we find the equation of the line of best fit, we need to decide whether or not the two variables appear to be linearly related. There are two ways to check on this: (1) draw a scatter diagram and see if the points suggest a linear relationship or (2) calculate the linear correlation coefficient and test the hypothesis $\rho = 0$, as in section 13-2. If the scatter diagram of the data or the calculated correlation coefficient suggests a linear correlation, you are justified in calculating the equation for the line of best fit. The scatter diagram of the CUMEs and costs of radio advertising is shown in figure 13-9. The scatter diagram suggests that a linear line of best fit is appropriate.

FIGURE 13-9
CUME Versus Cost

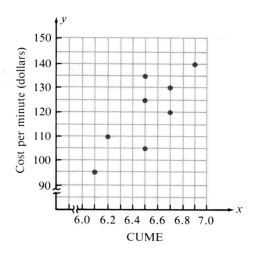

Let's calculate the equation of the line of best fit by using the summations from table 13-4 and formulas (13-5) and (13-6). By formula (13-6)

$$b_1 = \frac{6{,}275.0 - [(52.1)(960)/8]}{339.79 - [(52.1)^2/8]} = \frac{23}{0.48875} = 47.06 = \mathbf{47.1}$$

TABLE 13-4
Calculations for the Data of Illustration 13-4

Station	CUME, x	x^2	Cost, y	y^2	xy
1	6.5	42.25	105	11,025	682.5
2	6.5	42.25	125	15,625	812.5
3	6.2	38.44	110	12,100	682.0
4	6.7	44.89	120	14,400	804.0
5	6.9	47.61	140	19,600	966.0
6	6.5	42.25	135	18,225	877.5
7	6.1	37.21	95	9,025	579.5
8	6.7	44.89	130	16,900	871.0
	52.1	339.79	960	116,900	6,275.0

By formula (13-5)

$$b_0 = \frac{1}{8}[960 - (47.06)(52.1)] = -186.478 = \mathbf{-186.5}$$

Therefore

$$\hat{y} = -186.5 + 47.1x$$

To draw the line of best fit on the scatter diagram, we need to locate two points. Substitute two values of x, for example, 6.0 and 7.0, into the regression equation and obtain two corresponding values of $\hat{y}$:

$$\hat{y} = -186.5 + (47.1)(6.0) = -186.5 + 282.6 = 96.1 = \mathbf{96}$$

and

$$\hat{y} = -186.5 + (47.1)(7.0) = -186.5 + 329.7 = 143.2 = \mathbf{143}$$

Linear Regression: Fitting the Equation   Section 13-3

The values (6, 96) and (7, 143) represent two points (shown by + on figure 13-10) that enable us to draw the line of best fit.

NOTE: $(\bar{x}, \bar{y})$ is also on the line of best fit. It is the point shown by ◆ on figure 13-10. The use of a third point serves as a check on your work.

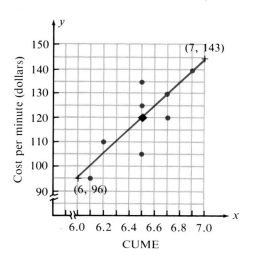

FIGURE 13-10
Line of Best Fit for Illustration 13-4

One of the main purposes for obtaining a regression equation is to make predictions. Once a linear relationship has been determined and the value of the input variable x is known, we can predict a value of y ($\hat{y}$). Suppose the client radio station in illustration 13-4 has a CUME of 6.6. What would we predict that a station with a 6.6 CUME would charge per minute for advertising? The equation was found to be $\hat{y} = -186.5 + 47.1x$. We can use the equation with $x = 6.6$ to predict the average cost per minute of all stations with 6.6 CUMEs. Thus

$$\hat{y} = -186.5 + (47.1)(6.6) = -186.5 + 310.86 = 124.36 = \$124.40$$

This equation can be used to determine if a client's advertising rate per minute is high, low, or average, given its CUME.

When making predictions based on the line of best fit, you must observe a few restrictions:

1. The equation should be used to make predictions only about the population from which the sample is drawn. For example, the use of the relationship between the CUMEs and costs per minute of advertising on radio stations in Tampa, Florida, to predict the costs per minute of New York City stations given their CUMEs would be questionable, since these two cities represent very different markets.

2. The equation should use only the sample domain of the input variables. For example, in illustration 13-4 you should not use CUMEs

outside the sample domain of 6.1 to 6.9 to predict advertising rates. The prediction that a station with a CUME of 0 would charge −$186.50 for advertising is nonsense.

3. If the sample was taken in 1984, do not expect the results to have been valid in 1928 or to hold in 1988. Present cost structures are different from past or future cost structures. On occasion you might wish to use the line of best fit to estimate values outside the domain interval of the sample. This procedure can be done; however, you should do it with caution and only for values close to the domain interval.

Exercises

13-12 Would you be justified in using the techniques of linear regression on the data given in exercise 3-10 to find the line of best fit? Explain.

13-13 If you fail to reject the hypothesis $\rho = 0$, would you be justified in using the techniques of linear regression on the data to find the line of best fit? Explain.

13-14 Use the data given in exercise 3-7.
(a) Calculate the equation of the line of best fit.
(b) Use the answer to part (a) to find how many applicants you would expect to get if your advertisement had six lines.
(c) What exactly does your answer to part (b) mean?
(Retain these solutions for use in answering exercise 14-3.)

13-15 (a) Calculate the equation of the line of best fit for the credit data in exercise 3-8, letting x = age of account.
(b) Calculate the equation of the line of best fit for the credit data in exercise 3-8, letting x = percentage delinquent.
(c) If you wanted to predict the percentage of 7-year-old accounts that are delinquent, which of the equations, part (a) or part (b), would you use? What would be your prediction?
(d) If you wanted to predict the percentage of 25-year-old accounts that are delinquent, what equation, if any, would you use? Explain.

13-16 (a) Calculate the line of best fit for the data in exercise 3-9 to predict sales based on price. (Retain this solution for use in answering exercise 13-23.)
(b) If a pen is priced at $0.89, how much sales would you predict?

13-17 (a) Calculate the line of best fit for the relationship between number of TV commercials, x, and sales units, y, for the data given in exercise 3-15.
(b) What does the value calculated for r in exercise 3-15 tell you about the answer you found in part (a) here?

13-18 A record of maintenance costs is kept for each electric typewriter throughout a company. A sample of 12 typewriters gave the data shown in the accompanying table.

Age, x (years)	Maintenance Cost, y (dollars)
6	62
7	151
1	0
3	10
6	96
4	52
5	77
2	19
1	20
9	121
3	90
8	75

(a) Draw a scatter diagram that shows these data.
(b) Calculate the equation of the line of best fit.
(c) A particular typewriter is 8 years old. How much maintenance (cost) do you predict it will require this year?
(d) Interpret your answer to part (c).
(e) Calculate r.
(f) What meaning does r have? How does it affect your answer in part (d)?

(Retain these solutions for use in answering exercise 13-24.)

Section 13-4 Coefficient of Determination: R Square

coefficient of determination

Having fit the regression line to the set of data, we now ask the next logical question: How well does the line fit the data? The measure used for this question is the **coefficient of determination** R^2, commonly referred to as R square.

$$R^2 = 1 - \frac{\sum (y - \hat{y})^2}{\sum (y - \bar{y})^2} \qquad (13\text{-}7)$$

or, equivalently,

$$R^2 = \frac{\sum (\hat{y} - \bar{y})^2}{\sum (y - \bar{y})^2} \qquad (13\text{-}8)$$

R^2 can take on any value between 0 and 1. For R^2 to equal 1, the numerator in formula (13-7), $\sum (y - \hat{y})^2$, must equal 0. That means each predicted value of y ($\hat{y}$) must equal the actual observed value of y. In other words, every observation must fall exactly on the regression line, and we would have a perfect fit. When R^2 equals 0, the numerator in formula (13-8), $\sum (\hat{y} - \bar{y})^2$, must equal 0, that is, each $\hat{y} = \bar{y}$. This situation occurs when we predict the same value of y for all values of x. Thus the slope of the line b_1 is 0; the regression line is horizontal. Hence extreme values of R^2 are easy to interpret. **When $R^2 = 1$, there is a perfect linear fit. When $R^2 = 0$, $b_1 = 0$ and there is no linear relationship between x and y.**

Now let us interpret values of R^2 between 0 and 1. Recall the problem of illustration 3-7 concerning the 10 salespeople, where we had each salesperson's sales y and entertainment expenditures x. (See table 13-2.) Suppose that we want to predict the amount of sales for 1 salesperson who is to be randomly selected from the set of 10. The best estimate that we could give, since we do not know which one is to be selected, is $36,500 (the mean amount of sales for all 10, $\bar{y}$). Suppose salesperson 10 was selected. Then the **error in our prediction** is the difference between the predicted value 36.5 and the actual value y of 43.0, namely, 6.5. This difference, $y - \bar{y}$, **is a measure of the accuracy with which we can predict y *without* knowledge of x.**

error in prediction

Now if we had been told that the salesperson selected had spent 40 (hundred dollars) on entertainment, then we could have used the line of fit to make a prediction: $\hat{y} = 15.8 + (0.61)(40) = 40.2$. That is, knowing that the salesperson spent $x = 40$ (hundred dollars), we can predict sales of $\hat{y} = 40.2$ (thousand dollars). The error in this prediction would be 2.8 ($43.0 - 40.2$), the difference between the actual sales y and the prediction from the regression equation, $\hat{y}$. This difference, $y - \hat{y}$, **is a measure of the accuracy for predictions made *with* knowledge of x.**

The difference between our initial estimate without knowledge of x (36.5) and our estimate with knowledge of x (40.2) is explained by what the regression line tells us about the predictive relationship between x and y. That is, we expect the selected salesperson's sales to be 3.7 ($40.2 - 36.5$) above the average of the 10 salespeople, since we know this salesperson spent 40 (hundred dollars) on entertainment. However, the 2.8 difference between the estimate from the regression and what salesperson 10 sold is not explained by the regression. As figure 13-11 shows, the initial error in predicting the sales of salesperson 10 can be separated into two parts: that which we would not have made if we knew x and that which we would have made even if we knew x:

Total deviation = unexplained deviation + explained deviation

or

Error in prediction without knowledge of x = error in prediction with knowledge of x + error that would not have occurred if regression had been used

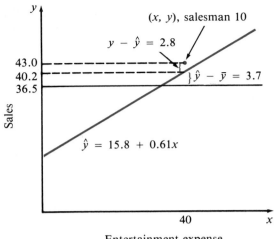

FIGURE 13-11
Entertainment Expense Versus Sales

That is,

$$y - \bar{y} = (y - \hat{y}) + (\hat{y} - \bar{y}) \tag{13-9}$$

$$(43.0 - 36.5) = (43.0 - 40.2) + (40.2 - 36.3)$$

$$6.5 = 2.8 + 3.7$$

To place the resulting reduction in error (the explained deviation) of 3.7 in perspective, it would make sense to divide the 3.7 reduction by the original error 6.5 (the total deviation) and talk in terms of the "percentage of reduction in error due to the regression." Thus in this case the percentage of reduction in error due to using regression is 3.7/6.5, or 57%.

Thus far we have analyzed only the usefulness of regression in predicting salesperson 10's sales. We can use the same approach over all 10 salespeople, with one major change. We cannot simply add up the total of the explained deviations and divide by the total of the deviations, because some deviations are positive and some are negative and they will cancel out. [Recall that $\sum (y - \bar{y}) = 0$.] That is, to say we have no error in prediction because we overestimated one salesperson's sales by 10 and underestimated another salesperson's sales by 10 is meaningless. To solve this problem, we choose to deal with **squared error**, or **squared deviation**. The sum of squared deviations is called **variation**. We can divide the variation that occurs without knowledge of x into two parts, just as we did for deviations in formula (13-9):

Total variation = unexplained variation + explained variation

$$\sum (y - \bar{y})^2 = \sum (y - \hat{y})^2 + \sum (\hat{y} - \bar{y})^2 \tag{13-10}$$

The proportion of the total variation about the mean (our best estimate without knowledge of x) that can be explained by a regression relationship of x and y is

$$\frac{\sum (\hat{y} - \bar{y})^2}{\sum (y - \bar{y})^2}$$

and is called R^2. Therefore

R^2 is the percentage of the variation in the dependent variable from the mean (its predicted value ignoring the independent variable) that can be explained by the relationship between y and x that is expressed by the regression equation.

Let's calculate R^2 for our sales forecast regression, which uses entertainment expenditures as the independent variable. The computations are shown in table 13-5.

NOTE: $\sum (y - \bar{y})$ and $\sum (\hat{y} - \bar{y})$ must both equal 0, except for rounding error. You can use this to check your calculations.

$$R^2 = \frac{366.59}{848.50} = 0.43$$

TABLE 13-5
Calculations Needed to Find R^2

Salesperson	y	$y - \bar{y}$	$(y - \bar{y})^2$	x	$\hat{y}(15.8 + 0.61x)$	$\hat{y} - \bar{y}$	$(\hat{y} - \bar{y})^2$
1	30	−6.5	42.25	27	32.3	−4.2	17.64
2	26	−10.5	110.25	22	29.2	−7.3	53.29
3	25	−11.5	132.25	15	25.0	−11.5	132.25
4	36	−0.5	0.25	35	37.2	+0.7	0.49
5	33	−3.5	12.25	33	35.9	−0.6	0.36
6	36	−0.5	0.25	52	47.5	+11.0	121.00
7	32	−4.5	20.25	35	37.2	+0.7	0.49
8	54	+17.5	306.25	40	40.2	+3.7	13.69
9	50	+13.5	182.25	40	40.2	+3.7	13.69
10	43	+6.5	42.25	40	40.2	+3.7	13.69
		0 ⒸⓀ	848.50			−0.1 ⒸⓀ	366.59

Thus the variation between the sales and our best prediction of the sales without knowledge of a salesperson's entertainment expenditures (the mean) for the 10 salespeople can be reduced by 43% by knowing each salesperson's entertainment expenditures and using the regression equation between sales and entertainment expenditures.

In two-variable regression (one x variable and one y variable), R^2 equals the square of the correlation coefficient for x and y. If you have calculated the correlation coefficient of your data first, you can tell how well the regression line will fit before you actually calculate b_0 and b_1 by simply squaring the value of the correlation coefficient.

Optional Formula

R^2 can be calculated directly from the computations for the regression line. As an alternative to formulas (13-7) and (13-8), you can use the following formula:

$$R^2 = \frac{b_0 \sum y + b_1 \sum xy - [(\sum y)^2/n]}{\sum y^2 - [(\sum y)^2/n]} \quad (13\text{-}11)$$

Exercises

13-19 Calculate R^2 for the following sets of data:
(a) $\sum (y - \bar{y})^2 = 210$ and $\sum (\hat{y} - \bar{y})^2 = 140$
(b) $\sum (y - \bar{y})^2 = 100$ and $\sum (y - \hat{y})^2 = 10$
(c) $\sum (\hat{y} - \bar{y})^2 = 75$ and $\sum (y - \hat{y})^2 = 25$

13-20 If the correlation between sales and price is -0.7 and between sales and advertisement expenditures is 0.5, which independent variable will better predict sales? Explain.

13-21 Calculate R^2 for the regression in illustration 13-4.

13-22 Calculate R^2 by using formula (13-11) for the sales and entertainment expenditure data, and compare your result with that found on page 475 using formula (13-8) and table 13-5.

13-23 How well does the regression line calculated in exercise 13-16 fit the data?

13-24 How well does the regression line predicting maintenance costs from the age of the electric typewriter, calculated in exercise 13-18, fit the data?

In Retrospect

In this chapter we have made a more thorough investigation of linear correlation. You should now have a better insight into what is measured by the correlation coefficient.

Also in this chapter we have applied some of the topics we learned about previously. For example, hypothesis testing and confidence interval estimates were applied to the sample correlation coefficient. These tests permit us to make inferences about the population correlation coefficient.

The news article in the beginning of this chapter presents a correlation analysis between success in graduate business school and certain undergraduate predictors. What does the statement "significantly correlated (at the 0.05 level)" mean to you now? Given the sizes of the correlations, would you feel confident in predicting someone's success (maybe yours) in graduate business school on the basis of one of the variables, say undergraduate grade point average?

Once we determine that a linear relationship exists between two variables, it is often useful to find the mathematical expression that will best predict one variable given the value of the other. For this purpose we have learned how to calculate the regression equation. We have learned how to compute the b_0 and b_1 values of the line of best fit, $\hat{y} = b_0 + b_1 x$, from a set of data and to measure how well the line fits the data.

In the next chapter we will explore the use of confidence intervals and tests of hypotheses to make inferences by using the computed regression line.

Chapter Exercises

13-25 Do the data in exercise 3-30 support Dr. Engel's economic theory? Test by using $\alpha = 0.05$.

13-26 One would expect that the number of hours a salesperson spends with a client, x, would have direct (positive) correlation with the size of the client's account, y. The hours spent and the size of the clients' accounts were recorded (see the following table) for eight randomly selected clients.

	Client							
	1	2	3	4	5	6	7	8
Hours spent, x	10	6	15	11	7	19	17	3
Account size, y ($\times \$100$)	51	36	67	63	44	89	80	26

(a) Draw the scatter diagram and estimate r.
(b) Calculate r. (How close was your estimate?)
(c) Is there significant positive correlation shown by this data? Test at $\alpha = 0.01$.
(d) Find the 95% confidence interval estimate for the true population value of ρ.

13-27 When buying almost any item, it is often advantageous to buy in as large a quantity as possible. The unit price is usually less for the larger quantities.

The data shown in the table were obtained to test this theory:

Number of Units, x	1	3	5	10	15
Cost per Unit, y	55	52	48	32	25

(a) Calculate r.

(b) Can we conclude that the number of units ordered and the cost per unit are correlated? Use $\alpha = 0.05$.

(c) Find the 95% confidence interval estimate for the true value of ρ.

13-28 When two dice are rolled simultaneously, the result on each roll is expected to be independent of the other. To test this expectation, roll a pair of dice 12 times. Identify one die as x and the other as y. Record the number observed on each of the two dice.

(a) Draw the scatter diagram of x versus y.

(b) If the two dice do behave independently, what value of r can be expected?

(c) Calculate r.

(d) Test for independence of these dice at $\alpha = 0.10$.

13-29 Use the market survey data given in exercise 2-79.

(a) Calculate the linear correlation coefficient between age and income for the 50 persons surveyed.

(b) Using the random number table, identify 16 persons randomly selected from the 50 and record their age and income. Save this sample for answering the remaining parts of this question.

(c) Calculate the linear correlation coefficient for your sample.

(d) Construct a 95% confidence interval based on your sample correlation coefficient.

(e) Compare your results in parts (a), (c), and (d).

13-30 Use the data in exercise 3-25.

(a) Calculate the equation to predict assets (y) based on interest income (x).

(b) Calculate R^2.

(c) If in 1983 PSFS's interest income were $1,200,000, what would you predict its assets to be?

13-31 If we are determining when it is best to replace a piece of equipment, it is important to know the relationship between the age of the machinery and the maintenance costs. The accompanying table shows the age and year's maintenance costs for 10 randomly selected lathe machines.

Number of Years Old, x	Maintenance Costs, y
4	45
2	20
3	28
4	55
7	62
5	45
7	40
10	80
12	90
1	20

(a) Calculate the equation of the regression line.
(b) Calculate R^2 and interpret it in light of the data.
(c) If a lathe is 4 years old, what would you estimate its maintenance costs to be?

(Retain these solutions for use in answering exercise 14-22.)

13-32 The accompanying table shows the man-hours charged (x) by Sun Docks in the construction of six cruise ships and the total cost (y) for a ship.

Ship	Man-Hours, x ($\times 1,000$)	Total Cost, y ($\times \$1$ million)
1	19	66
2	23	74
3	25	72
4	24	76
5	26	78
6	21	72

(a) Calculate the equation for the regression line.
(b) Calculate R^2 by using formula (13-7).
(c) Calculate R^2 by using formula (13-11).
(d) Interpret the value of b_0 calculated.
(e) Interpret the value of b_1 calculated.

(Retain these solutions for use in answering exercise 14-21.)

13-33 The variance in commercial credit institutions' rates of interest is claimed to reflect the different standards the institutions use in screening applications. The lower the rate, the higher the standards and thus the less likely the account is to default in payment. The data in the following table represent the interest rates charged by eight randomly selected credit institutions and their default rates:

Interest Rates, x (%)	Number of Defaults, y (per 1,000 loans)
7.0	39
6.5	38
5.5	18
6.0	36
8.0	36
8.5	45
6.0	39
6.5	36

(a) Calculate the equation of the regression line.
(b) Calculate the coefficient of determination.
(c) Calculate the correlation coefficient.
(d) Does the claim seem justified by the data? Use $\alpha = 0.05$.

(Retain these solutions for use in answering exercise 14-23.)

13-34 Use the data from exercise 3-31.

(a) Calculate the equation of the regression line to predict the price of beef based on the production of beef and veal.
(b) How well does the line fit the data?
(c) If production is 15 billion tons, what would you predict the price of beef per 100 pounds to be?

13-35 The accompanying set of data represents the occupancy rates (percent occupied) in summer (x) and winter (y) for 25 randomly selected hotels in Sea Isle City.

Hotel	Occupancy Rate		Hotel	Occupancy Rate	
	Summer, x	Winter, y		Summer, x	Winter, y
1	75	43	14	73	41
2	86	44	15	78	45
3	68	36	16	71	41
4	83	38	17	86	50
5	57	41	18	71	41
6	66	40	19	96	51
7	55	27	20	96	54
8	84	46	21	59	28
9	61	38	22	81	50
10	68	35	23	58	37
11	76	42	24	90	46
12	76	42	25	92	54
13	71	45			

(a) Draw a scatter diagram for these data.

(b) Calculate the equation for the regression line to predict winter occupancy rates from summer occupancy rates.

(c) How well does the regression line fit the data?

(d) If a hotel had an occupancy rate of 90 (%) in the summer, what would you predict that its occupancy rate would be in the winter?

(e) If a hotel had an occupancy rate of 10 (%) in the summer, what would you predict that its rate would be in the winter?

13-36 A small-town car dealer believed that the amount of money he spent on advertising in the local paper had a direct effect on how many cars he sold that month. The accompanying data show the amount of money spent in advertising per month and the number of cars sold per month for a 2-year period.

Advertising, x	Cars Sold, y	Advertising, x	Cars Sold, y
30	5	70	19
30	9	70	23
30	14	70	31
40	6	80	24
40	14	80	32
40	18	80	35
50	12	90	27
50	14	90	32
50	23	90	38
60	18	100	34
60	24	100	35
60	28	100	39

(a) Is there sufficient reason to conclude that the amount of advertising and the number of cars sold are correlated? Use $\alpha = 0.05$.

(b) Calculate the equation for the regression line.

(c) What is R^2 for the regression line?

Challenging Problem

13-37 Given the following random sample data:

y	log y	x	y	log y	x
10,600	4.025	2	15,938	4.202	9
11,236	4.051	3	17,908	4.253	11
10,000	4.000	1	15,036	4.177	8
11,236	4.051	3	14,185	4.152	7
11,910	4.076	4	17,908	4.253	11
13,382	4.127	6	11,236	4.051	3
15,938	4.202	9	11,910	4.076	4
16,895	4.228	10	12,625	4.101	5

where

$$y = \text{yearly salary of steelworker at Lupen Steel}$$
$$\log y = \text{the logarithm of } y \text{ to the base 10}$$
$$x = \text{years of service at Lupen Steel}$$

(a) Draw a scatter diagram for (i) x versus y and (ii) x versus $\log y$.
(b) Compute r for (i) x and y and (ii) x and $\log y$.
(c) Is there sufficient evidence to conclude that there is a correlation between (i) x and y? (ii) x and $\log y$? Use $\alpha = 0.05$.
(d) Find the line of best fit and calculate R^2 for the regression of (i) x predicting y and (ii) x predicting $\log y$.
(e) Compare the preceding results. Which regression fits better? How are the two regressions and correlations related? (*Hint*: Look at equation 3-5. What do you get if you take the log of y as the dependent variable.)

Hands-On Problems

Consider this problem: What is the relationship between the age x of a Volkswagen Rabbit and its selling value y? Define "age" to be the age in terms of the selling market. That is, $x = (\text{present year}) - (\text{year of manufacture}) + 1$. During the year 1984, a 1983 car will be considered to be 2 years old; a 1979 car will be 6 years old. Define the selling value as you prefer. It could be the advertised price by a dealer, the advertised price by an individual, the value that an owner thinks the car is worth, or whatever. Collect 20 pieces of bivariate data.

13-1 Define "selling value."

13-2 Describe your sampling plan.

13-3 Draw a scatter diagram of your data.

13-4 Calculate r and test to see if your data fit a linear model.

13-5 Obtain the equation for the line of best fit and determine R^2.

13-6 A friend has a Rabbit that is 5 years old. What is your estimate of its worth?

13-7 Another friend has a Rabbit that is 3 years old. What is your estimate of its worth?

13-8 What does the slope of the regression line represent?

13-9 What is the meaning of the y-intercept? Is it the value you expected it to be?

If you prefer, this problem set may be modified. Choose an item of your own interest—a different make of car, trucks, whatever; there are many problems for regression analysis. However, it is intended that you use a set of data that is expected to be linear. (Check with your instructor before proceeding.)

The calculations for this problem set can most easily be accomplished with the assistance of an electronic calculator or a packaged program on a computer. A list of the available programs can be obtained from your computer center. There are a variety of packaged programs available: Minitab, Biomed (Biomedical Programs), SAS (Statistical Analysis System), IBM Scientific Subroutine Packages, and SPSS (Statistical Package for the Social Sciences) program libraries. Your local computer center will assist you.

14 Linear Regression Analysis

Chapter Outline

14-1 Basic Concepts Needed for Inference in Regression
*To analyze a regression equation, we use the concepts of **random experimental error** and **standard error of regression**.*

14-2 Inferences Concerning the Slope
*We judge the **usefulness of the equation of best fit** to predict the value of one variable given the value of another variable.*

14-3 Confidence Interval Estimates for Regression
***If the line of best fit is usable**, we can establish confidence interval estimates.*

14-4 Common Mistakes Made in Using Regression
*Some of the **common pitfalls** to avoid in using regression analysis.*

Risk/Return: U.S. Industry Pattern

Reduction of this elusive relationship to mathematics provides a tool for judging industry and company performance.

In the affairs of corporations, what is the relationship between expected risk and the level of return on investment?...

Measuring the Relationship

A corporate manager bases his risk evaluation and investment decisions to a large degree on the experience of his industry. A way of picturing and quantifying an important part of that experience is by considering the scatter of returns on investment earned by the companies in the industry. We contend that industries characterized by highly dispersed profit distributions are judged by management and investors to be riskier than those characterized by compact distributions of profit rates.

We developed our measure of risk for each of 59 major S.I.C. fields of business (primarily industrial and nonfinancial) by calculating the dispersion (or variance, in mathematical terms) of return on capital of individual companies around the average return for that industry....

The average of the yearly dispersions then became the "typical risk quantity" for that industry for that period of time. A high dispersion of return for individual companies around the mean indicates a high industry risk quantity; a low dispersion, a low risk quantity.

Since the results depicted suggest the existence of a truly quantitative and predictable relationship between risk and return in the U.S. economy, they can be useful to solve problems that involve the important elements of risk and return.

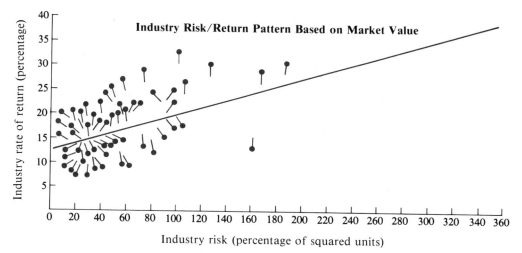

Regression equation: Industry return = 14.4 + 0.00724 (industry risk)

$$\text{Industry return} = \frac{\text{fixed charges} + \text{dividends} + \text{change in market value}}{\text{market value}}$$

Industry risk = average intercompany dispersion

Summary regression statistics: $R^2 = 0.59$ with 57 degrees of freedom

$F = 82.36$ with (1, 57) degrees of freedom for the slope coefficient

From Gordon R. Conrad and Irving H. Plotkin, "Risk/Return: U.S. Industry Pattern," *Harvard Business Review*, March–April 1968, pp. 90–99. Copyright © 1968 by the President and Fellows of Harvard College; all rights reserved. Reprinted by permission.

Chapter Objectives

In the previous chapter we saw how to fit a regression line to data and how to measure the goodness of the fit. Now we will explore how to analyze the results of regression analysis. We will test the hypothesis "the line is of no real use" against the alternative "the line is of real use." Previously, we used the equation of the line of best fit to make point predictions. Now we will make confidence interval estimations.

In short, this chapter will explore tests of hypotheses and confidence intervals associated with linear regression analysis. In this chapter we will restrict our work to two-variable regression. In the next chapter we will expand our study to cases in which there are more than two related variables.

Section 14-1 Basic Concepts Needed for Inference in Regression

Recall that the line of best fit results from an analysis of a situation in which we have two related variables. When two variables are studied jointly, we often would like to control one variable by means of controlling the other. Or we might want to predict the value of a variable based on knowledge of another variable. In both cases we want to find the line of best fit, provided one exists, that will best predict the value of the dependent, or output, variable. Recall that the variable we know or can control is called the **independent, or input, variable**; the variable resulting from the use of the equation of the line of best fit is called the **dependent, or predicted, variable**.

When the line of best fit is plotted, it does more than just show us a pictorial representation. It tells us two things: (1) there really is a functional (equational) relationship between the two variables and (2) the line expresses the quantitative relationship between the two variables. Recall that the line of best fit is useless when a change in the independent variable does not have a definite effect on the dependent variable. When there is no relationship between the variables, a horizontal line of best fit is plotted. A horizontal line has a slope of 0, which implies that a change in the independent variable has no effect on the dependent variable. This idea about the slope being 0 will be amplified later in this chapter.

As we saw in the last chapter, regression analysis on sample data results in a mathematical equation of the line of best fit, $\hat{y} = b_0 + b_1 x$. The **equation** used to explain the behavior of linear bivariate data **in the population** is

$$y = \beta_0 + \beta_1 x + \varepsilon \qquad (14\text{-}1)$$

experimental error

This equation represents the linear relationship between two variables in a population. β_0 is the y-intercept, and β_1 is the slope. ε is the **random experimental error** in the observed value of y at a given value of x.

The regression line from the sample data gives us b_0, which is **our estimate of** β_0, and b_1, which is **our estimate of** β_1. The error ε is ap-

arranged so that the x values are in numerical order.) Find the line of best fit, the coefficient of determination (R^2), and the variance of y about the line of best fit (s_e^2).

Solution The extensions and summations needed for this problem are also shown in table 14-1. We can now calculate the line of best fit, using formulas (13-4) and (13-5), and the measure of the goodness of fit, R^2, using formula (13-11).

$$b_1 = \frac{\sum xy - [(\sum x)(\sum y)/n]}{\sum x^2 - [(\sum x)^2/n]} = \frac{5{,}623 - [(184)(403)/15]}{2{,}616 - [(184)^2/15]}$$

$$= \frac{679.533}{358.933} = 1.8932 = 1.89$$

TABLE 14-1
Data for Illustration 14-1

Delivery	Miles, x	Minutes, y	x^2	xy	y^2
1	3	7	9	21	49
2	5	20	25	100	400
3	7	20	49	140	400
4	8	15	64	120	225
5	10	25	100	250	625
6	11	17	121	187	289
7	12	20	144	240	400
8	12	35	144	420	1,225
9	13	26	169	338	676
10	15	25	225	375	625
11	15	35	225	525	1,225
12	16	32	256	512	1,024
13	18	44	324	792	1,936
14	19	37	361	703	1,369
15	20	45	400	900	2,025
	184	403	2,616	5,623	12,493

$$b_0 = \frac{1}{n}\left(\sum y - b_1 \sum x\right) = \frac{1}{15}[403 - (1.8932)(184)]$$

$$= 3.6434 = 3.64$$

$$\hat{y} = 3.64 + 1.89x$$

$$R^2 = \frac{b_0 \sum y + b_1 \sum xy - [(\sum y)^2/n]}{\sum y^2 - [(\sum y)^2/n]}$$

$$= \frac{(3.6434)(403) + (1.8932)(5{,}623) - [(403)^2/15]}{(12{,}493) - [(403)^2/15]}$$

$$= \frac{1{,}286.487}{1{,}665.734} = 0.772 = \mathbf{0.77}$$

The variance of y about the regression line is calculated by using formula (14-5).

$$s_e^2 = \frac{\text{SSE}}{n-2} = \frac{(\sum y^2) - (b_0)(\sum y) - (b_1)(\sum xy)}{n-2}$$

$$= \frac{12{,}493 - (3.6434)(403) - (1.8932)(5{,}623)}{15 - 2}$$

$$= \frac{379.2462}{13} = 29.1728 = \mathbf{29.17}$$

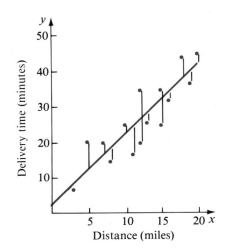

FIGURE 14-5
Distance Versus Time,
Illustration 14-1

NOTE: Extra decimal places are often needed for these types of calculations. Notice that b_1 (1.8932) was multiplied by 5,623. If 1.89 had been used instead, that one product would have changed the numerator by approximately 18. That, in turn, would have changed the final answer by almost 1.5, which is a sizable round-off error.

$s_e^2 = 29.17$ is the variance of the 15 e's shown in figure 14-5 as vertical line segments.

NOTE: The variance of e will be used in the following sections of this chapter in much the same way as the variance of x (as calculated in chapter 2) was used in chapters 8, 9, and 10.

Exercises

14-1 Use the data and the scatter diagram from exercise 13-2. Let x represent the East Coast amounts and y the West Coast amounts.
(a) Find the equation of the line of best fit, $\hat{y} = b_0 + b_1 x$, and graph it on the scatter diagram drawn in answering exercise 13-2.
(b) Find the ordinates y for the points on the line of best fit whose abscissas are $x = 2, 4, 6$, and 8.
(c) Find the value of e for each of the points in the given data. ($e = y - \hat{y}$.)
(d) Find the variance s_e^2 of those points about the line of best fit by using formula (14-3).
(e) Find the variance s_e^2 by using formula (14-5). [Answers to parts (d) and (e) should be the same.]

(Retain these solutions for use in answering exercise 14-4.)

14-2 Use the data given in exercise 13-4 to answer the questions of exercise 14-1. In part (b) use $x = 0, 1, 2, \ldots, 7$.

14-3 Use the data given in exercise 3-7 and the results obtained in exercise 13-14.
(a) Draw a scatter diagram of the data.
(b) Find the equation of the line of best fit and graph it on the scatter diagram.

(c) Find the ordinates y on the line of best fit that correspond to $x = 2, 3, 4, 5, 6,$ and 7.

(d) Find the four values of e that are associated with the points where $x = 3$ and $x = 6$.

(e) Find the variance s_e^2 of all points about the line of best fit.

(Retain these solutions for use in answering exercises 14-5, 14-7, and 14-9.)

Section 14-2 Inferences Concerning the Slope

Now that the equation of the line of best fit has been determined and the linear model has been verified (by inspection of the scatter diagram and by a high value of R^2), we are ready to determine if we can use the equation to predict y. We will test the null hypothesis "the equation of the line of best fit is of no value in predicting y given x." That is, the null hypothesis to be tested is β_1 (the slope of the relationship in the population) is 0. If $\beta_1 = 0$, the linear equation will be of no use in predicting y. To test this hypothesis, we will use a t test.

Before we look at the hypothesis test, let's discuss the sampling distribution of the slope. If random samples of size n are repeatedly taken from a bivariate population, the calculated slopes, the b_1's, would form a sampling distribution that is approximately normally distributed with a mean of β_1 and a variance of $\sigma_{b_1}^2$, where

$$\sigma_{b_1}^2 = \frac{\sigma_\varepsilon^2}{\sum (x - \bar{x})^2} \tag{14-6}$$

An appropriate estimate of the variance $\sigma_{b_1}^2$ is obtained by replacing σ_ε^2 with s_e^2, the estimate of the variance of the error about the regression line:

$$s_{b_1}^2 = \frac{s_e^2}{\sum (x - \bar{x})^2} \tag{14-7}$$

Formula (14-7) may be rewritten in the following, more manageable form:

$$s_{b_1}^2 = \frac{s_e^2}{\sum x^2 - [(\sum x)^2 / n]} \tag{14-8}$$

NOTE: The "standard error of—" is the standard deviation of the sampling distribution of—. Therefore the standard error of regression (slope) is σ_{b_1} and is estimated by s_{b_1}.

We are now ready to **test the hypothesis** $\beta_1 = 0$. Let's use the line of best fit determined in illustration 14-1, $\hat{y} = 3.64 + 1.89x$. That is, we want to determine if this equation is of any use in predicting delivery time y. In this type of hypothesis test, the null hypothesis is always $H_0: \beta_1 = 0$.

STEP 1: $H_0: \beta_1 = 0$. (This hypothesis implies that x is of no use in predicting y; that is, $\hat{y} = \bar{y}$ would be as effective.)

The alternative hypothesis can be either one tailed or two tailed. If we suspect that the slope is positive, as in illustration 14-1 [we would expect delivery time (y) to increase as the distance (x) increases], a one-tailed test is appropriate.

STEP 2: $H_a: \beta_1 > 0$.

STEP 3: The test statistic is t. The number of degrees of freedom for this test is $n - 2$, **df = $n - 2$**. Thus for our example df = $15 - 2 = 13$. If we use $\alpha = 0.05$, the critical value of t is $t(13, 0.05) = 1.77$; see table 7 of appendix D and the accompanying figure here.

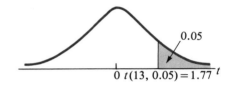

The formula used to calculate the value of the **test statistic t** for inferences about the slope is

$$t = \frac{b_1 - \beta_1}{s_{b_1}} \qquad (14\text{-}9)$$

STEP 4: In our illustration of delivery times and distances, the variance among the b_1's is estimated by use of formula (14-8):

$$s_{b_1}^2 = \frac{s_e^2}{\sum x^2 - (\sum x)^2/n} = \frac{29.1728}{2{,}616 - (184)^2/15}$$

$$= \frac{29.1728}{358.9333} = 0.08128 = 0.0813$$

If we use formula (14-9), the observed value of t becomes

$$t = \frac{b_1 - \beta_1}{s_{b_1}} = \frac{1.89 - 0}{\sqrt{0.0813}} = 6.629$$

$$t^* = 6.63$$

STEP 5: *Decision*: Reject H_0 ($t*$ is in the critical region; see the figure).

Conclusion: The slope of the line of best fit in the population is greater than 0. The evidence indicates that there is a linear relationship and that the delivery time (y) can be predicted based on the distance (x).

The slope β_1 of the regression line of the population can be estimated by means of a confidence interval. The **confidence interval** is written as

$$b_1 \pm t(n-2, \alpha/2) \cdot s_{b_1} \qquad (14\text{-}10)$$

The 95% confidence interval for the estimate of the population's slope, β_1, for illustration 14-1 is

$$1.89 \pm (2.16)(\sqrt{0.0813})$$
$$1.89 \pm 0.6159$$
$$1.89 \pm 0.62$$
$$1.27 \quad \text{to} \quad 2.51$$

Thus 1.27 to 2.51 is the 0.95 confidence interval for β_1. That is, we can say that the slope of the line of best fit of the population from which the sample was drawn is between 1.27 and 2.51 with 95% confidence.

Exercises

14-4 Calculate the variance $s_{b_1}^2$ for the data in exercise 13-2 (see exercise 14-1).
(a) Use formula (14-7).
(b) Use formula (14-8).

14-5 (a) Calculate the standard error of the slope, s_{b_1}, of the data in exercise 3-7 (see exercise 14-3).
(b) Is the observed value of the slope b_1 large enough to reject the null hypothesis $\beta_1 = 0$ in favor of the alternative hypothesis $\beta_1 > 0$ at $\alpha = 0.05$?

(Retain these solutions for use in answering exercises 14-7 and 14-9.)

14-6 An economist feels that the percentage of a product's total cost that is represented by advertising expenses (y) is a direct linear function of the number of directly competitive brands (x) on the market. A sample of 12 products was selected, and the number of competing brands (x) and the average percentage of total costs spent on advertising were recorded, as shown in the following table. (The data are listed with the x values in numerical order for your convenience.)

Number of Competing Brands, x	1	3	5	5	7	8	8	9	10	10	10	12
Percentage Spent on Advertising, y	5	10	15	20	15	20	25	25	25	30	35	35

(a) Draw a scatter diagram of the data.
(b) Find the equation of the regression line of the data.
(c) Does the value of b_1 show sufficient strength to conclude that the population's slope is greater than 0 at the $\alpha = 0.01$ level?

(Retain these solutions for use in answering exercises 14-8 and 14-10.)

14-7 Use the information found in exercises 14-3 and 14-5 to construct a 0.95 confidence interval for the estimation of β_1, the slope of the population.

14-8 Use the information found in exercise 14-6 to calculate the 0.99 confidence interval for the estimation of β_1.

Section 14-3 Confidence Interval Estimates for Regression

Once the equation of the line of best fit has been obtained and determined usable, we are ready to use the equation to make predictions. We can estimate two different quantities: (1) the mean of the population of y values at a given value of x, written as $\mu_{y|x_0}$, and (2) the individual y value selected at random that will occur at a given value of x, written as y_{x_0}. The best point estimate or prediction of both $\mu_{y|x_0}$ and y_{x_0} is $\hat{y}$. This estimate is the y value obtained when an x value is substituted into the equation of the line of best fit. Like other point estimates it is seldom right. The actual values of $\mu_{y|x_0}$ and y_{x_0} vary above and below the calculated value of $\hat{y}$.

Before developing confidence intervals for $\mu_{y|x_0}$ and y_{x_0}, recall the development of confidence intervals for the population mean μ in chapter 8 when the variance was known and in chapter 9 when the variance was estimated. The sample mean $\bar{x}$ was the best point estimate of μ. We used the fact that $\bar{x}$ is normally distributed with a standard error of $\sigma/\sqrt{n}$ to construct formula (8-2) for the confidence interval for μ. When σ had to be estimated, we used formula (9-2) for the confidence interval.

The confidence intervals for $\mu_{y|x_0}$ and y_{x_0} are constructed in a similar fashion. $\hat{y}$ replaces $\bar{x}$ as our point estimate. If we were to take random samples from the population, construct the line of best fit for each sample, calculate $\hat{y}$ for a given x using each regression line, and plot the various $\hat{y}$ values (they would vary, since each sample would yield a slightly different regression line), we would find that the $\hat{y}$ values form a normal distribution. That is, the sampling distribution of $\hat{y}$ is normal, just as the sampling distribution of $\bar{x}$ is normal. What about the appropriate standard error of $\hat{y}$? The standard error

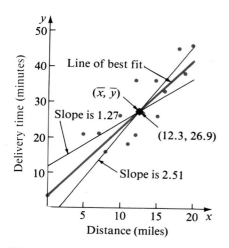

FIGURE 14-6
Three Lines That Pass Through the Centroid

in both cases ($\mu_{y|x_0}$ and y_{x_0}) is calculated by multiplying the square root of the variance of the error by an appropriate correction factor. Recall that the variance of the error, s_e^2, is calculated by means of formula (14-5).

Before looking at the correction factors for the two cases, let's see why they are necessary. Recall that the line of best fit passes through the point $(\bar{x}, \bar{y})$, the centroid. Earlier in this chapter we constructed a confidence interval estimate for the slope β_1 by using formula (14-10). If we draw lines with slopes equal to the extremes of that confidence interval, 1.27 to 2.51, through the point $(\bar{x}, \bar{y})$ [which is (12.3, 26.9)] on the scatter diagram, we will see that the value of $\hat{y}$ fluctuates considerably for different values of x (figure 14-6). Therefore we should suspect a need for a wider confidence interval as we select values of x that are further away from $\bar{x}$. Hence a correction factor is needed to adjust for the distance between x_0 and $\bar{x}$. This factor must also adjust for the variation of the y values about $\hat{y}$.

First let's estimate the **mean value of** y at a given value of x, $\mu_{y|x_0}$. The **confidence interval** estimate formula is

$$\hat{y} \pm t(n-2, \alpha/2) \cdot s_e \cdot \sqrt{\frac{1}{n} + \frac{n \cdot (x_0 - \bar{x})^2}{\sum (x - \bar{x})^2}} \qquad (14\text{-}11)$$

NOTE: The numerator of the second term under the radical sign is n times the square of the distance of x_0 from $\bar{x}$. The denominator is closely related to the variance of x and has a "standardizing" effect on this term.

Formula (14-11) can be modified to avoid having $\bar{x}$ in the denominator. The new form is

$$\hat{y} \pm t(n-2, \alpha/2) \cdot s_e \cdot \sqrt{\frac{1}{n} + \frac{(x_0 - \bar{x})^2}{\sum x^2 - (\sum x)^2/n}} \qquad (14\text{-}12)$$

Let's compare formula (14-11) with formula (9-2). $\hat{y}$ replaces $\bar{x}$, and

$$s_e \cdot \sqrt{\frac{1}{n} + \frac{n(x_0 - \bar{x})^2}{\sum (x - \bar{x})^2}} \qquad \text{(the standard error of } \hat{y}\text{)}$$

the estimated standard error of y in predicting $\mu_{y|x_0}$, replaces $s/\sqrt{n}$, the standard error of $\bar{x}$. The degrees of freedom are now $n-2$ instead of $n-1$ as before.

All these ideas are explored in the next illustration.

Illustration 14-2

To send a messenger (illustration 14-1), it costs the delivery firm $3.60 an hour, or $0.06 a minute. In developing its rate schedule, the firm used the regression equation calculated in illustration 14-1 to predict the time it would take to make a delivery of a given number of miles. To ensure a profit, the estimated delivery time was charged at a rate of $0.10 a minute. For example, the labor (messenger time) charge for a 7-mile delivery would be

$$\text{Labor charge} = (\$0.10)(\hat{y}_{x_0 = 7}) = (\$0.10)[3.64 + 1.89(7)]$$
$$= (\$0.10)(16.87) = \$1.687$$

or $1.70, rounded for simplicity. The actual cost for a particular 7-mile delivery would be $(\$0.06)(y)$, where y is the actual time. The average cost for all 7-mile deliveries would be $(\$0.06)(\mu_{y|x_0 = 7})$.

The regression equation value of 16.87 minutes is our best estimate of the average delivery time for all deliveries of 7 miles. Like any point estimate it is seldom exact. However, since the firm charges $1.70 for the 7-mile delivery (based on $0.10 per minute), as long as

$$(\$0.06)(\mu_{y|x_0 = 7}) < \$1.70$$

the firm will show a profit. By dividing both sides of the inequality by 0.06, we find that whenever

$$\mu_{y|x_0 = 7} < 28\tfrac{1}{3} \text{ minutes}$$

the firm makes a profit. Our estimate for $\mu_{y|x_0 = 7}$ was 16.87 minutes. How likely is it that the true value of $\mu_{y|x_0 = 7}$ is less than $28\tfrac{1}{3}$ minutes?

Construct a 95% confidence interval estimate for the mean travel time for all deliveries of 7 miles and determine if it includes $28\tfrac{1}{3}$ minutes.

Solution

STEP 1: Find $\hat{y}_{x_0}$ when $x_0 = 7$:

$$\hat{y} = 3.64 + 1.89x = 3.64 + (1.89)(7) = \mathbf{16.87}$$

STEP 2: Find s_e:

$$s_e^2 = 29.17 \quad \text{(found previously)}$$
$$s_e = \sqrt{29.17} = \mathbf{5.40}$$

STEP 3: Find $t(13, 0.025) = 2.16$ (from table 7 in appendix D).

STEP 4: Use formula (14-12):

$$16.87 \pm (2.16) \cdot (5.40) \cdot \sqrt{\frac{1}{15} + \frac{(7 - 12.27)^2}{2{,}616 - (184)^2/15}}$$

$$16.87 \pm (2.16)(5.40)\sqrt{0.06667 + 0.07738}$$

$$16.87 \pm (2.16)(5.40)\sqrt{0.14405}$$

$$16.87 \pm (2.16)(5.40)(0.38)$$

$$16.87 \pm 4.43$$

$$12.44 \quad \text{to} \quad 21.30$$

This interval is the 0.95 confidence interval for $\mu_{y|x_0=7}$. The price appears to ensure that charges will cover the costs for deliveries of 7 miles, since the value $28\frac{1}{3}$ minutes is not in the confidence interval estimate.

The confidence interval estimate is depicted in figure 14-7 by the heavy vertical line. The **confidence interval belt** showing the upper and lower boundaries of all interval estimates at 95% confidence is shown also. Notice

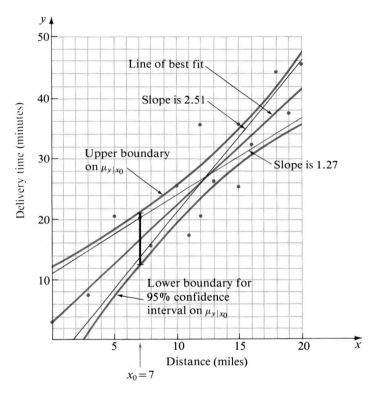

FIGURE 14-7
Confidence Interval Belts for $\mu_{y|x_0}$

that the boundary lines for x values far away from $\bar{x}$ become close to the two lines that represent the equations having slopes equal to the extreme values of the 95% confidence interval estimate for the slope (see figure 14-6). ☐

Often when a prediction is being made, we want to predict the value of an individual y. For example, before we set a rate based on the estimated time it will take to make a delivery, we would want to know the confidence interval for the delivery time to determine the risks that a particular delivery will take too long. The formula for the **confidence interval** for the value of a **single randomly selected** y is

$$\hat{y} \pm t(n-2, \alpha/2) \cdot s_e \cdot \sqrt{1 + \frac{1}{n} + \frac{(x_0 - \bar{x})^2}{\sum x^2 - (\sum x)^2/n}} \qquad (14\text{-}13)$$

Illustration 14-3

What is the 95% confidence interval estimate of the time it will take to make any randomly chosen delivery of 7 miles? Does this estimate include $28\frac{1}{3}$ minutes?

Solution

STEP 1: $\hat{y}$ at $x_0 = 7$ is 16.87, as found previously.

STEP 2: $s_e = 5.40$, as found previously.

STEP 3: $t(13, 0.025) = 2.16$.

STEP 4: Use formula (14-13). (Notice that some of the calculations needed were found in illustration 14-2.)

$$16.87 \pm (2.16)(5.40)\sqrt{1 + 0.14405}$$
$$16.87 \pm (2.16)(5.40)(1.07)$$
$$16.87 \pm 12.48$$
$$4.39 \quad \text{to} \quad 29.35$$

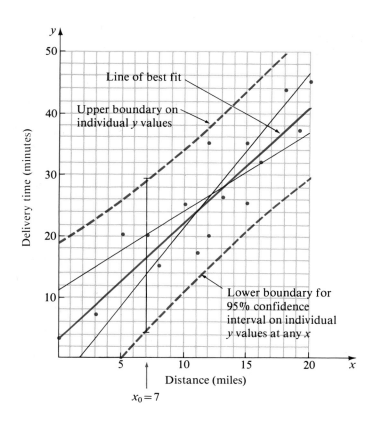

FIGURE 14-8
Confidence Interval Belts for y_{x_0}

This interval is the 0.95 confidence interval for $y_{x_0=7}$. Note that this wider interval includes $28\frac{1}{3}$ minutes.

The confidence interval is shown in figure 14-8 as the vertical line segment at $x_0 = 7$. The interval for y_{x_0} (at $x_0 = 7$) is much longer than the confidence interval for $\mu_{y|x_0}$ (at $x_0 = 7$). The dashed lines represent the upper and lower boundaries of the confidence intervals for individual y values for all given x values. □

Can you justify the fact that the confidence interval for individual values of y is wider than the confidence interval for mean values of y? Think of illustrations 14-2 and 14-3. Would you expect there to be a greater risk of losing money on travel time charges on a single trip than on the average over many trips? Think about the distributions of "individual values" and "mean values," and study figure 14-9.

The techniques of coding (for large numbers) and the techniques of frequency distributions (for large sets of numbers) are adaptable to the calculations presented in this chapter. Information about these techniques and about models other than linear models may be found in many other textbooks. It is worth noting here that if the data suggest a logarithmic or an exponential functional relationship, a simple change of variable will allow for the use of linear analysis.

FIGURE 14-9
Confidence Belts for Individual y's and for the Mean Value of y

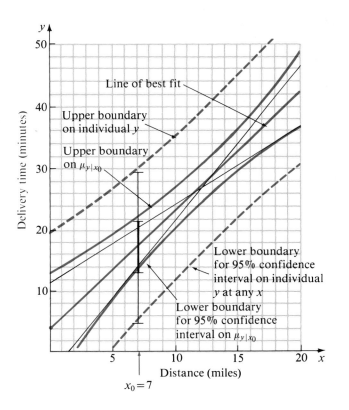

Exercises

14-9 Use the results obtained in answering exercises 14-3 and 14-5 to make the following estimates. (Exercise 3-7 has the original data.)

(a) Give a point estimate for the mean number of applicants that one might expect from an advertisement with six lines.

(b) Find the 0.95 confidence interval estimate for the mean number of applicants when $x_0 = 6$.

(c) Give a point estimate for the number of applicants that might occur from a particular advertisement of six lines.

(d) Find a 0.95 confidence interval estimate for the number of applicants from a particular advertisement of six lines.

14-10 Use the data and the answers found in exercise 14-6 to make the following estimates:

(a) Give a point estimate for the mean proportion of total costs spent on advertising for all products when there are four competing brands.

(b) Give a 0.90 confidence interval estimate for the mean proportion of total costs spent on advertising for all products when there are four competing brands.

(c) Give a 0.90 confidence interval estimate for the proportion of total costs spent on advertising for a particular product with four competing brands.

Section 14-4 Common Mistakes Made in Using Regression

There are some basic precautions that you need to be aware of before you work with regression analysis. Remember that the regression equation is meaningful only in the domain of the x variable studied. Prediction outside the range of given x values is extremely dangerous; it requires that we know or assume that the relationship between x and y remains the same outside the domain of the sample data. For example, suppose we want to know how long it will take to make a delivery of 75 miles (illustration 14-1). We can use $x = 75$ in all the formulas, but we do not expect the answers to carry the confidence or validity of the x values between 3 and 20, which were in the sample. Seventy-five miles may represent a delivery in the heart of a nearby major city. Do you think the estimated times, which were based on local distances of 3 to 20 miles, would be good predictors in this situation? Also, at $x = 0$ the equation has no real meaning. However, although projections outside the interval of sample values may be somewhat suspect, they may be the best predictors available.

Don't get caught by the common fallacy of applying the regression results inappropriately. For example, this fallacy would include applying the results of illustrations 14-1 through 14-3 to another company. But suppose the second company had a city location, whereas the first company had a rural location, or vice versa. Do you think the results for a rural location would also be valid for a city location? Basically, the results of one sample should not be used to make inferences about a population other than the one from which the sample was drawn.

Perhaps the most common fallacy is to jump to the conclusion that the results of the regression prove that *x causes y* to change. Regressions only measure movement between *x* and *y*; they **never prove causation**. A judgment of causation can be made only when it is based on theory or knowledge of the relationship separate from the regression results. The most common difficulty in this regard occurs because of what is called the *missing variable, or third-variable, effect*. That is, we observe a relationship between *x* and *y* because a third variable, one that is not in the regression, affects both *x* and *y*.

In Retrospect

In this chapter we have made a more thorough inspection of the linear relationship between two variables. Although it was not directly emphasized, we have applied many of the topics of earlier chapters in this chapter. The ideas of hypothesis testing and confidence interval estimates were applied to the regression problem. Reference was made to the sampling distribution of the sample slope b_1, which allowed us to make inferences about β_1, the slope of the population from which the sample was drawn. Reference was made to the sampling distribution of $\hat{y}$ and the role of the normal distribution in constructing confidence intervals. Finally, we measured the variance of y values and estimated values of y at given values of x.

As this chapter ends, you should be aware of the basic concepts of regression analysis and correlation analysis. You should now be able to design an elementary experiment, collect the data, and analyze a two-variable linear relationship. (The Hands-On Problems at the end of the chapter present an opportunity for you to do this.)

Reread the news article at the beginning of this chapter. The authors state, "the results depicted suggest the existence of a truly quantitative and predictable relationship between risk and return." Is the R^2 given in the article statistically significant? If so, does that support the authors' claim? What does the value of R^2, 0.59, tell you? Do you think other factors might affect industry return?

There are many problems where more than one variable is required to predict another. In the next chapter we will explore multiple regression, regression that uses more than one independent variable.

Chapter Exercises

14-11 Explain why a 95% confidence interval estimate for the mean value of *y* at a particular *x* is narrower than a 95% confidence interval for an individual *y* value at the same value of *x*.

14-12 When $x_0 = \bar{x}$, is the formula for the standard error of $\hat{y}$ ($s_e \cdot \sqrt{1/n}$) what you might have expected it to be?

14-13 If $R^2 = 0.95$, $n = 11$, and $\sum (y - \bar{y})^2 = 100$, what is s_e^2?

14-14 A personnel director collected the data shown in the following table from a sample of eight employees:

Years Employed, x	5	3	7	5	3	9	8	1
Days Absent, y	19	6	16	12	14	10	5	16

(a) What is the equation of the line of best fit?
(b) Test the null hypothesis that β_1 equals 0 against the alternative that β_1 is positive. Use $\alpha = 0.05$.
(c) Do the data indicate that years employed will be useful in predicting days absent?

14-15 If we fail to reject the null hypothesis that $\rho = 0$, do you think we would fail to reject or we would reject the hypothesis $\beta_1 = 0$? Explain.

14-16 The data shown in the following table resulted from an experiment performed for the purpose of regression analysis. The input variable x was set at five different levels, and three observations were made at each level.

(a) Draw a scatter diagram of the data.

x	0.5	1.0	2.0	3.0	4.0
y	3.8	3.2	2.9	2.4	2.3
	3.5	3.4	2.6	2.5	2.2
	3.8	3.3	2.7	2.7	2.3

(b) Draw the regression line by eye.
(c) Place an asterisk, *, at each level, approximately where the mean of the observed y values is located. Does your regression line look like the line of best fit for these five mean values?
(d) Calculate the equation of the regression line.
(e) Find the standard deviation of y about the regression line.
(f) Construct a 95% confidence interval estimate for the true value of β_1.
(g) Construct a 95% confidence interval estimate for the mean value of y at $x_0 = 3.0$; at $x_0 = 3.5$.
(h) Construct a 95% confidence interval estimate for an individual value of y at $x_0 = 3.0$; at $x_0 = 3.5$.

14-17 Commerce Credit believes that the longer the loan repayment schedule, the greater the risk of default. To test its claim, the firm collected the sample data shown in the following table:

Length of Loan (years)	15	10	20	5	15	20	10	25
Number of Defaults per Sample of 1,000 Loans	26	16	41	3	27	45	12	58

(a) Calculate the equation of best fit.

(b) Test the hypothesis that $\beta_1 = 0$. Use $\alpha = 0.05$.

(c) Construct a 95% confidence interval for β_1.

(d) Construct a 95% confidence interval for the mean number of defaults per 1,000 loans made with a repayment schedule of 5 years.

(e) Construct a 95% confidence interval for the mean number of defaults per 1,000 loans made with a repayment schedule of 25 years.

14-18 An economist developed the following regression equation based on a random sample of 100 lawyers in the Dallas area:

$$\hat{y} = 2{,}000 + 0.22x$$

where

x = income in 1983 (including tax-sheltered and tax-deferred income)

y = payments into tax-sheltered pension or tax-deferred funds

Based on these results, can the following conclusions be drawn?

(a) You would estimate that a lawyer in Dallas who earns $100,000 would tax shelter $24,000.

(b) You would estimate that a lawyer in New York who earns $100,000 would tax shelter $24,000.

(c) You would estimate that a lawyer in Houston who earns $5,000 would tax shelter $3,100.

14-19 The Health Service Agency constructed the following regression equation based on a sample taken in 1970 of 15 large hospitals in New York:

$$\hat{y} = 275{,}000 + 16{,}800x$$

where

y = total revenue of the hospital

x = total beds in the hospital

Explain why the following statements cannot be made based on the regression result:

(a) Hospitals in New York with 10 beds will have an estimated average revenue of $443,000.

(b) We would expect a hospital in Salt Lake City with 100 beds to have revenues that are $168,000 greater than a hospital in New York with 90 beds.

(c) If we increased the size of each hospital in New York by 10 beds, the total revenue of all New York hospitals would increase by (16,800) × (10).

14-20 You want to predict sales of a company. The value of R^2 using the variable GNP in a regression is 0.70, and the value of R^2 using the variable advertisement expenditures is 0.65.

(a) Which regression has the smaller s_e^2? Explain.

(b) Can we conclude that the GNP has a greater effect on sales than advertising does? Explain.

14-21 Using your results from exercise 13-32, construct a 0.95 confidence interval for β_1.

14-22 Use your results from exercise 13-31.

(a) Test the hypothesis that the age of the equipment is not useful in predicting maintenance costs. Use $\alpha = 0.10$.

(b) Calculate a 0.95 confidence interval for β_1.

14-23 Use the results from exercise 13-33.

(a) Calculate a 95% confidence interval for the number of defaults you would expect in a sample of 10,000 loans made at 7.0% interest.

(b) Calculate a 95% confidence interval for the mean number of defaults per 1,000 loans you would expect for all loans made at 7.0% interest.

14-24 (a) Using all the data from the market study in exercise 2-79, calculate the equation of the line of best fit for predicting the number of flights based on a person's income.

(b) Use the random number table to select 16 people from the data in exercise 2-79. Calculate the equation of the line of best fit for your sample of 16 people for predicting the number of flights based on a person's income.

(c) Construct a 0.90 confidence interval for β_1 based on your sample of 16.

(d) Compare your results in parts (a) and (c).

14-25 Recall that s_e^2 is the variance of individual values about the regression line, that

$$(s_{\bar{x}})^2 = \frac{s^2}{n}$$

is the variance of the sample means, and that

$$(s_{b_1})^2 = \frac{s_e^2}{(\sum x^2) - (\sum x)^2/n}$$

is the variance of the sample slopes. Can you justify the following two formulas?

(a) The standard error of estimate for $\mu_{y|x}$:

$$s_e \cdot \sqrt{\frac{1}{n} + \frac{(x_0 - \bar{x})^2}{(\sum x^2) - (\sum x)^2/n}}$$

which can be written as

$$\sqrt{\frac{s_e^2}{n} + \frac{s_e^2}{(\sum x^2) - (\sum x)^2/n} \cdot (x_0 - \bar{x})^2}$$

(b) The standard error of estimate for y_{x_0}:

$$s_e \cdot \sqrt{1 + \frac{1}{n} + \frac{(x_0 - \bar{x})^2}{(\sum x^2) - (\sum x)^2/n}}$$

which can be written as

$$\sqrt{s_e^2 + \frac{s_e^2}{n} + \frac{s_e^2}{(\sum x^2) - (\sum x)^2/n} \cdot (x_0 - \bar{x})^2}$$

(*Hint*: When have we added variances before?)

14-26 The following computer printout represents a regression analysis of the data presented in exercise 2-81. It predicts percent converted (A) on the basis of the variable difference (B).

```
DEPENDENT VARIABLE..      A              % CONVT

MEAN RESPONSE         38.60000      STD. DEV.       26.82430

VARIABLE(S) ENTERED ON STEP NUMBER      1..  B          DIFF

MULTIPLE R              0.63579
R SQUARE                0.40423
ADJUSTED R SQUARE       0.35841
STD DEVIATION          21.48613
```

```
---------------------- VARIABLES IN THE EQUATION -----------

VARIABLE              B              STD ERROR B                    t
                                                              ---------------
                                                              SIGNIFICANCE

B                 0.89457350          0.30120708                  2.969961
                                                                  0.011
(CONSTANT)        28.461500           6.5138424                   4.369387
                                                                  0.001


ALL VARIABLES ARE IN THE EQUATION
```

State, compute, or give the value of:

(a) the regression coefficient b_1

(b) the constant b_0

(c) the t^* for testing the hypothesis concerning β_1

(d) the prob-value for testing the hypothesis concerning β_1

(e) R^2

(f) s_e

(g) s_{b_1}

(h) a 95% confidence interval estimate for β_1

14-27 The following computer printout represents a regression analysis using the data presented in exercise 2-81. This study predicts percent converted (A) on the basis of stock price volatility. Answer parts (a) through (h) of exercise 14-26 and state which model you would use, the one in exercise 14-26 or the one in this exercise. Explain.

```
DEPENDENT VARIABLE..       A                % CONVT

MEAN RESPONSE         38.60000         STD. DEV.        26.82430

VARIABLE(S) ENTERED ON STEP NUMBER       1..    C             INDEX

MULTIPLE R              0.31495
R SQUARE                0.09919
ADJUSTED R SQUARE       0.02990

STD DEVIATION          23.42024

---------------------- VARIABLES IN THE EQUATION -----------
```

VARIABLE	B	STD ERROR B	t SIGNIFICANCE
C	-70.444121	58.877667	1.1964489 0.253
(CONSTANT)	110.26985	60.289315	1.8290115 0.090

ALL VARIABLES ARE IN THE EQUATION.

Hands-On Problems

Suppose the relationship between the yearly earnings y of a person at United Press Incorporated and the years he or she has been employed x is $y = 8{,}000 + 1{,}000x + \varepsilon$.

Employee	Employee's Weight	Employee's Years of Service, x	ε	Employee's Earnings, y
1	150			
2	165			
3	190			
4	145			
5	185			
6	205			
7	145			
8	185			
9	175			
10	190			
11	160			
12	170			
13	185			
14	210			
15	175			

Fifty random values of ε, with mean 0 and standard deviation 1,000, are

−1,088	−1,055	774	−731	55
109	−694	1,329	−309	365
1,181	984	−600	−588	842
239	279	56	−895	1,505
915	−1,010	−1,201	−884	272
826	2,286	−60	239	544
−2,064	−429	497	−571	191
−182	−1,456	1,054	1,112	1,853
−1,098	−291	−80	1,372	−1,675
354	785	−625	−603	−320

14-1 Using the random number table, randomly assign years of service between 1 and 20 years to each of the 15 employees and enter your results in the table. The number of years of service may repeat.

14-2 Using a random number table and the list of ε's presented on the preceding page, select, with replacement, an ε for each of the 15 employees. List the ε values you selected in the table.

14-3 Calculate an earnings for each employee, using the formula $y = 8{,}000 + 1{,}000x + \varepsilon$, by using the x and ε values you have selected for each employee.

NOTE: What you have done in problems 14-1, 14-2, and 14-3 is to *simulate* what a random sample of 15 employees might look like if drawn from a population where the regression relationship was $y = 8{,}000 + 1{,}000x + \varepsilon$. In the remaining problems you will analyze the sample data to infer the population relationships. This practice should give you some insight into the properties of inference, since you know, in this case, the "correct" answers. Of course, in real life we do not know the "correct" answers. If we did, we wouldn't need to use sample estimates. This technique—called *simulation, or Monte Carlo*, studies—is often used by statisticians to evaluate the goodness of inferential procedures.

14-4 Calculate the regression line for earnings y and length of service x by using the data you calculated for the 15 employees.

14-5 Calculate the error of the regression $(y - \hat{y})$ for each employee.

14-6 Calculate s_e for the regression.

14-7 Compare your findings in problems 14-5 and 14-6 with the ε values and σ_ε (1,000). Are they the same? Why or why not?

14-8 Test the hypothesis $\beta_1 = 0$. Use $\alpha = 0.01$. Are you surprised by the result?

14-9 Calculate a 95% confidence interval for β_1. Does it include the value 1,000? Should it?

14-10 Calculate the regression between earnings and the weight of each employee. Use the employee's weight as the independent variable.

14-11 Test the hypothesis that $\beta_1 = 0$ for your regression results in problem 14-10. Use $\alpha = 0.01$. Are you surprised by the result?

14-12 Using the results from the regression in problem 14-4, construct a 0.90 confidence interval for what you expect a particular person with 10 years of service to earn.

14-13 Using the random number table, select 50 values from the ε list (sample with replacement). For each ε value selected, compute $y = 8{,}000 + 1{,}000(10) + \varepsilon$, so that you have 50 earnings for people with 10 years of service. How many fall in the interval you calculated in problem 14-12? How many would you expect to fall in the interval?

15 Multiple Regression

Chapter Outline

15-1 Partial Correlation
Measures the linear relationship between two variables by removing the effect of a third variable.

15-2 Model Building and Multiple Regression
Analysis for more than one independent variable.

15-3 How Well Does the Model Fit: Multiple R^2
The measure of goodness of fit of a multiple regression model.

15-4 Inferences Concerning the Goodness of Fit of a Multiple Regression Model
We judge the usefulness of a multiple regression model in predicting one variable based on other variables.

15-5 Inferences Concerning Individual β's in the Multiple Regression Model
We judge the ability of each particular independent variable in a multiple regression model to predict the dependent variable.

15-6 Further Considerations in Modeling
The selection and construction of the independent variables to be introduced into a multiple regression model.

How to Get a Better Forecast

Regression analysis can predict more accurately than less scientific methods can....

While use of regression still remains outside the normal corporate forecasting process, it has come into increasing usage by many large companies in recent years. They are generally those companies that have made the greatest strides in using the computer as a management tool, rather than merely as an accounting device, for instance:

- The American Can Company for several years has used a regression technique to estimate sales on the basis of numerous external factors. For example, to forecast beer-can demand, the company has used an equation that correlates sales to income levels, number of drinking establishments per thousand persons, and age distribution of the population.

- Through regression analysis, Eli Lilly and Company has correlated the sale of pharmaceuticals with disposable income.

- The RCA Sales Corporation uses mathematical techniques for forecast sales of television sets, radio sets, and phonographs. Analysis at this company has concentrated on identifying the most important of more than 300 economic variables that might logically be connected with sales of RCA products.

- Armour & Company has found that it can accurately predict the number of cattle to be slaughtered in future months by using such explanatory variables as range-grass conditions and steer-corn ratios....

Forecasting Sales

Beginning with an attempt to predict sales, clearly the first step is to pinpoint those factors that are assumed to affect sales and to be associated with sales....

In the example used in this discussion we develop a forecast of sales and earnings for an actual (but disguised) company in the home furnishings industry, which we shall call the Cherryoak Company....

It is safe to say that most experienced observers, both within and outside the home furnishings field, have certain preconceived notions about the effects of these variables on sales of home furnishings. Through the regression technique, however, one can measure precisely how large and how significant each of them is in its historical relationship to total sales. These various relationships can be described by means of this industry sales regression equation:

$$S = B + B_m(M) + B_h(H) + B_i(I) + B_t(T),$$

where

S = Gross sales for year;

B = Base sales, or starting point from which other factors have influence;

M = Marriages during the year;

H = Housing starts during the year;

I = Annual disposable personal income;

T = Time trend (first year = 1, second year = 2, third year = 3, etc.)....

Results via the computer in an initial regression of the data show this pattern:

$$S = 49.85 - 0.068M + 0.036H + 1.22I - 19.54T$$

Chapter Objectives

In chapters 13 and 14 we restricted our analysis to bivariate data. Recall that bivariate data are ordered pairs of related variables, such as a person's height and weight. We explored in depth the use of correlation and regression to analyze the association between two variables.

In this chapter we will explore the concept of regression analysis as a model for representing the relationship between an independent variable and the dependent variable. We will expand the ideas of correlation, regression, and modeling to consider ordered groups of data in which three or more variables are related as the result of some common element. We will explore the logical extension of correlation analysis—partial correlation—and expand the simple regression models into multiple regression models. Although the calculations become more complex here, the basic concepts introduced in the previous two chapters apply.

Section 15-1 Partial Correlation

Thus far we have restricted our analysis to bivariate data. Now let's consider the case in which each observation yields a set of three variables rather than two variables.

In illustration 3-7 the amount of sales y and the entertainment expenses x were collected for a set of 10 salespeople. Now let's consider a third variable, the length of employment of each salesperson. The data are shown in table 15-1.

TABLE 15-1
Data for Entertainment Expenses, Years of Employment, and Sales

Salesperson	Entertainment Expense, x_1 ($\times \$100$)	Years of Employment, x_2	Sales, y ($\times \$1,000$)
1	27	6	30
2	22	4	26
3	15	7	25
4	35	6	36
5	33	5	33
6	52	8	36
7	35	7	32
8	40	13	54
9	40	11	50
10	40	9	43

Using formula (3-2) we can calculate the correlation between each pair of variables:

$$r = \frac{\sum xy - [(\sum x)(\sum y)/n]}{\sqrt{\sum x^2 - [(\sum x)^2/n]}\sqrt{\sum y^2 - [(\sum y)^2/n]}}$$

The results are (you could check the calculations here as a review):

$$r_{x_1y} = 0.660 = 0.66$$
$$r_{x_2y} = 0.901 = 0.90$$
$$r_{x_1x_2} = 0.518 = 0.52$$

NOTE: The subscripts used with r simply identify the variables involved.

Notice that sales has a better linear relationship (as evidenced by higher correlation) with years of employment (x_2) than with entertainment expenditures (x_1). Moreover, entertainment expenditures (x_1) is also correlated with years of employment (x_2). The accompanying diagram depicts the interrelationships.

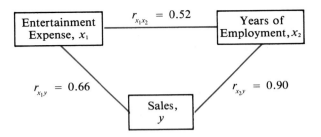

The next logical question is, how much of the observed correlation between variables x_1 and y is due to the fact that both are correlated with variable x_2?

---

partial correlation

The correlation between x_1 and y, after the effect of x_2 on both variables has been removed, is called the **partial correlation** of x_1 and y holding x_2 constant and is written $r_{x_1y \cdot x_2}$.

The partial correlation coefficient of any two variables x_1 and y holding variable x_2 constant, $r_{x_1y \cdot x_2}$, can easily be calculated by using the three correlation coefficients (r_{x_1y}, r_{x_2y}, and $r_{x_1x_2}$) and the following formula:

$$r_{x_1y \cdot x_2} = \frac{r_{x_1y} - (r_{x_1x_2} \cdot r_{x_2y})}{\sqrt{(1 - r_{x_1x_2}^2)(1 - r_{x_2y}^2)}} \qquad (15\text{-}1)$$

The formulas for the other two partial correlations can be found by interchanging the appropriate subscripts.

For our example the set of partial correlation coefficients are

$$r_{x_1y \cdot x_2} = \frac{r_{x_1y} - (r_{x_1x_2} \cdot r_{x_2y})}{\sqrt{(1 - r_{x_1x_2}^2)(1 - r_{x_2y}^2)}}$$

$$= \frac{0.660 - (0.518)(0.901)}{\sqrt{[1 - (0.518)^2][1 - (0.901)^2]}}$$

$$= \frac{0.193}{0.371} = 0.5202 = \mathbf{0.52}$$

$$r_{x_2y \cdot x_1} = \frac{0.901 - (0.518)(0.660)}{\sqrt{[1 - (0.518)^2][1 - (0.660)^2]}}$$

$$= \frac{0.559}{0.643} = 0.8694 = \mathbf{0.87}$$

The correlation between sales (y) and entertainment expenditures (x_1) of *salespeople who have the same employment seniority* is 0.52. The correlation between sales (y) and years of employment (x_2) of *salespeople who spend the same amount on entertainment* is 0.87.

The concept of partial correlation coefficients can be extended to consider any number of variables ordered together. For example, $r_{x_1y \cdot x_2 x_3}$ is the correlation between variables x_1 and y when the effect of variables x_2 and x_3 is removed.

A word of caution here: Partial correlation, like simple correlation, only measures the linear association between variables. It does not prove causation.

Exercises

15-1 Given $r_{x_1y} = 0.5$, $r_{x_1x_2} = 0.7$, and $r_{x_2y} = 0.4$, find the following:
(a) $r_{x_1y \cdot x_2}$ (b) $r_{x_1x_2 \cdot y}$ (c) $r_{x_2y \cdot x_1}$

15-2 Given $r_{x_1y} = -0.4$, $r_{x_1x_2} = 0.6$, and $r_{x_2y} = 0.8$, find the following:
(a) $r_{x_1y \cdot x_2}$ (b) $r_{x_1x_2 \cdot y}$ (c) $r_{x_2y \cdot x_1}$

15-3 Given $r_{x_1y} = 0.7$, $r_{x_1x_2} = 0$, and $r_{x_2y} = 0$, find the following:
(a) $r_{x_1y \cdot x_2}$ (b) $r_{x_1x_2 \cdot y}$ (c) $r_{x_2y \cdot x_1}$

15-4 Suppose the correlation between profits and advertising expenses is positive, but the partial correlation between profits and advertising expenses holding total expenses constant is negative.

(a) What do you predict would happen to profits if total expenses were kept constant, but the proportion of the budget allocated to advertising was decreased? Explain.

(b) What do you think would happen to profits if both advertising and total expenses were increased? Explain.

Section 15-2 Model Building and Multiple Regression

Read the article at the beginning of the chapter carefully. What is each company trying to do? They are attempting to construct realistic and reasonable *models* in order to use one or more input, or independent, variables to predict the dependent variable, sales.

model

Models

A model is a mathematical equation or set of equations that relates a set of independent variables to a dependent variable or dependent variables.

deterministic and probabilistic models

There are two types of models: **deterministic** and **probabilistic**. **In a deterministic model the independent variables determine exactly the value of the dependent variable.** Consider the compound interest model, which predicts how much money you will have at the end of n years if you put \$1,000.00 in the bank at 8% interest compounded yearly.

$$A = B(I)^n \qquad (15\text{-}2)$$

where

$B =$ amount invested

$I =$ interest rate

$n =$ number of years invested

Equation 15-2 is a deterministic model, since the values for the independent variables in the equation always produce the actual value of the predicted variable amount. **In a probabilistic model, in addition to the independent variables, there is a random variable that affects the value of the dependent variable.** Thus for a given set of values for the independent variables, the value of the dependent variable will not always be the same because of the value of the random variable. An example of a probabilistic model is the **simple linear regression model** presented in chapter 14, equation 14-1. Many different types of models are built and used in practice, but all model building consists of the same four steps:

1. Choosing the variables to use.
2. Choosing the form of model (equations) to use to relate the independent variables to the dependent variables.
3. Given the form of the model, estimating the parameters of the model.
4. Evaluating how well the model works.

In the Eli Lilly case cited in the article, the independent variable chosen to predict pharmaceutical sales was disposable income. (Disposable income means take-home income, the actual dollars a person has to spend.) The model chosen was a probabilistic model, a simple linear regression model:

$$y = \beta_0 + \beta_1 x + \varepsilon$$

where

y = pharmaceutical sales

x = disposable income

ε = random error

Lilly would then use the formulas given in chapter 13 to estimate the values of the parameters β_0 and β_1 and use the R^2 test procedures given in section 13-4 to evaluate the model.

Look now at what Armour & Company and Cherryoak did. They considered in their regression models more than one independent variable. In our problem of trying to predict sales in section 15-1, we began exploring what variables to use. Initially, we used a simple linear regression model using the one variable "entertainment expense." Now we find that when we consider the variable "years of employment of each salesperson," it had a stronger linear relationship to sales than entertainment expenditures did. Thus we could obtain a better-fitting regression equation if we used years of employment rather than entertainment expenditures to predict sales. In fact, since in two-variable regression the square of the correlation coefficient equals R^2, we know that "years of employment" explains 81% (0.901^2) of the variation in sales, whereas "entertainment expenses" explains 44% (0.660^2). However, we also found that sales and entertainment expenditures are correlated after the effect of years of service is removed (recall the partial correlation $r_{x_1 y \cdot x_2} = 0.52$).

We might wonder if we can use both independent variables in a regression to predict sales. Of course we can. In fact, we can build a regression model with three, four, or more independent variables used to predict the dependent variable. Such a model is called the *general linear model*.

general linear model

General Linear Model

$$y = \beta_0 + \beta_1 x_1 + \beta_2 x_2 + \cdots + \beta_p x_p + \varepsilon$$

where y is the dependent variable and $x_1, x_2, \ldots, x_p$ are the independent variables. ε is the random error and assumed to be normally distributed with a mean of 0 and a standard deviation of σ_ε.

multiple regression analysis

Given that we have selected the variables $x_1, x_2, \ldots, x_p$ and the general linear model to relate the independent x variables to the dependent y variable, the next step is to estimate the parameters of the model, $\beta_0, \beta_1, \ldots, \beta_p$. The process for estimating the parameters, determining how good the model fits, and specifying the inference tests associated with this model is called **multiple regression analysis**. We perform this analysis by finding the values of β_0 through β_p that make the model best fit the actual data. This task is done in the same manner as in simple linear regression—by using the method of least squares to find the best-fitting equation.

We must find the values of b_0, b_1, and b_p such that the sum of the squares of the error, $\sum (y - \hat{y})^2$, is as small as possible. Formulas for finding the coefficients b_0, b_1, and b_p exist. However, the formulas are very complicated and are tedious to solve. In practice, the solution for the coefficients is almost always done with the aid of a computer. Consequently, in this chapter we will not present the formulas and will assume you have a computer available to you.

A computer would calculate the best multiple regression model for our sales problem as

```
THE REGRESSION EQUATION IS
Y = 7.75    + (0.245      )X1 + ( 2.69       )X2
```

The computer does not print $\hat{y}$ or subscripts. Thus the answer translated to the form we have been using is

$$\hat{y} = 7.75 + 0.245x_1 + 2.69x_2$$

To predict sales using this linear model, we would substitute values of x_1 and x_2 into this equation and compute $\hat{y}$. For example, if a salesperson was employed 9 years (x_2) and spent 23 (hundred dollars) on entertainment (x_1), we would predict that he or she would sell 37.6 (thousand dollars) worth of merchandise (y):

$$\hat{y} = 7.75 + 0.245(23) + 2.69(9) = 37.595 = 37.6$$

When we interpret the regression equation, b_0 (7.75) is the intercept, that is, the value of y when x_1 and x_2 are *both* 0. b_1 (0.245), called the **net regression coefficient for x_1**, is the predicted change in y per unit increase in x_1, holding x_2 constant. It is the difference between two salespeople's predicted sales ($245) if they have the same years of employment, but one spends 1 (hundred dollars) more than the other on entertainment. b_2 (2.69), called the **net regression coefficient for x_2**, is the predicted change in y per unit increase in x_2, holding x_1 constant. It is the difference between two salespeople's predicted sales ($2,690) if they spend the same amount on entertainment, but one has been employed 1 year longer than the other.

Although the model in our example deals with only two independent variables, there is no reason why a third or fourth independent variable could not be introduced. A general linear multiple regression model may include some independent variables that appear at higher orders, such as x^2 and x^3, or that appear as the product of two independent variables, such as $x_1 x_2$. The model

$$y = \beta_0 + \beta_1 x_1 + \beta_2 x_1^2 + \beta_3 x_1 x_2 + \varepsilon \qquad (15\text{-}3)$$

is a linear model. If we let $x_3 = x_1^2$ and $x_4 = x_1 x_2$, we can rewrite model 15-3 as

$$y = \beta_0 + \beta_1 x_1 + \beta_2 x_3 + \beta_3 x_4 + \varepsilon \qquad (15\text{-}4)$$

and clearly model 15-4 fits the general linear model. A requirement for a model to be a linear model is that each term contains only one parameter and that the parameter is linear; that is, the parameter is β not β^2.

The exponential regression model, $y = \beta_0 \beta_1^x \varepsilon$, is *not* a linear regression model, since it contains two parameters $\beta_0 \beta_1$ together. A regression model that contains only one variable but at different higher orders is a special case of general linear model referred to as the *polynomial regression model*.

polynomial regression model

Polynomial Regression Model

$$y = \beta_0 + \beta_1 x_1 + \beta_2 x_1^2 + \cdots + \beta_p x_1^p + \varepsilon$$

quadratic and cubic regression models

Two specific examples of polynomial regression models are the **quadratic regression model**

$$y = \beta_0 + \beta_1 x_1 + \beta_0 x_2^2 + \varepsilon$$

and the **cubic regression model**

$$y = \beta_0 + \beta_1 x_1 + \beta_2 x_1^2 + \beta_3 x_1^3 + \varepsilon$$

curvilinear regression

Polynomial regression models are special cases of multiple regression and are referred to as **curvilinear regression**. The difference between regular, multiple, and curvilinear regression is in interpreting the β coefficients.

Illustration 15-1

According to general economic theory, the relationship between the number of items produced and the average cost (AC) of production items can be expressed as a quadratic regression model. That is, $AC = \beta_0 + \beta_1 x_1 + \beta_2 x_1^2 + \varepsilon$, where x represents units produced. Fifteen production runs were randomly sampled, and the average cost per unit was calculated. The resulting data are shown in table 15-2. Fit the model and graph the results.

TABLE 15-2
Production Lists

Units ($\times 1{,}000$)	Average Cost per Unit	Units ($\times 1{,}000$)	Average Cost per Unit
15	6.42	32	2.10
18	4.88	34	2.29
21	3.75	35	2.42
26	2.28	38	3.75
26	2.37	39	3.70
29	2.00	40	4.00
30	2.06	42	4.69
30	1.94		

Solution AC is the dependent variable, usually identified as y. Using a multiple regression program on a computer and considering x to be x_1 and x^2 to be x_2, we obtain

```
THE REGRESSION EQUATION IS
Y = 19.8      + (-1.18    )X1 + (0.197D-01)X2
```

Table 15-3 shows part of a computer printout for this problem. There are three columns of values: **X IN, Y IN, PRED Y**. The columns **X IN** and **Y IN** are the original data. **PRED Y** is the set of $\hat{y}$ values that are calculated by using the regression equation with each x value. The values 0, 1, and 2 under **X** in the **ESTIMATES FROM LEAST SQUARES FIT** part of the table are the subscripts corresponding to the coefficients (found under the heading **COEFFICIENTS**) and the independent variables. Thus column 3 is the column that contains the values of x^2 that were used for x_2 in this application (table 15-3 does not show column 3). The coefficient **0.19737D-01** is expressed in scientific notation. The **D-01** indicates that the decimal in **0.19737** should be moved one place to the left for it to be in its normal position.

The equation written in a format that we are accustomed to seeing is

$$AC = 19.8 - 1.18x + 0.0197x^2$$

A graph of the curvilinear regression is presented in figure 15-1. The original data are shown by the points, and the curve represents the regression model. It is determined by drawing a smooth curve through the points $(x, \hat{y})$. These values are listed in table 15-3 (**X IN, PRED Y**).

TABLE 15-3
Computer Printout for
Illustration 15-1

ROW	X IN COL. 1	Y IN COL. 2	PRED Y VALUES
1	15.000	6.420	6.460
2	18.000	4.880	4.864
3	21.000	3.750	3.624
4	26.000	2.280	2.347
5	26.000	2.370	2.347
6	29.000	2.000	2.055
7	30.000	2.060	2.036
8	30.000	1.940	2.036
9	32.000	2.100	2.118
10	34.000	2.290	2.357
11	35.000	2.420	2.536
12	38.000	3.750	3.309
13	39.000	3.700	3.648
14	40.000	4.000	4.022
15	42.000	4.690	4.893

ESTIMATES FROM LEAST SQUARES FIT

X	COLUMN	COEFFICIENTS
0	**	19.764
1	1	-1.1830
2	3	0.19737D-01

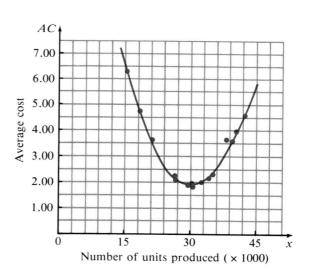

FIGURE 15-1
Curvilinear Regression,
Illustration 15-1

Exercises

15-5 Using the results of illustration 15-1, what do you predict the average cost per unit would be for the following production runs?
(a) 24 (b) 34 (c) 0

15-6 To predict sales of a certain microwave oven, the marketing department developed the following model

$$\hat{y} = 275 + 4x_1 - 250x_2$$

where

y = sales (thousands of units)

x_1 = number of retail outlets

x_2 = price of product minus price of competitor's product

(a) Explain what the coefficient 4 represents.
(b) Explain what the coefficient -250 represents.
(c) If the firm has the 100 retail outlets and it prices the product the same as its competitor does, what would you forecast its sales to be?
(d) If the firm raises its price $10 while its competitor's price remains the same, what do you predict would happen to the units sold?

15-7 Which of the following are general linear models:
(a) $y = \beta_0 + \beta_1 x_1 + \beta_2 + x_1^2 x_2 + \beta_3 x_2 + \varepsilon$
(b) $y = \beta_0 + \beta_1 x_1 + \beta_2 x_1^{\beta_3} + \varepsilon$
(c) $y = \beta_0 + \beta_1(1/x_1) + \varepsilon$
(d) $y = \dfrac{\beta_0}{1 + 10^{\beta_1 - \beta_2 x_1 + \varepsilon}}$

Section 15-3 How Well Does the Model Fit: Multiple R^2

multiple R^2 If we want to measure how well the multiple regression model fits the data, we can calculate R^2 by using formula (13-7). When R^2 is calculated for a multiple regression, it is called **multiple R^2** rather than simply R^2.

Illustration 15-2

Recall the multiple regression problem in section 15-1 predicting sales based on years of employment and entertainment expenses (see table 15-1). In section 15-2 we found the computer solution to be

$$\hat{y} = 7.75 + 0.245x_1 + 2.69x_2$$

How well does this multiple regression equation fit the data?

Solution Let's calculate R^2 by use of formula (13-7):

$$R^2 = 1 - \frac{\sum(y - \hat{y})^2}{\sum(y - \bar{y})^2}$$

For our example the necessary calculations for R^2 are presented in table 15-4.

TABLE 15-4
Calculations Needed to Find R^2

Salesperson	y	$(y - \bar{y})$	$(y - \bar{y})^2$	$\hat{y}$	$(y - \hat{y})$	$(y - \hat{y})^2$
1	30	−6.5	42.25	30.5	−0.5	0.25
2	26	−10.5	110.25	23.9	2.1	4.41
3	25	−11.5	132.25	30.3	−5.3	28.09
4	36	−0.5	0.25	32.5	3.5	12.25
5	33	−3.5	12.25	29.3	3.7	13.69
6	36	−0.5	0.25	42.0	−6.0	36.00
7	32	−4.5	20.25	35.2	−3.2	10.24
8	54	17.5	306.25	52.5	1.5	2.25
9	50	13.5	182.25	47.1	2.9	8.41
10	43	6.5	42.25	41.8	1.2	1.44
Total	365	0 ⓒⓚ	848.5		−0.1 ⓒⓚ	117.03
	$\bar{y} = 36.5$					

NOTES FOR TABLE 15-4:

1. The y values come from table 15-1.
2. $\bar{y}$ is the mean of the 10 y values.
3. $\hat{y}$ is the predicted amount of sales and must be calculated for each salesperson. For example, salesperson 1 spent 27 (hundred dollars) and has 6 years with the company. Therefore

$$\hat{y} = 7.75 + 0.245(27) + 2.69(6) = 30.5$$

The others are calculated in the same manner.

4. −0.1 is not exactly 0 for the check of the column labeled $(y - \hat{y})$. This discrepancy is due to the round-off error in the $\hat{y}$ values.

Using the calculations in the table with formula (15-7), we have

$$R^2 = 1 - \frac{117.03}{848.5} = 0.862 = \mathbf{0.86}$$

Thus 86% of the variation in the sales can be explained by a salesperson's entertainment expenditures and his or her years of employment. ☐

Section 15-4 Inferences Concerning the Goodness of Fit of a Multiple Regression Model

After determining the equation of best fit of the multiple regression model and measuring its goodness of fit, we are ready to determine if the regression model can be used to predict y values and whether a particular independent variable is of any use. First we need to test the null hypothesis "the linear multiple regression model is of no value in predicting y, given the set of independent variables." We do this by testing the null hypothesis $\rho^2 = 0$.

NOTE: If the multiple R^2 is 0, then all the β's are 0.

The test statistic that is used for this hypothesis test is F. The calculated value of F is found by using the formula

$$F = \frac{\sum (\hat{y} - \bar{y})^2 / p}{\sum (y - \hat{y})^2 / (n - p - 1)} \tag{15-5}$$

where p is the number of independent variables.

or its equivalent form

$$F = \frac{R^2 / p}{(1 - R^2)/(n - p - 1)} \tag{15-6}$$

This F statistic belongs to the F distribution with p and $n - p - 1$ degrees of freedom. Recall that each F distribution has two degrees of freedom: degrees of freedom for the numerator and degrees of freedom for the denominator. Notice that p is associated with the numerator and $n - p - 1$ is associated with the denominator for the calculated value of F.

Now let's test the null hypothesis $R^2 = 0$ for the sales example we considered in the previous section.

STEP 1: $H_0: \rho^2 = 0$ (all β's = 0).

STEP 2: $H_a: \rho^2 > 0$ (at least one $\beta \neq 0$, or at least one variable is useful in predicting y).

NOTE: R^2 must have a value between 0 and 1; that is, $0 \leq R^2 \leq 1$.

STEP 3: Let $\alpha = 0.05$. The critical value of F is $F(p, n - p - 1, \alpha)$, where p is the number of independent variables and n is the number of sets of data in the sample. For our illustration $p = 2$ and $n = 10$. Therefore the critical value of F is F(2, 7, 0.05) = 4.74, as found in table 9a of appendix D. See the figure.

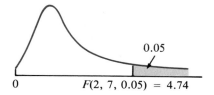

STEP 4: Use formula (15-6) to calculate F^*.

$$F = \frac{R^2/p}{(1 - R^2)/(n - p - 1)} = \frac{0.86/2}{(1 - 0.86)/7} = 21.50$$

$$F^* = 21.5$$

STEP 5: *Decision*: Reject H_0 (F^* is in the critical region).

Conclusion: $\rho^2 > 0$. The evidence indicates that *at least one* of the variables, "years of employment" or "entertainment expenses," is useful in predicting sales.

Exercises

15-8 Given the following results from a multiple regression model using three independent variables,

$$R^2 = 0.45 \quad n = 20$$

test the hypothesis $\rho^2 = 0$. Use $\alpha = 0.01$.

15-9 A linear multiple regression model using 25 sample pieces of data is calculated to predict the length of time it will take a client to pay his or her bill based on the variables "amount of bill," "credit rating of client," and "rate of discount for 30-day payment." The calculated R^2 is 0.60.

(a) Test the hypothesis that $\rho^2 = 0$. Use $\alpha = 0.05$.

(b) Can you conclude that the rate of discount for 30-day payment is useful in predicting time for payment? Explain.

15-10 A hospital-planning commission is interested in predicting the number of beds a hospital will allocate to obstetric services. It randomly selects 10 hospitals and builds a general linear multiple regression predicting the number of obstetric beds on the basis of "the number of members of the obstetric staff" and "the female population in the area between ages 14 and 45." The results of the multiple regression model are shown in the following table:

Hospital	Obstetric Beds, y	Predicted Obstetric Beds, ŷ
1	25	27
2	50	53
3	55	51
4	75	74
5	80	80
6	85	87
7	95	92
8	95	98
9	105	110
10	105	98

(a) How well does the regression fit the data?
(b) Test the hypothesis $\rho^2 = 0$. Use $\alpha = 0.05$.
(c) What can you conclude based on the results in (a) and (b)?

Section 15-5 Inferences Concerning Individual β's in the Multiple Regression Model

If we reject the hypothesis $\rho^2 = 0$, then we can conclude that at least one of the independent variables is useful in predicting y. The next step is to examine each of the independent variables separately to determine which one, or ones, is useful in predicting y. We do this by testing the null hypothesis "the slope of each variable is equal to 0." The test procedure is identical to that presented for β_1 in section 14-2 except for two things. First, instead of using formula (14-3) to calculate s_e^2, we use

$$s_e^2 = \frac{\sum (y - \hat{y})^2}{n - p - 1} \qquad (15\text{-}7)$$

NOTE: This formula is the same as formula (14-3) except for the denominator, the degrees of freedom. Compare the two formulas when $p = 1$.

Second, in calculating the critical value, the test statistic t has $n - p - 1$ degrees of freedom rather than $n - 2$.

In our sales prediction problem in section 15-4, we rejected the null hypothesis $\rho^2 = 0$. Now let's look at the independent variables, entertainment expenditures x_1 and years of service x_2, and determine which one is, or if both are, useful in the model for predicting sales.

Illustration 15-3

Test the hypothesis that the entertainment expenditures are not useful in the model for predicting sales. Use $\alpha = 0.05$.

Solution

STEP 1: $H_0: \beta_1 = 0$ (expenditures are not useful in the model for predicting sales).

STEP 2: $H_a: \beta_1 > 0$ (expenditures are useful in the model for predicting sales).

NOTE: A one-tailed test seems appropriate, since such expenditures would not be likely to lower the amount of sales.

STEP 3: The degrees of freedom for the test statistic t are

$$df = n - p - 1$$

Since $n = 10$ and $p = 2$ (there are two independent variables in our example), df = 7. Thus for $\alpha = 0.05$, the critical value of t is $t(7, 0.05) = 1.89$ (from table 7, appendix D). See the figure.

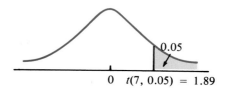

STEP 4: The t statistic is calculated by using formula (14-9):

$$t = \frac{b_1 - \beta_1}{s_{b_1}}$$

The value for b_1 is found from the regression equation in section 15-2 and is 0.245. The value for s_{b_1} is calculated by using formula (14-8):

$$s_{b_1}^2 = \frac{s_e^2}{\sum x^2 - (\sum x)^2/n}$$

First we find s_e^2 by using formula (15-7) and the calculations in table 15-4. Then we have

$$s_e^2 = \frac{\sum (y - \hat{y})^2}{n - p - 1} = \frac{117.03}{7} = 16.72$$

The sums needed to calculate $s_{b_1}^2$ are computed in table 15-5. Thus we have

$$s_{b_1}^2 = \frac{16.72}{12{,}481 - (339)^2/10} = \frac{16.72}{988.9} = 0.017$$

Therefore

$$t = \frac{0.245 - 0}{\sqrt{0.017}} = \frac{0.245}{0.130} = 1.8846$$

$$t^* = 1.88$$

STEP 5: *Decision*: Fail to reject H_0 (t^* is not in the critical region).

Conclusion: The evidence is not strong enough to reject the null hypothesis $\beta_1 = 0$. Hence entertainment expenses are not useful in the model for predicting sales holding years of employment constant.

TABLE 15-5 Calculations Needed for Computing $s_{b_1}^2$, Illustration 15-3

x_1	x_1^2	x_2	x_2^2
27	729	6	36
22	484	4	16
15	225	7	49
35	1,225	6	36
33	1,089	5	25
52	2,704	8	64
35	1,225	7	49
40	1,600	13	169
40	1,600	11	121
40	1,600	9	81
Total 339	12,481	76	646

Now let's consider years of employment.

Illustration 15-4

Test the hypothesis that the number of years of employment is not useful in the model for predicting sales. Use $\alpha = 0.05$.

Solution

STEP 1: $H_0: \beta_2 = 0$ (years of employment are not useful in the model for predicting sales).

STEP 2: $H_a: \beta_2 > 0$ (years of employment are useful in the model for predicting sales).

NOTE: As the number of years of employment increases, the experience gained should increase the amount of sales. Therefore a one-tailed test is used.

STEP 3: The critical value again is $t(7, 0.05) = 1.89$ (from table 7, appendix D).

STEP 4: The value for s_e^2 is the same as in illustration 15-3. Thus using formula (14-8), the calculated s_e^2, and the calculations in table 15-5, we have

$$s_{b_2}^2 = \frac{s_e^2}{\sum x^2 - (\sum x)^2/n} = \frac{16.72}{646 - (76)^2/10} = \frac{16.72}{68.4} = 0.244$$

From the regression equation found earlier, $b_2 = 2.69$. So using formula (14-9), we find the test statistic value to be

$$t = \frac{2.69 - 0}{\sqrt{0.244}} = \frac{2.690}{0.494} = 5.4453$$

$$t^* = 5.45$$

STEP 5: *Decision*: Reject H_0 (t^* is in the critical region).

Conclusion: $\beta_2 > 0$. The evidence indicates that "years of employment" is a useful variable for predicting sales in this regression model.

As a result of the two hypothesis tests completed in illustrations 15-3 and 15-4, we will disregard variable x_1 (entertainment expenses) from the regression model and keep variable x_2 (years of employment). With x_2 as the only useful independent variable for predicting sales, the regression model becomes $\hat{y} = 12.39 + 3.17x_2$. [This model is found by using the 10 ordered pairs of data (x_2, y) and formulas (13-5) and (13-6).] □

Exercises

15-11 Calculate R^2 for the regression determined in illustration 15-1 and test the hypothesis $\rho^2 = 0$. Use $\alpha = 0.05$.

15-12 If we fail to reject the hypothesis that $\rho^2 = 0$, should we test the hypothesis that each $\beta = 0$? Explain.

15-13 (a) For illustration 15-1 test the hypothesis $\beta_1 = 0$; use $\alpha = 0.05$.
(b) For illustration 15-1 test the hypothesis $\beta_2 = 0$; use $\alpha = 0.05$.
(c) Interpret your findings of (a) and (b) considered together.

15-14 You are given the following information from fitting a multiple regression model with three variables:

$$b_1 = 1.3 \qquad b_2 = 10.0 \qquad b_3 = 25.2 \qquad n = 30$$
$$s_{b_1} = 0.3 \qquad s_{b_2} = 2.0 \qquad s_{b_3} = 15.0$$

Test the following hypotheses. Use $\alpha = 0.05$.
(a) $\beta_1 = 0$ (b) $\beta_2 = 0$ (c) $\beta_3 = 0$

15-15 You are given the following information from fitting a multiple regression model with two independent variables to 18 sample data points:

$$\sum (y - \hat{y})^2 = 150 \quad \sum x_1 = 150 \quad \sum x_2 = 250$$
$$\sum x_1^2 = 2{,}250 \quad \sum x_2^2 = 6{,}255 \quad b_1 = 2.9 \quad b_2 = 4.0$$

Test the following hypotheses. Use $\alpha = 0.01$.

(a) $\beta_1 = 0$ (b) $\beta_2 = 0$

Section 15-6 Further Considerations in Modeling

In studying multiple regression models thus far, we have focused on steps 2, 3, and 4 in model building and have ignored step 1, choosing the variables to use. Choosing the variables to use consists of two parts: (1) measuring the factors we want to use as independent variables and (2) deciding which variables to use. In the previous section we learned the procedures in hypothesis testing to determine whether a variable we used in the model was useful or not. This determination, however, requires that we first build the regression model and then delete variables from the model based on evaluating how good it is.

Because multiple regression models deal with more than one independent variable, they create special problems. One problem is encountered when we try to enter variables containing the same information into a regression model. For example, suppose we attempt to build a model to predict sales of a consumer product using the male population (x_1), the female population (x_2), and the total population (x_3) as independent variables. Clearly, x_1 and x_2 contain all the information that x_3 contains. In fact, $x_1 + x_2 = x_3$. In this situation the regression model will not work, and no estimates of β_1, β_2, and β_3 are possible. This situation is known as **perfect multicollinearity**.

perfect and near multicollinearity

A more common problem is **near multicollinearity**. This situation occurs when we use a variable that is almost the same as some other variables or combinations of other variables. The result is an unreliable estimate of the regression model. In both cases (perfect and near multicollinearity), the variances of the slopes will be extremely large.

We often can think of more independent variables than we have sets of paired data (e.g., 10 observations and 25 independent variables). There are many techniques that can be used to select independent variables from a list of possible variables, the most common being stepwise regression. The question of variable selection, however, is beyond the scope of this basic text.

A word of caution is needed here. We cannot compare multiple R^2's of multiple regression models that do not have the same number of variables. R^2 will increase every time we add a new independent variable to the regression model, regardless of whether or not the variable is related to the dependent variable. If we were to add the shoe size of the salespeople to our regression model for predicting sales, R^2 would be higher than if just entertainment expenses and years of experience were used. Yet common sense tells us that shoe size cannot be used in a model to predict sales.

Let us now discuss the problem of how to measure the factor we want to use as an independent variable. There are two types of variables, *quantitative variables* and *dummy variables*.

quantitative and dummy variables

A **quantitative variable** represents counts or measurements. A **dummy variable** takes on a value of 0 or 1 to represent the presence or absence of a particular attribute quality.

For example, in a regression predicting a person's earnings, the variable equaling the person's age would be a quantitative variable, whereas the variable equaling a person's sex, represented by 1 if male and 0 if female, would be a dummy variable. (*Note*: The choice of whether male is given the value of 1 or female the value of 1 is totally arbitrary and affects only the sign of the regression coefficient, not the numerical value.)

Some qualitative factors may need to be represented by a series of dummy variables. Consider measuring the factor "market the stock is traded on." Assume, for simplicity, a stock can be traded only on either the American, the New York, or the Over-the-Counter Stock Exchange. In this case we would represent the factor by two dummy variables:

$$x_1 = 0 \text{ if not on New York Exchange}$$
$$= 1 \text{ if on New York Exchange}$$
$$x_2 = 0 \text{ if not on American Exchange}$$
$$= 1 \text{ if on American Exchange}$$

Notice that we do *not* use a third dummy variable, x_3, to represent traded on the Over-the-Counter Exchange. We exclude x_3, because, as discussed previously, we cannot enter variables containing the same information into a regression model. If we know the values of x_1 and the total, x_2, we already know the value of x_3; hence if we include x_3, we have perfect multicollinearity. Thus the regression will not work, and no estimates of β_1, β_2, and β_3 are possible. Which variable, x_1, x_2, or x_3, we leave out is arbitrary and does not affect the regression model.

Many factors can be measured as either a quantitative variable or a dummy variable in the regression model. How we measure it is a choice we must make. Suppose we are interested in building a linear multiple regression model to predict the earnings of people, and one of the factors we want to consider using is educational attainment. We could measure educational attainment as the number of years of schooling completed. For example, if a person completed high school, we would let $x_1 = 12$; if a person completed 3 years of college, $x_1 = 15$; and if a person completed college, $x_1 = 16$. In this case the factor "education" would be represented in the regression model by the quantitative variable x_1. You should realize, however, that by doing this type of measurement, we have made an important assumption: the effect of the education is linear; that is, each additional year of education is worth the same. Do you think that having completed 3 years of college rather than 2 years of college will increase your earnings potential as much as having completed 4 years of college and graduated rather than just 3 years? If not, then the choice of measuring education as years of schooling completed would not give us a good regression model. As an alternative we could measure education as a series of dummy variables: $x_1 = 1$ if high school graduate, $x_2 = 1$ if some college attended, $x_3 = 1$ if college graduate. Perhaps you can think of other ways to measure educational attainment.

In Retrospect

In this chapter we have expanded the concepts of bivariate correlation and regression analysis and considered situations in which we have more than two variables. We found that essentially all we had to do was modify the mathematical models and formulas; the basic concepts remained the same.

When a third variable is introduced into a regression model, the concept of partial correlation is encountered. You should be able to distinguish clearly between a simple and a partial correlation.

The concepts of modeling the general linear model, multiple regression, and curvilinear regression were presented. You should know the four basic steps of modeling, understand the difference between a probabilistic and deterministic model, and be able to distinguish the general linear regression model from the polynomial regression model. We discussed how we can interpret multiple and curvilinear regressions, how we calculate the multiple R^2, and how we test the usefulness of the regression model as a whole and the usefulness of each of the separate variables. However, the computation of the equation of best fit was left to the computer and standard multiple regression computer programs. In the next chapter we will explore the forecasting problem and models used for forecasting.

Chapter Exercises

15-16 In a curvilinear regression model, explain why it is not meaningful to interpret each b_i as the predicted change in y per unit change in x_i, holding all other x's constant.

15-17 Could you predict a company's stock price by using a multiple regression model with the three independent variables "the company's assets," "liabilities," and "net worth"? Explain.

15-18 Could you predict net sales of a company by using as independent variables "last year's sales," "sales 2 years ago," and "the average sales over the last 2 years"? Explain.

15-19 In studying the food price index, an economist calculated the following set of correlations:

$$r_{x_1 y} = 0.90 \qquad r_{x_1 x_2} = 0.74$$
$$r_{x_2 y} = 0.84 \qquad r_{x_1 x_3} = 0.62$$
$$r_{x_3 y} = 0.78 \qquad r_{x_2 x_3} = 0.50$$

where

$$y = \text{food price index}$$
$$x_1 = \text{per-capita-income index}$$
$$x_2 = \text{per-capita food consumption}$$
$$x_3 = \text{time}$$

(a) If you were trying to build a simple linear regression model to predict the food price index, which x variable would you use? Explain.

(b) After fitting the simple linear regression model, you decide to use a multiple regression model with a second variable. Which x variable would you add?

15-20 The accompanying table gives the results for a sample of six stocks.

Stock	Price-Earnings Ratio, x_1	Price, x_2	Volume, y (hundred shares)
A	7	36	84
B	5	28	74
C	8	39	89
D	7	30	78
E	10	45	92
F	3	22	70

(a) Calculate $r_{x_1 y}, r_{x_1 x_2}, r_{x_2 y}$.
(b) Calculate $r_{x_1 y \cdot x_2}, r_{x_1 x_2 \cdot y}, r_{x_2 y \cdot x_1}$.

15-21 A firm was interested in predicting the usage of computer time (y), measured in CPU (central processing units), that would be required by the firm's research department in a given month. Basing its study on the last 28 months, the firm built a multiple regression model by using the "number of projects" (x_1) and "monthly research budget" (x_2), measured in thousands of dollars, as independent variables. The results were as follows:

$$\hat{y} = 2.0 + 8x_1 + 0.02x_2$$

$$s_e^2 = 10, \qquad R^2 = 0.90, \qquad s_{b_1} = 3.0, \qquad s_{b_2} = 0.005$$

(a) Test the hypothesis that the regression model is useful. Use $\alpha = 0.05$.
(b) Test the hypothesis that the "number of projects" is a useful variable for predicting CPU usage. Use $\alpha = 0.10$.

(c) Test the hypothesis that "the monthly research budget" is a useful variable for predicting CPU usage. Use $\alpha = 0.10$.

(d) If there are five projects and the research budget is \$100,000, what would you predict CPU usage to be for the month?

15-22 A hospital administrator was interested in the relationship between hospital revenues (y) and Blue Cross payments (x_1) and the income of the community (x_2). He collected data on 15 hospitals in different counties of California and calculated the following multiple regression model:

$$\hat{y} = 3.45 + 0.495x_1 + 0.0092x_2$$

He also found the following results:

$$\sum (y - \hat{y})^2 = 560.8 \qquad \sum (y - \bar{y})^2 = 53{,}901$$
$$\sum (x_1 - \bar{x}_1)^2 = 129{,}328 \qquad \sum (x_2 - \bar{x}_2)^2 = 505{,}752$$

(a) Calculate R^2.
(b) Test the hypothesis $\rho^2 = 0$. Use $\alpha = 0.05$.
(c) Calculate s_e^2.
(d) Calculate s_{b_1} and s_{b_2}.
(e) Test the hypothesis $\beta_1 = 0$. Use $\alpha = 0.10$.
(f) Test the hypothesis $\beta_2 = 0$. Use $\alpha = 0.10$.

15-23 A firm was interested in the relationship between net income y (measured in \$10,000 units) and the number of salespeople x_1 and the number of products produced x_2. It collected the data shown in the accompanying table and built the following multiple regression model:

$$\hat{y} = -17.793 + 0.9121x_1 + 0.1589x_2$$

x_1	x_2	y
20	11	7
40	16	26
35	19	13
30	12	5
50	26	33

(a) If the company were to employ 40 salespeople and to carry 20 products, what would you predict the net income to be?
(b) Calculate R^2.
(c) Test the hypothesis $\rho^2 = 0$. Use $\alpha = 0.05$.
(d) Calculate s_e^2.
(e) Calculate s_{b_1} and s_{b_2}.
(f) Test the hypothesis $\beta_1 = 0$. Use $\alpha = 0.05$.
(g) Test the hypothesis $\beta_2 = 0$. Use $\alpha = 0.05$.

15-24 An economist was interested in the relationship between the consumer price index (CPI), the gross national product (GNP), and the prime interest rate. She collected the data shown in the accompanying table and built the following multiple regression model:

$$\hat{y} = 1.67 + 0.0029x_1 + 8.1115x_2$$

Year	CPI, x_1	GNP, x_2	Prime Interest Rate, y
1	100	1.05	7.1
2	103	1.21	8.2
3	112	1.26	9.1
4	104	1.35	9.8
5	116	1.47	10.4

(a) Calculate R^2.
(b) Test the hypothesis $\rho^2 = 0$. Use $\alpha = 0.05$.
(c) Calculate s_e^2.
(d) Calculate s_{b_1} and s_{b_2}.
(e) Test the hypothesis $\beta_1 = 0$. Use $\alpha = 0.05$.
(f) Test the hypothesis $\beta_2 = 0$. Use $\alpha = 0.05$.

15-25 Consider the accompanying data concerning hotel occupancy. The multiple regression model is

$$\hat{y} = -13.82 + 0.564x_1 + 1.099x_2$$

	Visitors to City, x_1	Number of Conventions, x_2	Vacancy Rate, y (%)
January	45	16	29
February	42	14	24
March	44	15	27
April	45	13	25
May	43	13	26
June	46	14	28
July	44	16	30
August	45	16	28
September	44	15	28
October	43	15	27

(a) Using x_1 as the only independent variable, calculate the simple regression model $\hat{y} = b_0 + b_1 x_1$.
(b) For the model in part (a), calculate each error e and R^2.
(c) Test the hypothesis $\rho^2 = 0$ for your answer in part (b). Use $\alpha = 0.05$.
(d) Using the given multiple regression model, calculate each error e and the multiple R^2.

(e) Test the hypothesis that the multiple $\rho^2 = 0$. Use $\alpha = 0.05$.
(f) Test the hypothesis $\beta_1 = 0$. Use $\alpha = 0.10$.
(g) Test the hypothesis $\beta_2 = 0$. Use $\alpha = 0.10$.
(h) Compare your results from the simple regression model with the results from the multiple regression model.

15-26 The following printout presents the computer output from a multiple regression analysis predicting percent of outstanding bonds converted, based on the data presented in exercise 2-81.

```
DEPENDENT VARIABLE..        A              % CONVT

MEAN RESPONSE           38.60000    STD. DEV.         26.82430

VARIABLE(S) ENTERED ON STEP NUMBER   1..   B           DIFF
                                           C           INDEX
                                           D           CONV TERMS


MULTIPLE R              0.93397    ANOVA                DF
R SQUARE                0.87229    REGRESSION           3.
                                   RESIDUAL            11.
STD DEVIATION          10.81445    COEFF OF            28.0 PCT
                                   VARIABILITY


SUM OF SQUARES      MEAN SQUARE         F         SIGNIFICANCE
   8787.12422         2929.04141    25.04474           .000
   1286.47578          116.95234


-------------------- VARIABLES IN THE EQUATION -----------

VARIABLE            B          STD ERROR B             t
                                                  -----------
                                                  SIGNIFICANCE

B             1.0516058        0.15416369           6.821358
                                                    0.000
C          -107.41302         24.986935             4.298767
                                                    0.001
D            24.932290         5.1278473            5.642190
                                                    0.000
(CONSTANT)   99.316241        24.740959             4.014244
                                                    0.002

VAR LABELS      A, % CONVT/B, DIFF/C, INDEX/D, CONV TERMS/
VALUE LABELS    D(1) NO(2) YES/
```

State or calculated the value of the following:

(a) The estimate of the linear multiple regression model.
(b) Which variable(s) is quantitative and which is dummy?
(c) Multiple R^2.
(d) F^* for testing hypothesis concerning ρ^2.
(e) Prob-value for testing hypothesis concerning ρ^2.
(f) Does the model fit better than chance (use $\alpha = 0.01$)?
(g) t^* associated with each variable.
(h) Prob-value for testing each variable.
(i) Which β's are not equal to 0 (use $\alpha = 0.01$)?
(j) If the difference is 10.00, the index is 1.05, and there is periodic reduction, what would you predict is the percent converted?

Hands-On Problems

Suppose you wish to determine whether the price of a stock and the price-earnings ratio can be used in a multiple regression model to predict the volume of trade in a stock. Go to a newspaper and randomly select 20 stocks and record their closing price (x_1), their price-earnings ratio (x_2), and the daily volume (y).

15-1 Draw a scatter diagram showing each pair—y and x_1, y and x_2, and x_1 and x_2.

15-2 Go to your computer center and find a packaged multiple regression program. (There are a variety available: Minitab, SPSS, SAS, BMDP, and IBM Scientific Subroutines. Your computer center can assist you.) Run your data and obtain a multiple regression equation.

15-3 What does b_1 mean?

15-4 What does b_2 mean?

15-5 Calculate and interpret R^2.

15-6 Test the significance of ρ^2. Use $\alpha = 0.05$.

15-7 If you reject the hypothesis $R^2 = 0$, test each individual coefficient. Use $\alpha = 0.05$.

15-8 What is your best prediction of the volume of a stock selling for $20 with a price-earnings ratio of 11?

16 Forecasting

Chapter Outline

16-1 Forecasting
We forecast the future in order to make decisions today.

16-2 Types of Models
Forecasting methods can be classified as either time series or causal.

16-3 Basic Elements of a Forecast
All forecasts have three basic elements in common: time, historical data, and uncertainty.

16-4 Measuring the Error of the Forecast: Mean Absolute Deviation and Mean Squared Error
Mean absolute deviation and mean squared error are the most common measures of the accuracy of forecasts.

16-5 Defining What Is Meant by the Best Forecasting Technique
The best forecast is the one that meets the needs at the minimum cost and inconvenience.

Disclosure of Projections of Future Performance

One of the major objectives of financial reporting is to supply information for the use of decision making. While traditional accounting statements are based on past transactions, most user decisions require the prediction of future events.

Investors are important users of financial statements; they must rely upon them as a basis for evaluating stock market prices and dividend paying ability. Since investment decisions reflect the judgment of investors concerning the future economic performance of a firm, it is not surprising that projections of future performance are sought by both individual and institutional investors. This concern with future events has resulted in an increasing interest in the possibility of including forecasts of future operations in the published financial statements of a firm....

Several individuals concerned with financial reporting have taken the view that the attest function should be extended to cover forecasted information. Roth commented that if the practice of including forecast data in published reports does develop, it is likely that the independent auditor will be expected to attest as to their fairness. He contends that the profession currently possesses the capability of performing this service. Willingham, Smith and Taylor stated that the extension of the auditor's opinion to include forecasts is not only within the competence of the profession, but that as the needs of society evolve, the practice of auditing must evolve to meet these changing needs. Others, however, disagree with the idea of reports by independent accountants attesting to forecasts. For example, Harvey Kapnick, the Chairman of Arthur Andersen & Co., contends that the audit of published forecasts raises certain serious questions regarding the competence and independence of the accountant.

In an attempt to provide answers to the questions raised above and to consider related issues, the present study was undertaken....

In order to determine the opinions of financial analysts, the authors sent a mail questionnaire to a random sample of two hundred twenty-five chartered financial analysts (CFAs). Of the 225 individuals selected, a total of 53 usable responses were received for an overall response rate of 23.6 percent....

TABLE 4

Potential Involvement of the CPAs with Published Forecast Data ("Agree" Replies Only)

	Number	Percentage
There is, or within ten years will be, a need for CPAs to attest to financial forecasts or other disclosures of estimates of future included in financial statements	29	54.7
Auditors' attestation would add to reliability of forecast data	17	32.0
SEC, or some other regulatory agency, *should* require an opinion of a CPA with regard to published financial forecasts within next ten years	19	35.8
SEC, or some other regulatory agency, *will* require an opinion of a CPA with regard to published financial forecasts within next ten years	21	39.6
If auditor's opinion is extended to published financial forecasts, opinion should specifically cover:		
Soundness of the accounting method used	53	100.0
Accuracy of the compilation of the data	44	83.0
Assumptions underlying the forecast	31	58.4

From James J. Benjamin and Robert H. Strawser, "Disclosure of Projections of Future Performance," *Journal of Business* (Seton Hall University), vol. 12, no. 2. (May 1974): 14–24. Reprinted by permission.

Chapter Objectives

We live in the present. We can study the past. But often what we really need to know is the future. Many of the decisions we face every day are difficult to make because what should be done depends on what will happen at some future time. For instance, choosing college courses would be easier if you knew exactly what you would learn in each course and exactly how useful the learned material would be in future endeavors (and perhaps also what your grade would be). However, since we must make these decisions in the present, we are forced to predict the future.

In this chapter we will present the basic concepts of forecasting. We will see that in business there is a wide diversity of forecasting situations. Similarly, there are numerous forecasting techniques. Unfortunately, there is no single, best forecasting technique. The forecasting problem, therefore, is to match the appropriate technique to the given forecasting situation. Thus in this chapter we will focus on the first step in forecasting, understanding the forecasting situation. We will discuss what all forecasting situations have in common. We will investigate how we can evaluate the accuracy of a forecasting technique and we will discuss what is meant by the "best" method. We will also define the different patterns of data with which we must deal. We will see that the various quantitative forecasting techniques can be classified into one of two categories, causal and time series, and we will outline the advantages and disadvantages of using each type.

Section 16-1 Forecasting

The name **forecasting** has been given to the problem of **trying to predict what the situation will be at some time in the future**. If you stop and think for a minute, you will realize how often you make decisions based on what you think will happen in the future. The decision of whether or not to carry an umbrella depends on what you forecast the weather to be like during the day. When you purchase something on extended time payments, intuitively (or formally, if you are a systematic budgeter) you have forecasted your future income and expenses and have concluded that you will be able to make the payments. When you choose a college major, your choice is probably based in part on an estimate of future labor market conditions.

In the business world, forecasts are prepared daily in all phases of operations. Some examples of the use of forecasts are the following:

> Marketing departments need forecasts of demand to plan sales strategies. Such forecasts of demand are needed on a variety of bases. For example, total demand is needed to plan total sales effort; demand by market regions is needed to allocate sales effort; demand by consumer characteristics (e.g., age, income groups, sex, etc.) is needed to plan effective advertising strategy.

> Production scheduling requires detailed estimates of demand by specific lines of production (e.g., color, style, etc.) and usually by the week

or month. Moreover, forecasts of trends in availability and prices of raw materials are needed to plan purchases.

Personnel departments need forecasts of labor turnover, labor supply, absenteeism, and trends in wages.

Top management requires forecasts of technological change, economic conditions, and future company growth in order to plan capital expenditure for new plant and equipment and to plan the long-term general future course of the company.

As you can see, there is a diversity of situations where forecasts are made in business. There also is a diversity of forecasting techniques available for the businessperson to use. They run the gamut from simply relying on personal judgment to the use of very complex quantitative models that simulate and forecast the whole economy. The following list presents some of the major quantitative forecasting techniques.

Simple exponential smoothing
Linear exponential smoothing
Linear moving average
Seasonal exponential smoothing
Classical time series decomposition
Simple regression
Multiple regression
Census II
Foran II
Adaptive filtering
Generalized adaptive filtering
Box–Jenkins
Econometric models
Business indicators
Growth curves

This list is presented here simply to show that there are many forecasting techniques available from which to choose. Only a few of the forecasting techniques will be discussed in this text. Regression, a very popular method used in forecasting, has been discussed already in chapters 13, 14, and 15; classical time series and business indicators will be discussed in subsequent chapters. In this chapter we will restrict our attention to understanding the basic concepts of the forecasting problem.

It is important to understand the basic concepts of the forecasting problem. Also, it is important to understand the elements that all forecasting situations have in common and how forecasting situations differ, because *there is no single best forecasting technique*. The reason that there is a diversity of forecasting techniques is that different techniques work best in different situations. **The forecasting problem, then, is that of trying to match the appropriate technique to the particular situation.** Thus the first step is to analyze the forecasting situation.

As we mentioned, there are a multitude of different forecasting techniques. These techniques can be divided into two basic categories. These two classifications are discussed in the next section.

Section 16-2 Types of Models

time series model

causal model

The multitude of different forecasting techniques can be classified into two basic types. The first and most common is a time series model. In a **time series model** the variable to be predicted is analyzed historically over time, and this historical pattern, or patterns, is modeled or estimated. A time series model assumes that the time-sequenced pattern that has occurred in the past will continue into the future. This model can be identified solely from the historical pattern of the variable to be predicted. Consequently, any time series model will give the same forecast no matter what management decisions are currently made. Such models are not useful for predicting what changes will result from a management policy change, such as a price reduction or a new advertising campaign.

The second type of forecasting technique is called a causal model. In a **causal model** other variables that are related to what is being predicted are used, and then their predictive relationship is used in making the forecast. For example, sales can be related to advertising expenditures, sales efforts (number of man-hours the salespeople have expended), price of the item, competitors' prices, and so forth. Sales would then be considered as a function of these independent variables, and the corresponding algebraic relationship would be estimated. Thus by using expected future values of the independent variables, we can predict the dependent variable. In causal models we seek relationships that generate the patterns on the data. The advantage of such models is that they allow us to incorporate into the model any changes in management policies or shifts in the environment. The disadvantages are that they are more difficult to develop, that they require much more historical data (i.e., they require data on various variables and not simply on the item to be predicted), and that the ability to predict requires knowledge of, or the ability to accurately predict, the future values of the independent variables. However, such models generally are more useful in evaluating the impact of management policy than are time series models, especially for a medium-range or long-range prediction.

The regression techniques that we have studied in chapters 13, 14, and 15 belong to the class of causal models. We will be studying the classical time series analysis in chapter 17.

Exercises

16-1 What are the advantages and disadvantages of time series and causal models?

16-2 Consider the following sales forecasting models and state whether they are causal or time series models.

(a) $\text{Sales}(t + 1) = \sum_{i=1}^{t} w_i \times \text{sales}(i)$, where w_i is a weight (constant).

(b) $\text{Sales}(t + 1) = 100 + 5$ (advertising dollars spent in year t) $+ 25$ (number of salespeople employed in year $t + 1$).

(c) $\text{Sales}(t + 1) = 0.8 \, [\text{sales}(t) - \text{forecasted sales}(t)] + \text{forecasted sales}(t)$.

Note: Sales($t + 1$) represents the amount of sales in time period $t + 1$. Sales(t) represents the sales in time period t.

16-3 Why are there many different forecasting techniques?

Section 16-3 Basic Elements of a Forecast

Although there are many types of forecasts, all forecasts have three common basic elements: **time, the reliance on historical data, and a degree of uncertainty.** We will deal with each of these three elements in the following subsections.

Time Frame

Forecasts deal with situations in the future, so time is directly and importantly involved. Every forecast must be made for a specific point in time. The length of time in the future over which the forecast is made is called the **time frame** or **time horizon** of the forecast. Generally, changing the time frame affects the choice of the forecasting technique and the ability to forecast accurately. Normally, *the longer the time frame, the more difficult it is to forecast accurately.* The length of time can be a day, a week, a month, a quarter, a year, several years, or any other desired period.

Depending on their length of time, time frame forecasts are generally classified as follows:

immediate, short-term, medium-term, and long-term time frames

Immediate: less than 1 month
Short-term: 1 to 3 months
Medium-term: more than 3 months to less than 2 years
Long-term: 2 years or more

Patterns of Data

All forecasts must, to some extent, rely on information that is contained in historical data. As the famous revolutionary hero Patrick Henry stated, "I know of no way of judging the future but by the past."

assumption of consistency

The basic assumption in using historical data is that some pattern(s) exists that can be identified and extrapolated to prepare predictions. This is known as the **assumption of consistency**.

The key to good forecasting is to select the techniques that are appropriate for the pattern of data that exists and for the time frame of the forecast. The **four basic types of patterns** are:

1. trend
2. seasonal
3. cyclical
4. irregular

trend pattern

Trend Patterns

Patterns that exist when the data follow a generally increasing or decreasing pattern over a period of time.

Trend patterns are generally very significant in medium-term and long-term forecasts. Most business and economic data (e.g., gross national product, stock prices, sales, etc.) exhibit a trend pattern. Figure 16-1 illustrates a

FIGURE 16-1
Example of a Trend Pattern

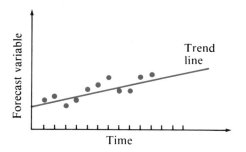

trend pattern. The straight line through the middle of the pattern represents the idea of trend. If there is a trend, this line will have either a positive or a negative slope. If the slope is 0, then there is no trend pattern. When no trend pattern is present, the data are said to be **stationary**. This classification does not mean that values of the data do not vary, but that they vary about a horizontal line displaying no overall growth or decline. Figure 16-2 illustrates a stationary pattern.

seasonal pattern

Seasonal Patterns

Fluctuations in the data that occur during a time period of 1 year or less and repeat themselves over each consecutive time period.

FIGURE 16-2
Example of a Stationary Pattern

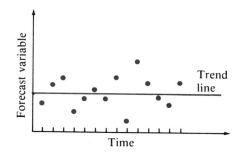

Although we tend to think of seasonal patterns in terms of a yearly time period, some seasonal patterns may be monthly or even weekly. For example, department store sales usually vary by month of the year; cash receipts may vary by the week or the month because of billing cycles; a firm's output may vary by the day of the week because of weekly production schedules. Figure 16-3 shows a seasonal pattern in data. The seasonal

FIGURE 16-3
Example of a Seasonal Pattern

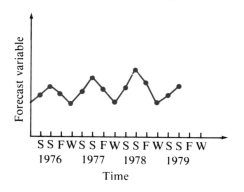

pattern is identified by the pattern of low points for W(winter), high points for S(summer), and in-between points for S(spring) and F(fall). The same basic pattern repeats over the years.

cyclical pattern

Cyclical Patterns

Similar to seasonal patterns except that the time period over which the pattern repeats itself exceeds 1 year. Cyclical patterns generally are induced by the long-term cyclical patterns in the economy and are explained by general business cycle theory.

For example, it seems that periods of prosperity and recession follow each other, and sales of commodities of major consumer goods (cars, televisions, steel, etc.) tend to follow the economy. Data taken in these situations would display a cyclical pattern. Figure 16-4 illustrates such a cyclical pattern (about a stationary trend line) in a set of data. Figure 16-4 shows a repetitive pattern much like the seasonal pattern, except that the length of time

FIGURE 16-4
Example of a Cyclical Pattern

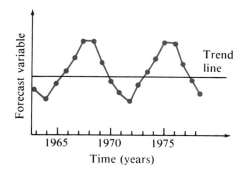

for each repeated pattern is 8 years. Recall that the length of time *for each repeated pattern* in the seasonal pattern in *less* than 1 year.

irregular pattern

Irregular Patterns

The unexplained or unpredictable fluctuations in the data.

Figure 16-5 shows an example of an irregular pattern—that is, there is no identifiable pattern in the plotted data.

FIGURE 16-5
Example of an Irregular Pattern

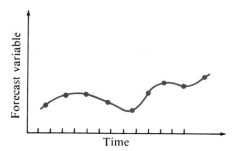

The four patterns outlined here do not necessarily occur alone. In many situations they can be found operating in various combinations or all together. In classical forecasting, which is discussed in the next chapter, all four patterns are assumed to exist simultaneously. The technique then focuses on decomposition of the data into each component, estimating each pattern separately, and combining the projected impact of each component in the future to produce the final forecast.

It should be pointed out that analyzing historical data to determine the underlying pattern or patterns can yield quite useful management information, aside from its use in forecasting. Such analysis allows the manager to better understand what has happened and hence to evaluate current management policies. For instance, the determination of whether a drop in sales is due to seasonal variation, cyclical variation, irregular fluctuation, or a trend of declining sales (or a determination of how much can be attributed to each) can be vitally important to management in evaluating its current policies and indicating needed action.

Error of the Forecast

forecast error

All forecasts deal with a certain level of uncertainty. This uncertainty is the third aspect common to every forecast. That is, we are never certain what the situation will actually be in the future. If we were, the forecasting problem would be trivial. Because of the uncertainty, it is necessary to make assumptions or judgments about relevant conditions in the future that underlie the forecast and, as a result, **some error in forecasting must be expected**. Therefore the irregular pattern must be included in all forecasts.

The magnitude of the irregular element will significantly affect the ability to forecast. If the irregular element is a very large portion of the data, then, in effect, the data are unpredictable. On the other hand, if the irregular element is small and if we can determine the trend, seasonal, and cyclical effects, we should be able to forecast fairly accurately.

Exercises

16-4   What are the three aspects that every forecasting situation has in common?

16-5   Explain the following statement: Whereas all forecasting situations have three basic aspects in common, it is the examination of these three aspects that helps separate one forecasting situation from another and that gives guidance in selecting a forecasting technique.

16-6   In quantitative forecasting, what is the basic assumption we make when using past data?

16-7   Explain the differences among trend, seasonal, cyclical, and irregular patterns of data.

Section 16-4 Measuring the Error of the Forecast: Mean Absolute Deviation and Mean Squared Error

To evaluate the forecasting error, let us denote the predicted value of y_i as $\hat{y}_i$. ($\hat{y}$ represents the predicted value of a forecasted variable. Essentially, this role is the same that $\hat{y}$ plays in regression analysis.) $\hat{y}_i$ includes the estimate of the trend, cyclical, and seasonal patterns of the data. If we subtract the predicted value $\hat{y}_i$ from the actual value of y_i, we have the **error of a particular forecast**, e_i:

$$e_i = y_i - \hat{y}_i \qquad (16\text{-}1)$$

The errors of forecast are the key to evaluating how accurately we can forecast. However, before we discuss the common ways to measure the accuracy of a given forecast, we stress one important point. If we could *perfectly* predict the trend, seasonal, and cyclical components, then e_i would simply represent the irregular element of the data. Hence our measure of error would tell us how predictable the data are and whether or not we can forecast with any accuracy. Realistically, however, we can never perfectly predict the trend, seasonal, and cyclical components. Thus our measure of error is a combination measure of how predictable the data are and how well the particular technique predicts the trend, seasonal, and cyclical patterns. Hence large errors may mean either that no forecasting technique will be accurate or that the particular technique we are using is inappropriate.

There are several alternative ways of measuring the predictive capabilities of any forecasting method. One approach might be simply to compute each error of forecast and sum these errors over all the data, $\sum (y_i - \hat{y}_i)$. The problem with this approach, of course, is that if errors tend to be irregular, we will find positive errors offsetting negative errors. Hence the sum will always be close to 0 regardless of how accurate the forecasts really are.

However, examination of the *pattern* of errors over time can be most useful in seeing if some trend, seasonal, or cyclical pattern is being ignored. For example, let's examine the errors of forecast presented in figure 16-6.

Looking at the graph of the errors in figure 16-6a, we can see that the errors exhibit an upward pattern. The errors are not random or unpredictable, since they exhibit a pattern. This pattern would indicate that the forecasting technique used is inadequate and another technique should be tried. In particular, the graph of the errors indicates that there is a trend pattern that is not accounted for. Figure 16-6b shows that a seasonal pattern has not been adequately identified in the forecasting model. (Basically, the points follow a pattern of two positive errors, two negative, two positive, etc.) Figure 16-6c shows that a cyclical pattern has not been adequately measured in the forecasting model.

FIGURE 16-6
Examples of Errors in Forecast That Are Not Irregular

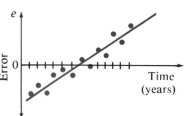

(a) Trend pattern not fully accounted for by model

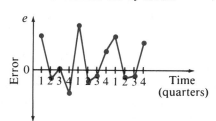

(b) Seasonal pattern not fully accounted for by model

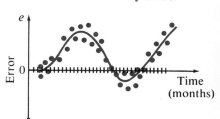

(c) Cyclical pattern not fully accounted for by model

On the other hand, if the errors were to appear purely irregularly, as in figure 16-7, we would conclude that the forecasting technique is adequate. In this case we would examine the magnitude of the errors to determine if accurate forecasting is possible.

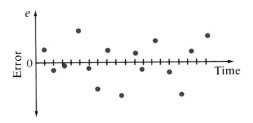

FIGURE 16-7
Errors in Forecast That Are Irregular

To determine the magnitude of the forecasting error, we can avoid the problem of positive and negative errors canceling by simply dealing with the absolute value of the error terms.

absolute deviation
mean absolute deviation

The sum of the absolute values of the error terms is referred to as the **absolute deviation (AD)**. When the absolute deviation is divided by the number of terms n, the result is called the **mean absolute deviation (MAD)**.

$$\text{Absolute deviation} = \sum |y_i - \hat{y}_i| = \sum |e_i| \quad (16\text{-}2)$$

$$\text{Mean absolute deviation} = \frac{\sum |y_i - \hat{y}_i|}{n} = \frac{\sum |e_i|}{n} \quad (16\text{-}3)$$

The accompanying table shows the calculations of the absolute deviation (AD) and the mean absolute deviation (MAD) for some data.

Actual y_i	−	Predicted $\hat{y}_i$	=	Error e_i	Absolute Error, $|e_i|$
20		18		2	2
35		39		−4	4

$$AD = 6$$
$$MAD = 6/2 = 3$$

An alternative to using the absolute value of the error is to use its square. Squaring each of the errors also eliminates the problem of positive and negative errors canceling.

squared error
mean squared error

The sum of the squares of the error terms is referred to as the **squared error**. When the squared error is divided by the number of terms n, the result is called the **mean squared error (MSE)**.

$$\text{Squared error} = \sum (y_i - \hat{y}_i)^2 = \sum e_i^2 \qquad (16\text{-}4)$$

$$\text{Mean squared error} = \frac{\sum (y_i - \hat{y}_i)^2}{n} = \frac{\sum e_i^2}{n} \qquad (16\text{-}5)$$

The accompanying table shows the calculations of the squared error and the mean squared error (MSE) for some data.

Actual y_i	− Predicted $\hat{y}_i$	= Error e_i	Error Squared, e_i^2
20	18	2	4
35	39	−4	16

Squared error = **20**
MSE = 20/2 = **10**

The major difference between the MSE and the MAD is that the MSE, unlike the MAD, penalizes a forecast method much more for extreme errors than it does for small errors. That is, if we consider errors of 4 and 2 when using MAD, an error of 4 is twice that of an error of 2. But if we use MSE, an error of 4 is four times that of an error of 2, since $4^2 = 16$ and $2^2 = 4$. (See the two preceding tables.) Thus if we would rather have several small errors and few large errors, we should use MSE rather than MAD.

Illustration 16-1

Consider a situation in which we have two possible forecasting techniques, A and B. The accompanying table gives the forecasts for each method in

(1) Actual, y_i	(2) Predicted A	(3) Predicted B
45	42	46
49	46	48
52	55	58

columns 2 and 3 and gives the actual values in column 1. Which forecast method has the smaller MSE? The smaller MAD? If the two measures are different, explain why.

Solution The calculations for the MAD and the MSE are shown in the next table. Forecast B has the smaller MAD, whereas forecast A has the

(1) Actual, y_i	(2) Predicted A	(3) Predicted B	Error A	Error B	Absolute Error A	Absolute Error B	Squared Error A	Squared Error B
45	42	46	3	−1	3	1	9	1
49	46	48	3	1	3	1	9	1
52	55	58	−3	−6	3	6	9	36
Sum					9	8	27	38
MAD					3	2.67	—	—
MSE					—	—	9	12.67

smaller MSE. The difference is due to the large penalty the MSE imposes for the error of −6 in forecast B. □

Exercises

16-8 Why is the sum of the forecasting error *not* a good measure of accuracy?

16-9 Suppose you are given the forecasts shown in the accompanying table. Compute the following:
(a) the forecast error
(b) MAD
(c) MSE

Date	Cash on Hand	Predicted Cash on Hand
1/31/83	$220	$205
2/28/83	210	198
3/31/83	195	197
4/30/83	184	184
5/31/83	163	175
6/30/83	152	165

16-10 The accompanying table presents the forecast errors associated with three sales models, each of a different quality.
(a) Graph the forecast errors for each model.
(b) Do the patterns of error appear to be random (i.e., irregular) for each model? If not, why not?
(c) Which model(s) would you conclude inadequately fits the pattern of the data?

		Forecast Error		
Quarter		A	B	C
1978	I	−25	−20	−19
	II	+5	+5	−15
	III	+15	−20	−9
	IV	−10	+6	−8
1979	I	−12	−5	−4
	II	+7	−1	−2
	III	+9	−12	+6
	IV	−7	+3	+10
1980	I	−14	+9	+15
	II	+12	+26	+22
	III	+21	−12	+27
	IV	−10	−8	+31
1981	I	−6	+20	+26
	II	+5	−2	+14
	III	+7	−4	+9
	IV	−2	−5	+4
1982	I	−9	+16	−2
	II	+8	−14	−6
	III	+13	+9	−2
	IV	−3	+5	+8

(d) Compute MSE for each model.
(e) Compute MAD for each model.

Section 16-5 Defining What Is Meant by the Best Forecasting Technique

In illustration 16-1 we saw that forecast method A had a larger MAD but a smaller MSE. This was because method B was heavily penalized for its large error in the last forecast when using the measure MSE. You should recognize, however, that neither of these accuracy evaluations is totally sufficient in evaluating a forecasting technique or in comparing different techniques. Other factors, such as costs and level of complexity, are directly related to accuracy. Since decisions vary considerably in scope and importance and since the need for accuracy varies, it should be recognized that the best forecasting technique is not necessarily the most accurate technique but is the one that meets the needs at the minimum cost and inconvenience.

For instance, suppose we develop a sophisticated forecasting model to predict sales 1 month in advance and find that it has a mean absolute deviation of $3,000. (For simplicity we have decided MAD is the best measure of error for our purposes.) What does this really tell us? First of all, the

$3,000 must be placed in perspective. If monthly sales average $6,000, then the MAD is extremely large and our forecasts aren't very good. (An error of $3,000 on a $6,000 figure is quite large.) Conversely, if monthly average sales are $300,000, our forecasts on the average are quite accurate.

However, even if average sales were $300,000, this information is still not enough to tell us that our forecasting method is worthwhile. Suppose that monthly sales never differ from $300,000 by more than $4,000. Then if we were simply to forecast $300,000 every month, we would do almost as well, with a lot less effort and cost, as the sophisticated model.

To evaluate any error measurement, some basic standard of comparison is needed to place the errors in perspective. A common practice is to use various so-called naive models. A **naive model** is an extremely simple model that can be implemented inexpensively and easily without formal statistical analysis and without the aid of a computer. A common naive model is to forecast that the next value will equal the last observed value (i.e., $\hat{y}_{t+1} = y_t$, which is read, "the predicted value of y in time period $t + 1$ equals the observed value of y in time period t"). Then the ease and cost of applying the naive model can be compared against the greater accuracy but higher cost of a more sophisticated and complex model.

In the following illustrations we investigate three forecasting methods to determine which one is best for the situation.

Illustration 16-2

Suppose that for production-scheduling purposes we need to forecast sales 1 month in advance. We are considering the use of a sales-forecasting technique called *exponential smoothing*. This technique is fairly simple and can easily be developed and used; hence the costs in developing and using it are not high when compared with most other methods. A yearly record of the results of using the exponential smoothing method is given in table 16-1

TABLE 16-1 Forecasting Record, Illustration 16-2

Time Period (months)	Actual Sales	Predicted Sales, A
1	1,500	—
2	1,725	1,500
3	1,510	1,573
4	1,720	1,522
5	1,330	1,542
6	1,535	1,521
7	1,740	1,522
8	1,810	1,544
9	1,760	1,571
10	1,930	1,590
11	2,000	1,624
12	1,850	1,662

under the heading labeled *Predicted Sales, A*. (Since a starting point is needed to use this method, the first month is not forecast.) Calculate the MAD and MSE for forecast A.

Solution The calculations for the errors ($e_i = y_i - \hat{y}_i$), the absolute value of the error $|e_i|$, the square of the errors e_i^2, and their sums are shown in table 16-2. The values of MAD and MSE are then calculated by using formulas (16-3) and (16-5), respectively.

$$\text{MAD} = \frac{\sum |e_i|}{n} = \frac{2{,}289}{11} = 208.09 = \mathbf{208}$$

$$\text{MSE} = \frac{\sum e_i^2}{n} = \frac{585{,}259}{11} = 53{,}205.36 = \mathbf{53{,}205}$$

TABLE 16-2
Calculations Needed to Find MAD and MSE for Prediction A

Time Period, i (months)	Actual, y_i	Predicted A, $\hat{y}_i$	Error, $e_i = y_i - \hat{y}_i$	$|e_i|$	e_i^2
1	1,500	—	—	—	—
2	1,725	1,500	225	225	50,625
3	1,510	1,573	−63	63	3,969
4	1,720	1,522	198	198	39,204
5	1,330	1,542	−212	212	44,944
6	1,535	1,521	14	14	196
7	1,740	1,522	218	218	47,524
8	1,810	1,544	266	266	70,756
9	1,760	1,571	189	189	35,721
10	1,930	1,590	340	340	115,600
11	2,000	1,624	376	376	141,376
12	1,850	1,662	188	188	35,344
Total				2,289	585,259

We see that forecasting method *A* has an MAD of 208 and an MSE of 53,205.

average absolute error of forecast

Average Absolute Error of Forecast

Found by dividing MAD by $\bar{y}$:

$$\text{Average absolute error of forecast} = \frac{\text{MAD}}{\bar{y}} \qquad (16\text{-}6)$$

Let's consider the data for illustration 16-2. The average sales $\bar{y}$ over the last 11 months (the same months for which the errors were calculated, namely, months 2 through 12) is 1,719.1 (18,910 ÷ 11). Thus the average absolute error of forecast for forecast A is approximately 12% (208 ÷ 1,719.1). Let's assume that this average absolute error of 12% shows that the forecast A technique is adequate for our needs in the situation described in illustration 16-2.

Now suppose someone else suggests using a more complex and costly forecasting technique called *Box–Jenkins* for forecasting the sales in illustration 16-2.

Illustration 16-3

Table 16-3 gives the Box–Jenkins sales forecasts (listed as *Predicted Sales, B*) for the sales of illustration 16-2. Calculate the MAD, MSE, and average absolute error of forecast for prediction B.

TABLE 16-3
Box–Jenkins Sales Forecasts

Time Period (months)	Predicted Sales, B	Time Period (months)	Predicted Sales, B
2	1,525	8	1,600
3	1,532	9	1,610
4	1,650	10	1,640
5	1,683	11	1,730
6	1,735	12	1,800
7	1,702		

Solution The calculations for the errors e_i, the absolute value of error $|e_i|$, the square of the error, and their sums are shown in table 16-4. The MAD and MSE are

$$\text{MAD} = \frac{\sum |e_i|}{n} = \frac{1,853}{11} = 168.45 = \mathbf{168}$$

$$\text{MSE} = \frac{\sum e_i^2}{n} = \frac{437,537}{11} = 39,776.09 = \mathbf{39,776}$$

Method B for forecasting sales has an MAD of 168 and an MSE of 39,776. The average absolute error of forecast for method B is 10% ($\bar{y} = 1,719.1$, as before; MAD/$\bar{y}$ = 168/1,719.1 = 0.098 = 0.10). Thus by using the more costly and complex technique, we reduce the average absolute error from 12% to 10%. □

If the first technique (illustration 16-2) adequately meets our needs, then increasing the complexity and cost of the technique for the slight improvement indicated in illustration 16-3 does not seem worthwhile. These facts alone might tempt us to go ahead and use the exponential smoothing

TABLE 16-4
Calculations Needed to Find MAD and MSE for Prediction B

Time Period, i (months)	Actual, y_i	Predicted B, $\hat{y}_i$	Error, e_i	$\|e_i\|$	e_i^2
1	1,500	—	—	—	—
2	1,725	1,525	200	200	40,000
3	1,510	1,532	−22	22	484
4	1,720	1,650	70	70	4,900
5	1,330	1,683	−353	353	124,609
6	1,535	1,735	−200	200	40,000
7	1,740	1,702	38	38	1,444
8	1,810	1,600	210	210	44,100
9	1,760	1,610	150	150	22,500
10	1,930	1,640	290	290	84,100
11	2,000	1,730	270	270	72,900
12	1,850	1,800	50	50	2,500
Total				1,853	437,537

method (method A). However, suppose we compare method A with the easier, cheaper naive model.

Illustration 16-4

Calculate the MAD, MSE, and average absolute error of forecast for the naive model, $\hat{y}_{t+1} = y_t$, for the sales in illustration 16-2.

Solution Table 16-5 lists the actual sales for the 12-month period and the 11 predicted values (*Predicted N*). It also gives the calculations we need to determine MAD and MSE.

$$\text{MAD} = \frac{\sum |e_i|}{n} = \frac{1,960}{11} = 178.18 = \mathbf{178}$$

$$\text{MSE} = \frac{\sum e_i^2}{n} = \frac{440,800}{11} = 40,072.73 = \mathbf{40,073}$$

$$\text{Average absolute error of forecast} = \frac{\text{MAD}}{\bar{y}} = \frac{178}{1,719.1}$$

$$= 0.1035 = \mathbf{0.10}$$

The naive model results in an MAD of 178, an MSE of 40,073, and an average absolute error of approximately 10%.

TABLE 16-5
Calculations Needed to Find MAD and MSE for Prediction N

Time Period, i (months)	Actual, y_i	Predicted N, $\hat{y}_i$	Error, e_i	$|e_i|$	e_i^2
1	1,500	—	—	—	—
2	1,725	1,500	225	225	50,625
3	1,510	1,725	−215	215	46,225
4	1,720	1,510	210	210	44,100
5	1,330	1,720	−390	390	152,100
6	1,535	1,330	205	205	42,025
7	1,740	1,535	205	205	42,025
8	1,810	1,740	70	70	4,900
9	1,760	1,810	−50	50	2,500
10	1,930	1,760	170	170	28,900
11	2,000	1,930	70	70	4,900
12	1,850	2,000	−150	150	22,500
Total				1,960	440,800

□

Let's compare the MAD, the MSE, and the average absolute error of forecast for the three forecast methods used in illustrations 16-2, 16-3, and 16-4; see table 16-6. If, as assumed before, the exponential smoothing model (method A) is adequate for forecasting the sales on a month-by-month basis, then certainly the naive model will be adequate also. (The values of all three measures of error for the naive model are less than the corresponding measures for the exponential smoothing model.) Since the naive model is very simple to use and is the least costly, it would be the best of these three methods to use for this situation.

TABLE 16-6
A Comparison of Methods A, B, and N

Method of Forecast	MAD	MSE	Average Absolute Error
Method A	208	53,205	12%
Method B	168	39,776	10%
Method N	178	40,073	10%

Exercises

16-11 Comment on the following statement: The best forecast is the most accurate one.

16-12 What is the purpose of a naive model?

16-13 Consider the results shown in the accompanying table of a model predicting the average dividend yield of common stocks.

Month		Actual Yield (%)	Forecasted Yield
1982	June	3.16	3.13
	July	3.13	3.10
	August	3.08	3.02
	September	3.00	3.01
	October	2.98	3.03
	November	3.18	3.08
	December	3.16	3.10
1983	January	3.17	3.15
	February	3.26	3.20
	March	3.36	3.22
	April	3.34	3.26
	May	3.49	3.36
	June	3.59	3.46
	July	3.64	3.56

(a) Compute the MAD and average absolute error of forecast for the model.

(b) Construct the naive forecast for the model (this month's yield = last month's yield) and the naive forecast error. (Yield in May 1982 was 3.14.)

(c) Compute the MAD and average absolute error of forecast of the naive model and compare it with the given model's results.

In Retrospect

In business we must plan for the future, and to plan effectively we must forecast effectively. In this chapter you have seen that there is a diversity of forecasting techniques available. The reason that there are alternative techniques is that different techniques work best in different situations. You have learned that forecasting situations can differ both in their time frame and in their patterns of data and that the forecasting problem is really one of finding the appropriate technique for the particular situation.

We must remember that all quantitative forecasting must rely on past data. This means we must assume that the data patterns—four typical patterns were discussed in this chapter—and relationships that exist can be identified and will continue to exist in the future. Realistically, however, we accept the fact that some data are unpredictable and that we cannot perfectly identify and predict existing patterns. Hence we must expect some forecast error for these two reasons. Consequently, it is the amount of error that determines the usefulness of the forecast.

While you have learned in this chapter how to measure forecast accuracy (you should be able to calculate MAD and MSE), you have also *learned that the best technique is not necessarily the most accurate. The best technique for a particular situation is found by comparing the trade-offs of cost, level of effort, and accuracy and thereby selecting the technique that meets the need for accuracy at the minimum cost and inconvenience. In certain situations where the rewards for accuracy are very great, a tremendous amount of effort and cost can be justified in developing a forecast. Conversely, there are also situations where a "ball park" estimate is all that is needed.*

The news article at the beginning of this chapter discusses the importance of forecasting. Forecasting techniques are becoming valuable tools in many areas of business—management, accounting, market research, sales, and so on. Thus it is important that forecasts be constructed and interpreted correctly.

Three specific forecasting techniques are discussed in this text: regression, classical time series, and business indicators. Regression is the topic of chapters 13, 14, and 15. Classical time series will be studied in the next chapter, chapter 17. Business indicators is the topic of chapter 18. Several books are available that survey the full range of forecasting techniques, and many books deal with a single technique in depth.

Chapter Exercises

16-14 In the following forecasting situations, what would be the most appropriate time frame?

(a) The corporate planning department is considering the construction of a new plant.

(b) The production manager is considering leasing another bottling machine to increase production capabilities. The lease must be a 1-year lease.

(c) The corporate treasurer is trying to decide how much money currently held in cash can be invested in 90-day treasury notes.

(d) The purchasing manager of a large grocery places orders for milk on Monday and Friday of each week for delivery 3 days later. Milk has a 7-day shelf life.

(e) The personnel manager is responsible for the recruitment and hiring of new employees. On the average it takes 45 days from the beginning of the hiring process to get a qualified employee on the job.

16-15 Why should it generally be easier to forecast accurately within the shorter time frame?

16-16 List the last five decisions you had to make that required a forecast, either formally or intuitively. For each decision state (a) the frame and (b) what information affected your evaluation of the future.

16-17 Which of the sales data in the following table displays a trend pattern?

	Sales ($\times 1,000$)		
Time Period	Company A	Company B	Company C
1	100	150	200
2	110	152	220
3	150	148	190
4	140	131	186
5	170	132	205
6	174	125	207
7	200	110	200
8	294	100	195
9	210	96	204

16-18 If the forecast error is large, what two things might that indicate?

16-19 Why would we want to examine the actual forecast errors for patterns?

16-20 In comparing two forecasting techniques, what is implied if one method has a smaller MAD but a larger MSE?

16-21 According to cash management theory, a firm should keep as little money in cash as possible. It is usually the treasurer's task to determine how much cash must be kept on hand so that daily expenditures can be met from current receipts and cash in the checking account. Excess cash should be invested either in interest-bearing, 90-day treasury notes or in short-term, low-interest-bearing investments. For this forecasting situation discuss the following:

(a) specifically, what must be forecasted
(b) the data that might be useful in forecasting
(c) the alternative of using a time series versus a causal model
(d) the more appropriate measure of error, MAD or MSE
(e) the costs of error for this problem

16-22 The accompanying data represent the government's forecast of per-capita personal income in the United States and the actual per-capita income in the United States.

(a) Compute the forecast error.
(b) Plot the forecast error.
(c) Compute the MAD and the average absolute error of forecast.
(d) Compute the MSE.

	Per-Capita Income (dollars)	
Year	Actual	Predicted
1976	5,068	5,267
1977	5,161	5,195
1978	5,215	5,210
1979	5,264	5,224
1980	5,368	5,275
1981	5,455	5,330
1982	5,579	5,405
1983	5,746	5,528

Source: U.S. Department of Commerce, Office of Business Economics.

(e) What do the results of (b) tell you about the model?

(f) Do you think a different model can be found that has a smaller MAD than this model? Why or why not?

16-23 Consider the two forecasts of yearly company sales shown in the accompanying table.

	Sales ($\times 1,000,000$)		
Year	Actual	Predicted, A	Predicted, B
1976	7	8	8.5
1977	11	10.5	9.5
1978	13	13.0	11.0
1979	15	15.5	12.0
1980	9	18.0	14.0

(a) Compute the MAD and MSE for each forecast, A and B.

(b) Explain the difference in the results, using MAD and MSE as measures of accuracy.

16-24 Assume that the monthly sales of a department store are comprised of a trend, a seasonal, a cyclical, and an irregular component and that actual sales are the sum of each component (i.e., $y_i = T_i + S_i + C_i + I_i$). The data in the accompanying table represent the component elements for the years 1982 and 1983.

(a) Plot the trend component.

(b) Plot the seasonal component.

Time		Trend, T	Seasonal, S	Cyclical, C	Irregular, I
1982	Jan.	5,000	−500	0	+136
	Feb.	5,200	−700	+2	−363
	Mar.	5,400	−300	16	+302
	Apr.	5,600	+400	36	+49
	May	5,800	+100	64	+436
	June	6,000	−600	100	−188
	July	6,200	−1000	144	+369
	Aug.	6,400	−400	196	−447
	Sept.	6,600	+300	324	−411
	Oct.	6,800	+300	400	+62
	Nov.	7,000	+900	484	+526
	Dec.	7,200	+1500	596	−7
1983	Jan.	7,400	−500	484	−328
	Feb.	7,600	−700	400	−31
	Mar.	7,800	−300	324	+426
	Apr.	8,000	+400	196	+92
	May	8,200	+100	144	+20
	June	8,400	−600	100	−191
	July	8,600	−1000	64	+179
	Aug.	8,800	−400	36	−396
	Sept.	9,000	+300	16	+66
	Oct.	9,200	+300	2	+344
	Nov.	9,400	+900	0	−285
	Dec.	9,600	+1500	−2	−380

(c) Plot the cyclical component.

(d) Plot the irregular component.

(e) Calculate and plot the actual sales.

(f) If we build a model that perfectly explains the patterns of data, what would be the MAD?

Hands-On Problems

Examine a particular newspaper for each of 14 consecutive days. Look at its weather forecast and record the predicted high temperature for that day or, if it is an evening paper, for the next day. Then record the high temperatures that actually occurred.

16-1 Construct a table with the predicted temperatures and actual temperatures and find each error of forecast.

16-2 Plot the errors of forecast as a function of time and comment on whether they look irregular. State the implications of your conclusions.

16-3 Compute the MAD and the MSE of the forecast error.

16-4 Use the naive model that predicts that the next day's high will be equal to today's high, and compare the forecast accuracy of the naive model with the newspaper's accuracy. Comment on the implications of your results.

NOTE: If you have competing newspapers in your area, you might wish to use the different papers' forecasts and compare them for accuracy.

17 Classical Time Series Analysis

Chapter Outline

17-1 Classical Time Series Decomposition
*A time series is expressed as the **product of four components**.*

17-2 Estimating Trend and Cyclical Components Together
*A **centered moving average** is used to estimate the combined trend and cyclical components.*

17-3 Separating the Trend and Cyclical Components: Trend
*We fit a **trend line** to the combined trend and cyclical parts of the data to estimate the trend component.*

17-4 Separating the Trend and Cyclical Components: Cyclical
*The **ratio of T × C and trend** is computed to estimate the cyclical part of the data.*

17-5 Seasonal and Irregular Components
*A **seasonal index** is calculated to represent the stable seasonal component, from which the irregular component can be separated.*

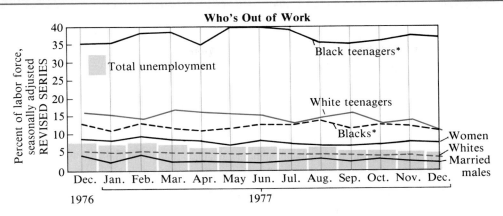

*Includes other minorities.
From *Newsweek*, 30 January 1978, p. 23. Reproduced by permission of Leckner Design Associates.

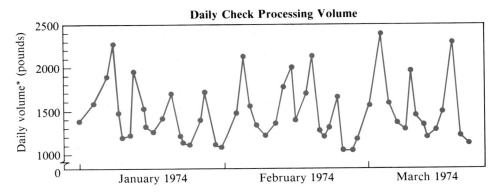

*There are between 300 to 375 transactions per pound.
From *Journal of Bank Research* (Summer 1977), published by Bank Administration Institute. Reprinted by permission.

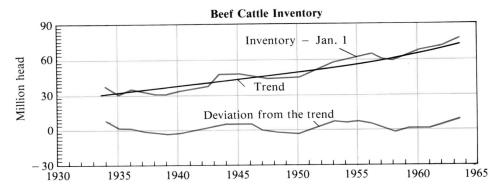

From Frederick V. Waugh, *Graphic Analysis*: *Applications in Agricultural Economics*, Agriculture Handbook no. 326, prepared for U.S. Department of Agriculture, Economic Research Service (Washington, D.C.: Government Printing Office, 1966), p. 21.

Chapter Objectives

In the last chapter we learned that the key to forecasting is recognizing the pattern of data. We also learned that different forecasting techniques work best for different patterns of data. Consider the simple case where only a trend component is present. If the trend component is linear, then the forecasting technique of simple linear regression that we learned about in chapters 13 and 14 will work well. However, if the trend looks like the graph in figures 3-13 or 3-14, simple linear regression will not work.

We have discussed four different types of patterns of data: trend, cyclical, seasonal, and irregular. Although each pattern was described in isolation, in practice, all patterns usually occur simultaneously. When the patterns occur together, it is often very difficult to recognize the patterns. In this chapter we will take a time series and separate, or **decompose***, the data into four separate series, one representing the trend component, one representing the seasonal component, one representing the cyclical component, and one representing the irregular component. This decomposition is one of the initial steps in forecasting. By decomposing the data, we should be better able to recognize the existing patterns of data and hence select the appropriate forecasting technique. The selection of the appropriate forecasting technique given the pattern of the data will not be discussed in this text, however.*

In this chapter you will learn how to (a) calculate a moving average to separate out the combined trend and cyclical patterns, (b) fit a trend line to the moving average results to isolate the trend pattern, (c) separate out the trend line from the moving average to isolate the cyclical pattern, and (d) compute a seasonal index to represent the seasonal pattern.

Section 17-1   Classical Time Series Decomposition

In our discussion of forecasting in chapter 16, we defined four distinct types of patterns that time series data may follow: trend, cyclical, seasonal, and irregular. An example of each pattern was illustrated in figures 16-1 through 16-5. Recognition of the patterns that are present in the data is a crucial part of forecasting. All forecasting techniques assume that some patterns exist and that these patterns can be identified and extrapolated to prepare forecasts.

Although each pattern is recognizable in figures 16-1 through 16-5, often things are not so simple. Examine figure 17-1, which shows 4 years of sales history for a firm. Is the pattern obvious? No. Why? Because in figure 17-1, unlike figures 16-1 through 16-5, all four types of patterns are occurring together.

In classical time series analysis, we start with the assumption that each observation is made up of the combination of the four patterns of data.

decomposition

The basic classical time series problem, referred to as **decomposition**, is to separate the observed y values into each of their component parts, trend (T), cyclical (C), seasonal (S), and irregular (I).

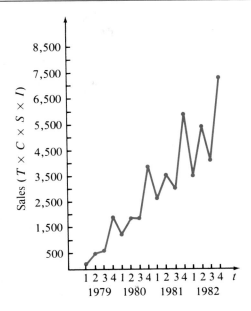

FIGURE 17-1
Four Years' Sales Figures of Texot Corporation

Decomposition allows us to identify the past patterns to better understand what has happened and it helps us in eventually choosing an appropriate forecasting technique.

multiplicative model

The most common classical time series model is called the **multiplicative model** and is represented by the formula

$$y = T \times C \times S \times I \qquad (17\text{-}1)$$

The only value directly observable is the y value. **The trend component T is expressed in terms of the actual units of y.** That is, if we are studying sales, we say that the trend component in a given month is a certain amount of dollars. The remaining three components are expressed as **percentage adjustments**.

The cyclical component C is expressed as a percentage of the trend component. Thus a C value of 110 means that the effect of the cyclical component will be to increase the value of y 10% above the trend. (*Note*: 100 means 100%, or 1.00, or no change.)

The seasonal component S is expressed as a percentage of the trend and cyclical components. An S value of 90 means that the effect of the seasonal component will be to decrease the value of y 10% below that of the combined trend and cyclical components.

The irregular component I is expressed as a percentage of the trend, cyclical, and seasonal components. Thus an irregular component of 105 means that the actual y value will be 5% above the combined trend, cyclical, and seasonal components.

Illustration 17-1

In the first quarter the trend component of sales was $15,000, the cyclical component was 102, the seasonal component was 90, and the irregular component was 103. What was the actual sales value? How much of the sales can be attributed to each component?

Solution   According to formula (17-1),

$$\text{Sales} = y = T + C \times S \times I$$
$$= (15{,}000)(1.02)(0.90)(1.03) = \$14{,}183.10$$

The trend component is $T = 15{,}000$.

The cyclical component is 102; thus the cyclical component accounts for 2% above the trend amount. The dollar amount of sales attributable to the cyclical component is thus

$$(15{,}000)(0.02) = 300$$

The seasonal component is 90; thus the seasonal effect is to reduce the trend and cyclical component by 10%:

$$[(15{,}000)(1.02)] \times (-0.10) = -1{,}530$$

The irregular component is 103; thus the irregular factors increase the trend, cyclical, and seasonal components by 3%:

$$[(15{,}000)(1.02)(0.90)] \times (0.03) = 413.10$$

Notice that the sum of the dollar amounts attributable to each component is the actual sales (y):

$$\$15{,}000 + \$300 + \$(-1{,}530) + \$413.10 = \$14{,}183.10$$

Hence we have decomposed the first quarter's sales into each component. □

Recall the sales data presented in figure 17-1. No clear pattern was observed. The data in figure 17-1 were generated by combining the four patterns of data presented in table 17-1 using formula (17-1). Figure 17-2 shows the graph of the trend component, figure 17-3 shows the cyclical components, figure 17-4 shows the seasonal component, and figure 17-5 shows the irregular component.

TABLE 17-1

Sales of Texot Corporation Decomposed by Trend, Cyclical, Seasonal, and Irregular Components

Year	Quarter	Trend	Cyclical (%)	Seasonal (%)	Irregular (%)	Sales
1979	1	1,150	95.6	85	102	953
	2	1,495	96.6	100	95	1,371
	3	1,840	97.4	85	98	1,493
	4	2,185	98.2	130	101	2,817
1980	1	2,530	98.9	85	100	2,127
	2	2,875	99.5	100	96	2,746
	3	3,220	100.0	85	99	2,710
	4	3,565	100.4	130	102	4,746
1981	1	3,910	100.8	85	105	3,518
	2	4,255	101.0	100	102	4,384
	3	4,600	101.2	85	99	3,917
	4	4,945	101.3	130	104	6,773
1982	1	5,290	101.2	85	97	4,414
	2	5,635	101.0	100	110	6,261
	3	5,980	100.8	85	98	5,021
	4	6,325	100.4	130	99	8,173

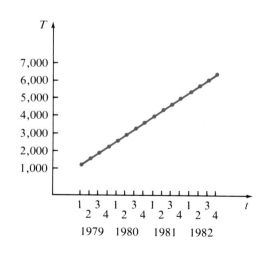

FIGURE 17-2

Trend Component of Sales, Texot Corporation

FIGURE 17-3
Cyclical Component of Sales,
Texot Corporation

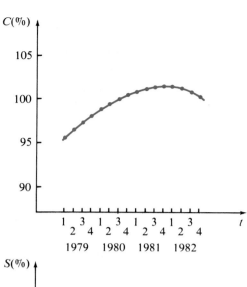

FIGURE 17-4
Seasonal Component of Sales,
Texot Corporation

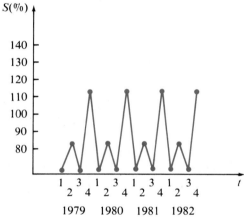

FIGURE 17-5
Irregular Component of Sales,
Texot Corporation

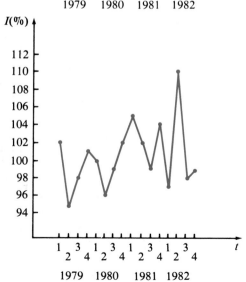

Exercises

17-1 Identify the four components of a time series and explain the kind of pattern over time to which each applies.

17-2 Which component of time series would account for the following patterns in sales?
  (a) general growth of demand
  (b) a strike
  (c) the effect of Christmas sales

17-3 For a forecast of housing starts in the next month, the forecast's components are as follows: trend, 86.3 thousand units; seasonal, 83; and cyclical, 102. What is the forecast for housing starts?

17-4 A sales amount for a given month has been decomposed into the following dollar amounts:

Trend:	70,000
Cyclical:	1,000
Seasonal:	−2,000
Irregular:	1,000

If we have a multiplicative model, how would we express each of these components?

Section 17-2 Estimating Trend and Cyclical Components Together

moving average

The first step in decomposition is to isolate the combined trend and cyclical components. This isolation is done by calculating what is called a **moving average**. The idea behind a moving average is that if we average a year's worth of data, the seasonal effect will tend to average out. Moreover, the irregular component will also tend to average out. Consequently, what will be left should be a good estimate of the combined trend and cyclical components.

The procedure for calculating a moving average is relatively straightforward. The 4-year Scotch whiskey sales history of Debow Industries is presented in table 17-2, and the calculations for the moving averages are presented in table 17-3.

The first step of the calculations is to add together the sales for the first four quarters. This total is placed in the column marked *Four-Quarter Moving Total* of table 17-3 and is positioned in the middle of the four quarters for which the data are summed. Since there are four quarters, the middle is between the second and third quarters. The first entry is thus 4,559 (the sum of the sales for the first four quarters of table 17-2) and it is placed between the second and the third quarter of 1979 in table 17-3. The

TABLE 17-2
Monthly Sales of Scotch, Debow Industries

Year	Quarterly Sales (× 10,000 bottles)			
	Q1	Q2	Q3	Q4
1979	805	1,028	839	1,887
1980	1,017	1,145	836	2,207
1981	1,219	1,357	1,004	2,439
1982	1,207	1,352	1,074	2,831

second entry in column (2) is calculated by dropping sales for the first quarter of 1979 and adding the next quarter, the first quarter of 1980. Sales for these four quarters are then totaled and the sum placed between the third and fourth quarter of 1979. The second entry is 4,771 (4,559 − 805 + 1,017).

This procedure of dropping the first quarter and adding the next quarter is continued until the last sum is calculated. The last sum will be found by adding the last four quarters and placing the answer between the second and third quarter of the last year. The term "moving" comes from the idea that we are always summing four consecutive quarters together and moving the starting point down the list, one quarter at a time.

The next step is to compute the four-quarter moving average. This computation is done by dividing each four-quarter moving total in column (2) by 4. The result is placed in column (3), headed *Four-Quarter Moving Average*. For example, the first entry, the four-quarter moving average between the third and fourth quarter, is 4,559 ÷ 4 = 1,139.75. (*Note*: The four-quarter moving averages should be accurate to at least two decimal places more than are contained in the original data.)

The final step is to **center the average**. Each figure in column (3) falls between two quarters. We must align the average to a particular quarter to do further analysis. This task is done by simply averaging each consecutive pair of four-quarter moving averages found in column (3) and aligning the result on the appropriate quarter. For example, referring to the table, you will notice that we averaged two figures in column (3) and aligned the new average with the quarter between those two figures.

	Four-Quarter Moving Average	Four-Quarter Center Moving Average
Q2		
	1,139.75	
Q3		$\dfrac{2{,}332.5}{2} = 1{,}166.2$
1979	1,192.75	
Q4		$\dfrac{2{,}414.75}{2} = 1{,}207.4$
	1,222.00	
Q1		

TABLE 17-3

Calculation of Moving Average to Estimate Trend and Cyclical Components for Scotch Sales, Debow Industries

Year	Quarter	(1) Sales, $T \times C \times S \times I$	(2) Four-Quarter Moving Total	(3) Four-Quarter Moving Average [column (2) ÷ 4]	(4) Four-Quarter Centered Moving Average
1979	1	805			
	2	1,028			
			4,559	1,139.75	
	3	839			1,166.3
			4,771	1,192.75	
	4	1,887			1,207.4
			4,888	1,222.00	
1980	1	1,017			1,221.6
			4,885	1,221.25	
	2	1,145			1,261.3
			5,205	1,301.25	
	3	836			1,326.5
			5,407	1,351.75	
	4	2,207			1,378.3
			5,619	1,404.75	
1981	1	1,219			1,425.8
			5,787	1,446.75	
	2	1,357			1,475.8
			6,019	1,504.75	
	3	1,004			1,503.3
			6,007	1,501.75	
	4	2,439			1,501.1
			6,002	1,500.50	
1982	1	1,207			1,509.3
			6,072	1,518.00	
	2	1,352			1,567.0
			6,464	1,616.00	
	3	1,074			
	4	2,831			

The centering is done consecutively for each pair, and the result is placed in column (4), headed *Four-Quarter Centered Moving Average*. (*Note*: The four-quarter centered moving average should be accurate to at least one decimal place more than is contained in the original data.)

The four-quarter centered moving averages represent our estimates of the trend and cyclical components of the data. Note that we have no estimate for the first two quarters nor the last two quarters.

Illustration 17-2

Potter Industries is interested in understanding its inventory level to better schedule production and coordinate marketing efforts to control inventory costs. Table 17-4 presents the quarterly levels of inventory for the past 2 years. Separate out the trend and cyclical components by using a four-quarter centered moving average.

TABLE 17-4
Quarterly Inventory Levels of Potter Industries

Year	Quarterly Inventory ($\times 10{,}000$)			
	Q1	Q2	Q3	Q4
1982	7	12	23	14
1983	10	14	27	17

Solution The first step is to calculate the four-quarter moving totals, as described in the preceding pages. Add the first four quarters:

$$7 + 12 + 23 + 14 = 56$$

and place that figure in its proper slot in column (2) in table 17-5, between the second and third quarter of 1982. To calculate the next four-quarter moving total, we add the next quarter (first quarter of 1983) and subtract the first quarter in the previous addition (first quarter of 1982) from the total just calculated:

$$56 + 10 - 7 = 59$$

and place it between the third and fourth quarter of 1982. This addition and subtraction are repeated until all the totals are computed.

The completed calculations are presented in table 17-5. Column (3) was computed by dividing each entry in column (2) by 4. Column (4), the four-quarter centered moving average, was found by averaging each consecutive pair of figures in column (3) and aligning the result between the two figures that were averaged.

TABLE 17-5

Calculation of $T \times C$ Component Using the Four-Quarter Moving Average, Potter Industries

Year	Quarter	(1) Inventory, $T \times C \times S \times I$	(2) Four-Quarter Moving Total	(3) Four-Quarter Moving Average [column (2) ÷ 4]	(4) Four-Quarter Centered Moving Average
1982	1	7			
	2	12			
			56	14.00	
	3	23			14.4
			59	14.75	
	4	14			15.0
			61	15.25	
1983	1	10			15.8
			65	16.25	
	2	14			16.6
			68	17.00	
	3	27			
	4	17			

Illustration 17-3

In this illustration we wish to compare a set of calculated $T \times C$ values with a set of actual $T \times C$ values. The sales data for Texot Corporation, which are presented in table 17-1, include the actual trend and cyclical components for each quarter, from 1979 through 1982. These data are repeated in table 17-6, and the actual combined $T \times C$ components are calculated. If, however, the actual *sales* figures (which are also shown in table 17-1) were the only data available, as is usually the case, we would have to estimate the $T \times C$ component using the four-quarter centered moving average method, which we have just learned. Table 17-7 shows the calculation of the four-quarter centered moving averages and compares those estimates with the actual $T \times C$ components from table 17-6. Notice that the estimates do not exactly match the actual values. Is this what you expected? It should be. As with most estimates, you should have expected some error. The reason for the error here is that over four quarters the seasonal and irregular components do not usually average out perfectly; hence the discrepancy.

TABLE 17-6
Trend and Cyclical Components of Sales of Texot Corporation

Year	Quarter	Trend	Cyclical	T × C
1979	1	1,150	95.6	1,099.4
	2	1,495	96.6	1,444.2
	3	1,840	97.4	1,792.2
	4	2,185	98.2	2,145.7
1980	1	2,530	98.9	2,502.2
	2	2,875	99.5	2,860.6
	3	3,220	100.0	3.220.0
	4	3,565	100.4	3,579.3
1981	1	3,910	100.8	3,941.3
	2	4,255	101.0	4,297.6
	3	4,600	101.2	4,655.2
	4	4,945	101.3	5,009.3
1982	1	5,290	101.2	5,353.5
	2	5,635	101.0	5,691.4
	3	5,980	100.8	6,027.8
	4	6,325	100.4	6,350.3

TABLE 17-7
Calculation of Four-Quarter Moving Average Estimate of T × C for Texot Corporation

Year	Quarter	(1) Sales	(2) Four-Quarter Moving Total	(3) Four-Quarter Moving Average	(4) Four-Quarter Centered Moving Average (T × C)	(5) Actual T × C (table 17-6)
1979	1	953				
	2	1,371				
			6,634	1,658.50		
	3	1,493			1,805.3	1,792.2
			7,808	1,952.00		
	4	2,817			2,123.9	2,145.7
			9,183	2,295.75		
1980	1	2,127			2,447.9	2,502.2
			10,400	2,600.00		
	2	2,746			2,841.1	2,860.6
			12,329	3,082.25		
	3	2,710			3,256.1	3,220.0
			13,720	3,430.00		

TABLE 17-7
(continued)

Year	Quarter	(1) Sales	(2) Four-Quarter Moving Total	(3) Four-Quarter Moving Average	(4) Four-Quarter Centered Moving Average (T × C)	(5) Actual T × C (table 17-6)
	4	4,746			3,634.8	3,579.3
			15,358	3,839.50		
1981	1	3,518			3,990.4	3,941.3
			16,565	4,141.25		
	2	4,384			4,394.6	4,297.6
			18,592	4,648.00		
	3	3,917			4,760.0	4,655.2
			19,488	4,872.00		
	4	6,773			5,106.6	5,009.3
			21,365	5,341.25		
1982	1	4,414			5,479.3	5,353.5
			22,469	5,617.25		
	2	6,261			5,792.3	5,691.4
			23,869	5,967.25		
	3	5,021				
	4	8,173				

□

Exercises

17-5 Complete the following table:

Quarter	Four-Quarter Moving Total	Four-Quarter Moving Average	Four-Quarter Centered Moving Average
1980 1			
	110		
2			
	113		
3			
	115		
4			
	117		
1981 1			

17-6 The data in the accompanying table represent the last 2 years' sales of Xtra Corporation. Compute the four-quarter centered moving average for these data. (Retain this solution for use in answering exercise 17-9.)

	Quarterly Sales			
Year	Q1	Q2	Q3	Q4
1982	964	868	973	911
1983	1,129	905	1,075	997

17-7 The data in the accompanying table represent the number of two-unit (duplex) housing starts for a 2-year period. Compute the four-quarter centered moving average for these data. (Retain this solution for use in answering exercise 17-10.)

	Quarterly Two-Unit Housing Starts ($\times 1{,}000$)			
Year	Q1	Q2	Q3	Q4
1981	9.7	12.5	8.4	9.9
1982	9.9	14.2	13.7	12.5

Section 17-3 Separating the Trend and Cyclical Components: Trend

After we estimate the combined trend and cyclical components, the next step is to separate the trend and cyclical components. This separation is done by **fitting a trend equation** to the $T \times C$ data. The trend equation is a line through the bivariate data, time and $T \times C$. To determine the trend equation, we must first code the time variable. We do this coding by sequentially numbering the time periods, starting with $t = 1$ for the first quarter that has a four-quarter center moving average value.

The idea behind fitting a trend line is that the line through the combined trend and cyclical data will represent the trend, and the deviation from the line $(y - \hat{y})$ of the combined trend and cyclical line will represent the cyclical component about the trend. The trend line is nothing more than a regression line, where time (t) is the independent variable and $T \times C$ is the observed dependent variable. Different types of trend formulas can be used, for example, linear, quadratic, or exponential. In section 3-3 we illustrated some of the types of trend (regression) lines that can be fit to the data. [The equation for trend is obtained by regression analysis, with t replacing x, $T \times C$ replacing y, and T (the trend component) being equal to $\hat{y}$.] Although the symbols for the variables used in this chapter are different from the symbols we used in our discussion of regression in chapters 3, 13, 14, and 15, what we discussed about fitting a regression line in those chapters still holds for fitting a trend line. The concept of least squares and the formulas for calculating the coefficients of the line remain the same and can be used to estimate the trend line.

Let's return to the data of Debow Industries, table 17-2. Figure 17-6 shows the graph of the $T \times C$ component estimated in table 17-3 by the four-quarter centered moving average. If we look at the points plotted in figure 17-6, it appears that a linear trend line would be appropriate.

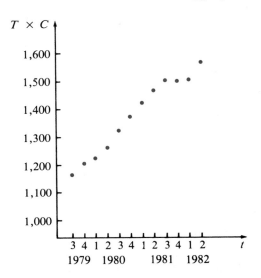

FIGURE 17-6
$T \times C$ Component of Sales, Debow Industries

To find the coefficients b_0 and b_1, we use the formulas for linear regression from chapter 13, formulas (13-4) and (13-5), writing t in place of x, $T \times C$ in place of y, and $\overline{T \times C}$ in place of $\bar{y}$.

$$b_1 = \frac{\sum (t - \bar{t})(T \times C - \overline{T \times C})}{\sum (t - \bar{t})^2} \tag{17-2}$$

$$b_0 = \frac{\sum T \times C - b_1 \sum t}{n} \tag{17-3}$$

Formula (17-2) can be rewritten [just as was done for formula (13-4)] in a form that is easier to apply:

$$b_1 = \frac{\sum ty - \sum t \sum y / n}{\sum t^2 - \sum t^2 / n} \tag{17-4}$$

NOTE: y is used for $T \times C$ for simplicity.

Separating the Trend and Cyclical Components: Trend   Section 17-3

NOTE: b_0 should be calculated to the nearest hundredth and b_1 to the nearest thousandth.

The formula for the trend line becomes

$$\text{Trend} = b_0 + b_1 t$$

Two useful formulas when t (time) is numbered sequentially from 1 to n are

$$\sum t = \frac{(n)(n+1)}{2} \qquad (17\text{-}5)$$

and

$$\sum t^2 = \frac{(n)(n+1)(2n+1)}{6} \qquad (17\text{-}6)$$

Table 17-8 presents the necessary calculations for using formulas (17-3) and (17-4) with the data of Debow Industries.

Using formula (17-5), we have

$$\sum t = \frac{(12)(13)}{2} = 78$$

Using formula (17-6), we have

$$\sum t^2 = \frac{(12)(13)(25)}{6} = 650$$

Using formulas (17-4) and (17-3), we have

$$b_1 = 112{,}927.4 - \frac{(78)(16{,}543.7)}{12} = 37.716 = 37.72$$

$$b_0 = \frac{16{,}543.7 - (37.716)(78)}{12} = 1{,}133.49$$

Therefore the formula for the trend line is

$$\text{Trend} = 1{,}133.49 + 37.72t$$

Once we have the trend formula, we can compute the trend component for each quarter. The trend component T is found by substituting the t value corresponding to the quarter into the trend formula. The trend

TABLE 17-8
Calculation for Linear Trend for Debow Industries Data

Year	Quarter	t	y = (T × C)	ty
1979	3	1	1,166.3	1,166.3
	4	2	1,207.4	2,414.8
1980	1	3	1,221.6	3,664.8
	2	4	1,261.3	5,045.2
	3	5	1,326.5	6,632.5
	4	6	1,378.3	8,269.8
1981	1	7	1,425.8	9,980.6
	2	8	1,475.8	11,806.4
	3	9	1,503.3	13,529.7
	4	10	1,501.1	15,011.0
1982	1	11	1,509.3	16,602.3
	2	12	1,567.0	18,804.0
Total		78	16,543.7	112,927.4

component for the third quarter of 1979 is 1,171.21 [1,133.49 + 37.72(1)]. For the first quarter of 1981, it is 1,397.53 [1,133.49 + 37.72(7)]. Table 17-9 shows the trend components for each quarter.

TABLE 17-9
Trend Component for Debow Industries

Year	Quarter	t	y = T × C	Trend T, 1,133.49 + 37.72t
1979	3	1	1,166.3	1,171.2
	4	2	1,207.4	1,208.9
1980	1	3	1,221.6	1,246.7
	2	4	1,261.3	1,284.4
	3	5	1,326.5	1,322.1
	4	6	1,378.3	1,359.8
1981	1	7	1,425.8	1,397.5
	2	8	1,475.8	1,435.3
	3	9	1,503.3	1,473.0
	4	10	1,501.1	1,510.7
1982	1	11	1,509.3	1,548.4
	2	12	1,567.0	1,586.1

Illustration 17-4

In illustration 17-3 we estimated the trend and cyclical components combined for the Texot Corporation data presented in table 17-1. We saw that the estimated combined trend and cyclical components did not perfectly match the actual values, but were close. Now let's do the next step and fit the trend line for the estimated $T \times C$ data. If we look closely at the actual trend data in table 17-1, we see that the slope (the change in trend per quarter) is 345. If we let $t = 1$ for the third quarter of 1979, the intercept (the value of trend at $t = 0$, the second quarter of 1979) is 1,495. Thus the real trend line is $1,495 + 345t$. Do you think the trend line we estimate will be the same? Will it be close?

Solution Table 17-10 presents the calculations necessary for finding the trend line formulas.

TABLE 17-10
Calculations for Trend Line Fit for Texot Corporation

Year	Quarter	Four-Quarter Centered Moving Average, $y = T \times C$	t	ty
1979	3	1,805.3	1	1,805.3
	4	2,123.9	2	4,247.8
1980	1	2,447.9	3	7,343.7
	2	2,841.1	4	11,364.4
	3	3,256.1	5	16,280.5
	4	3,634.8	6	21,808.8
1981	1	3,990.4	7	27,932.8
	2	4,394.6	8	35,156.8
	3	4,760.0	9	42,840.0
	4	5,106.6	10	51,066.0
1982	1	5,479.3	11	60,272.3
	2	5,792.3	12	69,507.6
Total		45,632.3		349,626.0

Using formula (17-5), we have

$$\sum t = \frac{(12)(13)}{2} = 78$$

Using formula (17-6), we have

$$\sum t^2 = \frac{(12)(13)(25)}{6} = 650$$

Using formulas (17-3) and (17-4), we get

$$b_1 = \frac{349,626 - (78)(45,632.3)/12}{650 - (78)^2/12} = \frac{53,016.05}{143} = 370.742$$

$$b_0 = \frac{45,632.3 - (370.742)(78)}{12} = 1,392.87$$

Thus the trend line is

$$\text{Trend} = 1{,}392.87 + 370.742t$$

As table 17-11 shows and as you should by now have expected, the trend component estimates are close but not perfect.

TABLE 17-11
Trend Component Estimates for Texot Corporation Compared with Actual Trend Components

Year	Quarter	t	Estimated Trend, $1{,}392.87 + 370.742t$	Actual Trend
1979	3	1	1,763.6	1,840
	4	2	2,134.4	2,185
1980	1	3	2,505.1	2,530
	2	4	2,875.8	2,875
	3	5	3,246.6	3,220
	4	6	3,617.3	3,565
1981	1	7	3,988.1	3,910
	2	8	4,358.8	4,255
	3	9	4,729.5	4,600
	4	10	5,100.3	4,945
1982	1	11	5,471.0	5,290
	2	12	5,841.8	5,635

□

Exercises

17-8 The trend formula found for the quarter sales levels (in units) of Quarto Electronic amplifiers is

$$T = 75 + 3t, \quad t = 1 \text{ in first quarter, 1979}$$

What would be the predicted trend by quarters for the years 1982 and 1983?

17-9 Find the linear trend line formula by using the moving average calculated for the sales of Xtra Corporation in exercise 17-6. List the quarterly trend components for the data. (Retain this solution for use in answering exercise 17-13.)

17-10 Find the linear trend line formula for the data on housing starts given in exercise 17-7.

Section 17-4 Separating the Trend and Cyclical Components: Cyclical

After we have fit a trend line by using the $T \times C$ data, the cyclical component is simply the deviation around the trend line. Recall that the cyclical component is expressed as a percentage of the trend component. Thus if we divide each quarter's estimated $T \times C$ value by the estimated trend value, we get the cyclical component, $(T \times C)/T = C$.

Illustration 17-5

Estimate the cyclical component for the Debow Industries problem we analyzed previously.

Solution Previously, we estimated the combined trend and cyclical components by using a four-quarter moving average (see illustration 17-2). In section 17-3 we fitted a trend line through the $T \times C$ data and estimated the trend component. The results of both analyses were summarized in table 17-9 and are reproduced here in table 17-12.

TABLE 17-12
Estimation of Cyclical Component for Debow Industries Data

Year	Quarter	(1) $T \times C$	(2) Trend, T	(3) Cyclical (%) [column (1) ÷ column (2) × 100]
1979	3	1,166.3	1,171.2	99.6
	4	1,207.4	1,208.9	99.9
1980	1	1,221.6	1,246.7	98.0
	2	1,261.3	1,284.4	98.2
	3	1,326.5	1,322.1	100.3
	4	1,378.3	1,359.8	101.4
1981	1	1,425.8	1,397.5	102.0
	2	1,475.8	1,435.3	102.8
	3	1,503.3	1,473.0	102.1
	4	1,501.1	1,510.7	99.4
1982	1	1,509.3	1,548.4	97.5
	2	1,567.0	1,586.1	98.8

To estimate the cyclical component, we divide the $T \times C$ estimates, column (1) in table 17-12, by the trend estimates, column (2) in table 17-12. The result, $(T \times C)/T = C$, is our estimate of the cyclical component. These results are given in column (3) in table 17-12. □

Exercises

17-11 The cyclical component can be viewed as a deviation about the trend line. Explain this statement.

17-12 Given the accompanying table containing the $T \times C$ component and the trend component, compute the cyclical component.

Year	Quarter	$T \times C$	Trend
1977	1	834	1,000
	2	953	1,100
	3	1,675	1,200
	4	1,201	1,300
1978	1	1,330	1,400
	2	1,461	1,500
	3	1,594	1,600
	4	1,727	1,700
1979	1	1,861	1,800
	2	1,995	1,900
	3	2,128	2,000
	4	2,260	2,100
1980	1	2,389	2,200
	2	2,516	2,300
	3	2,640	2,400
	4	2,760	2,500
1981	1	2,876	2,600
	2	2,986	2,700
	3	3,090	2,800
	4	3,190	2,900
1982	1	3,282	3,000
	2	3,367	3,100
	3	3,443	3,200
	4	3,511	3,300
1983	1	3,570	3,400
	2	3,619	3,500
	3	3,658	3,600
	4	3,685	3,700

17-13 Using the results of exercises 17-6 and 17-9, calculate the cyclical component for the sales of Xtra Corporation.

Section 17-5 Seasonal and Irregular Components

Thus far we have learned how to estimate the combined trend and cyclical components and how to separate the trend and cyclical components. Now let us turn our attention to the seasonal and irregular components.

Remember that the observations y are equal to $T \times C \times S \times I$. In section 17-2 we learned how to use a four-quarter moving average to estimate the combined trend and cyclical component, $T \times C$. If we divide each of our observed y values by its associated $T \times C$ value, the result will be a measure of the combined seasonal and irregular components.

Now we have to separate the seasonal component from the irregular component, which is done by first grouping all the $S \times I$ values by quarter. That is, we group all the first quarter values together, all the second quarter values together, and so forth. We assume that the seasonal effect of each quarter is the same each year. This assumption is known as **stable (or constant) seasonal effects**. Thus the variation of the $S \times I$ values for the same quarter in different years should be due to the irregular component.

stable seasonal effects

Next we find the mean of the quarterly values for each quarter. The mean of the quarterly values (all first quarters together, all second quarters together, etc.) for each of the four quarters will give us an estimate of the seasonal components that is relatively free of the irregular components. (The irregular components tend to average out.)

The final step is to adjust the seasonal components so that the sum equals 400. This adjustment is done so that the average seasonal component is 100. The seasonal effect will therefore average out over the total year's data, as it should. (*Note*: Since this model is multiplicative, a component of 100% represents multiplication by 1.0. Consequently, it has no effect on the product.) The adjustment is computed by dividing the number 400 by the sum of the quarterly average over the four quarters and then multiplying each quarterly average by its adjustment factor. The resulting adjusted seasonal component is called the **seasonal index**. The irregular component can then be separated from the combined seasonal and irregular data by dividing each combined value by the corresponding seasonal index.

seasonal index

Illustration 17-6

To illustrate the calculation of a seasonal index and the determination of irregular components, let's use the data for the sales of Scotch whiskey of Debow Industries. The 4-year sales history was presented in table 17-2, and $T \times C$ combined components were estimated by using a four-quarter moving average in table 17-3. The first step in calculating the seasonal index is to divide each sales figure by its associated $T \times C$ figure and multiply by 100. This calculation gives us the combined seasonal and irregular ($S \times I$) components for each quarter. See table 17-13. Note that we have no values for the first two and last two quarters of data, since there is no moving average estimate for $T \times C$ for these quarters.

TABLE 17-13
Calculation of Seasonal
and Irregular Components
for Debow Industries
Scotch Sales

Year	Quarter	(1) Sales	(2) T × C, Four-Quarter Centered Moving Average	(3) S × I (%) [column (1) ÷ column (2) × 100]
1979	3	839	1,166.3	71.9
	4	1,887	1,207.4	156.3
1980	1	1,017	1,221.6	83.3
	2	1,145	1,261.3	90.8
	3	836	1,326.5	63.0
	4	2,207	1,378.3	160.1
1981	1	1,219	1,425.8	85.5
	2	1,357	1,475.8	92.0
	3	1,004	1,503.3	66.8
	4	2,439	1,501.1	162.5
1982	1	1,207	1,509.3	80.0
	2	1,352	1,567.0	86.3

The next step is to separate the $S \times I$ values by quarter, as shown in table 17-14. We then compute the mean for each quarter. (*Note*: Keep extra decimal places for the quarterly averages. Round off the adjusted averages to the nearest 10th of a percent.) Finally, we compute the adjustment factor by summing all the averages and dividing that into 400:

$$\text{Adjustment} = \frac{400}{82.933 + 89.7 + 67.233 + 159.633} = \frac{400}{399.499} = 1.00125$$

We now multiply each quarterly average by the adjustment (1.00125) to get the adjusted quarterly averages. *Note*: As a check, sum the adjusted quarterly averages and make sure the sum equals 400.

The seasonal indexes for Scotch sales of Debow Industries are as follows:

Quarter	Seasonal Index
First	83.04
Second	89.81
Third	67.32
Fourth	159.83

TABLE 17-14
Calculation of Seasonal Index for Scotch Sales, Debow Industries

	Quarter			
Year	1	2	3	4
1979	—	—	71.9	156.3
1980	83.3	90.8	63.0	160.1
1981	85.5	92.0	66.8	162.5
1982	80.0	86.3	—	—
Average	82.933	89.7	67.233	159.633
Adjusted Average	83.04	89.81	67.32	159.83

Sum of quarterly average = 399.499

$$\text{Adjustments} = \frac{400}{399.499} = 1.00125$$

Adjusted average = average × adjustment

TABLE 17-15
Calculation of Irregular Components for Debow Industries Scotch Sales

Year	Quarter	(1) S × I	(2) Seasonal Index	(3) Irregular Component [column (1) ÷ column (2) × 100]
1979	3	71.9	67.32	106.8
	4	156.3	159.83	97.8
1980	1	83.3	83.04	100.3
	2	90.8	89.81	101.1
	3	63.0	67.32	93.6
	4	160.1	159.83	100.2
1981	1	85.5	83.04	103.0
	2	92.0	89.81	102.4
	3	66.8	67.32	99.2
	4	162.5	159.83	101.7
1982	1	80.0	83.04	96.3
	2	86.3	89.81	96.1

We can now separate the irregular components from the $S \times I$ component. The calculations and resulting irregular components are shown in table 17-15.

For many common economic series, such as gross national product, unemployment, and consumer prices, the seasonal index already has been calculated by the government. For many government series the seasonal effects are removed before the data are published. This procedure is done by dividing each month of the original data by its respective seasonal index. Such data are said to be **deseasonalized, or seasonally adjusted**.

deseasonalized, or seasonally adjusted, data

Exercises

17-14 Explain what is meant by the statement, "the series is seasonally adjusted, or deseasonalized."

17-15 Explain what is meant by the statement, "the seasonal effects are stable."

17-16 The accompanying table gives the averages of the ratios of $y/(T \times C)$ by the quarter. Compute the adjustment factor and the seasonal index.

	Quarter		
1	2	3	4
93.7	89.1	92.2	127.3

17-17 The accompanying table gives the ratios of $y/(T \times C)$ for the quarterly sales of Richet Brothers Department Stores. Calculate the seasonal index.

Year	Quarter			
	1	2	3	4
1978	—	—	86.3	124.1
1979	107.2	93.8	88.6	107.9
1980	110.9	97.2	90.6	113.4
1981	110.8	95.2	—	—

17-18 Suppose the quarterly sales of Richet Brothers (see exercise 17-17) for 1982 were 1,900, 1,645, 1,816, and 2,010. Use the seasonal index calculated in exercise 17-17 to compute the quarterly seasonally adjusted sales.

17-19 The accompanying table gives the historical sales of long-playing records of Atwic Records and the estimated trend and cyclical components of sales (from a four-quarter centered moving average). Compute the seasonal index of sales.

Year	Quarter	Sales	T × C
1978	3	832	965.5
	4	1,249	1,006.7
1979	1	1,129	1,018.8
	2	997	1,047.2
	3	940	1,065.2
	4	1,310	1,066.9
1980	1	1,131	1,065.9
	2	973	1,067.6
	3	901	1,059.3
	4	1,284	1,104.5
1981	1	1,198	1,108.0
	2	1,122	1,196.3
	3	1,125	1,269.7
	4	1,477	1,369.4
1982	1	1,598	1,382.0
	2	1,280	1,369.4

In Retrospect

In this chapter we have learned how to take a quarterly time series and decompose it into its four component parts. You should realize why we would want to do this. By isolating the distinct patterns, we can better understand what has happened in the past, and we can recognize the past patterns better. This recognition of past patterns is essential in being able eventually to select an appropriate forecasting technique.

The basic assumption of decomposition is that the time series values are made up of the combination of four components: trend, cyclical, seasonal, and irregular. We have dealt with the multiplicative model, which says that the values are the product of the four components, $y = T \times C \times S \times I$. You should review chapter 16 to make sure you fully understand the different types of patterns.

Reviewing the decomposition analysis, we see that it is a step-by-step process. First we isolate the combined trend and cyclical components by calculating a four-quarter centered moving average. This calculation gives us a $T \times C$ estimate. Next we fit a trend line to separate out the trend component T. The deviation about the trend line is the cyclical component. The cyclical component C is calculated by dividing the $T \times C$ estimate by the trend estimate T. (Remember that the cyclical, seasonal, and irregular components are expressed as percentages of the other components. Only the trend component is expressed in units.) The next calculation, the seasonal index, uses the $T \times C$ estimates to determine the $S \times I$ components, which are then averaged and adjusted. Keep in mind that the *procedure we have discussed assumes stable seasonal effects, that is, that the seasonal effects are the same each year. The deviation of the $S \times I$ components about the seasonal index is calculated by dividing the $S \times I$ estimates by the seasonal indexes.*

Look at the various time series presented at the beginning of the chapter. Notice that the graph of daily check-processing volumes seems to indicate a strong seasonal pattern, but the seasonal effect here is a daily effect, not a monthly effect. The graph of beef cattle inventory is decomposed into the trend and cyclical pattern. The graph of unemployment data is deseasonalized (seasonally adjusted) so as to highlight trend patterns. These graphs show just a few of the many examples of time series found in business. Pick up your local newspaper and turn to the business section; you will probably find at least one time series graph there.

A seasonal index is really a special type of time series. It represents that pattern of change due to the effect of the season, but it does not by itself tell you the actual change. You probably can think of other similar time series, such as the consumer price index. Such series are called index numbers. In the next chapter we will look at the concept of index numbers in more detail. We will also look at a special type of series particularly useful in business—business indicators. These indicators are specially designed time series that denote the present and future state of the economy.

Chapter Exercises

17-20 Explain why we would want to decompose a time series into its four component parts.

17-21 The following table gives the weekly average unemployment claims within each quarter made in Billings, Delaware, in 1982:

Quarter			
1	2	3	4
340	218	286	248

The seasonal index for this series is given in the next table:

Quarter			
1	2	3	4
137.6	81.1	101.5	79.8

(a) Plot the unemployment claims time series.
(b) Deseasonalize the series.
(c) Plot the deseasonalized series on the same graph as the original series.

17-22 The accompanying table gives the number of new checking account customers per quarter at Lincoln Bank.

Quarter	1980	1981	1982
First	918	1,018	1,138
Second	879	983	1,094
Third	837	948	1,053
Fourth	928	1,038	1,085

(a) Compute the four-quarter centered moving average.
(b) Calculate the linear trend line formula and the quarterly trend components.
(c) Calculate the cyclical components.
(d) Compute the seasonal index.

17-23 To better understand and schedule emergency room services, a hospital administrator collected data on the quarterly amount of hours of patient services rendered. The data are shown in the following table:

	Quarter			
Year	1	2	3	4
1	8,824	8,948	9,355	9,170
2	9,164	8,615	9,603	9,907
3	9,816	10,019	9,940	9,565
4	10,166	10,081	11,897	11,454
5	10,291	11,400	12,654	12,194

(a) Compute the four-quarter centered moving average.
(b) Fit a trend line formula and calculate the trend components.
(c) Compute the cyclical components.
(d) Compute the seasonal indexes.
(e) Determine the irregular component for each of the quarters.

17-24 The data in the accompanying table represent the number of bankruptcy claims filed in a certain county in Florida for 5 years. Answer parts (a) through (e) of exercise 17-23 using these data.

Quarter	1978	1979	1980	1981	1982
1	1,938	1,985	2,063	2,121	2,176
2	1,876	1,938	1,999	2,066	2,122
3	1,873	1,821	1,995	2,042	2,102
4	1,917	1,968	2,062	2,094	2,149

17-25 The accompanying data represent the average quarterly product costs for 1,000 pounds of coffee beans for Xefco Foods. Answer parts (a) through (e) of exercise 17-23 for these data.

	Quarter			
Year	1	2	3	4
1	236	190	213	209
2	416	328	303	262
3	356	238	228	187
4	307	234	254	181
5	340	250	276	259

17-26 The accompanying data represent the number of new mortgage loans in Wayne, Iowa, for the last 5 years. Answer parts (a) through (e) of exercise 17-23 for these data.

	Year				
Quarter	1	2	3	4	5
1	410	476	553	620	696
2	406	478	550	618	696
3	454	519	599	671	738
4	511	581	653	724	799

17-27 The data in the accompanying table represent the monthly unemployment rates in Philadelphia, both seasonally adjusted and not seasonally adjusted.

Month	Not Seasonally Adjusted	Seasonally Adjusted
January	6.37	6.50
February	5.85	6.50
March	6.27	6.60
April	6.70	6.70
May	6.90	6.70
June	8.71	6.70
July	8.50	6.80
August	8.50	6.80
September	6.23	7.00
October	6.80	6.80
November	5.10	6.80
December	4.69	6.70

(a) What are the seasonal indexes for unemployment?

(b) Did you expect the values for November and December to be below 100? Explain. Did you expect the values for June, July, and August to be above 100? Explain.

Challenging Problems

17-28 Show that if you do the following calculation:

	(1) Four-Quarter Moving Total		(2) Sum of Consecutive Four-Quarter Totals	(3) Column (2) ÷ 8
1				
	110			
2		(+)	240	30
	130			
3				

(i.e., add the 2 four-quarter moving totals and divide by 8), you get the same result as that from calculating the four-quarter moving average and then centering it.

17-29 What changes in the procedures for calculating the moving average to isolate trend and cycle and for calculating the seasonal index do we have to make if the data are monthly rather than quarterly?

Hands-On Problems

Go to the library and investigate one of the business publications (e.g., *Survey of Current Business, Federal Reserve Bulletins, Business Week, Monthly Labor Review*). Find a quarterly time series of at least 6 years in length that is not seasonally adjusted. For problems 17-1 through 17-5, use only the latest 5 years of data.

17-1 Plot the time series.

17-2 Calculate a four-quarter centered moving average.

17-3 Compute a trend line through the moving averages and graph the trend line on a separate graph.

17-4 Calculate the cyclical components and graph them on a separate graph.

17-5 Calculate the seasonal indexes.

17-6 Deseasonalize the series and plot the deseasonalized series on the same graph as the original series.

The calculations for this problem set can most easily be accomplished with the assistance of an electronic calculator or a packaged program on a computer. A list of the available programs can be obtained from your computer center. There are a variety of packaged programs available: Minitab, Biomed (Biomedical Programs), SAS (Statistical Analysis System), IBM Scientific Subroutine Packages, and SPSS (Statistical Package for the Social Sciences) program libraries. Your local computer center will assist you.

18 Business Indicators and Index Numbers

Chapter Outline

18-1 Business Indicators
Used to measure or predict changes in particular business activities.

18-2 Indexes and Index Numbers
Special types of business indicators.

18-3 Uses of Index Numbers
Index numbers can be used to remove the effect of price changes from a time series.

18-4 Construction of a Weighted Price Index
The two most common types of price indexes are the weighted aggregate and the weighted average of relatives.

18-5 The Consumer Price Index
The most widely used price index.

G.N.P. (Yawn) at $2 Trillion!

It should happen at 2:36 this afternoon, give or take a few minutes....

The country's gross national product will cross the $2 trillion mark. That is $2,000,000,000,000, a lot of money.

Does anyone really care? To be sure, the gross national product is not an inconsequential number. It represents after all the nation's total output of goods and services. But what difference does it make that G.N.P. has hit $2 trillion?

• • •

In this case, not much. The big number doesn't mean a whole lot, because most of it is merely a result of inflation. It seems difficult to believe, but after it took the economy a full 200 years to get to the $1 trillion juncture, the second trillion is upon us after a mere seven years.

Something like two-thirds of the second trillion was "inflation, inflation, inflation," in the words of one economist. He reasons that inflation represented only about a third of the first trillion. Small wonder, then, that economists instead pay attention to "real" G.N.P., which is a measure adjusted to discount inflation.

• • •

The whole calculation of G.N.P.—the handiwork of the Department of Commerce—is an inexact and somewhat arbitrary undertaking to begin with. After all, the thought of taking a huge and hopelessly complex economy and expressing it as a single number is a bit unnerving.

• • •

So is the man in the street better off with a $2 trillion G.N.P. than he was with a measly $1 trillion G.N.P.?

Well, the important thing is what's happening with real G.N.P. Real G.N.P. is figured in terms of the purchasing power of 1972 dollars, to discount inflation. When G.N.P. hits $2 trillion, according to Mr. Stallings' math, real G.N.P. will approach about $1.37 trillion....

From N. R. Kleinfield, *New York Times*, 27 January 1978, pp. A1, D3. © 1978 by the New York Times Company. Reprinted by permission.

Chapter Objectives

The United States economy is highly complex and constantly changing. To make rational business and economic policy decisions, we must be able to assess the changes in current business conditions and to predict future changes.

In this chapter we will discuss business indicators, which attempt to measure the change in business activity. You will learn that there are economic time series that change along with changes in the economy as a whole or with changes in specific business conditions. You probably are familiar with many business indicators already. To quote Dr. Julius Shiskin, the last commissioner of the Bureau of Labor Statistics: "The unemployment rate and the change in the consumer price index are as well known to the public today as the World Series and Super Bowl scores. It is no wonder. They provide simple measures of the performance of the economy and the basis for major policy decisions by the President and the Congress."

Indexes, especially price indexes, play an important role in business. Although the businessperson seldom constructs an index number, he or she does constantly use index numbers published by the government. We will illustrate many of the uses of index numbers and how to construct price indexes in general and the consumer price index in particular, so that you can better understand and use index numbers.

Section 18-1 Business Indicators

It is crucial to any businessperson to know what is happening in business activity. For example, are prices rising? Buying and selling decisions often hinge on the answer to that question. Or are we in a recession or heading toward one? Major government policies and actions frequently depend on the answer to that question.

Finding the answers to these types of questions is not as simple as it may seem, because today's economy is highly complex. For example, consider the question of prices. During inflationary periods, not all prices go up. During a boom, not all industries grow. During a bull market, some stocks go down (especially, it seems, the ones that we buy).

In the study of business conditions, two major statistical questions arise: (1) how do we define and measure business conditions and (2) how can we predict business conditions? The major statistical tools available to the businessperson to be used in studying business conditions are *business indicators* (also called *economic indicators*).

business indicator

Business Indicator
An economic time series that has a reasonably stable relationship to the average of the whole economy or some particular business condition or economic process.

For example, the gross national product is considered to reflect the general state of the economy. The consumer price index is considered to reflect the inflationary trend.

It is important to realize that many business indicators do not specifically attempt to measure business conditions, such as a stock price measures the current value of a stock. Many indicators simply try to measure or indicate the change or fluctuation in the business activity. For example, the consumer price index does not give us the "average price." Instead, it gives us a number that can be compared over time with measured changes in prices. If the consumer price index in 1982 is 110.0 and in 1983 is 112.2, then we can say that prices generally are increasing (112.2 is greater than 110.0). More specifically, we can say that prices generally are 2% higher $[(112.2 - 110.0)/110.0 = 0.02]$ in 1983 than in 1982.

Business indicators are classified into one of three types—leading, coincident, and lagging—depending on their relationship to the economic activity.

leading indicator

Leading Indicators
Indicators for which a change in the indicator normally precedes a change in the economic activity.

For example, unemployment is considered to be a leading indicator of the level of overall economic activity. This means that when unemployment rises, the economy will slow down, whereas when it falls, the economic strength will pick up. Leading indicators are useful in forecasting changes in business activities.

coincident indicator

Coincident Indicators
Indicators in which peaks and troughs in the indicator normally occur at the same time as peaks and troughs in the business condition.

For example, gross national product is considered to be a coincident indicator of the general state of the economy. Coincident indicators are useful in assessing the current state of business conditions.

lagging indicator

Lagging Indicators
Indicators for which change in the indicator normally occurs after the change in the business condition.

For example, labor cost per unit of output in manufacturing industries is considered to be a lagging indicator of the general state of the economy. Lagging indicators are useful in confirming past assessments of business activities.

Figure 18-1 shows what these three types of indicators might look like if they were graphed over a period of time.

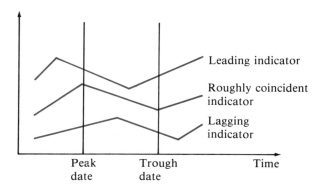

FIGURE 18-1
Economic Indicators

There are a multitude of business indicators. The major publisher of business indicators is the government. (The *Survey of Current Business* and *Business Conditions Digest* are the major business indicator publications.) Moreover, there are several indicators published that are considered to measure the same business condition. Pick up the financial section of almost any newspaper and you will find reported the Dow-Jones stock averages, the Standard and Poor index, and the New York Stock Exchange index. All three are supposed to measure the fluctuations in the New York Stock Exchange securities' market. Why three and not one, if they are all supposed to measure the same thing? Because indicators, like any other *descriptive* statistic, are not perfect. No indicator perfectly measures the fluctuations in business conditions. This is especially true of leading indicators, which are used to attempt to forecast future business change. It is not uncommon to find leading indicators that are supposed to relate to the same business activity giving conflicting signals. One may be signaling an upturn, while others are signaling a downturn.

Although business indicators are far from perfect, they are widely used by economists and businesses. They play an important role in shaping government economic policy and in influencing business decisions.

Exercises

18-1   What are the primary functions of business indicators?

18-2   Suppose we had the true measure of a business condition over time. We let x_t be that measure for time period t and y_t be the business indicator of that activity. Which of the following three sets of correlations do you think would be highest if the business indicator is a (a) lagging indicator, (b) leading indicator, or (c) coincident indicator?

(i) the correlation between x and y in the same time period

(ii) the correlation between x in one time period and y in a time period a few time periods later

(iii) the correlation between x in one time period and y in a time period a few time periods before

18-3 If we could actually measure directly the business condition, which business indicators do you think would be of no value? Which might still be of some value? Explain.

Section 18-2 Indexes and Index Numbers

Business indicators are time series. Two different types of time series are used in business indicators, depending on whether an absolute change or a relative change is reported. Consider the following series:

	1980	1981	1982
Retail sales	$375,527	$408,850	$448,379

This series reports absolute amounts.

Now suppose we were interested only in the relative change over the 3-year period. We could calculate the percentage change in retail sales between 1981 and 1980 and also between 1982 and 1980. The result would be as follows:

1980	1981	1982
$\frac{375,527}{375,527} = 100\%$	$\frac{408,850}{375,527} = 108.9\%$	$\frac{448,379}{375,527} = 119.4\%$

NOTE: $408,850/375,527 = 1.089$, which is a ratio. Indexes are expressed as percentages; thus the ratio of 1.089 must be multiplied by 100 to be expressed as a percentage.

This series (leaving out the percent sign, as is always done) would be called an *index*.

	1980	1981	1982
Retail sales index	100.0	108.9	119.4

The index 108.9 would mean that retail sales in 1981 were 8.9% greater than sales in 1980. The index 119.4 would mean that retail sales in 1982 were 19.4% greater than sales in 1980.

index

Indexes
Business indicators that report a percentage change compared with a specific point in the series.

In the preceding example we could have just as easily made our comparisons with 1981 instead of 1980. The index would then have been as follows:

	1980	1981	1982
Retail sales index	$91.8 \left(\dfrac{375,527}{408,850}\right)$	$100.0 \left(\dfrac{408,850}{408,850}\right)$	$109.7 \left(\dfrac{448,379}{408,850}\right)$

In this case we have used 1981 as the base period of the index.

base period of an index

Base Period of an Index
The specific point from which each percentage change is calculated.

Whenever an index is presented, the base year must also be given to interpret the index value.

In the preceding example we could have used either the actual amount or an index to present the time series of retail sales. In some situations, however, no meaningful absolute amounts exist. This situation would occur, for example, when we are dealing with a business activity that is being related to a *composite* of individual items to study quantity or price changes. Consider the problem of trying to measure general food production. How can we add up pounds of beef and quarts of milk? Do we use 1 pound of beef and 1 quart of milk or 10 pounds of beef and 1 gallon of milk? What does the price of 1 pound of beef and 1 quart of milk represent? Although there are problems in construction, such composites of items can be and are formed. The composite is called an **index number**. The index number by itself has little meaning. However, as a series it allows us to measure relative change. This measurement is extremely useful as a business indicator of price and production movement.

index number

In the next section we will explore some of the major uses of index numbers—in particular, price indexes. Then in section 18-4 we will tackle the problem of constructing index numbers.

Exercises

18-4 Explain the difference between an index and an absolute measure.

18-5 Convert the business indicator in the accompanying table to an index with a base year of 1978. Now convert it to an index with a base year of 1980.

Exports of Passenger Cars ($\times \$1$ million)

1976	1977	1978	1979	1980
837	1,183	1,322	1,825	2,334

18-6 If the index for automobile production is 129.1 in 1983, what does this index mean if the base year is 1978? If the base year is 1979?

18-7 Do you agree or disagree with the statement, "all indexes are simply absolute values, where each year's value is divided by the value in the base year." Explain.

Section 18-3 Uses of Index Numbers

The gross national product, the total dollar value of the nation's output, passed $2 trillion on 27 January 1978. It took the country 200 years to get to the first trillion, but it took only 7 years to reach the second trillion. Does this mean that the physical output of the United States doubled between 1971 and 1978? Hardly. As the *New York Times* reported, "something like two-thirds of the second trillion was 'inflation, inflation, inflation' in the words of one economist."

Time series that are measured in dollars will change because of price changes as well as quantity changes. Consider two time series: (1) gallons of gasoline sold and (2) dollars of gasoline sold. During the oil shortage of the 1970s, the first series showed a drop while the second showed an increase due to the fast rise in prices.

deflating a time series

constant dollars

The process of removing the effect of price changes from a time series is called **deflation** or **deflating a time series**. The resulting deflated time series is said to be measured in **constant dollars**, or the series is prefixed with the word **"real."**

For example, deflated gross national product is called real gross national product. To deflate a series, you simply divide each year's entry by the appropriate price index for that year.

Illustration 18-1

The president of Rextron Corporation is upset because in reviewing the company's balance sheets, he notes that over the last 5 years the value of inventory has gone up 25%, as shown in table 18-1. He asks the manager in charge of inventory control to explain why this increase has happened. The manager responds that the physical volume of inventory has changed only slightly, but that prices have risen sharply. She presents the data in table

18-2 to demonstrate her argument. The real inventory, or deflated inventory, is calculated by dividing the inventory values in column (1) by the price index in column (2).

TABLE 18-1
Value of Inventory, Rextron Corporation

Year	Inventory ($\times \$1,000$)
1979	400
1980	415
1981	445
1982	475
1983	500

What do the deflated figures represent? They represent what the dollar value of the inventory would have been in each of the years if the prices had been the same as in the base year 1975. Thus all the inventories are measured in 1975 dollars (referred to as *constant dollars*).

TABLE 18-2
Value of Real Inventory

Year	(1) Inventory ($\times \$1,000$)	(2) Price Index of Product Mix (1975 = base year)	(3) Real Inventory ($\times \$1,000$)
1979	400	1.10	363.6
1980	415	1.12	370.5
1981	445	1.18	377.1
1982	475	1.20	395.8
1983	500	1.24	403.2

Illustration 18-2

The data in table 18-3 represent the sales of Litown Clothing Stores for the past 5 years and the number of salespersons employed. Has the productivity of the sales force increased?

TABLE 18-3
Sales and Sales Force for Litown Clothing Stores

Year	Sales ($\times \$1,000$)	Sales Force	Sales per Employee
1979	3,100	310	10,000
1980	3,350	315	10,635
1981	3,800	315	12,063
1982	4,400	320	13,750
1983	4,950	330	15,000

In terms of average dollar sales per employee, yes, the sales force's productivity has increased. But how about in terms of units sold? Are they selling the same number of items, but at increasing prices? To answer this question, the firm obtained the apparel and upkeep price index series from the government's Bureau of Labor Statistics. It then deflated the sales series (divided each sales figure by the index) and calculated the average constant dollar sales per employee (divided the deflated sales figures by the number of employees). See table 18-4.

From the table we see that constant dollar sales also went up, indicating that the sales increase per employee was not all due to increased prices. However, the percentage increase is a lot less in constant dollars than in actual dollars—only a 10% increase in constant dollar sales over the 4 years as opposed to a 50% increase in actual dollar sales.

TABLE 18-4
Deflated Sales and Sales per Employee

Year	Sales (× $1,000)	Apparel and Upkeep Price Index	Constant Dollar Sales (÷ $1,000)	Sales Force	Constant Dollar Sales per Employee
1979	3,100	128.2	2,418.1	310	7,800
1980	3,350	140.0	2,392.9	315	7,597
1981	3,800	150.4	2,526.6	315	8,021
1982	4,400	165.8	2,653.8	320	8,293
1983	4,950	174.0	2,844.8	330	8,621

Illustration 18-3

Last summer you worked for $3.50 an hour. This summer you are working for $3.85 an hour. You got a 10% raise, but in terms of purchasing power, are you 10% better off? For example, suppose tuition also has increased by 10%. To remove the effect of price changes in your purchasing power, you would divide your income by the consumer price index (CPI).

Illustration 18-4

Suppose the consumer price index in 1979 is 205.0 and in 1984 it is 305.0. The base year is 1967. You earned $20,000 in 1979 and $30,000 in 1984. How much did your real income increase?

Solution The $20,000 income in 1979 is the same (in terms of buying power) as a $9,756 (20,000/2.05 = 9,756) income in 1967. The $30,000 income in 1984 is the same as a $9,836 (30,000/3.05 = 9,836) income in 1967. The percentage increase in real income over the 3 years is

$$\left(\frac{9,836}{9,756} \times 100\right) - 100\% = 101\% - 100\% = 1\%$$

NOTE: Remember to place the decimal point in the index. An index represents a percentage. Thus an index of 120 is 120% or 1.20.

In general, to calculate the percentage change in real income from period 1 to period 2, you can use this formula:

$$\text{Percentage increase in real income between periods 1 and 2} = \frac{\text{dollars in period 2}}{\text{dollars in period 1}} \times \frac{\text{CPI in period 1}}{\text{CPI in period 2}} \times 100 - 100\% \qquad (18\text{-}1)$$

Thus using formula (18-1), we calculate the solution for illustration 18-4:

$$\text{Percentage increase in real income between 1979 and 1984} = \left[\left(\frac{30{,}000}{20{,}000}\right) \times \left(\frac{205.0}{305.0}\right) \times (100)\right] - 100$$

$$= 1\% \qquad \square$$

Illustration 18-5

If the consumer price index in 1984 is 305, how much money must you earn to have the same real income as in 1979? Refer to illustration 18-4.

Solution   In illustration 18-4 we found that a $20,000 income in 1979 is the same as an income of $9,756 in the base year 1967. In 1984 the consumer price index is 305. Hence prices in 1984 are 3.05 more than they were in 1967. Correspondingly, you will need 3.05 more income in 1984 than you needed in 1967 to buy the same goods and services. Thus an income of $9,756 in the base year is equivalent to a 1984 income of

$$3.05 \times \$9{,}756 = \$29{,}756 \qquad \square$$

In general, to calculate what an amount in dollars at one point in time is worth at another point in time, you can use this formula:

$$\frac{\text{Dollar value in period 2}}{} = \frac{\text{dollar value in period 1}}{} \times \frac{\text{CPI in period 2}}{\text{CPI in period 1}} \qquad (18\text{-}2)$$

Thus if we use formula (18-2), the solution for illustration 18-5 would be

$$\text{Dollar value in 1984 of 1979's } \$20{,}000 = \$20{,}000 \times \frac{3.05}{2.05} = \$29{,}756$$

Illustration 18-6

The two columns in table 18-5 were taken from the June 1975 *Survey of Current Business*. Column (1) represents the average weekly take-home income of U.S. workers with three dependents, adjusted for seasonal variations (i.e., the earnings each month were divided by the seasonal index). It shows a general increase in take-home pay. Column (2) was calculated by

TABLE 18-5
Spendable Income

Month, 1974	(1) Current Dollars, Seasonally Adjusted	(2) 1967 Dollars, Seasonally Adjusted
April	131.27	91.16
May	133.28	91.62
June	134.41	91.55
July	134.98	91.13
August	136.11	90.90
September	137.52	90.78
October	138.04	90.31
November	136.98	88.79
December	138.50	89.08

dividing column (1) by the consumer price index. It says "1967 dollars" because the base year for the consumer price index is 1967. The figures in column (2) represent the income needed in 1967 to purchase the same amount of goods that could be purchased in 1974 with the income given in column (1). For example, in April 1974, $131.27 would purchase the same amount of goods that could have been purchased with $91.16 in 1967. Comparing real take-home income shows a different situation than that revealed when comparing actual income. The average wage earnings in terms of buying power were actually declining during 1974. □

Illustration 18-7

In estate planning it is important to recognize the problem of inflation. It is fairly common for a person to prepare his or her will so that a trust is set up from which his or her surviving mate can draw a fixed yearly income; upon the mate's death the trust passes to the children. But suppose that the person making the will lives a long period of time after the will is drawn up and a high rate of inflation occurs. What seemed like a livable income when the will was prepared may not be when that person dies. Consider a fixed yearly income of $20,000. When the will was drawn, $20,000 was an adequate income. Suppose, however, that in 15 years the consumer price index goes from 305.0 to 450.0. Let's calculate, using formula (18-2), what this income would mean 15 years later.

$$\frac{\text{Dollar value}}{\text{(in 15 years)}} = \frac{\text{dollar value}}{\text{(when will drawn)}} \times \frac{4.50}{3.05}$$

$$\$20{,}000 = \text{dollar value} \times \frac{4.50}{3.05}$$

$$\text{dollar value} = \$20{,}000 \times \frac{3.05}{4.50}$$

$$= 13{,}560$$

In terms of when the will was drawn, 15 years ago, the $20,000 is equivalent to an income of only $13,560. □

Today, many pensions, Social Security, for instance, are tied to the consumer price index. When the CPI increases, the pension payments increase to ensure that those receiving the pension do not suffer a drop in real income (purchasing power). Many union contracts are also tied to the CPI. When the CPI increases, the wage ratio increases. This increase is generally known as a *cost of living adjustment*, or *escalator clause*. Again, this adjustment is done so that real wages do not decline.

Price indexes are also used in assessing "real" depreciation to aid in financial planning in business. In tax law, depreciation over the life of an asset is equal to the cost of the asset minus the salvage value. Depreciation is charged against income and reduces taxable income and reported profits. However, a business that pays dividends or makes financial decisions based on the reported profits after depreciation deductions have been taken may get into financial trouble, as we will see in the next illustration.

Illustration 18-8

Ajax Products built its plant in 1950. The plant cost $70,000 and is expected to have to be replaced in 35 years with no salvage value. The firm uses a straight-line depreciation basis. That is, total depreciation is assumed to be $70,000 and is allocated equally over each year for 35 years, giving a deduction of $2,000 a year. In 1983 the firm reported a net profit of $25,000. The firm is considering paying this out in a dividend. Should it?

Solution   When the building was built, the construction index (a measure of the change in the cost of building) was 110.0. In 1984 it is 1,100.0. If we use formula (18-2), the cost of replacing the building in 1984 is

$$\$70{,}000\left(\frac{1{,}100.0}{110.0}\right) = \$700{,}000$$

Thus the real yearly replacement cost is $700,000/35 = $20,000 rather than the $2,000 taken as depreciation. Reflecting the replacement cost of the asset depreciated, profits in 1983 were therefore really $7,000 [i.e., 25,000 + (2,000 − 20,000)]. Without awareness of this situation and the proper financial planning for replacing the plant when it has to be replaced, the firm will be in real financial trouble. □

Another common use of indexes is for a firm to compare changes in its business with the average changes in other business.

Illustration 18-9

Aton Home Products increased its advertising budget by 15%. The advertising manager wanted to know if this increase was in line with other firms' increases. Examining the *Survey of Current Business*, she noted that the McCann-Erickson national advertising index went from 142 to 150 during this period. Hence nationally the percentage change in advertising expenditures was 5.6% {[(1.50/1.42) × 100] − 100}. Thus her increase was considerably larger. □

Exercises

18-8   What is meant by the term *real wages*?

18-9   What is meant by the term *constant dollars*?

18-10   Give an example of four different uses of index numbers.

18-11   Suppose you earned $3.00 an hour in 1982 and $3.20 an hour in 1983. The CPI in 1982 was 250 and in 1983 it was 280.
   (a) What was your real wage rate in 1982?
   (b) What was your real wage rate in 1983?
   (c) What is the percentage change in your real wage?
   (d) If your real wage rate in 1983 was the same as it was in 1982, what would your wage rate have been in 1983?

18-12   The CPI in April 1983 was 295.5. The base of the CPI is 1967. In April 1983 $1.00 will buy the same amount of goods that could have been purchased with how many dollars in 1967? (This amount is called the *purchasing power of the dollar*.)

18-13   A firm buys a car for $10,000 in 1979. The car has a 5-year life and a $1,000 salvage value. The firm uses a straight-line depreciation basis. The 1979 CPI for new cars is 100. The 1984 CPI for new cars is 115.0. What is the real yearly replacement cost in 1984? Compare the yearly replacement cost with depreciation.

18-14   Between 1983 and 1984 Porter Brewery increased its magazine advertising budget by 5%. The McCann-Erickson national advertising index for magazines was 126 in 1983 and 131 in 1984. Did Porter Brewery increase its magazine advertising budget more than, less than, or the same as other companies, on the average?

Section 18-4 Construction of a Weighted Price Index

Construction of any index requires four decisions: (1) the choice of the items to be used in the index, (2) the choice of a base year, (3) the choice of importance or weights to be given each item, and (4) the choice of the index formula to be used.

The choice of items to be used depends on what the index is supposed to represent. The items selected are either a census of all relevant items or a judgment sample of the important items. The New York Stock Exchange index uses all stocks listed on the exchange, whereas the Dow-Jones index uses only 65 selected stocks. There is no "right" or "wrong" selection of items, but when you use an index, be sure to find out exactly which items are included so that you can properly interpret what is being measured. For example, the Dow-Jones index basically selects blue-chip stocks. Hence it has been questioned whether it really measures total market change, particularly of the more speculative stocks.

Any year can be chosen for the base year. However, since each index number represents a comparison to the base year, it is important to choose a base year so that the comparison is meaningful. This usually means choosing a base year that is not too far in the past (generally, no more than 10 to 15 years) and choosing a year in which nothing abnormal happened. For example, you would not want to choose a base year in which there was a price freeze.

weight The hardest choice is the selection of weights. The **weights** represent the importance we are going to place on the price change of each item. Consider a food index that is to be constructed based on the items and prices given in table 18-6. The only price that has risen is that of caviar. If we were simply to compare the sum of the prices of the items in each year, we would see a substantial increase ($\sum P_1 = \$7.10$ versus $\sum P_0 = \$5.80$). But are food prices in general really rising sharply? Looking at table 18-6, we would have to say that food prices are rising only if you are one of the few people addicted to caviar.

To avoid this kind of distortion in a food index, we *weight* each item by the quantity a "typical" consumer would buy in a given period of time.

TABLE 18-6
Items and Prices for Food Index

Item	Base Year 1967, P_0	1968, P_1
Milk	$0.35 (qt)	$0.35 (qt)
Eggs	1.15 (doz)	1.15 (doz)
Beef	1.10 (lb)	1.10 (lb)
Caviar	3.20 (lb)	4.50 (lb)

market basket

The weights given to the items constitute what is referred to as the typical **market basket**. You should realize, however, that the "typical" consumer may not be very typical of certain people's buying habits. This atypical behavior may be especially true of persons in the extremely high- or low-income categories.

You should be thinking now, do I compute a market basket for each year or for just 1 year, and if for just 1 year, which year? We cannot change the market basket each year. If we did change it each year, then we would not know if the change in the index number was due to price increases or simply changes in consumer preference to more or less expensive items. Thus to construct a food index, we use the typical market basket of the base year. This method of index construction is known as the **Laspeyres method**, after Etienne Laspeyres, who first suggested it in 1864. **The price index number then represents the percentage of change in the cost of purchasing the typical market basket bought in the base year.**

Laspeyres method

Let's use the data in table 18-7 to construct a food price index. However, we still need an index formula to be able to use these data to construct an index. The two most commonly used formulas are given next.

TABLE 18-7
Data for Constructing Food Index

Item	Prices (per unit) 1967, P_0	1975, P_{75}	Market Basket,[†] Q_0
Milk	$0.35 (qt)	$0.45	5 (qt)
Eggs	1.15 (doz)	1.45	2 (doz)
Beef	1.10 (lb)	1.65	5 (lb)
Bread	0.40 (loaf)	0.44	3 (loaves)

† Typical weekly purchase of a consumer in 1967.

weighted price aggregate formula

The **weighted price aggregate formula** is

$$I_i = \frac{\sum (P_i \times Q_0)}{\sum (P_0 \times Q_0)} \times 100 \qquad (18\text{-}3)$$

This formula involves the calculation of an aggregation of prices (weighted by their respective quantities).

weighted average of price relatives formula

The **weighted average of price relatives formula** is

$$I_i = \frac{\sum [(P_i/P_0) \times P_0 \times Q_0]}{\sum (P_0 \times Q_0)} \times 100 \qquad (18\text{-}4)$$

This formula involves the relative or ratio of prices (weighted by the item's value, price × quantity).

The symbols in these formulas are defined as follows:

$$I_i = \text{index value in year } i$$
$$P_i = \text{price of item in year } i$$
$$P_0 = \text{price of item in base year}$$
$$Q_0 = \text{amount of item in market basket in base year}$$
$$\frac{P_i}{P_0} = \text{relative prices}$$
$$\frac{P_i}{P_0} \times (P_0 \times Q_0) = \text{weighted relatives}$$

Let's use formula (18-3) to calculate the 1975 price index. The preliminary calculations are given in table 18-8.

TABLE 18-8
Calculations for Weighted Aggregate

Item	Prices		Weights, Q_0	Price × Weight	
	P_0	P_{75}		$P_0 Q_0$	$P_{75} Q_0$
Milk	$0.35	$0.45	5	1.75	2.25
Eggs	1.15	1.45	2	2.30	2.90
Beef	1.10	1.65	5	5.50	8.25
Bread	0.40	0.44	3	1.20	1.32
				10.75	14.72

Using formula (18-3), we have

$$I_{75} = \frac{14.72}{10.75} \times 100 = 1.369 \times 100 = \mathbf{136.9}$$

(Remember, indexes are percentages.) Thus we can say that the purchase of the typical 1967 market basket in 1975 will cost 36.9% more money.

Now let's use formula (18-4) to calculate the 1975 price index. The preliminary calculations are shown in table 18-9. Using formula (18-4), we get

$$I_{75} = \frac{14.72}{10.75} \times 100 = 136.9$$

TABLE 18-9
Calculations for Weighted Average of Relatives

	(1)	(2)	(3)	(4)	(5)	(6)
	Prices		Weights,		Relatives,	Weighted Price Relatives,
Item	P_0	P_{75}	Q_0	$P_0 Q_0$	P_{75} / P_0	$(P_{75} \div P_0) \times (P_0 Q_0)$
Milk	$0.35	$0.45	5	1.75	1.286	2.25
Eggs	1.15	1.45	2	2.30	1.261	2.90
Beef	1.10	1.65	5	5.50	1.500	8.25
Bread	0.40	0.44	3	1.20	1.100	1.32
				10.75		14.72

The results of using formulas (18-3) and (18-4) are the same. In fact, they will always be the same. (The proof of this formula is left for you, exercise 18-31.) Although the weighted average of price relatives formula is more difficult to calculate, it does have an advantage. The intermediate step, the calculation of price relatives [column (5)], gives us an index number for each item. From these individual indexes it is an easy task to select any subset and weight them by their $P_0 Q_0$ values and construct an index for that subset. For example, if the price relatives for all food prices were published, you could select those items that you purchase normally and construct a price index for the items you buy.

CAUTION: Price indexes are far from perfect measures of price change. The market basket may not perfectly represent the mix in which you are interested. Also, price changes may, to some extent, represent quality changes in the item over time and not true price increases. In addition, price indexes do not truly reflect expenditure increases, since you may substitute items in the market basket as a result of a price increase. For example, if butter increased 10% but margarine decreased 5%, you might switch from butter to margarine and reduce your market basket costs despite the rise in the food index. Price indexes at best are imperfect measures of change in a complex economic activity. These shortcomings do not mean, however, that they are not useful—only that they should be used with care. If we always waited for perfect information before making a business decision, few business decisions would ever be made.

As an example of an actual price index, the next section describes the consumer price index, the most widely used index published by the government.

Exercises

18-15 (a) Does a price index measure changes in consumption patterns?

(b) Does a price index measure changes in product quality?

(c) Does a price index measure the change in the cost of the market basket for all consumers?

18-16 Suppose you are given the market basket data in the accompanying table for college expenses. Calculate, using formula (18-3) (the weighted price aggregate formula), the index for college costs for 1975 and 1980. Interpret your results.

Item	1970, P_0	1975, P_{75}	1980, P_{80}	Market Basket, Q_0
Tuition	$ 25 (credit)	40	50	30
Fees	2 (credit)	4	8	30
Books	8 (book)	12	20	10
Room	300 (semester)	350	400	2
Board	300 (semester)	375	425	2

18-17 Using formula (18-4) (the weighted average of price relatives), calculate a college cost index for the data in exercise 18-16. Are your results the same as in exercise 18-16? Interpret the individual price relatives you calculated.

Section 18-5 The Consumer Price Index

The consumer price index, first published in 1919, is the most widely used price index in the United States. As of 1978, CPI's have been published for two populations: (1) a new CPI-U for All Urban Consumers, which covers about 80% of the civilian population, and (2) the revised CPI-W for Urban Wage Earners and Clerical Workers. The CPI-W represents about half the population of the CPI-U. The CPI attempts to measure changes in prices in general. It often is referred to as the *cost of living index*, but in reality it does not actually indicate how much families spend on living costs, since it does not consider that people often shift their market basket by switching their purchases of high-priced items to lower-priced items.

Let us explore how the CPI deals with the four decisions outlined in the previous section.

Items: The CPI includes approximately 400 items, covering everything people buy, from food, clothing, housing, and transportation to health and recreation. It deals with actual prices people pay, including sales taxes, excise taxes, and real estate taxes. (It does not include income tax.)

Base year: The CPI uses a 1967 base year.

Weights: The market basket is based on the average expenditures of urban wage earners and clerical workers and now includes professional, managerial, technical, and other workers who were previously excluded. The data are obtained from a national survey of consumer expenditures.

Formula: The basic formula used in the construction of the index is the weighted average of price relatives, formula (18-4). The CPI also has a special method of attempting to correct for quality changes and the introduction of major new products.

Aside from the overall CPI's, the government publishes a number of separate consumer price indexes for specific subgroups of items—for example, a rent index, a medical care index, and a dairy product index. The index is also published for 28 large metropolitan areas, such as New York, Boston, Atlanta, Milwaukee, and San Diego.

In Retrospect

Business indicators are time series that measure or predict change in business conditions. We learned how to distinguish among a leading, a lagging, and a coincident indicator and we learned their respective uses. When the indicator is reported in relative terms, we call it an index.

Of special importance to businesspeople are composite indexes, particularly price indexes. There are four decisions that must be made in constructing price indexes: (1) choice of items, (2) choice of base year, (3) choice of weights, and (4) choice of index formula. Although most businesspeople never compute an index, they do use published indexes frequently. When using a published index, make sure you know what choices were used in the construction of the index.

Index numbers can be classified as descriptive statistics. Like most summary figures they do not perfectly measure the complete process. In addition, many problems are involved in constructing index numbers.

Reread the news article at the beginning of this chapter. You should now have a better appreciation for the statement that a GNP of $2 trillion is comparable with a real GNP of $1.37 trillion.

Despite their problems, indexes are widely used in business. You should now be able to use a price index to remove price change effects from a time series, to convert income into real income, and to calculate real depreciation and purchasing power. The meanings of terms like "constant dollars" and "real wages" should be clear to you now.

Chapter Exercises

18-18 Increases in the consumer price index probably overstate the true cost of living increase. Do you agree or disagree? Explain.

18-19 In 1985, tuition, fees, room, and board at a certain eastern college cost $6,000 per year, and the price index for college education costs is 115 (base year 1980). You estimate that in 1990 it will be 145. What do you estimate the cost of the college education will be in 1990?

18-20 You rent an apartment and currently pay $300 per month. Suppose the CPI currently is 115. Three years later your rent is raised to $325. The CPI at that time is 130. You complain that the rent raise is too large. In light of general price raises, is your complaint justified?

18-21   With respect to exercise 18-20, do you think the CPI is the appropriate index to use? Explain.

18-22   Packard Toys evaluates its raw materials inventory at the end of each year. For the past 5 years, this inventory has been increasing. Do the data in the accompanying table indicate a problem in inventory control?

Year	Inventory	Raw Material Price Index
1978	$280,000	115.0
1979	295,000	116.5
1980	310,000	118.0
1981	330,000	119.5
1982	370,000	121.0

18-23   Suppose the data given in the accompanying table represent the consumer price index for two major metropolitan areas.

Year	New York	Atlanta
1982	115.6	120.3
1983	120.8	124.8
1984	130.0	132.4

(a)   Can you conclude that prices generally are higher in Atlanta? Explain.
(b)   Which area has undergone a higher rate of inflation?

18-24   How do you think large increases in the prices of the following items would affect the CPI? Explain.
(a)   yachts
(b)   beef
(c)   zinc
(d)   tractors
(e)   scientific calculators
(f)   sales tax
(g)   income tax

18-25   The pay rate for student assistants for the last 5 years has been $3.00, $3.10, $3.15, $3.20, and $3.30 per hour. The CPIs for the periods were 115.1, 123.0, 133.0, 139.0, 139.1, and 150.0, respectively.
(a)   How much would a student in 1967 (base year for the index) have had to be paid to be able to purchase the same amount of goods as the student currently earning $3.30?
(b)   If wage rates had kept up with inflation, how much would a student currently be paid?

18-26 Explain the statement, "By paying later, I can pay with cheaper dollars." What does that statement assume will happen to the CPI?

18-27 Given the accompanying data on beverage prices and the market basket of a typical urban wage-earning family of four, and using 1974 as base year, construct the weighted average of price relatives index for these beverage prices:

(a) alcoholic beverage prices

(b) nonalcoholic beverage prices

	Price per Unit		Market Basket
Item	1974	1984	in 1974
Soda	$0.10	$0.45	12
Milk	0.49	0.89	5
Coffee	1.89	3.99	1
Tea	1.00	2.00	1
Beer	1.69	3.50	2
Liquor	5.85	9.85	1

18-28 Given the data in the accompanying table, construct an index of raw materials costs for Ace Toys. Use the Laspeyres weighted aggregate method and 1979 as the base year.

	Price per Unit			Quantity		
Material	1979	1980	1981	1979	1980	1981
A	$0.60	$0.63	$0.65	3	4	4
B	1.00	1.90	1.70	2	1	1
C	0.75	0.90	0.90	6	6	6
D	6.00	6.25	6.50	1	1	1

18-29 For the data in exercise 18-28, construct an index of raw materials cost.

(a) Use 1980 as the base year.

(b) Use 1981 as the base year.

18-30 Use the data and the answer to exercise 18-28.

(a) Compute the percentage change in the index number between 1980 and 1979.

(b) Compute the cost of the 1979 market basket in 1979.

(c) Compute the cost of the 1980 market basket in 1980.

(d) Compute the percentage increase in expenditures between 1980 and 1979 to purchase the year's market basket.

(e) Compare answer (d) with answer (a). Explain your findings.

18-31 Show algebraically that formula (18-4) is equivalent to formula (18-3).

Challenging Problem

18-32 Go to the library and get a list of the component indexes of the CPI. Which component or components do you think are
(a) most likely to measure quality rather than price changes
(b) least likely to measure quality rather than price changes
(c) most likely to reflect a cost of living change to you as a college student
(d) least likely to reflect a cost of living change to you as a college student

Explain your answers.

Hands-On Problems

18-1 Go to the library and find the annual reports for the past 5 years of any major company. Record their sales for the past 5 years.

18-2 Go to the library and get a copy of the recent *Survey of Current Business*. Record the CPI and three other series (do not choose all index series) that you think might be related to sales of the company you selected in problem 18-1. For example, you might choose GNP, retail sales, and industrial sales. Study the indicators published and choose whichever ones you think might be relevant.

18-3 Plot dollar sales of the firm and sales in constant dollars on the same graph. Do you think the firm is selling more or less items over time?

18-4 Plot the three indicators you chose on the same graph as dollar sales and constant dollar sales. Do any of the indicators appear to be reasonable indicators of sales? If so, what type?

18-5 Find a sales index for the same industry as the firm you selected. Plot that index and the dollar sales on the same graph. What does this graph indicate about the firm's sales relative to that of the industry?

19 Elements of Nonparametric Statistics

Chapter Outline

19-1 Nonparametric Statistics
Distribution-free, or nonparametric, methods provide test statistics from an unspecified distribution.

19-2 The Sign Test
Uses a simple **quantitative count** of plus or minus signs to tell us whether or not to reject the null hypothesis.

19-3 The Mann–Whitney U Test
The **rank number** for each piece of data is used to compare **two independent samples**.

19-4 The Runs Test
A **sequence of data** that possesses a common property, a run, is used to test the question of randomness.

19-5 Rank Correlation
The linear correlation coefficient's **nonparametric alternative**, rank correlation, uses only rankings to determine whether or not to reject the null hypothesis.

19-6 Comparing Statistical Tests
When choosing between parametric and nonparametric tests, we are interested primarily in the **control of error**, the relative **power** of the test, and **efficiency**.

Study Finds That Money Can't Buy Job Satisfaction

When it comes to getting workers to produce—do their level best—money is less than everything. Feeling appreciated—having a sense of being recognized—is more important.

That is the thesis of Kenneth A. Kovach, of George Mason University....

Kovach reproduces a consequential survey of 30 years ago in which workers and supervisors gave their opinions of "what workers want from their jobs and what management thinks they want." Here it is:

Managements today are more "behavioral" in their approach to managing. Nevertheless, Kovach contends that a wide gap still exists, a gap that is suggested by what workers rank as No. 1 in importance to them ("appreciation") and what supervisors consider No. 1 to workers ("good wages")....

	Worker Ranking	Boss Ranking
Full appreciation of work done	1	8
Feeling of being in on things	2	10
Sympathetic help on personal problems	3	9
Job security	4	2
Good wages	5	1
Interesting work	6	5
Promotion and growth in the organization	7	3
Personal loyalty to employees	8	6
Good working conditions	9	4
Tactful disciplining	10	7

Reprinted by permission of the *Philadelphia Inquirer*, 29 December 1976.

Chapter Objectives

Nonparametric methods of statistics dominate the success story of statistics in recent years. Unlike their parametric counterparts, many of the best-known nonparametric tests, also known as distribution-free tests, are founded on a basis of elementary probability theory. The derivation of most of these tests is well within the grasp of the student who is competent in high school algebra and understands binomial probability. Thus the nonmathematical statistics user is much more at ease with the nonparametric techniques.

This chapter is intended to give you a feeling for basic concepts involved in nonparametric techniques and to show you that nonparametric methods are extremely versatile and easy to use, once a table of critical values is developed for a particular application. The selection of the nonparametric methods presented here includes only a few of the common tests and applications. You will learn about the sign test, the Mann–Whitney U test, the runs test, and Spearman's rank correlation test. These tests will be used to make inferences corresponding to both one- and two-sample situations.

Section 19-1 Nonparametric Statistics

Most of the statistical procedures that we have studied in this book are known as *parametric* methods. For a statistical procedure to be parametric, we either assume that the parent population is at least approximately normally distributed or we rely on the central limit theorem to give us a normal approximation. This is particularly true of the statistical methods studied in chapters 8, 9, and 10.

nonparametric, or distribution-free, methods

The **nonparametric methods,** or **distribution-free methods**, as they are also known, do not depend on the distribution of the population being sampled. The nonparametric statistics are usually subject to much less confining restrictions than their parametric counterparts. Some, for example, require only that the parent population be continuous.

The recent popularity of nonparametric statistics can be attributed to the following characteristics:

1. Nonparametric methods require few assumptions about the parent population.
2. Nonparametric methods are generally easier to apply than their parametric counterparts.
3. Nonparametric methods are relatively easy to understand.
4. Nonparametric methods can be used in situations where the normality assumptions cannot be made.
5. Nonparametric methods appear to be wasteful of information in that they sacrifice the value of the variable for only a sign or a rank number. However, nonparametric statistics are generally only slightly less efficient than their parametric counterparts.

Section 19-2   The Sign Test

sign test The ordinary **sign test** is a very versatile and exceptionally easy-to-apply nonparametric method that **uses only plus and minus signs**. (There are several specific techniques.) The sign test is useful in two situations: (1) a hypothesis test concerning the value of the median for one population and (2) a hypothesis test concerning the median difference (paired difference) for two dependent samples. Both tests are carried out using the same basic procedures and are nonparametric alternatives to the t tests used with one mean (section 9-1) and the difference between two dependent means (section 10-5). Notice, however, that the sign test applies to the median, whereas the t test applies to the mean.

Single-Sample Hypothesis Test Procedure

The sign test can be used when a random sample is drawn from a population with an unknown **median** M and the population is assumed to be continuous in the vicinity of M. The null hypothesis to be tested concerns the value of the population median M. The test may be either one or two sided. We present the test procedure in the next illustration.

Illustration 19-1

An important number in economic theory and economic policy is the marginal propensity to save (MPS), or the proportion of a person's income he or she saves rather than spends. A random sample of 75 people, all earning $20,000, was selected, and they were asked to carefully determine the amount of money they saved. The data collected were used to test the hypothesis "the median MPS is 0.15" against the alternative that the median is unequal to 0.15. The 75 pieces of data were summarized as follows:

Under 0.15:	18
0.15:	15
Over 0.15:	42

Use the sign test to test the null hypothesis against the alternative hypothesis.

Solution   The data are converted to (+) and (−) signs. A plus sign will be assigned to each piece of data larger than 0.15, a minus sign to each piece of data smaller than 0.15, and a 0 to those data equal to 0.15. The sign test uses only the plus and minus signs; therefore the 0s are discarded and the usable sample size becomes 60. That is, $n(+) = 42$, $n(-) = 18$, and $n = n(+) + n(-) = 42 + 18 = 60$.

STEP 1: $H_0: M = 0.15$.

STEP 2: $H_a: M \neq 0.15$.

STEP 3: We will use $\alpha = 0.05$ for a two-sided test. The **test statistic** that will be used is the **number of the less frequent sign**, the smaller of $n(+)$ and $n(-)$, which is $n(-)$ for our illustration. We will want to reject the null hypothesis whenever the number of the less frequent sign is extremely small. Table 10 of appendix D gives the maximum allowable number k of the less frequent sign that will allow us to reject the null hypothesis. That is, if the number of the less frequent sign is less than or equal to the critical value in the table, we will reject H_0. If the observed value of the less frequent sign is larger than the table value, we will fail to reject H_0. n in the table is the total number of signs, not including 0s.

For our illustration $n = 60$ and the critical value from table 10 is 21. (See the diagram.)

Reject H_0		Fail to reject H_0
0	21	22

x, number of less frequent sign

STEP 4: The observed statistic is $x = n(-) = 18$ and it falls in the critical region.

STEP 5: *Decision*: Reject H_0.

Conclusion: The sample shows sufficient evidence at the 0.05 level to reject the claim that the median MPS is 0.15. □

Two-Sample Hypothesis Test Procedure

The sign test may also be applied to a hypothesis test dealing with the **median difference** between paired data that result from **two dependent samples**. A familiar application is the use of before-and-after testing to determine the effectiveness of some activity. In a test of this nature, the signs of the differences are used to carry out the test. Again 0s are disregarded. The following illustration shows the procedure.

Illustration 19-2

A new training program to reduce clerical errors is advertised by Haim Associates. According to Haim, "After 1 week of training you will see an immediate reduction in clerical errors on the part of your staff." To test the program, Penn Insurance randomly selects 18 premium clerks and enrolls them in the program. Each clerk records premium payments of clients by typing the information into a computer terminal. The number of errors per 10,000 keystrokes of each clerk was calculated for the week before entering the program and the week after completing the program. Table 19-1 lists the people, their error counts, and a − for those whose error rate was reduced (i.e., improved), a 0 for those whose error rate remained the same, and a + for those whose error rate actually increased.

The claim being tested is that the program increased quality of work. The null hypothesis that will be tested is that there is no quality gain (i.e., that the median change in error counts is 0), meaning that only a rejection of the null hypothesis will allow us to conclude in favor of the advertised claim. Actually, we will be testing to see if there are significantly more minus signs than plus signs. If the program is of absolutely no value, we

TABLE 19-1
Data for Keystroke Errors

Clerk	Keystroke Errors Before	Keystroke Errors After	Sign of Difference, Before to After
Mrs. Smith	146	142	−
Mr. Brown	175	178	+
Mrs. White	150	147	−
Mr. Collins	190	187	−
Mr. Gray	220	212	−
Miss Collins	157	160	+
Mrs. Allen	136	135	−
Mrs. Noss	146	138	−
Miss Wagner	128	132	+
Mr. Carroll	187	187	0
Mrs. Black	172	171	−
Mrs. McDonald	138	135	−
Miss Henry	150	151	+
Miss Greene	124	126	+
Mr. Tyler	210	208	−
Mrs. Williams	148	148	0
Mrs. Moore	141	138	−
Mrs. Sweeney	164	159	−

would expect to find an equal number of plus and minus signs. If it works, there should be significantly more minus signs than plus signs. Thus the test performed here will be a one-tailed test. (We will want to reject the null hypothesis in favor of the advertised claim if there are "many" minus signs.)

Solution

STEP 1: $H_0: M = 0$ (no quality gain).

STEP 2: $H_a: M < 0$ (quality gain).

STEP 3: Use $\alpha = 0.05$. $n(+) = 5$, $n(-) = 11$, and $n = 16$. The critical value from table 10 shows $k = 4$ as the maximum allowable number. (You must use the $\alpha = 0.10$ column, as the table is set up for two-tailed tests.)

Reject H_0	Fail to reject H_0
0 1 2 3 4	5

x, number of less frequent sign

STEP 4: $x = n(+) = 5$.

STEP 5: *Decision*: Fail to reject H_0 (we have too many plus signs).

Conclusion: The evidence observed is not sufficient to allow us to reject the no-quality-gain null hypothesis. □

The sign test may be carried out by means of a **normal approximation** using the standard normal variable z. The normal approximation will be used if table 10 does not show the particular critical value desired or if n is large. z will be calculated by using the formula

$$z = \frac{x' - (n/2)}{\frac{1}{2}\sqrt{n}} \qquad (19\text{-}1)$$

(See note 3 with regard to x'.)

NOTES:

1. x may be the number of the less frequent sign or the most frequent sign. You will have to determine this designation in such a way that your test is consistent with the interpretation of the situation.

2. x is really a binomial random variable, where $p = \frac{1}{2}$. The sign test statistic satisfies the properties of a binomial experiment (see p. 192). Each sign is the result of an independent trial. There are n trials, and each trial has two possible outcomes (+ or −). Since the median is used, the probabilities for each outcome are both $\frac{1}{2}$. Therefore the mean μ_x is equal to $n/2$ ($\mu = np = n \cdot \frac{1}{2} = n/2$), and the standard deviation σ_x is equal to $\frac{1}{2}\sqrt{n}$ ($\sigma = \sqrt{npq} = \sqrt{n(\frac{1}{2})(\frac{1}{2})} = \frac{1}{2}\sqrt{n}$).

3. x is a discrete variable. But recall that the normal distribution must be used only with continuous variables. However, although the binomial random variable is discrete, it does become approximately normally distributed for large n. Nevertheless, when using the normal distribution for testing, we should make an adjustment in the variable so that the approximation is more accurate. (See section 6-5 on the normal approximation.) This adjustment is illustrated in figure 19-1 and is referred to as a **continuity correction**. For this discrete variable the area that represents the probability is a rectangular bar. Its width is 1 unit wide, from $\frac{1}{2}$ unit below to $\frac{1}{2}$ unit above the value of interest.

FIGURE 19-1
Continuity Correction

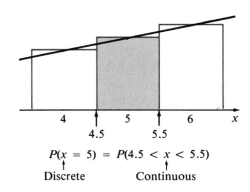

Therefore when z is to be used, we will need to make a $\frac{1}{2}$-unit adjustment before calculating the observed value of z. x' will be the adjusted value for x. If x is larger than $n/2$, $x' = x - \frac{1}{2}$. If x is smaller than $n/2$, $x' = x + \frac{1}{2}$. The test is then completed by the usual procedure.

Illustration 19-3

Use the sign test to test the hypothesis that the median number of days, M, that an assembly line employee was absent is at least 15 days. A survey of the time records of 120 randomly selected employees was taken, and a plus sign was recorded if the employee was absent 15 days or more and a minus sign if the employee was absent less than 15 days. Totals showed 80 minus signs and 40 plus signs.

Solution

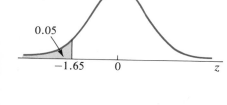

STEP 1: H_0: $M = 15$ ($\geq$) (at least as many plus signs as minus signs).

STEP 2: H_a: $M < 15$ (fewer plus signs than minus signs).

STEP 3: $\alpha = 0.05$; x is the number of plus signs. The critical value is shown in the figure.

STEP 4:

$$z = \frac{x' - n/2}{(\frac{1}{2})\sqrt{n}} = \frac{40.5 - 60}{(\frac{1}{2})\sqrt{120}} = \frac{-19.5}{(\frac{1}{2})(10.95)}$$

$$= \frac{-19.5}{5.475} = -3.562$$

$$z^* = -3.56$$

STEP 5: *Decision*: Reject H_0.

Conclusion: There are significantly more minus signs than plus signs, thereby implying that the median is less than the claimed 15 days. □

Confidence Interval Procedure

The sign test techniques can be applied to obtain a confidence interval estimate for the unknown population median M. To accomplish this, we will need to arrange the data in ascending order (smallest to largest). The data are identified as x_1 (smallest), $x_2, x_3, \ldots, x_n$ (largest). The critical value k for the maximum allowable number of signs, obtained from table 10, tells us the number of positions to be dropped from each end of the ordered data. The remaining extreme values become the bounds of the $1 - \alpha$ confidence interval. That is, the lower boundary for the confidence interval is x_{k+1}, the $(k + 1)$th piece of data; the upper boundary is x_{n-k}, the $(n - k)$th piece of data. The following illustration will clarify this procedure.

Illustration 19-4

We have 12 pieces of data in ascending order $(x_1, x_2, x_3, \ldots, x_{12})$ and we wish to form a 0.95 confidence interval for the population median. Table 10 shows a critical value of 2 ($k = 2$) for $n = 12$ and $\alpha = 0.05$ for a hypothesis test. This means that the critical region for a two-tailed hypothesis test would contain the last two values on each end (x_1 and x_2 on the left, x_{11} and x_{12} on the right). The noncritical region is then from x_3 to x_{10} inclusive. The 0.95 confidence interval is then x_3 to x_{10}, expressed as

$$x_3 \text{ to } x_{10}, \quad 0.95 \text{ confidence interval for } M$$

In general, the two data that bound the confidence interval will occupy positions $k + 1$ and $n - k$, where k is the value read from table 10. Thus

$$x_{k+1} \text{ to } x_{n-k}, \quad 1 - \alpha \text{ confidence interval for } M$$

If the normal approximation is to be used (including the continuity correction), the position numbers become

$$\tfrac{1}{2}n \pm [\tfrac{1}{2} + \tfrac{1}{2} \cdot z(\alpha/2) \cdot \sqrt{n}] \tag{19-2}$$

The interval is

$$x_L \text{ to } x_U, \quad 1 - \alpha \text{ confidence interval for } M$$

where

$$L = n/2 - \tfrac{1}{2} - \tfrac{1}{2} \cdot z(\alpha/2) \cdot \sqrt{n}$$

$$U = n/2 + \tfrac{1}{2} + \tfrac{1}{2} \cdot z(\alpha/2) \cdot \sqrt{n}$$

(L should be rounded down and U should be rounded up to be sure the level of confidence is at least $1 - \alpha$.) ☐

Illustration 19-5

Estimate the population median with a 0.95 confidence interval given a set of 60 pieces of data, $x_1, x_2, x_3, \ldots, x_{59}, x_{60}$.

Solution When we use formula (19-2), the position numbers L and U are

$$(\tfrac{1}{2})(60) \pm [\tfrac{1}{2} + \tfrac{1}{2}(1.96)\sqrt{60}]$$

$$30 \pm [0.50 + 7.59]$$

$$30 \pm 8.09$$

Thus

$$L = 30 - 8.09 = 21.91 \quad \text{(21st piece of data)}$$
$$U = 30 + 8.09 = 38.09 \quad \text{(39th piece of data)}$$

Therefore

$$x_{21} \text{ to } x_{39}, \quad 0.95 \text{ confidence interval } M \qquad \square$$

Exercises

19-1 Use the sign test to test the hypothesis that the median daily volume of Selen Stock in December was 48 ($\times 100$) shares. The following daily volumes ($\times 100$) were recorded on 14 randomly selected days in December:

45 49 50 55 47 48 52 65 40 39 30 46 52 45

(a) State the null hypothesis under test.
(b) State specifically what it is that you are actually testing when using the sign test.
(c) Complete the test for $\alpha = 0.05$ and carefully state your findings.

19-2 According to a time-motion study, the median time it should take a worker to make a certain weld on an assembly line is 24.8 seconds. The following times were recorded by a foreman for the production operation. Use the sample data drawn from the foreman's records and test the study's claim using the sign test. (Use $\alpha = 0.05$.)

24.7 24.7 24.6 25.5 25.7 25.8 26.5 24.5 25.3
26.2 25.5 26.3 24.2 25.3 24.3 24.2 24.2

(a) Perform the sign test. Use both techniques shown in the text: (1) the number of the least frequent sign and table 10 and (2) the number of signs and the standard normal z.
(b) Perform the hypothesis test using the parametric method, the t test.
(c) Compare the results of the three methods.

19-3 Use the sign test to test the null hypothesis that there is no difference in the amount of wear on the tires tested in illustration 10-15 on page 368. Use $\alpha = 0.05$.

19-4 Test the claim that the median difference before and after is 0 in the zero-defective study in exercise 10-39 on page 372. Use $\alpha = 0.05$.

19-5 Determine the 0.90 confidence interval for the median volume of Selen Stock in December based on the sample given in exercise 19-1.

19-6 Find the 0.95 confidence interval estimate for the median time for completion of the weld in exercise 19-2.

19-7 Find a 0.75 confidence interval for the median change in median number of defects in the zero-defective data in exercise 10-39 on page 372.

Section 19-3 The Mann-Whitney U Test

Mann–Whitney U test

The **Mann–Whitney U test** is a nonparametric alternative to the *t* test for the **difference between two independent means**. It can be applied when we have two independent random samples (independent within each sample as well as between samples) in which the random variable is continuous. This test is often applied in situations in which the two samples are drawn from the same population, but different "treatments" are used on each set. We will demonstrate the procedure in the following illustration.

Illustration 19-6

Pern Data Services hires temporary keypunch operators from two temporary help service companies. The firms' prices are the same, and it was always assumed by the personnel department that the quality of workers sent by each firm was the same. However, a newly hired personnel manager wishes to test this hypothesis: "The two temporary help agencies send employees whose keypunch performances, measured by the number of correct column entries per minute, are the same."

To test this hypothesis, the following two random samples were taken:

A	52	78	56	90	65	86	64	90	49	78
B	72	62	91	88	90	74	98	80	81	71

The size of the individual samples will be called n_a and n_b; actually, it makes no difference which group is assigned a or b. In our illustration they both have the value 10.

The first thing that must be done with the entire sample (all $n_a + n_b$ pieces of data) is to order it into one sample, smallest to largest:

49 52 56 62 64 65 71 72 74 78
78 80 81 86 88 90 90 90 91 98

rank number

Each piece of data is then assigned a **rank number**. The smallest (49) is assigned rank 1; the next smallest (52) is assigned 2; and so on up to the largest, which is assigned rank $n_a + n_b$ (20). Ties are handled by assigning to each of the tied observations the mean rank of those rank positions that they occupy. For example, in our illustration there are two 78s; they are the 10th and 11th pieces of data. The mean rank for each is then

$(10 + 11)/2 = 10.5$. In the case of the three 90s, the 16th, 17th, and 18th pieces of data, each is assigned 17, since $(16 + 17 + 18)/3 = 17$. The rankings are shown in table 19-2.

TABLE 19-2
Ranked Data and Ranks for Pern Data Services

Ranked Data	Rank	Source	Ranked Data	Rank	Source
49	1	A	78	10.5	A
52	2	A	80	12	B
56	3	A	81	13	B
62	4	B	86	14	A
64	5	A	88	15	B
65	6	A	90	17	A
71	7	B	90	17	A
72	8	B	90	17	B
74	9	B	91	19	B
78	10.5	A	98	20	B

The calculation of the test statistic U is a two-step procedure. We first determine the **sum of the ranks** for each of the two samples. Then using the two sums of ranks, we calculate a U score for each sample. **The smaller U score is the test statistic**.

The sum of ranks R_a for sample A is computed as

$$R_a = 1 + 2 + 3 + 5 + 6 + 10.5 + 10.5 + 14 + 17 + 17 = 86$$

The sum of ranks R_b for sample B is

$$R_b = 4 + 7 + 8 + 9 + 12 + 13 + 15 + 17 + 19 + 20 = 124$$

The U score for each sample is obtained by use of the following pair of formulas:

$$U_a = n_a \cdot n_b + \frac{(n_b)(n_b + 1)}{2} - R_b \qquad (19\text{-}3)$$

$$U_b = n_a \cdot n_b + \frac{(n_a)(n_a + 1)}{2} - R_a \qquad (19\text{-}4)$$

and U, the test statistic, will be the smaller of U_a and U_b.

For our illustration we obtain

$$U_a = (10)(10) + \frac{(10)(10+1)}{2} - 124 = 31$$

$$U_b = (10)(10) + \frac{(10)(10+1)}{2} - 86 = 69$$

Therefore

$$U = 31$$

Before we carry out the test for the illustration, let's try to understand some of the underlying possibilities. Recall that the null hypothesis is that the distributions are the same and that we will most likely want to conclude from this hypothesis that the means are approximately equal. Suppose for a moment that they are indeed quite different—say all of one sample comes before the smallest piece of data in the second sample when they are ranked together. This arrangement would certainly mean that we want to reject the null hypothesis. What kind of a value can we expect for U in this case? Suppose in illustration 19-6 that the 10 A values had ranks 1 through 10 and the 10 B values ranks 11 through 20. Then we would obtain

$$R_a = 55 \qquad R_b = 155$$

$$U_a = (10)(10) + \frac{(10)(10+1)}{2} - 155 = 0$$

$$U_b = (10)(10) + \frac{(10)(10+1)}{2} - 55 = 100$$

Therefore

$$U = 0$$

Suppose, on the other hand, that both samples were perfectly matched, that is, a score in each set identical to one in the other. Now what would happen?

54	54	62	62	71	71	72	72	...
A	B	A	B	A	B	A	B	...
1.5	1.5	3.5	3.5	5.5	5.5	7.5	7.5	...

$$R_a = R_b = 105$$

$$U_a = U_b = (10)(10) + \frac{(10)(10+1)}{2} - 105 = 50$$

Therefore

$$U = 50$$

If this case were to exist, we certainly would want to reach the decision "fail to reject the null hypothesis."

It should be noted that the sum of the two U's ($U_a + U_b$) will always be equal to the product of the two sample sizes ($n_a \cdot n_b$). For this reason we only need to concern ourselves with one of them, the smaller one.

Now let's return to the solution of illustration 19-6. To complete our hypothesis test, we need to be able to determine a critical value for U. Table 11 in appendix D gives us the critical value for some of the more common testing situations, as long as both sample sizes are reasonably small. Table 11 shows only the critical region in the left-hand tail, and the null hypothesis will be rejected if the observed value for U is less than or equal to the value read from the table. For our example $n_a = n_b = 10$; at $\alpha = 0.05$ in a two-tailed test, the critical value is 23. We observed a value of 31 and therefore we make the decision "fail to reject the null hypothesis." This decision means that we do not have sufficient evidence to reject the "equivalent" hypothesis. □

If the **samples are larger than size 20**, we may make the test decision with the aid of the standard normal variable z. This is possible, since the distribution of U is approximately normal with a mean

$$\mu_U = \frac{n_a \cdot n_b}{2} \tag{19-5}$$

and a standard deviation

$$\sigma_U = \sqrt{\frac{n_a n_b (n_a + n_b + 1)}{12}} \tag{19-6}$$

The null hypothesis is then tested by use of

$$z = \frac{U - \mu_U}{\sigma_U} \tag{19-7}$$

in the typical fashion. The test statistic z may be used whenever n_a and n_b are both greater than 20.

Illustration 19-7

First Oklahoma Bank has, in the past, required a bachelor's degree in business as a minimum requirement for entry into its management training program. The Equal Opportunity Commission has challenged the practice as discriminatory against females, claiming that female college graduates disproportionately have majors in nonbusiness fields and that a business degree is not necessary for the training program. The bank says that a business degree is job related to the training program, since a person with a business degree will be able to learn the banking procedures faster than a person with a major other than business. To prove its case, the bank randomly selects 27 job applicants, 15 with business degrees (B) and 12 with other degrees (NB). The training material is then divided into hour sessions, and at the end of each session the trainees are tested. The process continues until every session is passed. The accompanying table shows the number of sessions necessary for each employee to learn the required material.

B	29	27	32	25	27	28	23	31	37	28	22	24	28	31	34
NB	40	44	33	26	31	29	34	31	38	33	42	35			

Do the data show sufficient evidence to reject the claim that the average amount of training time required is the same regardless of degree type?

Solution

STEP 1: H_0: The average amount of training time required is the same regardless of degree type.

STEP 2: H_a: The business majors require less time, on the average.

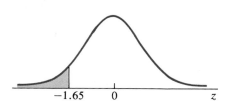

STEP 3: $\alpha = 0.05$; the critical value is shown in the figure.

STEP 4: The two sets of data are ranked jointly, and ranks are assigned, as shown in table 19-3. Then the sums are

$$R_B = 1 + 2 + 3 + 4 + 6.5 + \cdots + 20.5 + 23 = 151.0$$
$$R_{NB} = 5 + 11.5 + \cdots + 26 + 27 = 227.0$$

Now by using formulas (19-3) and (19-4), we can obtain the value of the U scores:

$$U_B = (15)(12) + \frac{(12)(12+1)}{2} - 227 = 180 + 78 - 227 = 31$$

$$U_{NB} = (15)(12) + \frac{(15)(15+1)}{2} - 151 = 180 + 120 - 151 = 149$$

Therefore

$$U = 31$$

TABLE 19-3
Rankings for Training Program

Number of Sessions	Group	Rank		Number of Sessions	Group	Rank	
22	B	1		31	NB	15	14.5
23	B	2		31	NB	16	14.5
24	B	3		32	B	17	
25	B	4		33	NB	18	18.5
26	NB	5		33	NB	19	18.5
27	B	6	6.5	34	B	20	20.5
27	B	7	6.5	34	NB	21	20.5
28	B	8	9	35	NB	22	
28	B	9	9	37	B	23	
28	B	10	9	38	NB	24	
29	B	11	11.5	40	NB	25	
29	NB	12	11.5	42	NB	26	
31	B	13	14.5	44	NB	27	
31	B	14	14.5				

Now we use formulas (19-5), (19-6), and (19-7) to determine the z statistic.

$$\mu_U = \frac{12 \cdot 15}{2} = 90$$

$$\sigma_U = \sqrt{\frac{(12)(15)(12 + 15 + 1)}{12}} = \sqrt{\frac{(180)(28)}{12}} = \sqrt{420} = 20.49$$

$$z = \frac{31 - 90}{20.49} = \frac{-59}{20.49} = -2.879$$

$$z^* = -2.88$$

STEP 5: *Decision*: Reject H_0.

Conclusion: The data show sufficient evidence to conclude that business majors require less training time. □

Exercises

19-8 Use the Mann–Whitney U test to test the claim that returns on investment of load and no-load mutual funds are the same. The data are shown in the accompanying table. Use $\alpha = 0.05$.

Fund Type	Investment Return (%)								
Load	6.5	6.8	3.0	6.0	5.8	6.1	3.3	3.4	5.9
No-load	7.6	6.0	6.1	7.4	7.3	6.0	6.2	6.3	6.0

19-9 To test if there is a difference in the labor force participation rate of women in northern and southern cities, an economist collected the sample data shown in the following table. Use the Mann–Whitney U test and $\alpha = 0.05$ to test if there is a difference.

Area	Labor Force Participation Rates for Women
Southern	60.4 59.8 63.5 64.5 65.1 64.0 63.5 63.1 61.0 62.8
Northern	62.7 61.7 60.5 63.5 65.3 64.2 61.3 62.3 63.4 61.5 63.2 63.0

19-10 The data shown in the accompanying table represent a random sample of the ages of unemployed persons in a small community. Do these data present sufficient evidence to conclude that there is a difference in the average age of unemployed men and women? Use a two-tailed test with $\alpha = 0.05$.

Men	70 60 77 39 36 28 19 40 24 46
	23 23 63 31 36 55 24 76 31 22
Women	62 46 43 28 21 22 27 42 41 27
	21 46 33 29 44 29 56 70 29 25

(a) State the null hypothesis that is being tested.
(b) Complete the test.

19-11 The management of a plastics firm, wishing to comply with the occupational health and safety laws, set out to test whether the chemicals it uses in production are affecting the systolic blood pressure of its employees. The firm took a random sample of blood pressures of employees that were not in contact with the chemical on a daily basis (x) and a sample of those that were in contact (y). The data are shown in the accompanying table. Do these data show sufficient evidence to conclude that workers in contact with the chemical have a higher systolic blood pressure at the 0.02 level of significance?

x	95 100 100 105 106 108 110 110
	115 118 120 122 124 125 130 130
	130 132 136 138 140 148 150 156
y	108 110 110 114 116 118 120 122 124
	126 126 128 130 130 132 136 136 140
	142 142 146 148 150 154 160 164 176

Section 19-4 The Runs Test

runs test The **runs test** is most frequently used to test a question involving the **randomness of data** (or lack of randomness). A **run** is a sequence of data that possesses a common property. One run ends and another starts when the piece of data does not display the property in question. The random variable that will be used in the test is V, the number of runs that are observed.

Illustration 19-8

To illustrate the idea of runs, let's draw a sample of 10 single-digit numbers from the telephone book using the next-to-last digit from each of the selected telephone numbers.

$$\text{Sample: } 2 \quad 3 \quad 1 \quad 1 \quad 4 \quad 2 \quad 6 \quad 6 \quad 6 \quad 7$$

Let's consider the property of "odd" (o) or "even" (e). The sample, as it was drawn, becomes $e, o, o, o, e, e, e, e, e, o$. Four runs are displayed.

$$e \quad o \quad o \quad o \quad e \quad e \quad e \quad e \quad e \quad o$$

So $V = 4$.

In illustration 19-8, if the sample contained no randomness, there would be only two runs—all the evens, then all the odds, or the other way around. We would also not expect to see them alternate: odd, even, odd, even, ... The maximum number of possible runs would be $n_1 + n_2$ or less (provided n_1 and n_2 are not equal), where n_1 and n_2 are the number of data that possess each of the two properties being identified.

We will often want to interpret the maximum number of runs as a rejection of a null hypothesis of randomness, as we often want to test randomness of the data in reference to how they were obtained. For example, if the data alternated all the way down the line, we might suspect that they had been tampered with. There are many aspects to the concept of randomness. The occurrence of odd and even as discussed in illustration 19-8 is one. Another aspect of randomness that we might wish to check is the ordering of fluctuations of the data above (a) or below (b) the mean or median of the sample.

Illustration 19-9

Consider the sequence that results from determining whether each of the data points in the sample of illustration 19-8 is above or below the median value. Test the null hypothesis that this sequence is random. Use $\alpha = 0.05$.

Solution

STEP 1: H_0: The numbers in the sample form a random sequence with respect to the two properties "above" and "below" the median value.

STEP 2: H_a: The sequence is not random.

$$\text{Sample: } 2 \quad 3 \quad 1 \quad 1 \quad 4 \quad 2 \quad 6 \quad 6 \quad 6 \quad 7$$

First we must rank the data and find the median. The ranked data are 1, 1, 2, 2, 3, 4, 6, 6, 6, 7. Since there are 10 pieces of data, the median is at the $i = 5.5$ position. Thus $\tilde{x} = (3 + 4)/2 = 3.5$. By comparing each number in the original sample with the value of the median, we obtain the following sequence of a's and b's:

$$b \quad b \quad b \quad b \quad a \quad b \quad a \quad a \quad a \quad a$$

We observe $n_a = 5$, $n_b = 5$, and 4 runs. So $V = 4$.

If n_1 and n_2 are both less than or equal to 20 and a two-tailed test at $\alpha = 0.05$ is desired, then table 12 of appendix D will give us the two critical values for the test. For our illustration with $n_a = 5$ and $n_b = 5$, table 12 shows critical values of 2 and 10. These values mean that if 2 or fewer, or 10 or more, runs are observed, the null hypothesis will be rejected. If between 3 and 9 runs are observed, we will fail to reject the null hypothesis.

STEP 3: $\alpha = 0.05$, and a two-tailed test is used. The critical values for V are found in table 12 (see the diagram).

Reject H_0	Fail to reject H_0	Reject H_0
0 2	3 9	10

V, number of runs

STEP 4: Four runs were observed; $V^* = 4$.

STEP 5: *Decision*: Fail to reject H_0.

Conclusion: We are unable to reject the hypothesis of randomness. □

To complete the hypothesis test about randomness when n_1 or n_2 is larger than 20 or when α is other than 0.05, we will use z, the standard normal variable. V is approximately normally distributed with a mean of μ_V and a standard deviation of σ_V. The formulas are as follows:

$$\mu_V = \frac{2n_1 \cdot n_2}{n_1 + n_2} + 1 \tag{19-8}$$

$$\sigma_V = \sqrt{\frac{(2n_1 \cdot n_2)(2n_1 \cdot n_2 - n_1 - n_2)}{(n_1 + n_2)^2(n_1 + n_2 - 1)}} \tag{19-9}$$

$$z = \frac{V - \mu_V}{\sigma_V} \tag{19-10}$$

Illustration 19-10

Test the null hypothesis that the sequence that results from classifying the sample data in illustration 19-8 as "odd" or "even" is a random sequence. Use $\alpha = 0.10$.

Solution

STEP 1: H_0: The sequence of odd and even occurrences is a random sequence.

STEP 2: H_a: The sequence is not random.

STEP 3: $\alpha = 0.10$, and a two-tailed test is to be used. The test criteria are shown in the figure.

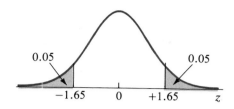

STEP 4: The sample and the sequence of odd and even properties are shown in illustration 19-8. $n_1 = n(\text{even}) = 6$, and $n_2 = n(\text{odd}) = 4$. There were four runs; so $V = 4$.

$$\mu_V = \frac{2n_1 \cdot n_2}{n_1 + n_2} + 1 = \frac{2(6)(4)}{6 + 4} + 1 = \frac{48}{10} + 1 = 5.8$$

$$\sigma_V = \sqrt{\frac{(2n_1 \cdot n_2)(2n_1 \cdot n_2 - n_1 - n_2)}{(n_1 + n_2)^2(n_1 + n_2 - 1)}}$$

$$= \sqrt{\frac{(2 \cdot 6 \cdot 4)(2 \cdot 6 \cdot 4 - 6 - 4)}{(6 + 4)^2(6 + 4 - 1)}}$$

$$= \sqrt{\frac{(48)(38)}{(10)^2(9)}} = \sqrt{\frac{1,824}{900}} = \sqrt{2.027} = 1.42$$

$$z = \frac{V - \mu_V}{\sigma_V} = \frac{4.0 - 5.8}{1.42} = \frac{-1.8}{1.42} = -1.268$$

$$z^* = -1.27$$

STEP 5: *Decision*: Fail to reject H_0.

Conclusion: We are unable to reject the hypothesis of randomness.

Exercises

19-12 Two automated teller machines are placed at opposite ends of a bus depot. The district supervisor of the bank gathered the following data to substantiate her claim that the machine chosen by a customer for a purchase was completely random. (A = machine at west side of depot; B = machine at east side of depot.) She observed 25 customers using the machines.

A A B B B B A A B A B B A A A B B A B A B A A B B

Use the runs test at a 5% level of significance to test the supervisor's claim that the results are random.

19-13 An advertising agency claims that when an advertisement is shown on television for phone sales, female (F) customers will respond at a different speed than male (M) customers. The following data are 28 responses to a randomly chosen ad, given in sequence of when the order was placed:

F F F F M F M M M M F F F F F M M F F F M M M M F M M F

Do these data show sufficient lack of randomness to support the agency's claim? Use $\alpha = 0.05$.

19-14 The manager of a garment factory senses that his workers are unhappy, as witnessed by the steadily increasing number of employees calling in sick. He collected the following data (number of people out per day) over the past 20 days in order of occurrence:

6 1 3 9 10 10 5 2 5 6 3 12 8 9 7 5 4 8 11 14

At $\alpha = 0.05$, do these data show sufficient lack of randomness to support the manager's suspicion?

19-15 Local union leaders were concerned about the lack of interest in union affairs, as shown by the lack of attendance at the weekly union meetings. They recorded the number of absences for each of the last 26 weeks. In order of occurrence they are:

5 16 6 9 18 11 16 21 14 17 12 14 10
6 8 12 13 4 5 5 6 1 7 18 26 6

Do these data show a lack of randomness about the median value at $\alpha = 0.05$? Conduct this test two ways: (1) by using critical values from table 12 and (2) by using the standard normal distribution.

Section 19-5 Rank Correlation

The rank correlation coefficient was developed by C. Spearman in the early 1900s. It is a nonparametric alternative to the linear correlation coefficient, which was discussed in chapters 3 and 13. It uses rankings only. If the data are quantitative, we will rank each of the two variables separately. The **Spearman rank correlation coefficient** r_s is found by using the formula

Spearman's rank correlation coefficient

$$r_s = 1 - \frac{6 \sum (d_i)^2}{n(n^2 - 1)} \qquad (19\text{-}11)$$

where d_i is the difference in the rankings and n is the number of pairs of data. The value of r_s will range from -1 to $+1$ and will be used in much the same manner as the linear correlation coefficient was used previously.

The null hypothesis that we will be testing is "there is no correlation between the two rankings." The alternative hypothesis may be either two tailed, "there is correlation," or one tailed, if we anticipate the existence of either positive or negative correlation. The critical region will be on the side(s) corresponding to the specific alternative that is expected. For example, if we suspect negative correlation, the critical region will be in the left-hand tail.

Illustration 19-11

Let's consider a hypothetical consumer panel in which four consumers rank five test products as to taste. The consumers are identified as A, B, C, and D and the products as a, b, c, d, and e. Table 19-4 lists the rankings that were awarded.

TABLE 19-4
Rankings of Five Test Products

	Panel			
Product	A	B	C	D
a	1	5	1	5
b	2	4	2	2
c	3	3	3	1
d	4	2	4	4
e	5	1	5	3

When we compare consumer judges A and B, we see that they ranked the products in exactly the opposite order—perfect disagreement (see table 19-5). From our previous work with correlation, we expect the calculated value for r_s to be exactly -1 for these data.

TABLE 19-5
Rankings of A and B

	A	B	$d_i = A - B$	$(d_i)^2$
a	1	5	-4	16
b	2	4	-2	4
c	3	3	0	0
d	4	2	2	4
e	5	1	4	16
				40

$$r_s = 1 - \frac{6[\sum (d_i)^2]}{n(n^2 - 1)} = 1 - \frac{(6)(40)}{(5)(5^2 - 1)} = 1 - \frac{240}{120} = -1$$

When judges A and C are compared, we see that their rankings of the products are identical (see table 19-6). We would expect to find a calculated correlation coefficient of $+1$ for these data.

TABLE 19-6
Rankings of A and C

	A	C	$d_i = A - C$	$(d_i)^2$
a	1	1	0	0
b	2	2	0	0
c	3	3	0	0
d	4	4	0	0
e	5	5	0	0
				0

$$r_s = 1 - \frac{(6)(0)}{(5)(5^2 - 1)} = 1 - 0 = 1$$

By comparing judge A with judge B and then with judge C, we have seen the extremes, total agreement and total disagreement. Now let's compare judge A with judge D (see table 19-7). There seems to be no real agreement or disagreement here. Let's compute r_s:

TABLE 19-7
Rankings of A and D

	A	D	$d_i = A - D$	$(d_i)^2$
a	1	5	-4	16
b	2	2	0	0
c	3	1	2	4
d	4	4	0	0
e	5	3	2	4
				24

$$r_s = 1 - \frac{(6)(24)}{(5)(5^2 - 1)} = 1 - \frac{144}{120} = 1 - 1.2 = -0.2$$

This value is fairly close to 0, which is what we should have suspected, since there was no real agreement or disagreement.

The test of significance will result in failing to reject the null hypothesis when r_s is close to 0 and will result in a rejection of the null hypothesis in cases where r_s is found to be close to $+1$ or -1, as appropriate. The critical values found in table 13 of appendix D are positive critical values only. Since the null hypothesis is "the population correlation coefficient is 0" (i.e., $\rho = 0$), we have a symmetric test statistic. Hence we need only add a plus or minus sign to the value found in the table, as appropriate. This sign will be determined by the specific alternative that we have in mind.

In our illustration the critical values for a two-tailed test at $\alpha = 0.10$ are ± 0.900 (remember that n represents the number of pairs). If the calculated value for r_s is between 0.9 and 1.0 or between -0.9 and -1.0, we will reject the null hypothesis in favor of the alternative that there is a correlation. If the calculated value for r_s is not between 0.9 and 1.0 or between -0.9 and -1.0, we will fail to reject the null hypothesis that there is no correlation. □

Illustration 19-12

Hay Corporation is interested in knowing if a proposed test to predict performance of managerial trainees will work. It takes 12 of its current first-level managers and ranks them on performance on the job and then gives them the test. The results are shown in the accompanying table. At the 0.05 level, do these data support the alternative hypothesis that the better managers had the higher grades?

Order of Performance	1	2	3	4	5	6	7	8	9	10	11	12
Exam Score	90	74	74	60	68	86	92	60	78	74	78	64

Solution

STEP 1: H_0: Order of performance has no relationship to exam score.

STEP 2: H_a: Better performance is related to the higher exam score.

STEP 3: $\alpha = 0.05$ and $n = 12$; the critical region is shown in the diagram.

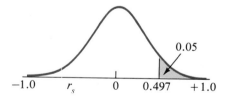

STEP 4: Rank the scores from highest to lowest, assigning the highest score the rank number 1, as shown.

92	90	86	78	78	74	74	74	68	64	60	60
1	2	3	4	5	6	7	8	9	10	11	12
			4.5	4.5	7	7	7			11.5	11.5

The rankings and the preliminary calculations are shown in table 19-8.

TABLE 19-8
Rankings for Test Scores and Preliminary Calculations

Test Score Rank	Order of Performance	Difference, d_i	$(d_i)^2$
1	7	−6	36.00
2	1	1	1.00
3	6	−3	9.00
4.5	9	−4.5	20.25
4.5	11	−6.5	42.25
7	2	5	25.00
7	3	4	16.00
7	10	−3	9.00
9	5	4	16.00
10	12	−2	4.00
11.5	4	7.5	56.25
11.5	8	3.5	12.25
			247.00

Using formula (19-11), we obtain

$$r_s = 1 - \frac{(6)(247.0)}{(12)(143)} = 1 - \frac{1,482}{1,716} = 1 - 0.864$$

$$= 0.136$$

STEP 5: *Decision*: Fail to reject H_0.

Conclusion: There is not sufficient evidence presented by these sample data to enable us to conclude that the better managers have the higher grades.

Exercises

19-16 The final test scores on two different business aptitude surveys given to 12 potential trainees are shown in the accompanying table. Is there sufficient evidence produced by this sample to reject the null hypothesis that the rank of the applicant on one survey is independent of the rank on the second survey? Use $\alpha = 0.05$.

	Testee											
	1	2	3	4	5	6	7	8	9	10	11	12
Survey 1	73	60	92	91	80	70	81	83	97	76	68	88
Survey 2	90	70	81	97	85	66	75	82	89	79	80	91

19-17 The sets of air pollution readings shown in the accompanying table were taken from 10 major industrial plants throughout the East Coast.

	Day									
	1	2	3	4	5	6	7	8	9	10
Sulfur dioxide	0.07	0.01	0.01	0.02	0.07	0.02	0.00	0.01	0.05	0.04
Carbon monoxide	0.9	2.0	0.8	0.8	0.9	1.3	3.8	2.0	1.3	2.5
Soiling or dust	0.3	0.5	0.3	0.3	0.3	0.5	0.3	0.5	0.3	0.5

(a) Draw a scatter diagram showing the sulfur dioxide and carbon monoxide readings as paired data. Use Spearman's rank correlation coefficient to test the correlation between the sulfur dioxide and the carbon monoxide readings. Complete a two-tailed test at $\alpha = 0.05$.
  (i) State the null hypothesis being tested.
  (ii) Complete the test and carefully state the conclusion reached.
(b) Repeat part (a) using the sulfur dioxide and soiling readings ($\alpha = 0.05$).
(c) Repeat part (a) using the carbon monoxide and soiling readings ($\alpha = 0.01$).

Section 19-6 Comparing Statistical Tests

Only a few nonparametric tests have been presented in this chapter. Many more nonparametric tests can be found in other books, and many of them seem to do the same job. In addition, most of them can be used in place of certain parametric tests. The question is, then, which test do we use, the parametric or nonparametric? Furthermore, when there is more than one nonparametric test, which one do we use?

The decision about which test to use must be based on the answer to the question, which test will do the job best? First let's agree that we are dealing with two or more tests that are equally qualified to be used. That is, each test has a set of assumptions that must be satisfied before they can be applied. From this starting point we will attempt to define "best" to mean the test that is best able to control the risks of error, keep the size of the sample to a number that is reasonable to work with, and be compatible with the data. (Sample size means cost, cost to you or your employer.)

Let's look first at the ability to **control the risk of error**. The risk that is associated with a type I error is controlled directly by the level of significance α. Recall that P(type I error) = α and P(type II error) = β. Therefore it is β that we must control. Statisticians talk about the **power of a statistical test**, which is defined to be $1 - \beta$. Thus the power of a test, $1 - \beta$, is the probability that we reject the null hypothesis when we should have rejected it. If two tests with the same α are equal candidates for use, the one with the greater power is the one you would want to choose.

power

efficiency   The next factor is the **sample size required** to do a job. Suppose that you set the levels of risk that you can tolerate, α and β, and then are able to determine the sample size that it would take to meet your specified challenge. The test that required the smaller sample size would then seem to have the edge. Statisticians usually use the term *efficiency* to talk about this concept. **Efficiency** is defined to be the ratio of the sample size of the best parametric test to the sample size of the best nonparametric test, when compared under a fixed set of risk values. For example, the efficiency rating for the sign test is approximately 0.63. This value means that a sample of size 63 with a parametric test will do the same job as a sample of size 100 with the sign test.

The power and the efficiency of a test cannot be used alone to determine the choice of test. Sometimes you will be forced to use a certain test because of the **data** you are given. When a decision is to be made, the final decision rests in a trade-off of three factors: (1) the power of the test, (2) the efficiency of the test, and (3) the data (and the number of data) available. Table 19-9 shows how the nonparametric tests discussed in this chapter compare with the parametric tests covered in previous chapters.

TABLE 19-9
Comparison of Parametric and Nonparametric Tests

Test Situation	Parametric Test	Nonparametric Test	Efficiency of Nonparametric Test
One mean	t test (p. 308)	Sign test (p. 623)	0.63
Two independent means	t test (p. 359)	U test (p. 630)	0.95
Two dependent means	t test (p. 367)	Sign test (p. 624)	0.63
Correlation	Pearson's (p. 461)	Spearman test (p. 640)	0.91
Randomness		Runs test (p. 636)	Not meaningful, there is no parametric test for comparison

In Retrospect

You have become acquainted in this chapter with some of the basic concepts of nonparametric statistics. While learning about the use of nonparametric methods and specific nonparametric tests of significance, you should have also come to realize and understand some of the basic assumptions that are needed when the parametric techniques of the earlier chapters are encountered. You now have seen a variety of tests, many of which somewhat duplicate the job done by others. What you must keep in mind is that you should use the best test for your particular needs. The power of the test and the cost of sampling, as related to size and *available data, will play important roles in determining the specific test to be used.*

The news article at the beginning of this chapter shows two sets of rankings. The ranks shown reflect the composite viewpoints of workers and bosses with regard to 10 work satisfaction factors. Does it appear that the two groups disagree on the relative importance of these factors?

Chapter Exercises

19-18 A local bottling company recently conducted a study to consider the efficiency of two different assembly line techniques. The new technique consisted of a new bottling machine and 12 workers. The old technique consisted of the old bottling machine and 14 workers. The number of cases of beer produced was recorded for a 10-day period; see the accompanying table.

Technique	Day									
	1	2	3	4	5	6	7	8	9	10
New	277	268	270	255	266	274	272	274	258	251
Old	270	267	253	260	261	253	267	271	248	271

Test the claim that the new technique is conducive to producing more cases of beer.

(a) State the null hypothesis and alternative hypothesis being tested. Complete the test with the sign test using $\alpha = 0.05$.

(b) State your conclusions.

19-19 The vice-president in charge of mergers and acquisitions asked her financial analyst to evaluate and rate the financial position of 18 corporations being considered for acquisition. She also asked her marketing analyst to evaluate and rate the same firms on their potential future sales strength. The data in the accompanying table represent their ratings, based on a scale of 0 to 100 (100 being best). Do the data present sufficient evidence to justify the statement, "the ranking of corporations by the financial and marketing analysts are correlated"? Use $\alpha = 0.05$.

Analyst	Corporation																	
	1	2	3	4	5	6	7	8	9	10	11	12	13	14	15	16	17	18
Financial	50	53	68	32	50	62	55	61	85	83	89	62	76	61	44	50	55	90
Marketing	40	52	69	45	60	58	65	30	90	70	75	70	63	68	50	55	48	79

19-20 The owner and manager of a delivery service company was concerned about the amount of overtime he had been paying his drivers per week. He would like to keep overtime under 10 hours a week, but he contends that most drivers work more than 15 hours a week overtime. He randomly chose 18 weekly pay checks from the previous year and found the following overtime hours (per week):

$$4 \quad 14 \quad 13 \quad 10 \quad 3 \quad 23 \quad 7 \quad 16 \quad 5$$
$$7 \quad 17 \quad 11 \quad 6 \quad 7 \quad 9 \quad 7 \quad 8 \quad 9$$

Use the sign test and check the owner's claim.

(a) State the specific null hypothesis that you wish to test.

(b) State the null hypothesis that is going to be tested in place of the null hypothesis stated in part (a).

(c) Complete the test and state a conclusion. Use $\alpha = 0.05$.

19-21 Two table tennis ball manufacturers have agreed that the quality of their products can be measured by the height to which the balls rebound. A test is arranged in which the balls are dropped from a constant height and the rebound heights measured. The results (in inches) are shown in the accompanying table. Manufacturer A claims, "The results show my product to be superior." Manufacturer B replies, "I know of no statistical test that supports this claim."

A	14.0	12.5	11.5	12.2	12.3	11.8	11.9	13.7	13.2
B	12.0	12.5	11.6	13.3	13.0	12.1	12.8	12.2	12.6

(a) What parametric test (or tests) is appropriate and what exactly does it show?

(b) Does the appropriate parametric test show that A's product is superior?

(c) Does the appropriate nonparametric test show that A's product is superior?

19-22 The average age in each of 32 advertising markets examined is as follows:

$$\begin{array}{cccccccc}
41 & 42 & 48 & 46 & 50 & 54 & 51 & 42 \\
51 & 50 & 45 & 42 & 32 & 45 & 43 & 56 \\
55 & 47 & 45 & 51 & 60 & 44 & 57 & 57 \\
47 & 28 & 41 & 42 & 54 & 48 & 47 & 32
\end{array}$$

(a) Does this sample show that the median age of the population from which this sample of 32 markets is drawn is different from 50? Use $\alpha = 0.05$.

(b) Does this sample show that the median age for the markets is less than 50? Use $\alpha = 0.05$.

19-23 From the bond underwriting data of exercise 2-80 (see p. 93), use the first eight bonds as your sample. Does this sample show sufficient evidence to reject the null hypothesis that gross proceeds and spread are independent?

(a) Use the Spearman rank correlation coefficient and test at $\alpha = 0.05$.

(b) Use the Pearson correlation coefficient (see chapter 13, p. 459) and test at $\alpha = 0.05$.

19-24 (a) Do the data collected for group A in exercise 10-67 (p. 385) show sufficient evidence to conclude that the special training made significant improvement in their ability to remember? Use a nonparametric test and $\alpha = 0.05$ to obtain your answer.

(b) Did group B show no gain during the experiment, as one would expect? (Recall that they were not given the special training.)

19-25 The accompanying data represent the number of Mercedes sold the first 15 working days of January. Last January sales averaged 110 cars per working day. Do these data support the claim that Mercedes are averaging more sales per day this year than last at $\alpha = 0.05$?

Day	1	2	3	4	5	6	7	8	9	10	11	12	13	14	15
Sales	109	108	110	112	108	111	112	113	110	112	115	114	116	116	118

19-26 For each of the sales-forecasting models in exercise 16-10 (p. 551), test the hypothesis that the pattern of error appears random. Use $\alpha = 0.05$.

19-27 While trying to decide on the best time to harvest his crop, a commercial apple farmer recorded the day on which the first apple on the top half and the first apple on the bottom half of 20 randomly selected trees were ripe. The variable x was assigned a value of 1 on the first day that the first ripe apple appeared on one of the 20 trees. The days were then numbered sequentially. The observed data are shown in the accompanying table. Do these data provide convincing evidence that the apples on the top of these trees start to ripen before the apples on the bottom half? Use $\alpha = 0.05$.

	Tree																			
	1	2	3	4	5	6	7	8	9	10	11	12	13	14	15	16	17	18	19	20
Top	5	6	1	4	5	3	6	7	8	5	8	6	4	7	8	10	3	2	9	7
Bottom	6	5	5	7	3	6	6	8	9	4	10	7	5	11	6	11	5	6	9	8

Challenging Problem

19-28 A small firm is considering two word-processing computer software packages, Wordwise and Wordsure. To test the packages, they randomly assign 14 secretaries to Wordwise and 14 to Wordsure. Each secretary was asked to type a copy of a letter, then make certain editorial changes. The length of time to complete the task was recorded, and the results are given in the following table:

Software Package	Minutes to Complete Job
Wordsure	29 31 31 35 20 28 30 32 28 28 33 37 30 31
Wordwise	34 43 29 31 33 34 33 36 31 32 27 30 34 31

(a) If you do not assume the distribution of scores to be normally distributed, is there enough evidence to conclude that the software programs do not perform equally well? Use $\alpha = 0.05$.

(b) If you do assume the scores to be normally distributed, is there enough evidence to conclude that the software programs do not perform equally well? Use $\alpha = 0.05$.

(c) Compare your results.

Hands-On Problems

Suppose that you had a large sum of money and were going to invest it in stocks. Obviously, many things are to be considered if you are to make such an investment. This problem set is only going to investigate one particular concern: "Would it be best to invest a fixed amount in low-cost stock or to invest that same amount in higher-priced stocks?" Let's say that you had $10,000. You could buy 50 shares at $200 each or you could buy 500 at $20 each. Which would be the better investment? Let's define low-cost stocks to be those stocks whose current selling price is less than $50 per share and higher-priced stocks to be those that are currently selling for more than $50 per share. From a newspaper that prints the daily stock market report, obtain a random sample of 50 daily net changes for each type of stock. Use two separate samples, each of size 50. Record the closing price and the net change for each stock selected in your sample. Use the data from these two samples to answer each of the following questions by performing an appropriate nonparametric test. Use $\alpha = 0.05$.

19-1 One concern that might be investigated is that of total earnings of the $10,000. Using the evidence found in the sample, can you expect the higher-priced stocks to show a larger gain?

19-2 Is there a relationship between the current selling price (closing price) and the absolute net change (the net change ignoring the sign, whether it was a gain or loss)? Plot the collected data as paired data (closing price, absolute net change) and test the claim that the stocks with the highest selling prices tend to have the greatest absolute net change.

20 Decision Theory

Chapter Outline

20-1 Decision Making Under Certainty and Uncertainty

*When all possible **outcomes** are **known and measurable**, we are making decisions under **certainty**; when the **outcomes** depend on the **unknown** state of nature, we are making decisions under **uncertainty**.*

20-2 Payoff and Opportunity Loss Tables

*Measure the **results of each action** under each possible state of nature.*

20-3 Decision Strategies

***Maximin, maximax,** and **minimax regret** are methods for selecting among alternative actions.*

20-4 Maximizing Expected Value of Payoff or Minimizing Expected Opportunity Loss

*Give us the **best decision on the average** over a period of time.*

20-5 Expected Value of Perfect Information

*We learn to measure the **cost of uncertainty** in decision making.*

20-6 Misuses of Maximizing Expected Value of Payoff

*Maximizing expected value of payoff is a long-run decision that **ignores the risks** associated with a single decision.*

The Valuation of Human Resources

It should be recognized that HRA [human resource accounting] is a generic term, including diverse techniques for evaluating human resources that yield quite different kinds of information. HRA theory is based primarily on the cost approach and the value approach....

One approach to HRA is the economic value model, most often represented as the present value of the stream of net future contributions of the individual to the organization. ...the present value of an employee is calculated as the present value of expected future salaries....

In Table 1 we go through all the steps in calculating the value of two hypothetical employees, one with a current salary of $9,500 who has been newly hired (Employee A) and one with a current salary of $8,500 who has two years of service (Employee B). To simplify the example, we will consider only the contributions to value from the next six years; ordinarily, each year until retirement would have a bearing on the calculation of present value.

Column 1 forecasts the employees' future salaries.... In this example we have ... assumed $500 annual increases.

... we add the probability of separation in Column 2. The probability that is of interest, though, is not the probability of leaving in a given year but the probability of staying through the end of that year. Thus, we derive Column 3, the cumulative probability of separation, from the subtotals of Column 2. The cumulative probability of staying (Column 4)....

The productivity factor (Column 6).... Multiplying this by the interim calculations in Column 5 yields the expected future contributions of each employee (Column 7).

The final step is to calculate the present value of each year's contribution and find the total present value. Any reasonable discount rate could be used, the cost of capital being one alternative. In this example, we have arbitrarily used a discount rate of 10 percent. Using this rate, the present value is calculated in Column 8, and values of 13,712 for employee A and 29,416 for employee B are obtained....

HRA may be particularly useful in large companies as a surrogate for the judgment of managers regarding the value of groups of employees....

We should point out that the Bank of America has not formally adopted HRA; the work being done is still exploratory and experimental. Although the model we have described is being used to evaluate turnover in several units of the bank....

Moreover, HRA emphatically points out that people are important assets of a firm, not just so many interchangeable parts. If all managers realize that employees generate future benefits and, therefore, realize the extent of their value to the organization, HRA will have served a useful purpose.

TABLE 1 *Calculation of Present Value*

Years of Service	(1) Salary Expected	(2) Annual Probability of Separation	(3) Cumulative Probability of Separation	(4) Cumulative Probability of Staying 1−(3)	(5) Expected Contribution Measured by Salary & Retention (4) × (1)	(6) Productivity Factor	(7) Contribution (5) × (6)	(8) Present Value (7)
				Employee A				
1	10,000	0.20	0.20	0.80	8,000	0.50	4,000	3,636
2	10,500	0.30	0.50	0.50	5,250	0.80	4,200	3,471
3	11,000	0.20	0.70	0.30	3,300	0.85	2,805	2,107
4	11,500	0.05	0.75	0.25	2,875	0.90	2,588	1,768
5	12,000	0.05	0.80	0.20	2,400	0.95	2,280	1,416
6	12,500	0.01	0.81	0.19	2,375	0.98	2,328	1,314
								13,712
				Employee B				
3	9,000	0.20	0.20	0.80	7,200	0.85	6,120	5,564
4	9,500	0.05	0.25	0.75	7,125	0.90	6,413	5,300
5	10,000	0.05	0.30	0.70	7,000	0.95	6,650	4,996
6	10,500	0.01	0.31	0.69	7,245	0.98	7,100	4,849
7	11,000	0.02	0.33	0.67	7,370	0.99	7,296	4,530
8	11,500	0.02	0.35	0.65	7,475	0.99	7,400	4,177
								29,416

From Richard B. Frantzreb, Linda L. T. Landau, and Donald P. Lundberg, "The Valuation of Human Resources," *Business Horizons* (Indiana University Graduate School of Business), vol. 27, no. 3 (June 1974): 73–80. Reprinted by permission.

Chapter Objectives

In this chapter we will explore the process of decision making. We will learn to distinguish if a decision problem is one of decision making under uncertainty or certainty.

Our discussion will deal primarily with strategies for making decisions under uncertainty. The following ingredients are necessary for evaluating a decision under uncertainty: to be able to list all the possible decisions we can make (actions); to be able to list all the possible conditions (states of nature) that will affect the outcome of each action; and to be able to measure the possible outcomes of each action under each state of nature. This preliminary analysis is presented in a payoff table or an opportunity loss table.

We will then discuss strategies for choosing the action to take. Three strategies, maximin, maximax, and minimax regret, do not consider the likelihoods of each state of nature occurring, but simply look at the best or worst results. Two other strategies, maximizing expected value of payoff or, equivalently, minimizing expected opportunity loss, offer us a formula for combining the probabilities of occurrence with the payoffs, to find the best action, on the average, over a period of time. We will see, however, that a long-run view is not always appropriate, since it ignores the risk involved in a single decision.

Section 20-1 Decision Making Under Certainty and Uncertainty

Do we market the new product or not? How much should we produce this week? Should we hold the stock or sell the stock? There are 10 clients that must be visited today; in what order should they be visited? In business the number and type of decisions a manager faces each day is almost endless.

The one characteristic common to all decision making is that there are a number of alternative things that can be done and the decision maker must choose between them. The alternatives a decision maker must decide among are called **actions**. If there is only one action, then obviously no decision need be made. However, there is no upper limit to the number of actions that are possible. Consider the decision about the order in which to call on 10 clients. There are 3,628,800 different possible ways to order the 10 visits and hence that many possible actions to choose among.

All decision problems can be categorized as either decisions under certainty or decisions under uncertainty.

actions

---

decision under certainty

If it is possible to calculate the outcome of each possible action before choosing a particular action, we are said to be making a **decision under certainty**.

In such a situation, making the best decision is conceptually simple. We simply calculate the outcome of each action, compare all of them, and select the one that gives the best result. However, in situations where there are a great many possible actions to consider, the task of predicting the outcome for every possible action can be extremely difficult, expensive, and time consuming. Consider the problem of scheduling the visits to 10 clients. It would be possible to calculate the time of travel between every pair of clients. We could therefore calculate the total time for any particular ordering of the 10 clients. However, this task is not easy, since there are 3,628,800 different orders for which we would have to calculate the travel time.

In decision problems under certainty, the major problem is to develop a method for quickly finding the best action without having to consider every single possible action. A number of mathematical approaches have been developed, the most common approach being *mathematical programming*, of which *linear programming* is the most important example. Although a knowledge of these techniques is very valuable for a business student, mathematical programming does not really fall under the heading of statistics and hence is not covered in this text.

Now let us consider the problem of whether or not to market a new product. The decision would be easy to make if we knew how well the product would sell. If it would sell enough to make a profit, we should market it; and if not, we shouldn't market it.

decision under uncertainty

In situations where the outcome of each action will differ depending on the conditions that prevail, and we cannot predict with certainty which condition *will* prevail, we are said to be making a **decision under uncertainty**.

We must decide whether or not to buy a stock without exact knowledge of its future price. A company must decide whether to expand without exact knowledge of the future of the economy. A publisher must decide whether to publish a book without exact knowledge of its future sales potential. A firm must decide whether to invest in research and development, uncertain of whether it will succeed or not. These are just a few of the almost unlimited examples of decision making under uncertainty that face a businessperson.

When a decision is made and an action is taken, how can we judge whether it was a "good" or a "bad" decision? In time we can always see how things turned out. But is this really the best way to judge? You have probably heard the common statement, "Hindsight makes us all great decision makers." Consider the situation where someone offers you the following bet: he will roll a fair die and if a 1 appears, he will give you $10; if any other number appears, you pay him $10. Having read the chapter on probability, you will turn down his offer. He rolls the die anyway and a 1 appears. Did you make a bad decision? In one sense yes and in another no.

The decision you made did not turn out, in this case, to give the best result, but the decision, viewed before the roll, was clearly the "best bet."

Statistical decision theory under uncertainty concerns itself with the "goodness" of the procedure for making decisions rather than with the "goodness" of any particular decision after the fact. The idea, simply, is this: if you use the "best" procedure, although it may not lead to the best result every time, in the long run you will make out better. For example, if you always take an umbrella when the weather forecaster says the probability of rain is 95%, undoubtedly there will be some days when you will carry the umbrella unnecessarily. But clearly, in the long run, carrying the umbrella will pay off (fewer colds, dryer clothes, etc.).

Exercises

20-1 State whether each of the following are examples of decision making under certainty or uncertainty. Explain.

(a) whether to invest in a 90-day treasury note or a 90-day bank savings certificate

(b) the order in which to make deliveries to seven different stores

(c) whether to hire Bill Wilson or Tom Blunt

(d) whether or not to purchase fire insurance

20-2 Give an example of a decision you made recently that would be called decision making under uncertainty.

20-3 Give an example of a decision you made recently that would be called decision making under certainty.

Section 20-2 Payoff and Opportunity Loss Tables

The first step in analyzing a decision under uncertainty is to list the choices, or actions, we have and to list the possible conditions that may exist that would affect the outcome of each action. The possible conditions that may exist are called the **states of nature**. We then calculate or estimate the worth of the outcomes, or **payoffs**, that would result from taking *each* action under *each* possible state of nature. The results are displayed in tabular form, as illustrated in table 20-1, and the table is called a **payoff table**.

states of nature

payoffs

payoff table

In setting up a payoff table, we usually use capital letters for the states of nature and lowercase letters for the actions. The numbers in the cells of the table represent the payoffs from taking each action under each possible state of nature. For example, in table 20-1 the 3 means that if we take action *a* and state of nature *A* occurs, we will gain 3. Similarly, if we take action *a* and state of nature *B* occurs, we will gain 2. If we take action *c* and state of nature *C* occurs, we lose 1. (*Note*: If the payoffs are costs rather

TABLE 20-1
A Payoff Table

	State of Nature		
Action	A	B	C
a	3	2	7
b	2	0	5
c	7	8	−1

than gains, simply use a negative sign in recording the numbers. But remember that when we deal with negative numbers, the larger the absolute value, the smaller the number—that is, −100 is less than −50.)

Illustration 20-1

In the past a bakery has found that on a Saturday it has sold between 8 and 11 birthday cakes. It costs $3 to make a cake that the bakery sells for $6. All cakes for sale on Saturday are baked Friday night, and all cakes not sold by closing on Saturday are given free to the local children's hospital. The bakery must decide how many cakes to bake on Friday night.

Let's construct the payoff table for this problem. There are four possible states of nature: sales demand is 8, 9, 10, and 11. There are four actions the bakery should consider: bake 8, 9, 10, or 11 cakes. The basic payoff table is shown in table 20-2.

TABLE 20-2
Payoff Table for the Bakery

Cakes Baked (action)	Demand for Cake (state of nature)			
	8(A)	9(B)	10(C)	11(D)
8(a)				
9(b)				
10(c)				
11(d)				

Now let's calculate the payoffs. If 8 cakes are made and 8 cakes demanded, then the bakery will make $8 \times (6 − 3) = \$24$. If 9, 10, or 11 cakes are demanded, the bakery will still make only $24, since it baked only 8 cakes. Thus all the entries in the first row of payoffs are $24. If 9 cakes are made and 8 cakes are demanded, then the bakery will make $8 \times (6 − 3) = \$24$ from the sale of the 8 cakes, but lose $3, since it has an extra cake it must give away. The payoff is therefore $21. If it bakes 9 cakes and the demand is 9, 10, or 11, it will sell 9 and make $27 ($9 \times 3$). Thus the second row of the payoff table will read 21, 27, 27, and 27. The third and fourth rows of the table are completed in a similar manner. The completed payoff table is shown in table 20-3.

TABLE 20-3
Payoff Table for the Bakery

Cakes Baked (action)	Demand for Cakes (state of nature)			
	8(A)	9(B)	10(C)	11(D)
8(a)	24	24	24	24
9(b)	21	27	27	27
10(c)	18	24	30	30
11(d)	15	21	27	33

We will discuss how the results are used in decision making in section 20-3.

Illustration 20-2

Retra Corporation has been sued for constraint of trade by Electron Corporation. The general counsel informs the president that Electron is willing to settle the case (i.e., drop the charges) if Retra Corporation will pay them $500,000. The counsel estimates the cost of defending the case will be $100,000 and if they lose the case, the damages will be $1,000,000. Construct the payoff table for this decision problem.

Solution   The actions that Retra Corporation can take are (*a*) try the case or (*b*) settle the case. The states of nature (the conditions that will affect the payoffs) are (*A*) they win the case if tried and (*B*) they lose the case if tried. The payoff table that results is shown in table 20-4.

TABLE 20-4
Payoff Table for Retra Corporation

Action	Case Disposition (state of nature)	
	(A) Win	(B) Lose
Try case (*a*)	−100,000	−1,100,000
Settle (*b*)	−500,000	−500,000

Illustration 20-3

Suppose you are the comptroller of a corporation. On 1 November you find that you have $100,000 in cash above what you need to meet your normal monthly expenses. However, you expect a bill for $85,000 to arrive from a contractor sometime this month. The bill is payable 30 days from receipt, but if it is paid within 10 days, you get a 2% discount. You maintain a $500,000 line of credit with a bank. The bank charges 1% monthly interest, with a minimum charge of 15 days of interest. That is, if you borrow money for from 1 to 15 days, you pay 15 days' interest (0.5%); for 16 to 30 days, you pay 30 days' interest (1%); and so forth. You can invest the excess cash in a 30-day, $100,000 treasury note that pays 0.75% interest for the month. As comptroller, what are your options and possible returns and what do your returns depend upon?

Solution   The uncertainty you face is just when the bill will arrive. If the bill arrives after 20 November, then you can invest the money, earn the interest, and pay the bill on 30 November, when the note matures, and still take the 2% discount for payment within 10 days. This arrangement would be the best of all possibilities. However, if the bill arrives earlier than 20 November and if you have invested the money, you will have to either borrow money to take advantage of the discount for prompt payment or wait until 30 November (when the note matures) to pay the bill and thus lose the discount.

Now let's list the possible actions you can take.

a. Do not invest. Keep the money and pay the bill within 10 days of receipt to get the discount.
b. Invest the money and pay the bill after the note matures.
c. Invest the money and borrow to pay the bill if that is necessary to get the discount.

The states of nature affecting the payoffs of these actions are as follows:

A. The bill arrives after 20 November (this means that the discount can be taken regardless of the action, and you do not have to borrow money in action c).
B. The bill arrives before 6 November [this means that under action b you lose the discount and under c you must pay 30 days' (1%) interest on the money borrowed to get the discount].
C. The bill arrives between 6 November and 20 November [under action b you lose the discount and under action c you must pay 15 days' (0.5%) interest on the money borrowed to get the discount].

Now let's calculate the payoffs for each action under each state of nature. These calculations are shown in table 20-5.

TABLE 20-5
Payoff Calculations for Comptroller's Problem

Action	Payoff, State of Nature A	
a	Save $1,700 (85,000 × 0.02) on discount	$1,700
b	Make $750 (100,000 × 0.0075) on investment and save $1,700 on discount	2,450
c	Make $750 on investment and save $1,700 on discount	2,450

Action	Payoff, State of Nature B	
a	Save $1,700 on discount	$1,700
b	Make $750 on investment	750
c	Make $750 on investment, save $1,700 on discount, and pay $850 (85,000 × 0.01) in interest	1,600

Action	Payoff, State of Nature C	
a	Save $1,700 on discount	$1,700
b	Make $750 on investment	750
c	Make $750 on investment, save $1,700 on discount, and pay $425 (85,000 × 0.005) in interest	2,025

TABLE 20-6
Payoff Table for
Comptroller's Problem

	Bill Arrives (state of nature)		
Action	After 20th (A)	Before 6th (B)	6th to 20th (C)
Do not invest (*a*)	1,700	1,700	1,700
Invest but not borrow (*b*)	2,450	750	750
Invest and borrow (*c*)	2,450	1,600	2,025

opportunity loss table

opportunity loss

From a payoff table we can construct an **opportunity loss table**. In an opportunity loss table we measure each outcome of an action under each state of nature against what we could have gained if we had made the best decision for that state of nature. In economic language this difference is called the **opportunity loss** from taking an action. For example, in table 20-1 if we take action *a* and state of nature *A* occurs, we make 3. However, if we had taken action *c* (when the state of nature is *A*), we would have made 7. Thus by taking action *a* we lost the opportunity of making an additional 4. The opportunity loss of action *a* under state of nature *A* is thus 4.

An opportunity loss table can easily be calculated from the payoff table. First you find the highest payoff in each column (this payoff is the best action under that state of nature). Then to convert a payoff to an opportunity loss, you subtract each payoff from the highest payoff in the column in which that payoff is located. The opportunity loss table for the payoffs in table 20-1 is shown in table 20-7. Note that the best action for each state of nature will have a 0 entry in an opportunity loss table.

TABLE 20-7
Opportunity Loss Table for
the Payoffs in Table 20-1

	State of Nature		
Action	A	B	C
a	4 (7 − 3)	6 (8 − 2)	0 (7 − 7)
b	5 (7 − 2)	8 (8 − 0)	2 (7 − 5)
c	0 (7 − 7)	0 (8 − 8)	8 [7 − (−1)]

Illustration 20-4

Construct the opportunity loss table for the decision problem in illustration 20-2. The payoffs are shown in table 20-4.

Solution   Under the condition that the case would win, the best action is "try case," and the payoff is −100,000. Under the condition that the case would lose, the best action is "settle," and the payoff is −500,000. Then the opportunity loss table is as shown in table 20-8.

TABLE 20-8
Opportunity Loss Table for Payoffs in Table 20-4

	Case Disposition (state of nature)	
Action	Win (A)	Lose (B)
Try case (*a*)	0 [−100,000 − (−100,000)]	600,000 [−500,000 − (−1,100,000)]
Settle (*b*)	400,000 [−100,000 − (−500,000)]	0 [−500,000 − (−500,000)]

NOTES:
1. The entries in an opportunity loss table will always be positive because of the way that opportunity loss is defined. However, remember that these entries *are* losses. Therefore the smaller the value, the better.
2. The entries in a payoff table may be either positive or negative.

Illustration 20-5

Construct the opportunity loss table for the decision problem in illustration 20-3 (the payoffs are listed in table 20-6).

Solution In table 20-6 the highest payoff in the first column is 2,450; the highest payoff in the second column is 1,700; the highest payoff in the third column is 2,025. Thus the opportunity loss table is as shown in table 20-9.

TABLE 20-9
Opportunity Loss Table for Payoffs in Table 20-6

	Bill Arrives (state of nature)		
Action	After 20th (A)	Before 6th (B)	6th to 20th (C)
Do not invest (*a*)	750 (2,450 − 1,700)	0 (1,700 − 1,700)	325 (2,025 − 1,700)
Invest but not borrow (*b*)	0 (2,450 − 2,450)	950 (1,700 − 750)	1,275 (2,025 − 750)
Invest and borrow (*c*)	0 (2,450 − 2,450)	100 (1,700 − 1,600)	0 (2,025 − 2,025)

□

Exercises

20-4 Doyles of London is trying to decide whether or not to underwrite a 1-year, $1 million policy on a 3-year-old champion thoroughbred horse. The premium on the policy is $100,000. If the horse is injured and cannot race again, the value of the policy must be paid to the insured. Construct the payoff table for this decision problem. (Use the solution to answer exercises 20-10 and 20-14.)

20-5 The vice-president of research at Rolon Drugs has a proposal for a new project to develop a new drug. The proposed research will cost $7.5 million. If the research is successful, he estimates the firm can expect to make $550 million profit on the drug. If the research is unsuccessful, the research costs are lost. Construct the payoff table for this decision problem. (Use the solution to answer exercises 20-11 and 20-17.)

20-6 Construct the opportunity loss table for the decision problem in illustration 20-1 (the payoffs are given in table 20-3).

20-7 An investment club meets once a month. The club is considering selling its 100 shares of Xtron stock at $50 per share and holding the cash until its next meeting. Assume that by the next meeting Xtron stock will be $40, $48, or $56. Construct the payoff table and opportunity loss table. (Use the solution to answer exercise 20-21.)

20-8 A publisher has a proposal from an author to publish a novel. From past records the publisher finds that a poor-selling novel will result in a $40,000 loss, an average-selling novel results in a $20,000 profit, a good-selling novel in an $80,000 profit, and a best-seller in a $500,000 profit. Construct the payoff and opportunity loss tables for this decision problem. (Use the solution to answer exercises 20-13 and 20-16.)

Section 20-3 Decision Strategies

After calculating the payoff table and the opportunity loss table, we now are ready to decide what action to take. Let's examine the payoffs given in table 20-1. What action should we take? Compare actions a and b. If state A exists, action a is better. If state B exists, action a is better. If state C exists, action a is better. Thus action a is always preferable (yields a higher payoff) to action b. When this situation occurs, we say that action a **dominates** action b, and we can draw a line through action b in the payoff table to indicate that this action is, in effect, to be eliminated. See table 20-10.

TABLE 20-10
Payoff Table with Line Through Action b to Indicate Action b Is Dominated by Action a

Action	State of Nature		
	A	B	C
a	3	2	7
~~b~~	~~2~~	~~0~~	~~5~~
c	7	8	−1

Now let's compare action *a* with *c*. We find that if the state of nature is *A* or *B*, action *c* is better (7 versus 3 and 8 versus 2). But if the state of nature is *C*, then action *a* is better (7 versus −1). Since we don't know the state of nature, how do we decide what to do?

There are several methods that can be used to help us to make a decision in cases like the one just presented. We will discuss three of the more commonly used alternatives in the following paragraphs.

Maximin

One approach, the pessimist's view, is to believe that nature will always play against us, so that if we take an action, the worst possible state of nature will exist. That is, if we take action *a*, the state of nature will be *B* and we will make $2, whereas if we take action *c*, the state of nature will be *C* and we will lose $1. If we take this pessimistic (or conservative) approach, we should then select action *a*. That is, we should select the action whose minimum payoff is a maximum when compared with the minimum payoffs of all other actions. Such a strategy is called **maximin** (maximize the minimum payoff).

maxamin

Maximax

The opposite strategy from maximin is to believe that nature is on our side and that, whatever action we take, the best possible state of nature will occur. If we take action *a*, the state of nature will be *C* and we will make $7. If we take action *c*, the state of nature will be *B* and we will make $8. If we follow this optimistic approach, we would take action *c*. That is, we should select the action whose maximum payoff is the highest payoff in the table. Such a strategy is called **maximax**. (Note that the maximax strategy is the easiest to follow. Simply find the largest entry in the payoff table and take the action associated with it.)

maximax

Minimax Regret

A third possible strategy uses the opportunity loss table. Recall that the opportunity loss measures the difference between what the payoff is and what it would have been if we had taken the best action for the given state of nature. The third approach considers that a decision maker could want to minimize his or her regret of what could have been. (This approach has been jokingly referred to as minimizing the "I could have, I should have hindsight syndrome.")

Let's look at the opportunity loss table for our problem, which is shown in table 20-11. Again we take a pessimistic view. That is, if we take *a*, we assume state of nature *B* will occur and we will suffer an opportunity loss of 6, whereas if we take action *c*, we assume state of nature *C* will occur and we will suffer an opportunity loss of 8 (these values are circled in table 20-11). Thus we select action *a* to minimize the maximum possible opportunity loss we can suffer. Such a decision strategy is called **minimax regret** (minimize the maximum regret).

minimax regret

TABLE 20-11
Opportunity Loss Table for the Payoffs in Table 20-1

| | State of Nature | | |
Action	A	B	C
a	4	⑥	0
c	0	0	⑧

Illustration 20-6

Considering the decision problem in illustrations 20-2 and 20-4, what would be the maximin, maximax, and minimax regret decisions?

Solution The circled values in table 20-12 represent the minimum payoff—the worst that could happen—for each action. (*Note*: We could also have circled the $-500,000$ in the second column, since the minimum payoff from settling is the same regardless of what would have happened in the trial.) Thus the **maximin decision is "settle"** ($-500,000$ is greater than $-1,100,000$).

TABLE 20-12
Payoff Table for Retra Corporation

| | Case Disposition (state of nature) | |
Action	Win	Lose
Try case	$-100,000$	$\boxed{-1,100,000}$
Settle	$\boxed{-500,000}$	$-500,000$

The best possible payoff is $-100,000$ and that occurs for the action "try case" (assuming, optimistically, that you win if you try the case). Thus the **maximax decision is "try case."**

The opportunity loss table for this problem is shown in table 20-8. From table 20-8 we see that if we settle, the worst opportunity loss is $400,000, whereas if we try the case, the worst opportunity loss is $600,000. Thus the **minimax regret decision is "settle"** ($400,000 < $600,000). □

Illustration 20-7

Considering the decision problem in illustrations 20-3 and 20-5, what would be the maximin, maximax, and minimax regret decisions?

Solution The circled values in table 20-13 represent the minimum payoff that could occur for each action. Thus the **maximin decision is "do not invest"** ($1,700 > 1,600 > 750$).

The best possible payoff is 2,450, which occurs with the actions "invest but not borrow" and "invest and borrow." However, since action c dominates b (it is always at least as good as b), b would be eliminated from consideration. Therefore the **maximax decision is "invest and borrow."**

TABLE 20-13
Payoff Table for Comptroller's Problem

	Bill Arrives (state of nature)		
Action	After 20th (A)	Before 6th (B)	6th to 20th (C)
Do not invest (a)	(1,700)	1,700	1,700
Invest but not borrow (b)	2,450	(750)	750
Invest and borrow (c)	2,450	(1,600)	2,025

The circled values in table 20-14 represent the maximum opportunity losses that could occur from each action. Thus the **minimax regret decision is "invest and borrow"** (100 < 750 < 1,275).

TABLE 20-14
Opportunity Loss Table for Comptroller's Problem

	Bill Arrives (State of Nature)		
Action	After 20th (A)	Before 6th (B)	6th to 20th (C)
Do not invest (a)	(750)	0	325
Invest but not borrow (b)	0	950	(1,275)
Invest and borrow (c)	0	(100)	0

Exercises

20-9 Explain why maximin and minimax regret are called conservative strategies.

20-10 For the decision problem in exercise 20-4, find each of these decisions:
(a) maximin
(b) maximax
(c) minimax regret

20-11 For the decision problem in exercise 20-5, find each of these decisions:
(a) maximin
(b) maximax
(c) minimax regret

20-12 Consider the accompanying payoff table. Are any of the actions dominated? Is this really a decision problem? Explain.

| | State of Nature | | | |
Action	A	B	C	D
a	25	35	45	55
b	15	35	55	70
c	70	70	70	70
d	65	25	45	70

20-13 For the decision problem in exercise 20-8, find each of these decisions:
(a) maximin
(b) maximax
(c) minimax regret

Section 20-4 Maximizing Expected Value of Payoff or Minimizing Expected Opportunity Loss

The decision strategies maximin, maximax, and minimax regret assume that nature is either playing against us or for us. In reality, the action we take has no effect on the state of nature that will occur. If it does, we call the problem one of game theory rather than of decision theory, and the approach is different. (We will not discuss game theory, but you will find references on game theory in other textbooks if you are interested.)

So let's assume that our actions do not affect the states of nature. Then it would make sense to consider the probability that each state of nature will occur. Consider illustration 20-2. If the counsel says that he is 99% certain that he will win the case, you probably would let him try the case. On the other hand, if he says that he is 99% certain that he cannot win the case, you would probably settle. What would you do if he said that the probability was 0.55 that he would win and 0.45 that he would lose? What to do is not so clear now. What we need is some method for weighting the payoff of each action by the likelihood of each payoff occurring. The likelihood of each payoff is just the probability that the payoff occurs. The weighting of each possible payoff of an action by the probability of the associated state of nature occurring is called the **expected value of the payoff of the action**. It is written in symbols as $E(\text{action})$ and is calculated by the formula

expected value of payoff

$$E(\text{action}) = \sum [A_l \cdot P(l)] \quad (20\text{-}1)$$

where

A_l = payoff of an action under state of nature l

$P(l)$ = probability of state of nature l

NOTE: The sum in formula (20-1) is taken across the states of nature (columns). There will be an expected payoff for each action (row).

For example, if the counsel were to assign a probability of 0.55 to winning the case (A) and 0.45 to losing the case (B), then applying formula (20-1) to the payoffs in table 20-4, we would get

$$E(\text{try case}) = (-100{,}000)(0.55) + (-1{,}100{,}000)(0.45) = -550{,}000$$

$$E(\text{settle}) = (-500{,}000)(0.55) + (-500{,}000)(0.45) = -500{,}000$$

Now let's see if we can interpret what the expected value of the payoff of an action actually represents. Let's consider the simple decision problem of whether or not to play the following game of chance. We flip a coin; if it is heads, we win $1.00, and if it is tails, we lose $1.10. The payoff table would be as shown in table 20-15.

TABLE 20-15
Payoff Table for Flipping Coin

	State of Nature	
Action	Heads	Tails
Bet	1.00	−1.10
Don't bet	0	0

If we decide to bet and play the game 100 times, we would expect to find close to 50% heads and 50% tails. If we get exactly 50 heads and 50 tails, we would lose $5 [(1.00)(50) + (−1.10)(50)]. The average payoff per bet would be

$$\frac{(50)(1.00) + (50)(-1.10)}{100} = \frac{-5}{100} = -\$0.05$$

NOTE: We can rewrite the equation above as

$$\left(\frac{50}{100}\right)(1.00) + \left(\frac{50}{100}\right)(-1.10) = -0.05$$

or, in general terms,

$$\left(\frac{\text{Number of heads}}{\text{Number of trials}}\right) \times \left(\begin{array}{c}\text{payoff}\\ \text{for heads}\end{array}\right)$$

$$+ \left(\frac{\text{number of tails}}{\text{number of trials}}\right) \times \left(\begin{array}{c}\text{payoff}\\ \text{for tails}\end{array}\right) = \text{expected payoff} \qquad (20\text{-}2)$$

We know that in 100 trials it is hardly a sure thing that we will get 50 heads and 50 tails. However, recall the law of large numbers from chapter 4. As the number of trials gets large, then

$$\left(\frac{\text{Number of heads}}{\text{Number of trials}}\right) \text{ becomes } P \text{ (head)}$$

and

$$\left(\frac{\text{Number of tails}}{\text{Number of trials}}\right) \text{ becomes } P \text{ (tail)}$$

Thus in the long run, formula (20-2) becomes the formula for expected payoff (20-1).

The **expected value of the payoff** of an action represents the average payoff per decision that will occur in the long run from continuously choosing that action.

---

maximizing expected value of payoff

If we are looking for a decision strategy that will, in the long run, give us the greatest average payoff, we should choose the action that has the highest expected value of payoff. This decision strategy is called **maximizing the expected value of payoff**. Using this criterion, we should advise counsel to settle the case if his assessment is 0.55 he will win, since the expected value of the action "settle" ($-500{,}000$) is greater than the expected value of the action "try case" ($-550{,}000$).

Illustration 20-8

Referring to the bakery problem in illustration 20-1, how many cakes should be baked if the bakery wants to maximize its expected profits? For the past 200 Saturdays, the demand for cakes was as given in table 20-16.

TABLE 20-16
Demand for Cakes for Past 200 Saturdays

Number of Cakes	Demand
8	40
9	50
10	60
11	50

Solution   Based on the past demand, our empirical probability assignments for each possible demand would be as shown in table 20-17.

TABLE 20-17
Probability Assignments for the Demand

Number of Cakes	Probability of Demand
8	40/200 = 0.20
9	50/200 = 0.25
10	60/200 = 0.30
11	50/200 = 0.25

Using the payoffs of table 20-3, the probabilities in table 20-17, and formula (20-1), we can calculate the expected value of each action:

$$E(\text{bake 8}) = (0.20)(24) + (0.25)(24) + (0.30)(24) + (0.25)(24)$$
$$= \$24.00$$

$$E(\text{bake 9}) = (0.20)(21) + (0.25)(27) + (0.30)(27) + (0.25)(27)$$
$$= \$25.80$$

$$E(\text{bake 10}) = (0.20)(18) + (0.25)(24) + (0.30)(30) + (0.25)(30)$$
$$= \$26.10$$

$$E(\text{bake 11}) = (0.20)(15) + (0.25)(21) + (0.30)(27) + (0.25)(33)$$
$$= \$24.60$$

Thus the action "**bake 10 cakes**" **maximizes the expected profits**. That means that if the bakery were to consistently bake 10 cakes every Friday night, its average profit for Saturday would be $26.10, in the long run. (Note that we are talking about *average* profit. On any given Saturday the bakery will never make $26.10. It can make $18, $24, or $30.) □

Illustration 20-9

Reconsider the cash management problem presented in illustration 20-3. Suppose the comptroller feels that the probabilities for the date of the bill's arrival are as follows: before 6 November, $\frac{1}{2}$; between 6 November and 20 November, $\frac{1}{4}$; after 20 November, $\frac{1}{4}$. What action maximizes the expected payoff?

Solution Using the payoffs in table 20-6, the probabilities $P(A) = \frac{1}{4}$, $P(B) = \frac{1}{2}$, and $P(C) = \frac{1}{4}$, and formula (20-1), we can calculate the expected value of the payoff of each action:

$$E(\text{do not invest}) = (\tfrac{1}{4})(1{,}700) + (\tfrac{1}{2})(1{,}700) + (\tfrac{1}{4})(1{,}700)$$
$$= \$1{,}700.00$$

$$E(\text{invest but not borrow}) = (\tfrac{1}{4})(2{,}450) + (\tfrac{1}{2})(750) + (\tfrac{1}{4})(750)$$
$$= \$1{,}175.00$$

$$E(\text{invest and borrow}) = (\tfrac{1}{4})(2{,}450) + (\tfrac{1}{2})(1{,}600) + (\tfrac{1}{4})(2{,}025)$$
$$= \$1{,}918.75$$

Thus the decision to **invest and borrow** money, if needed, to take the discount is the action that **maximizes the expected return**. ☐

In this section so far, we have been dealing with the payoff table and finding the action that maximizes the expected payoff. We could have used the opportunity loss table instead. If we did that, we would calculate the **expected opportunity loss of the action**, written as *EOL* (action) and calculated by the formula

expected opportunity loss

$$EOL(\text{action}) = \sum [\ell_l P(l)] \qquad (20\text{-}3)$$

where

ℓ_l = opportunity loss of an action for state of nature l
$P(l)$ = probability of state of nature l

minimize expected opportunity loss

Since we are dealing with losses rather than gains, we would select the action that **minimizes the expected opportunity loss**. Although the values calculated for the expected opportunity loss of each action will be different from the values calculated for the expected payoff of each action, *the action that has the maximum expected payoff will have the minimum expected opportunity loss*. Consequently, the two decision strategies will always lead to the same action

Illustration 20-10

Considering the cash management problem presented in illustration 20-3, 20-5 (the opportunity loss table), and 20-9 (the probabilities for the states of nature), find the decision that minimizes the expected opportunity loss.

Solution Using the opportunity losses in table 20-9, the probabilities $P(A) = \frac{1}{4}$, $P(B) = \frac{1}{2}$, and $P(C) = \frac{1}{4}$ for the states of nature, and formula (20-3), we can calculate the expected opportunity loss for each action:

$$EOL \text{ (do not invest)} = (\tfrac{1}{4})(750) + (\tfrac{1}{2})(0) + (\tfrac{1}{4})(325)$$
$$= \$268.75$$

$$EOL \text{ (invest but not borrow)} = (\tfrac{1}{4})(0) + (\tfrac{1}{2})(950) + (\tfrac{1}{4})(1{,}275)$$
$$= \$793.75$$

$$EOL \text{ (invest and borrow)} = (\tfrac{1}{4})(0) + (\tfrac{1}{2})(100) + (\tfrac{1}{4})(0)$$
$$= \$50.00$$

Thus **the action that minimizes the expected opportunity loss is "invest and borrow."** (Note that this action is the same as was chosen in illustration 20-9, where we were interested in the action that maximized the expected payoff.)

Exercises

20-14 In the decision problem presented in exercise 20-4, if the probability that a horse will be injured so that it cannot race again in 0.05, what is the expected payoff of each action? What action maximizes the expected payoff? (Use the solution to answer exercise 20-20.)

20-15 Would you agree with the statement, "The expected payoff of an action is the most likely payoff from taking that action"? Explain.

20-16 Consider the decision problem presented in exercise 20-8. Suppose the probabilities for the selling potential of the novel are as follows: poor seller, $\frac{1}{3}$; average seller, $\frac{1}{3}$; good seller, $\frac{1}{3}$. (Use the solution to answer exercise 20-19.)
 (a) What is the expected payoff for the action "publish"?
 (b) What is the expected payoff for the action "don't publish"?
 (c) What is the expected opportunity loss for the action "publish"?
 (d) What is the expected opportunity loss for the action "don't publish"?
 (e) What action minimizes the expected opportunity loss?
 (f) What action maximizes the expected payoff?

20-17 In the decision problem presented in exercise 20-5, if the probability of the research being successful is 0.02, should the research be undertaken if we use the decision strategy of maximizing expected profit? Use the solution to answer exercise 20-22.)

20-18 Consider the decision problem of whether to try or to settle the constraint-of-trade lawsuit presented in illustration 20-2. Which decision maximizes the expected payoff for the following probabilities?
 (a) The probability of winning is 0.58.
 (b) The probability of winning is 0.62.
 (c) The probability of winning is 0.60.

Section 20-5 Expected Value of Perfect Information

Recall the cash management problem we have been discussing. In illustration 20-9 we found that the maximum expected payoff was $1,918.75 by taking the action "invest and borrow." However, suppose that someone offers to sell us the information of when the bill will arrive, and hence we will not have to act without knowledge of when the bill will come in. How much is this information worth?

If we know in advance when the bill will arrive, then we can select the best action for that state of nature. If the bill arrives after 20 November (A), then we can take either action b, invest but do not borrow, or action c, invest and borrow if necessary. (Note that if the bill arrives after 20 November, the actions are equivalent, since we will not have to borrow.) If A

occurs, we would then make $2,450 (see table 20-6). If the bill arrives before 6 November (*B*), we would not invest (*a*), and we would make $1,700. If the bill arrives between 6 November and 20 November (*C*), we would invest and borrow (*c*), and we would make $2,025. Thus if we could choose the action after we know the state of nature, our possible payoffs would be as shown in table 20-18.

TABLE 20-18
Payoffs, Knowing State of Nature

State of Nature	Action to Take	Payoff
A	*b* or *c*	2,450
B	*a*	1,700
C	*c*	2,025

What can we expect our payoff to be if we had advance knowledge of the state of nature? As you should expect, we can calculate the expected payoff by weighting each payoff by the probability that each state of nature will occur. Thus the expected value of the payoff for taking the best action after knowing the state of nature would be

$$(\tfrac{1}{4})(2{,}450) + (\tfrac{1}{2})(1{,}700) + (\tfrac{1}{4})(2{,}025) = \$1{,}968.75$$

Recall that if we had to act *before* the state of nature was known, the best action was *c* (invest and borrow), and the expected value of the payoff for that action was $1,918.75 (illustration 20-9).

---

cost of uncertainty
expected value of perfect information

The difference between the expected payoff from taking the best action after knowing the state of nature and the maximum expected payoff from taking an action *before* knowing the state of nature is the **cost of uncertainty** and is called the **expected value of perfect information** (*EVPI*).

Thus the expected value of knowing beforehand when the bill would arrive (*EVPI*) is **$50.00** ($1,968.75 − $1,918.75).

We are now ready to give a more meaningful definition to the expected opportunity loss of an action. Recall that the opportunity loss measures the differences between the payoff of the action under a state of nature against the payoff that would occur if the best action for that state of nature had been taken. Hence opportunity loss measures the payoff of the action compared with the payoff that would occur if the state of nature were known beforehand. The expected opportunity loss of an action is the difference between the expected payoff of the action and the expected payoff from taking the best action after knowing the state of nature. The minimum expected opportunity loss is thus the expected value of perfect information. Look at illustration 20-10; what was the minimum expected opportunity loss? Compare that value with the expected value of perfect information ($EVPI$) calculated previously.

Illustration 20-11

Calculate the $EVPI$ for the bakery problem.

Solution If we knew in advance what demand would be, we would always bake the same number of cakes as demanded. (See tables 20-3 and 20-17.) The data given previously that are needed for this problem are summarized in table 20-19. Thus the expected payoff of always taking best action is

$$(0.20)(24) + (0.25)(27) + (0.30)(30) + (0.25)(33) = \$28.80$$

The maximum expected payoff was $26.10 (see illustration 20-8). Thus

$$EVPI = \$28.80 - \$26.10 = \mathbf{\$2.70}$$

TABLE 20-19
Summary of Bakery Problem Data

Cakes Demanded	Best Action, Cakes Baked	Best Payoff	Probability of Cake Demanded
8	8	24	0.20
9	9	27	0.25
10	10	30	0.30
11	11	33	0.25

Alternative Solution The opportunity loss for the problem is shown in table 20-20.

TABLE 20-20
Opportunity Loss Table for Bakery Problem

Cakes Baked (action)	Cakes Demanded (state of nature)			
	8	9	10	11
8	0	3	6	9
9	3	0	3	6
10	6	3	0	3
11	9	6	3	0

Thus each expected opportunity loss is

$$EOL(8) = (0.20)(0) + (0.25)(3) + (0.30)(6) + (0.25)(9) = \$4.80$$

$$EOL(9) = (0.20)(3) + (0.25)(0) + (0.30)(3) + (0.25)(6) = \$3.00$$

$$EOL(10) = (0.20)(6) + (0.25)(3) + (0.30)(0) + (0.25)(3) = \$2.70$$

$$EOL(11) = (0.20)(9) + (0.25)(6) + (0.30)(3) + (0.25)(0) = \$4.20$$

Thus the minimum expected opportunity loss is 2.70, which is the $EVPI$ calculated in the first solution. □

Exercises

20-19 For the publishing problem in exercise 20-16, what is the *EVPI*? If the publisher can hire a reviewer who is never wrong in her prediction of sales, what is the maximum she should be paid? Explain.

20-20 What is the *EVPI* for the insurance problem in exercise 20-14?

20-21 Consider the situation in exercise 20-7. Suppose that the club assesses the following probabilities of the stock's prices: 0.3 that it will be $40, 0.4 that it will be $48, and 0.3 that it will be $56.
  (a) Should the club sell the stock at $50 now?
  (b) If an investment advisor costs $200 and can correctly predict stock prices 70% of the time, should the club purchase this service? Explain.

20-22 What is the *EVPI* for the research and development investment problem in exercise 20-17?

Section 20-6 Misuses of Maximizing Expected Value of Payoff

Someone offers you the following bet. If the next car passing the intersection is not a bright red Rolls Royce, he will pay you $1; but if it is, you pay him $100,000. For discussion's sake, suppose only one in a million cars on the road are bright red Rolls Royces. Then the expected payoff of each action would be

$$E(\text{bet}) = (0.999999)(\$1) + (0.000001)(-100{,}000)$$
$$= 0.999999 - 0.10 = 0.899999 = \mathbf{\$0.90}$$
$$E(\text{not bet}) = (0.999999)(0) + (0.000001)(0) = \mathbf{0}$$

Thus the action "bet" is the best under maximizing expected return. But would you take the bet? What would you do if the car turns out to be a bright red Rolls Royce?

Consider a second bet. Put slips of paper numbered 1 through 100 in a hat and randomly select one slip. If 95 or less is selected, you win $1; whereas if 96 or more is selected, you lose $1 (−$1). The expected payoff of the bet is $0.90 [(0.95)(1) + (0.05)(−1)], exactly the same as the first bet. However, most people would take the second bet but not the first bet. Why? Because they cannot afford the *risk* of losing $100,000. Remember that expected payoff refers to the average payoff in the *long run*. However, in real life we seldom get the chance to repeat our decisions over and over again. Consequently, we must consider the risks associated with the possibility that the worst will happen when we make a decision.

Maximizing expected value does not consider risk. The two bets discussed above are considered to be the same under expected value theory, yet clearly one is much riskier than the other. (There are methods, such as utility analysis, that try to incorporate risk. We will not discuss them in this text, but references to the topic can be found in other textbooks.)

Maximizing expected payoff, however, is the best decision strategy to use when the decision can be made over and over again and when the risks involved with the worst occurrences are not significant. Consider the problem of term life insurance. The expected value of writing the policy (the insurance company's viewpoint) is positive. That is, the expected average return per policy is positive. The expected value of the return from buying a policy (the buyer's viewpoint) is negative. Yet people do buy insurance. This is to be expected, because the insurance company writes lots of policies and in a sense deals with the long run, whereas the individual buyer deals only with one policy and must consider the risks associated with the payoffs (paying a premium versus the cost of leaving a family without adequate financial protection). Would it surprise you to know that a small number of insurance companies underwrite almost all the insurance policies and that small insurance companies generally sell their large policies to the large insurance companies? If you understand expected values, it should not.

In Retrospect

In this chapter we have explored the basic concepts of statistical decision making. You should now be able to clearly distinguish between decisions under certainty and decisions under uncertainty and to measure the cost of uncertainty (EVPI).

In statistical decision making we are concerned with decisions under uncertainty and, in particular, with developing strategies for making such decisions. Note that when we evaluate a decision, we are referring to the evaluation of that decision before the outcomes are known, not after. Thus the "best" decision may turn out in a particular case to yield a poorer result than another decision.

In evaluating a decision problem, we must be able to list the actions, the states of nature, and the payoffs. These ideas are then combined in a payoff table or in its corresponding opportunity loss table.

In this chapter we discussed four different decision strategies: (1) minimax regret, (2) maximin, (3) maximax, and (4) maximizing the expected value of payoff or minimizing the expected opportunity loss. You should be able to find the action that is best under each strategy. We also discussed the problem that the first three ignore—the likelihoods associated with each state of nature—and their view that nature is playing for or against us (which it doesn't). The fourth approach considers the likelihood of the states of nature, but does not consider risks.

The news article at the beginning of this chapter presents an approach to human resource accounting using expected value criteria. It calculated the expected future salary the person will earn assuming the firm takes the action "do not fire" him today. Decision theory in all its aspects is an extremely popular topic in business research and practice today. This chapter has provided you with a brief introduction to the topic.

Chapter Exercises

20-23 State whether each of the following are examples of decisions under certainty or uncertainty:
 (a) whether to buy GM or XEROX stock
 (b) whether to computerize the payroll operation
 (c) how many items to keep in inventory to meet future demands

20-24 A firm is considering marketing a new product. If the economy is strong, it expects to make $3.7 million profit on the new product. However, if the economy is not strong, it expects to lose $1.7 million.
 (a) Construct the payoff table.
 (b) Construct the opportunity loss table. (Use the solution to answer exercise 20-32.)

20-25 Suppose you are given the accompanying opportunity loss table. If the payoff for action *a* under state of nature *A* is 100, for action *b* under state of nature *B* is 150, and for action *c* under state of nature *C* is 200, what is the payoff table?

	State of Nature		
Action	A	B	C
a	0	10	20
b	10	0	20
c	30	20	0

20-26 A newsstand historically orders between 100 and 105 (inclusive) Sunday papers. They make 10¢ on each paper sold and lose 5¢ on each paper not sold.
 (a) What is the payoff table?
 (b) What is the opportunity loss table?
 (Use the solution to answer exercises 20-31 and 20-38.)

20-27 An insurance adjustor is trying to decide whether to accept an offer of settlement from the plaintiff. The plaintiff is asking for $50,000 to settle the case. The adjustor feels that there is a 50% chance the insurance company can win the case; if they lose, there is an equal chance that the damages the jury will award will be either $100,000 or $50,000.
 (a) What is the payoff table?
 (b) What is the opportunity loss table?

(c) What is the minimax regret decision?
(d) What is the maximin decision?
(e) What is the maximax decision?
(f) What decision maximizes the expected payoff?

20-28 A businessman asks his accountant whether he should purchase a car or lease a car. The accountant responds that if the IRS rules that the car is a business necessity, leasing the car, after tax deductions, will cost him $7,000. If the IRS disallows it as a business necessity, leasing the car, after tax deductions, will cost him $10,000. Outright purchase will cost him, after tax deductions, $9,000, independent of an IRS ruling. (Use the solution to answer exercises 20-33, 20-36, and 20-44.)

(a) What is the payoff table?
(b) What is the opportunity loss table?
(c) What is the minimax regret decision?
(d) What is the maximin decision?
(e) What is the maximax decision?

20-29 Suppose you are given the accompanying payoff table. (Use the solution to answer exercises 20-34 and 20-37.)

	Broadway Show Runs (state of nature)		
Action	Less Than 1 Week	1 Week to 1 Month	More Than 1 Month
Do not invest	0	0	0
Invest in show	−20,000	0	+40,000

(a) Construct the opportunity loss table.
(b) What is the maximax decision?
(c) What is the minimax regret decision?
(d) What is the maximin decision?

20-30 If we find that action a dominates action b, why can we eliminate action b from consideration?

20-31 Suppose in exercise 20-26 that the demand for Sunday papers in the past is as shown in the accompanying table. (Use the solution to answer exercise 20-38.)

Papers Demanded	100	101	102	103	104	105
Number of Times	70	90	100	110	70	60

(a) What is the expected payoff from ordering 100 papers?
(b) What is the expected payoff from ordering 101 papers?

(c) What is the expected payoff from ordering 102 papers?
(d) What is the expected payoff from ordering 103 papers?
(e) What is the expected payoff from ordering 104 papers?
(f) What is the expected payoff from ordering 105 papers?
(g) How many papers should be ordered?

20-32 If in exercise 20-24 the firm assesses the probability of a strong economy as 0.3, what action should it take to maximize the expected value of payoff?

20-33 In reference to the tax problem in exercise 20-28, if the accountant feels that the probability of a favorable tax ruling on leasing is 0.8, should he advise the client to lease? (Use the decision strategy of maximizing expected payoff.)

20-34 In reference to the investment problem in exercise 20-29, if the probabilities of each occurrence are 0.4, 0.3, and 0.3, respectively, which decision maximizes expected payoff?

20-35 Show that the action that maximizes expected payoff in exercise 20-34 is also the action that minimizes expected opportunity loss.

20-36 Show that the action that maximizes expected payoff in exercise 20-28 also minimizes the expected opportunity loss.

20-37 If in exercise 20-29 a critic offers to review the script of a show and guarantees that he can perfectly predict its success, what is the expected value of this perfect information?

20-38 What is the expected value of perfect information in the inventory order problem in exercise 20-31?

20-39 What is the expected value of perfect information in the tax problem in exercise 20-33?

20-40 Suppose you are given the accompanying payoff table. Your production team feels that it has a 0.90 probability of fulfilling the contract.

	State of Nature	
Action	Fulfill Contract	Not Fulfill Contract
Take contract	+$5,000,000	−$20,000,000
Not take contract	0	0

(a) What action maximizes the expected payoff?
(b) Do you think a firm that has a net worth of $1 billion would bid? Do you think a firm with a net worth of $30 million would bid? Explain.

20-41 Would you agree with the statement that a firm that has large net worth, so that any single loss is not serious, would be wise to make all its decisions based on maximizing expected payoff? Explain.

22-42 A firm is trying to decide whether to raise the price of its product. It estimates that if the product is price inelastic (i.e., the drop in sales from a price increase will not be large; thus the total revenue will increase), it will increase profits by $800,000. If the product is price elastic (i.e., the drop in sales from a price increase will be so large that total revenues will decrease), it will decrease profits by $500,000. Its studies indicate that the probability that the price is elastic is 0.60.
  (a) Calculate the payoff table.
  (b) Calculate the opportunity loss table.
  (c) Find the maximax decision.
  (d) Find the maximin decision.
  (e) Find the minimax regret decision.
  (f) Find the decision that maximizes expected payoff.
  (g) Find the expected value of perfect information.

20-43 The president is considering two methods of stimulating the economy: a tax cut (fiscal policy) or increasing the money supply (monetary policy). The monetarists agree that the tax cut will increase the GNP by 10 billion and that increasing the money supply will increase the GNP by 30 billion. The fiscal economists argue that the tax cut will stimulate the economy by 50 billion and that increasing the money supply will increase the GNP by 5 billion. The president gives equal weight to each economic school of thought. Complete parts (a) through (g) of exercise 20-42 for this decision problem.

20-44 In the tax problem of exercise 20-28, how large a probability must the accountant assign to a favorable tax ruling for him to recommend leasing based on maximizing the expected value of payoff?

20-45 A manufacturing firm is trying to decide whether to raise the price of its pens. In evaluating what to do, it must concern itself with the reaction and price response that its main competitors will make to its action. Is this situation an example of decision theory? Explain.

Challenging Problem

20-46 Suppose you are given the accompanying payoff table.

	State of Nature	
Action	Interest Rates Down	Interest Rates Up
Invest	+7,000	−3,000
Don't invest	0	0

Your assessment of interest rates is that there is a 30% chance interest rates will drop. You now hear that Wharton Models predicts interest rates will rise. In the past when interest rates rose, Wharton correctly predicted it 80% of the time; and conversely, when interest rates dropped, Wharton incorrectly predicted they would rise 35% of the time.

(a) Based solely on your assessment, ignoring Wharton's prediction, what should you do if you follow a maximizing expected payoff strategy?

(b) How should you revise your prior probabilities about interest rate movement in light of Wharton's estimates and what should you do now? (*Hint*: What you want is the probability of interest rates going up or down given Wharton's prediction of a decline. Consider Bayes's formula.)

(c) Suppose your initial probability assessment that interest rates would drop had been 90%. Redo parts (a) and (b) and compare your results. Did Wharton's prediction have as big an effect when your initial opinion was stronger?

Appendixes

A	Summation Notation	**A-2**
B	Using the Random Number Table	**A-8**
C	Round-Off Procedure	**A-10**
D	Tables	**A-12**
1.	Random Numbers	A-12
2.	Factorials	A-14
3.	Binomial Coefficients	A-15
4.	Binomial Probabilities	A-16
5.	Poisson Probabilities	A-20
6.	Areas of the Standard Normal Distribution	A-21
7.	Critical Values of Student's t Distribution	A-22
8.	Critical Values of the χ^2 Distribution	A-23
9a.	Critical Values of the F Distribution ($\alpha = 0.05$)	A-24
9b.	Critical Values of the F Distribution ($\alpha = 0.025$)	A-26
9c.	Critical Values of the F Distribution ($\alpha = 0.01$)	A-28
10.	Critical Values for the Sign Test	A-30
11.	Critical Values of U in the Mann-Whitney Test	A-31
12.	Critical Values for Total Number of Runs (V)	A-32
13.	Critical Values of Spearman's Rank Correlation Coefficient	A-33
14.	Critical Values of r When $\rho = 0$	A-34
15.	Confidence Belts for the Correlation Coefficient ($1 - \alpha = 0.95$)	A-35

A Summation Notation

The capital Greek letter sigma (Σ) is used in mathematics to indicate the summation of a set of addends. Each of these addends must be of the form of the variable following Σ. For example:

1. Σx means sum the variable x.
2. $\Sigma (x - 5)$ means sum the set of addends that are each 5 less than the values of each x.

When large quantities of data are collected, it is usually convenient to index the response variable so that at a future time its source will be known. This indexing is shown on the notation by using i (or j or k) and affixing the index of the first and last addend at the bottom and top of the Σ. For example,

$$\sum_{i=1}^{3} x_i$$

means to add all the consecutive values of x's starting with source number 1 and proceeding to source number 3.

Illustration A-1

Consider the inventory in the accompanying table about the number of defective stereo tapes per lot of 100.

Lot Number, i	1	2	3	4	5	6	7	8	9	10
Number of Defective Tapes per Lot, x	2	3	2	4	5	6	4	3	3	2

(a) Find $\sum_{i=1}^{10} x_i$.

(b) Find $\sum_{i=4}^{8} x_i$.

Solution

(a) $\sum_{i=1}^{10} x_i = x_1 + x_2 + x_3 + x_4 + \cdots + x_{10}$

$= 2 + 3 + 2 + 4 + 5 + 6 + 4 + 3 + 3 + 2 = 34$

(b) $\sum_{i=4}^{8} x_i = x_4 + x_5 + x_6 + x_7 + x_8 = 4 + 5 + 6 + 4 + 3 = 22$

This index system must be used whenever only part of the available information is to be used. In statistics, however, we will usually use all the available information, and to simplify the formulas we will make an adjustment. This adjustment is actually an agreement that allows us to do away with the index system in situations where all values are used. Thus in our previous illustration, $\sum_{i=1}^{10} x_i$ could have been written simply as $\sum x$.

NOTE: The lack of the index indicates that all data are being used.

Illustration A-2

Given the following six values for x: 1, 3, 7, 2, 4, 5; find $\sum x$.

Solution

$$\sum x = 1 + 3 + 7 + 2 + 4 + 5 = 22 \qquad \square$$

Throughout the study and use of statistics you will find many formulas that use the $\sum$ symbol. Care must be taken so that the formulas are not misread. Symbols like $\sum x^2$ and $(\sum x)^2$ are quite different. $\sum x^2$ means "square each x value and then add up the squares," while $(\sum x)^2$ means "sum the x values and then square the sum."

Illustration A-3

Find (a) $\sum x^2$ and (b) $(\sum x)^2$ for the sample in illustration A-2.

Solution

(a)

x	1	3	7	2	4	5
x^2	1	9	49	4	16	25

$$\sum x^2 = 1 + 9 + 49 + 4 + 16 + 25 = 104$$

(b) $\sum x = 22$, as found in illustration A-2. Thus

$$(\sum x)^2 = (22)^2 = 484$$

As you can see, there is quite a difference between $\sum x^2$ and $(\sum x)^2$. $\qquad \square$

Likewise, $\sum xy$ and $\sum x \sum y$ are different. These forms will appear only when there are paired data, as shown in the following illustration.

Illustration A-4

Given the five pairs of data shown in the accompanying table, find (a) $\sum xy$ and (b) $\sum x \sum y$.

x	1	6	9	3	4
y	7	8	2	5	10

Solution

(a) $\sum xy$ means to sum the products of the corresponding x and y values. Therefore, we have

x	1	6	9	3	4
y	7	8	2	5	10
xy	7	48	18	15	40

$$\sum xy = 7 + 48 + 18 + 15 + 40 = 128$$

(b) $\sum x \sum y$ means the product of the two summations, $\sum x$ and $\sum y$. Therefore, we have

$$\sum x = 1 + 6 + 9 + 3 + 4 = 23$$
$$\sum y = 7 + 8 + 2 + 5 + 10 = 32$$
$$\sum x \sum y = (23)(32) = 736$$ □

There are three basic rules for algebraic manipulation of the $\sum$ notation.

NOTE: c represents any constant value.

Rule I

$$\sum_{i=1}^{n} c = nc$$

To prove this rule, we need only write down the meaning of $\sum_{i=1}^{n} c$:

$$\sum_{i=1}^{n} c = \underbrace{c + c + c + \cdots + c}_{n \text{ addends}}$$

Therefore,

$$\sum_{i=1}^{n} c = n \cdot c$$

Illustration A-5

Show that $\sum_{i=1}^{5} 4 = (5)(4) = 20$.

Solution

$$\sum_{i=1}^{5} 4 = \underbrace{4_{(\text{when } i=1)} + 4_{(\text{when } i=2)} + 4_{(i=3)} + 4_{(i=4)} + 4_{(i=5)}}_{\text{five 4s added together}}$$

$$= (5)(4) = 20$$ □

Rule 2

$$\sum_{i=1}^{n} cx_i = c \cdot \sum_{i=1}^{n} x_i$$

To demonstrate the truth of rule 2, we will need to expand the term $\sum_{i=1}^{n} cx_i$ and then factor out the common factor c.

$$\sum_{i=1}^{n} cx_i = cx_1 + cx_2 + cx_3 + \cdots + cx_n$$
$$= c(x_1 + x_2 + x_3 + \cdots + x_n)$$

Therefore,

$$\sum_{i=1}^{n} cx_i = c \cdot \sum_{i=1}^{n} x_i$$

Rule 3

$$\sum_{i=1}^{n} (x_i + y_i) = \sum_{i=1}^{n} x_i + \sum_{i=1}^{n} y_i$$

The expansion and regrouping of $\sum_{i=1}^{n} (x_i + y_i)$ is all that is needed to show this rule.

$$\sum_{i=1}^{n} (x_i + y_i) = (x_1 + y_1) + (x_2 + y_2) + \cdots + (x_n + y_n)$$
$$= (x_1 + x_2 + \cdots + x_n) + (y_1 + y_2 + \cdots + y_n)$$

Therefore,

$$\sum_{i=1}^{n} (x_i + y_i) = \sum_{i=1}^{n} x_i + \sum_{i=1}^{n} y_i$$

Illustration A-6

Show that $\sum_{i=1}^{3} (2x_i + 6) = 2 \cdot \sum_{i=1}^{3} x_i + 18$.

Solution

$$\sum_{i=1}^{3} (2x_i + 6) = (2x_1 + 6) + (2x_2 + 6) + (2x_3 + 6)$$
$$= (2x_1 + 2x_2 + 2x_3) + (6 + 6 + 6)$$
$$= (2)(x_1 + x_2 + x_3) + (3)(6)$$
$$= 2 \sum_{i=1}^{3} x_i + 18$$

Illustration A-7

Let $x_1 = 2, x_2 = 4, x_3 = 6, f_1 = 3, f_2 = 4,$ and $f_3 = 2$. Find $\sum_{i=1}^{3} x_i \cdot \sum_{i=1}^{3} f_i$.

Solution

$$\sum_{i=1}^{3} x_i \cdot \sum_{i=1}^{3} f_i = (x_1 + x_2 + x_3) \cdot (f_1 + f_2 + f_3)$$
$$= (2 + 4 + 6) \cdot (3 + 4 + 2)$$
$$= (12)(9) = 108 \qquad \square$$

Illustration A-8

Using the same values for the x's and f's as in illustration A-7, find $\sum (xf)$.

Solution   Recall that the use of no index numbers means "use all data."

$$\sum (xf) = \sum_{i=1}^{3} (x_i f_i) = (x_1 f_1) + (x_2 f_2) + (x_3 f_3)$$
$$= (2 \cdot 3) + (4 \cdot 4) + (6 \cdot 2) = 6 + 16 + 12 = 34 \qquad \square$$

Exercises

A-1 Write each of the following in expanded form (without the summation sign):

(a) $\sum_{i=1}^{4} x_i$ (b) $\sum_{i=1}^{3} (x_i)^2$ (c) $\sum_{i=1}^{5} (x_i + y_i)$

(d) $\sum_{i=1}^{5} (x_i + 4)$ (e) $\sum_{i=1}^{8} x_i y_i$ (f) $\sum_{i=1}^{4} x_i^2 f_i$

A-2 Write each of the following expressions as summations, showing the subscripts and the limits of summation:

(a) $x_1 + x_2 + x_3 + x_4 + x_5 + x_6$
(b) $x_1 y_1 + x_2 y_2 + x_3 y_3 + \cdots + x_7 y_7$
(c) $x_1^2 + x_2^2 + \cdots + x_9^2$
(d) $(x_1 - 3) + (x_2 - 3) + \cdots + (x_n - 3)$

A-3 Show each of the following to be true:

(a) $\sum_{i=1}^{4} (5x_i + 6) = 5 \cdot \sum_{i=1}^{4} x_i + 24$

(b) $\sum_{i=1}^{n} (x_i - y_i) = \sum_{i=1}^{n} x_i - \sum_{i=1}^{n} y_i$

A-4 Given $x_1 = 2, x_2 = 7, x_3 = -3, x_4 = 2, x_5 = -1,$ and $x_6 = 1$, find each of the following:

(a) $\sum_{i=1}^{6} x_i$ (b) $\sum_{i=1}^{6} x_i^2$ (c) $\left(\sum_{i=1}^{6} x_i\right)^2$

A–5 Given $x_1 = 4$, $x_2 = -1$, $x_3 = 5$, $f_1 = 4$, $f_2 = 6$, $f_3 = 2$, $y_1 = -3$, $y_2 = 5$, and $y_3 = 2$, find each of the following:

(a) $\sum x$ (b) $\sum y$ (c) $\sum f$ (d) $\sum (x - y)$
(e) $\sum x^2$ (f) $(\sum x)^2$ (g) $\sum xy$ (h) $\sum x \cdot \sum y$
(i) $\sum xf$ (j) $\sum x^2 f$ (k) $(\sum xf)^2$

A–6 Suppose you take out a $12,000 small-business loan. The terms of the loan are that each month for 10 years (120 months) you will pay back $100 plus accrued interest. The accrued interest is calculated by multiplying 0.005 (6%/12) times the amount of the loan still outstanding. That is, the first month you pay $12,000 × 0.005 in accrued interest, the second month ($12,000 − 100) × 0.005 in interest, the third month [$12,000 − (2)(100)] × 0.005, and so forth. Express the total amount of interest paid over the life of the loan by using summation notation.

The answers to these exercises are on page A-37.

B Using the Random Number Table

The random number table is a collection of "random" digits. The term *random* means that each of the ten digits (0, 1, 2, 3, ..., 9) has an equal chance of occurrence. The digits in table 1 of appendix D can be thought of as single-digit numbers (0–9), as two-digit numbers (00–99), as three-digit numbers (000–999), or as numbers of any desired size. The digits presented in table 1 are arranged in pairs and grouped into blocks of five rows and five columns. This format is used for convenience. Tables in other books may be arranged differently.

Random numbers are used primarily for one of two reasons: (1) to identify the source element of a population (the source of data) or (2) to simulate an experiment.

Illustration B-1

A simple random sample of 10 people is to be drawn from a population of 7564 people. Each person will be assigned a number, using the numbers from 0001 to 7564. We will view table 1 as a collection of four-digit numbers (two columns used together), where the numbers 0001, 0002, 0003, ..., 7564 identify the 7564 people. The numbers 0000, 7565, 7566, ..., 9999 represent no one in our population; that is, they will be discarded if selected.

Now we are ready to select our 10 people. Turn to table 1 (page A-12). We need to select a starting point and a "path" to be followed. Perhaps the most common way to locate a starting point is to look away and arbitrarily point to a starting point. The number we located this way was 3909. (It is located in the upper left corner of the block that is in the fourth large block from the left and the ninth large block down.) From here we will proceed down the column, then go to the top of the next set of columns, if necessary. The person identified by number 3909 is the first source of data selected. Proceeding down the column, we find 8869 next. This number is discarded. The number 2501 is next. Therefore, the person identified by 2501 is the second source of data to be selected. Continuing down this column, our sample will be obtained from those people identified by the numbers 3909, 2501, 7485, 0545, 6104, 3347, 6743, 1168, 3398, 3852. (The number 6104 is the second four-digit number at the top of the next set of columns. The numbers 9091 and 9074 were discarded.) □

Illustration B-2

Let's use the random number table and simulate 100 tosses of a coin. The simulation is accomplished by assigning numbers to each of the possible outcomes of a particular experiment. The assignment must be done in such a way as to preserve the probabilities.

Perhaps the simplest way to make the assignment for the coin toss is to let the even digits (0, 2, 4, 6, 8) represent heads and the odd digits (1, 3, 5, 7, 9) represent tails. The correct probabilities are maintained: $P(H) = P(0, 2, 4, 6, 8) = \frac{5}{10} = 0.5$ and $P(T) = P(1, 3, 5, 7, 9) = \frac{5}{10} = 0.5$. Once this assignment is complete, we are ready to obtain our sample.

Since the question asked for 100 tosses and there are 50 digits to a "block" in table 1, let's select two blocks as our sample of random one-digit numbers (instead of a column 100 lines long). Let's look away and point to one block on page A-12 and then do the same to select one block from page A-13. We picked the sixth block down in the first column of blocks on page A-12 (24 even and 26 odd numbers) and the fifth block down in the third column of blocks on page A-13 (23 even and 27 odd numbers). Thus we obtain a sample of 47 heads and 53 tails for our 100 simulated tosses. □

There are, of course, many ways to use the random number table. You must use your good sense in assigning the numbers to be used and in choosing the "path" to be followed through the table. One bit of advice is to make the assignments in as simple and easy a method as possible to avoid errors.

Exercises

B-1   A random sample of size 8 is to be selected from a population that contains 75 elements. Describe how the random sample of the eight objects could be made with the aid of the random number table.

B-2   A coin-tossing experiment is to be simulated. Two coins are to be tossed simultaneously and the number of heads appearing is to be recorded for each toss. Ten such tosses are to be observed. Describe two ways to use the random number table to simulate this experiment.

B-3   Simulate five rolls of three dice by using the random number table.

The answers to these exercises are on page A-37.

C Round-Off Procedure

When rounding off a number, we use the following procedure.

STEP 1   Identify the position where the round-off is to occur. This is shown by use of a vertical line that separates the part of the number to be kept from the part to be discarded. For example,

    125.267   to the nearest tenth is written as   125.2|67

    7.8890   to the nearest hundredth is written as   7.88|90

STEP 2   Step 1 has separated all numbers into one of four cases. (X's will be used as placeholders for number values in front of the vertical line. These X's can represent any number value.)

    Case I: XXXX|000...

    Case II: XXXX|---- (any value from 000...1 to 499...9)

    Case III: XXXX|5000...0

    Case IV: XXXX|---- (any value from 5000...1 to 999...9)

STEP 3   Perform the rounding off.

    Case I requires no round-off. It is exactly XXXX.

Illustration C-1

Round 3.5000 to the nearest tenth.

$$3.5|000 \quad \text{becomes} \quad 3.5 \qquad \square$$

Case II requires rounding. We will round down for this case. That is, just drop the part of the number that is behind the vertical line.

Illustration C-2

Round 37.6124 to the nearest hundredth.

$$37.61|24 \quad \text{becomes} \quad 37.61 \qquad \square$$

Case III requires rounding. This is the case that requires special attention. **When a 5 (exactly a 5) is to be rounded off, round to the even digit.** In the long run, half of the time the 5 will be preceded by an even digit (0, 2, 4, 6, 8) and you will round down, while the other half of the time the 5 will be preceded by an odd digit (1, 3, 5, 7, 9) and you will round up.

Illustration C-3

Round 87.35 to the nearest tenth.

$$87.3|5 \quad \text{becomes} \quad 87.4$$

Round 93.445 to the nearest hundredth.

$$93.44|5 \quad \text{becomes} \quad 93.44$$

(Note: 87.35 is 87.35000... and 93.445 is 93.445000....)

Case IV requires rounding. We will round up for this case. That is, we will drop the part of the number that is behind the vertical line and we will increase the last digit in front of the vertical line by one.

Illustration C-4

Round 7.889 to the nearest tenth.

$$7.8|89 \quad \text{becomes} \quad 7.9$$

NOTE: Cases I, II, and IV describe what is commonly done. Our guidelines for case III are the only ones that are different from typical procedure.

If the typical round-off rule (0, 1, 2, 3, 4 are dropped; 5, 6, 7, 8, 9 are rounded up) is followed, then $(n + 1)/(2n + 1)$ of the situations are rounded up. (n is the number of different sequences of digits that fall into each of case II and case IV.) That is more than half. You (as many others have) may say, "So what?" In today's world that tiny, seemingly insignificant amount becomes very significant when applied repeatedly to large numbers.

Exercises

C-1 Round each of the following to the nearest integer:

(a) 12.94   (b) 8.762   (c) 9.05   (d) 156.49
(e) 45.5   (f) 42.5   (g) 102.51   (h) 16.5001

C-2 Round each of the following to the nearest tenth:

(a) 8.67   (b) 42.333   (c) 49.666   (d) 10.25   (e) 10.35
(f) 8.4501   (g) 27.35001   (h) 5.65   (i) 3.05   (j) $\frac{1}{4}$

C-3 Round each of the following to the nearest hundredth:

(a) 17.6666   (b) 4.444   (c) 54.5454   (d) 102.055   (e) 93.225
(f) 18.005   (g) 18.015   (h) 5.555   (i) 44.7450   (j) $\frac{2}{3}$

The answers to these exercises are on page A-37.

D Tables

TABLE 1
Random Numbers

10	09	73	25	33	76	52	01	35	86	34	67	35	48	76	80	95	90	91	17	39	29	27	49	45
37	54	20	48	05	64	89	47	42	96	24	80	52	40	37	20	63	61	04	02	00	82	29	16	65
08	42	26	89	53	19	64	50	93	03	23	20	90	25	60	15	95	33	47	64	35	08	03	36	06
99	01	90	25	29	09	37	67	07	15	38	31	13	11	65	88	67	67	43	97	04	43	62	76	59
12	80	79	99	70	80	15	73	61	47	64	03	23	66	53	98	95	11	68	77	12	17	17	68	33
66	06	57	47	17	34	07	27	68	50	36	69	73	61	70	65	81	33	98	85	11	19	92	91	70
31	06	01	08	05	45	57	18	24	06	35	30	34	26	14	86	79	90	74	39	23	40	30	97	32
85	26	97	76	02	02	05	16	56	92	68	66	57	48	18	73	05	38	52	47	18	62	38	85	79
63	57	33	21	35	05	32	54	70	48	90	55	35	75	48	28	46	82	87	09	83	49	12	56	24
73	79	64	57	53	03	52	96	47	78	35	80	83	42	82	60	93	52	03	44	35	27	38	84	35
98	52	01	77	67	14	90	56	86	07	22	10	94	05	58	60	97	09	34	33	50	50	07	39	98
11	80	50	54	31	39	80	82	77	32	50	72	56	82	48	29	40	52	42	01	52	77	56	78	51
83	45	29	96	34	06	28	89	80	83	13	74	67	00	78	18	47	54	06	10	68	71	17	78	17
88	68	54	02	00	86	50	75	84	01	36	76	66	79	51	90	36	47	64	93	29	60	91	10	62
99	59	46	73	48	87	51	76	49	69	91	82	60	89	28	93	78	56	13	68	23	47	83	41	13
65	48	11	76	74	17	46	85	09	50	58	04	77	69	74	73	03	95	71	86	40	21	81	65	44
80	12	43	56	35	17	72	70	80	15	45	31	82	23	74	21	11	57	82	53	14	38	55	37	63
74	35	09	98	17	77	40	27	72	14	43	23	60	02	10	45	52	16	42	37	96	28	60	26	55
69	91	62	68	03	66	25	22	91	48	36	93	68	72	03	76	62	11	39	90	94	40	05	64	18
09	89	32	05	05	14	22	56	85	14	46	42	75	67	88	96	29	77	88	22	54	38	21	45	98
91	49	91	45	23	68	47	92	76	86	46	16	28	35	54	94	75	08	99	23	37	08	92	00	48
80	33	69	45	98	26	94	03	68	58	70	29	73	41	35	53	14	03	33	40	42	05	08	23	41
44	10	48	19	49	85	15	74	79	54	32	97	92	65	75	57	60	04	08	81	22	22	20	64	13
12	55	07	37	42	11	10	00	20	40	12	86	07	46	97	96	64	48	94	39	28	70	72	58	15
63	60	64	93	29	16	50	53	44	84	40	21	95	25	63	43	65	17	70	82	07	20	73	17	90
61	19	69	04	46	26	45	74	77	74	51	92	43	37	29	65	39	45	95	93	42	58	26	05	27
15	47	44	52	66	95	27	07	99	53	59	36	78	38	48	82	39	61	01	18	33	21	15	94	66
94	55	72	85	73	67	89	75	43	87	54	62	24	44	31	91	19	04	25	92	92	92	74	59	73
42	48	11	62	13	97	34	40	87	21	16	86	84	87	67	03	07	11	20	59	25	70	14	66	70
23	52	37	83	17	73	20	88	98	37	68	93	59	14	16	26	25	22	96	63	05	52	28	25	62
04	49	35	24	94	75	24	63	38	24	45	86	25	10	25	61	96	27	93	35	65	33	71	24	72
00	54	99	76	54	64	05	18	81	59	96	11	96	38	96	54	69	28	23	91	23	28	72	95	29
35	96	31	53	07	26	89	80	93	54	33	35	13	54	62	77	97	45	00	24	90	10	33	93	33
59	80	80	83	91	45	42	72	68	42	83	60	94	97	00	13	02	12	48	92	78	56	52	01	06
46	05	88	52	36	01	39	09	22	86	77	28	14	40	77	93	91	08	36	47	70	61	74	29	41
32	17	90	05	97	87	37	92	52	41	05	56	70	70	07	86	74	31	71	57	85	39	41	18	38
69	23	46	14	06	20	11	74	52	04	15	95	66	00	00	18	74	39	24	23	97	11	89	63	38
19	56	54	14	30	01	75	87	53	79	40	41	92	15	85	66	67	43	68	06	84	96	28	52	07
45	15	51	49	38	19	47	60	72	46	43	66	79	45	43	59	04	79	00	33	20	82	66	95	41
94	86	43	19	94	36	16	81	08	51	34	88	88	15	53	01	54	03	54	56	05	01	45	11	76
98	08	62	48	26	45	24	02	84	04	44	99	90	88	96	39	09	47	34	07	35	44	13	18	80
33	18	51	62	32	41	94	15	09	49	89	43	54	85	81	88	69	54	19	94	37	54	87	30	43
80	95	10	04	06	96	38	27	07	74	20	15	12	33	87	25	01	62	52	98	94	62	46	11	71
79	75	24	91	40	71	96	12	82	96	69	86	10	25	91	74	85	22	05	39	00	38	75	95	79
18	63	33	25	37	98	14	50	65	71	31	01	02	46	74	05	45	56	14	27	77	93	89	19	36

For specific details on the use of this table, see page A-8.

74	02	94	39	02	77	55	73	22	70	97	79	01	71	19	52	52	75	80	21	80	81	45	17	48
54	17	84	56	11	80	99	33	71	43	05	33	51	29	69	56	12	71	92	55	36	04	09	03	24
11	66	44	98	83	52	07	98	48	27	59	38	17	15	39	09	97	33	34	40	88	46	12	33	56
48	32	47	79	28	31	24	96	47	10	02	29	53	68	70	32	30	75	75	46	15	02	00	99	94
69	07	49	41	38	87	63	79	19	76	35	58	40	44	01	10	51	82	16	15	01	84	87	69	38
09	18	82	00	97	32	82	53	95	27	04	22	08	63	04	83	38	98	73	74	64	27	85	80	44
90	04	58	54	97	51	98	15	06	54	94	93	88	19	97	91	87	07	61	50	68	47	66	46	59
73	18	95	02	07	47	67	72	62	69	62	29	06	44	64	27	12	46	70	18	41	36	18	27	60
75	76	87	64	90	20	97	18	17	49	90	42	91	22	72	95	37	50	58	71	93	82	34	31	78
54	01	64	40	56	66	28	13	10	03	00	68	22	73	98	20	71	45	32	95	07	70	61	78	13
08	35	86	99	10	78	54	24	27	85	13	66	15	88	73	04	61	89	75	53	31	22	30	84	20
28	30	60	32	64	81	33	31	05	91	40	51	00	78	93	32	60	46	04	75	94	11	90	18	40
53	84	08	62	33	81	59	41	36	28	51	21	59	02	90	28	46	66	87	95	77	76	22	07	91
91	75	75	37	41	61	61	36	22	69	50	26	39	02	12	55	78	17	65	14	83	48	34	70	55
89	41	59	26	94	00	39	75	83	91	12	60	71	76	46	48	94	97	23	06	94	54	13	74	08
77	51	30	38	20	86	83	42	99	01	68	41	48	27	74	51	90	81	39	80	72	89	35	55	07
19	50	23	71	74	69	97	92	02	88	55	21	02	97	73	74	28	77	52	51	65	34	46	74	15
21	81	85	93	13	93	27	88	17	57	05	68	67	31	56	07	08	28	50	46	31	85	33	84	52
51	47	46	64	99	68	10	72	36	21	94	04	99	13	45	42	83	60	91	91	08	00	74	54	49
99	55	96	83	31	62	53	52	41	70	69	77	71	28	30	74	81	97	81	42	43	86	07	28	34
33	71	34	80	07	93	58	47	28	69	51	92	66	47	21	58	30	32	98	22	93	17	49	39	72
85	27	48	68	93	11	30	32	92	70	28	83	43	41	37	73	51	59	04	00	71	14	84	36	43
84	13	38	96	40	44	03	55	21	66	73	85	27	00	91	61	22	26	05	61	62	32	71	84	23
56	73	21	62	34	17	39	59	61	31	10	12	39	16	22	85	49	65	75	60	81	60	41	88	80
65	13	85	68	06	87	64	88	52	61	34	31	36	58	61	45	87	52	10	69	85	64	44	72	77
38	00	10	21	76	81	71	91	17	11	71	60	29	29	37	74	21	96	40	49	65	58	44	96	98
37	40	29	63	97	01	30	47	75	86	56	27	11	00	86	47	32	46	26	05	40	03	03	74	38
97	12	54	03	48	87	08	33	14	17	21	81	53	92	50	75	23	76	20	47	15	50	12	95	78
21	82	64	11	34	47	14	33	40	72	64	63	88	59	02	49	13	90	64	41	03	85	65	45	52
73	13	54	27	42	95	71	90	90	35	85	79	47	42	96	08	78	98	81	56	64	69	11	92	02
07	63	87	79	29	03	06	11	80	72	96	20	74	41	56	23	82	19	95	38	04	71	36	69	94
60	52	88	34	41	07	95	41	98	14	59	17	52	06	95	05	53	35	21	39	61	21	20	64	55
83	59	63	56	55	06	95	89	29	83	05	12	80	97	19	77	43	35	37	83	92	30	15	04	98
10	85	06	27	46	99	59	91	05	07	13	49	90	63	19	53	07	57	18	39	06	41	01	93	62
39	82	09	89	52	43	62	26	31	47	64	42	18	08	14	43	80	00	93	51	31	02	47	31	67
59	58	00	64	78	75	56	97	88	00	88	83	55	44	86	23	76	80	61	56	04	11	10	84	08
38	50	80	73	41	23	79	34	87	63	90	82	29	70	22	17	71	90	42	07	95	95	44	99	53
30	69	27	06	68	94	68	81	61	27	56	19	68	00	91	82	06	76	34	00	05	46	26	92	00
65	44	39	56	59	18	28	82	74	37	49	63	22	40	41	08	33	76	56	76	96	29	99	08	36
27	26	75	02	64	13	19	27	22	94	07	47	74	46	06	17	98	54	89	11	97	34	13	03	58
91	30	70	69	91	19	07	22	42	10	36	69	95	37	28	28	82	53	57	93	28	97	66	62	52
68	43	49	46	88	84	47	31	36	22	62	12	69	84	08	12	84	38	25	90	09	81	59	31	46
48	90	81	58	77	54	74	52	45	91	35	70	00	47	54	83	82	45	26	92	54	13	05	51	60
06	91	34	51	97	42	67	27	86	01	11	88	30	95	28	63	01	19	89	01	14	97	44	03	44
10	45	51	60	19	14	21	03	37	12	91	34	23	78	21	88	32	58	08	51	43	66	77	08	83
12	88	39	73	43	65	02	76	11	84	04	28	50	13	92	17	97	41	50	77	90	71	22	67	69
21	77	83	09	76	38	80	73	69	61	31	64	94	20	96	63	28	10	20	23	08	81	64	74	49
19	52	35	95	15	65	12	25	96	59	86	28	36	82	58	69	57	21	37	98	16	43	59	15	29
67	24	55	26	70	35	58	31	65	63	79	24	68	66	86	76	46	33	42	22	26	65	59	08	02
60	58	44	73	77	07	50	03	79	92	45	13	42	65	29	26	76	08	36	37	41	32	64	43	44
53	85	34	13	77	36	06	69	48	50	58	83	87	38	59	49	36	47	33	31	96	24	04	36	42
24	63	73	87	36	74	38	48	93	42	52	62	30	79	92	12	36	91	86	01	03	74	28	38	73
83	08	01	24	51	38	99	22	28	15	07	75	95	17	77	97	37	72	75	85	51	97	23	78	67
16	44	42	43	34	36	15	19	90	73	27	49	37	09	39	85	13	03	25	52	54	84	65	47	59
60	79	01	81	57	57	17	86	57	62	11	16	17	85	76	45	81	95	29	79	65	13	00	48	60

From tables of the RAND Corporation. Reprinted from Wilfred J. Dixon and Frank J. Massey, Jr., *Introduction to Statistical Analysis*, 3rd ed. (New York: McGraw-Hill, 1969), pp. 446–447. Reprinted by permission of the RAND Corporation.

TABLE 2
Factorials

n	n!
0	1
1	1
2	2
3	6
4	24
5	120
6	720
7	5,040
8	40,320
9	362,880
10	3,628,800
11	39,916,800
12	479,001,600
13	6,227,020,800
14	87,178,291,200
15	1,307,674,368,000
16	20,922,789,888,000
17	355,687,428,096,000
18	6,402,373,705,728,000
19	121,645,100,408,832,000
20	2,432,902,008,176,640,000

For specific details on the use of this table, see page 194.

TABLE 3
Binomial Coefficients

n	$\binom{n}{0}$	$\binom{n}{1}$	$\binom{n}{2}$	$\binom{n}{3}$	$\binom{n}{4}$	$\binom{n}{5}$	$\binom{n}{6}$	$\binom{n}{7}$	$\binom{n}{8}$	$\binom{n}{9}$	$\binom{n}{10}$
0	1										
1	1	1									
2	1	2	1								
3	1	3	3	1							
4	1	4	6	4	1						
5	1	5	10	10	5	1					
6	1	6	15	20	15	6	1				
7	1	7	21	35	35	21	7	1			
8	1	8	28	56	70	56	28	8	1		
9	1	9	36	84	126	126	84	36	9	1	
10	1	10	45	120	210	252	210	120	45	10	1
11	1	11	55	165	330	462	462	330	165	55	11
12	1	12	66	220	495	792	924	792	495	220	66
13	1	13	78	286	715	1287	1716	1716	1287	715	286
14	1	14	91	364	1001	2002	3003	3432	3003	2002	1001
15	1	15	105	455	1365	3003	5005	6435	6435	5005	3003
16	1	16	120	560	1820	4368	8008	11440	12870	11440	8008
17	1	17	136	680	2380	6188	12376	19448	24310	24310	19448
18	1	18	153	816	3060	8568	18564	31824	43758	48620	43758
19	1	19	171	969	3876	11628	27132	50388	75582	92378	92378
20	1	20	190	1140	4845	15504	38760	77520	125970	167960	184756

If necessary, use the identity $\binom{n}{k} = \binom{n}{n-k}$

From John E. Freund, *Statistics, A First Course*, Prentice-Hall, Inc., Englewood Cliffs, N.J., 1970, p. 313. Reprinted by permission.

For specific details on the use of this table, see page 194.

TABLE 4

Binomial Probabilities $\left[\binom{n}{x} \cdot p^x q^{n-x}\right]$

n	x	0.01	0.05	0.10	0.20	0.30	0.40	0.50	0.60	0.70	0.80	0.90	0.95	0.99	x
2	0	980	902	810	640	490	360	250	160	090	040	010	002	0+	0
	1	020	095	180	320	420	480	500	480	420	320	180	095	020	1
	2	0+	002	010	040	090	160	250	360	490	640	810	902	980	2
3	0	970	857	729	512	343	216	125	064	027	008	001	0+	0+	0
	1	029	135	243	384	441	432	375	288	189	096	027	007	0+	1
	2	0+	007	027	096	189	288	375	432	441	384	243	135	029	2
	3	0+	0+	001	008	027	064	125	216	343	512	729	857	970	3
4	0	961	815	656	410	240	130	062	026	008	002	0+	0+	0+	0
	1	039	171	292	410	412	346	250	154	076	026	004	0+	0+	1
	2	001	014	049	154	265	346	375	346	265	154	049	014	001	2
	3	0+	0+	004	026	076	154	250	346	412	410	292	171	039	3
	4	0+	0+	0+	002	008	026	062	130	240	410	656	815	961	4
5	0	951	774	590	328	168	078	031	010	002	0+	0+	0+	0+	0
	1	048	204	328	410	360	259	156	077	028	006	0+	0+	0+	1
	2	001	021	073	205	309	346	312	230	132	051	008	001	0+	2
	3	0+	001	008	051	132	230	312	346	309	205	073	021	001	3
	4	0+	0+	0+	006	028	077	156	259	360	410	328	204	048	4
	5	0+	0+	0+	0+	002	010	031	078	168	328	590	774	951	5
6	0	941	735	531	262	118	047	016	004	001	0+	0+	0+	0+	0
	1	057	232	354	393	303	187	094	037	010	002	0+	0+	0+	1
	2	001	031	098	246	324	311	234	138	060	015	001	0+	0+	2
	3	0+	002	015	082	185	276	312	276	185	082	015	002	0+	3
	4	0+	0+	001	015	060	138	234	311	324	246	098	031	001	4
	5	0+	0+	0+	002	010	037	094	187	303	393	354	232	057	5
	6	0+	0+	0+	0+	001	004	016	047	118	262	531	735	941	6
7	0	932	698	478	210	082	028	008	002	0+	0+	0+	0+	0+	0
	1	066	257	372	367	247	131	055	017	004	0+	0+	0+	0+	1
	2	002	041	124	275	318	261	164	077	025	004	0+	0+	0+	2
	3	0+	004	023	115	227	290	273	194	097	029	003	0+	0+	3
	4	0+	0+	003	029	097	194	273	290	227	115	023	004	0+	4
	5	0+	0+	0+	004	025	077	164	261	318	275	124	041	002	5
	6	0+	0+	0+	0+	004	017	055	131	247	367	372	257	066	6
	7	0+	0+	0+	0+	0+	002	008	028	082	210	478	698	932	7

For specific details on the use of this table, see page 197.

TABLE 4 (Continued)

n	x	0.01	0.05	0.10	0.20	0.30	0.40	p 0.50	0.60	0.70	0.80	0.90	0.95	0.99	x
8	0	923	663	430	168	058	017	004	001	0+	0+	0+	0+	0+	0
	1	075	279	383	336	198	090	031	008	001	0+	0+	0+	0+	1
	2	003	051	149	294	296	209	109	041	010	001	0+	0+	0+	2
	3	0+	005	033	147	254	279	219	124	047	009	0+	0+	0+	3
	4	0+	0+	005	046	136	232	273	232	136	046	005	0+	0+	4
	5	0+	0+	0+	009	047	124	219	279	254	147	033	005	0+	5
	6	0+	0+	0+	001	010	041	109	209	296	294	149	051	003	6
	7	0+	0+	0+	0+	001	008	031	090	198	336	383	279	075	7
	8	0+	0+	0+	0+	0+	001	004	017	058	168	430	663	923	8
9	0	914	630	387	134	040	010	002	0+	0+	0+	0+	0+	0+	0
	1	083	299	387	302	156	060	018	004	0+	0+	0+	0+	0+	1
	2	003	063	172	302	267	161	070	021	004	0+	0+	0+	0+	2
	3	0+	008	045	176	267	251	164	074	021	003	0+	0+	0+	3
	4	0+	001	007	066	172	251	246	167	074	017	001	0+	0+	4
9	5	0+	0+	001	017	074	167	246	251	172	066	007	001	0+	5
	6	0+	0+	0+	003	021	074	164	251	267	176	045	008	0+	6
	7	0+	0+	0+	0+	004	021	070	161	267	302	172	063	003	7
	8	0+	0+	0+	0+	0+	004	018	060	156	302	387	299	083	8
	9	0+	0+	0+	0+	0+	0+	002	010	040	134	387	630	914	9
10	0	904	599	349	107	028	006	001	0+	0+	0+	0+	0+	0+	0
	1	091	315	387	268	121	040	010	002	0+	0+	0+	0+	0+	1
	2	004	075	194	302	233	121	044	011	001	0+	0+	0+	0+	2
	3	0+	010	057	201	267	215	117	042	009	001	0+	0+	0+	3
	4	0+	001	011	088	200	251	205	111	037	006	0+	0+	0+	4
	5	0+	0+	001	026	103	201	246	201	103	026	001	0+	0+	5
	6	0+	0+	0+	006	037	111	205	251	200	088	011	001	0+	6
	7	0+	0+	0+	001	009	042	117	215	267	201	057	010	0+	7
	8	0+	0+	0+	0+	001	011	044	121	233	302	194	075	004	8
	9	0+	0+	0+	0+	0+	002	010	040	121	268	387	315	091	9
	10	0+	0+	0+	0+	0+	0+	001	006	028	107	349	599	904	10

TABLE 4 (Continued)

n	x	0.01	0.05	0.10	0.20	0.30	0.40	p 0.50	0.60	0.70	0.80	0.90	0.95	0.99	x
11	0	895	569	314	086	020	004	0+	0+	0+	0+	0+	0+	0+	0
	1	099	329	384	236	093	027	005	001	0+	0+	0+	0+	0+	1
	2	005	087	213	295	200	089	027	005	001	0+	0+	0+	0+	2
	3	0+	014	071	221	257	177	081	023	004	0+	0+	0+	0+	3
	4	0+	001	016	111	220	236	161	070	017	002	0+	0+	0+	4
	5	0+	0+	002	039	132	221	226	147	057	010	0+	0+	0+	5
	6	0+	0+	0+	010	057	147	226	221	132	039	002	0+	0+	6
	7	0+	0+	0+	002	017	070	161	236	220	111	016	001	0+	7
	8	0+	0+	0+	0+	004	023	081	177	257	221	071	014	0+	8
	9	0+	0+	0+	0+	001	005	027	089	200	295	213	087	005	9
	10	0+	0+	0+	0+	0+	001	005	027	093	236	384	329	099	10
	11	0+	0+	0+	0+	0+	0+	0+	004	020	086	314	569	895	11
12	0	886	540	282	069	014	002	0+	0+	0+	0+	0+	0+	0+	0
	1	107	341	377	206	071	017	003	0+	0+	0+	0+	0+	0+	1
	2	006	099	230	283	168	064	016	002	0+	0+	0+	0+	0+	2
	3	0+	017	085	236	240	142	054	012	001	0+	0+	0+	0+	3
	4	0+	002	021	133	231	213	121	042	008	001	0+	0+	0+	4
	5	0+	0+	004	053	158	227	193	101	029	003	0+	0+	0+	5
	6	0+	0+	0+	016	079	177	226	177	079	016	0+	0+	0+	6
	7	0+	0+	0+	003	029	101	193	227	158	053	004	0+	0+	7
	8	0+	0+	0+	001	008	042	121	213	231	133	021	002	0+	8
	9	0+	0+	0+	0+	001	012	054	142	240	236	085	017	0+	9
	10	0+	0+	0+	0+	0+	002	016	064	168	283	230	099	006	10
	11	0+	0+	0+	0+	0+	0+	003	017	071	206	377	341	107	11
	12	0+	0+	0+	0+	0+	0+	0+	002	014	069	282	540	886	12
13	0	878	513	254	055	010	001	0+	0+	0+	0+	0+	0+	0+	0
	1	115	351	367	179	054	011	002	0+	0+	0+	0+	0+	0+	1
	2	007	111	245	268	139	045	010	001	0+	0+	0+	0+	0+	2
	3	0+	021	100	246	218	111	035	006	001	0+	0+	0+	0+	3
	4	0+	003	028	154	234	184	087	024	003	0+	0+	0+	0+	4
13	5	0+	0+	006	069	180	221	157	066	014	001	0+	0+	0+	5
	6	0+	0+	001	023	103	197	209	131	044	006	0+	0+	0+	6

TABLE 4 (Continued)

n	x	0.01	0.05	0.10	0.20	0.30	0.40	0.50	0.60	0.70	0.80	0.90	0.95	0.99	x
	7	0+	0+	0+	006	044	131	209	197	103	023	001	0+	0+	7
	8	0+	0+	0+	001	014	066	157	221	180	069	006	0+	0+	8
	9	0+	0+	0+	0+	003	024	087	184	234	154	028	003	0+	9
	10	0+	0+	0+	0+	001	006	035	111	218	246	100	021	0+	10
	11	0+	0+	0+	0+	0+	001	010	045	139	268	245	111	007	11
	12	0+	0+	0+	0+	0+	0+	002	011	054	179	367	351	115	12
	13	0+	0+	0+	0+	0+	0+	0+	001	010	055	254	513	878	13
14	0	869	488	229	044	007	001	0+	0+	0+	0+	0+	0+	0+	0
	1	123	359	356	154	041	007	001	0+	0+	0+	0+	0+	0+	1
	2	008	123	257	250	113	032	006	001	0+	0+	0+	0+	0+	2
	3	0+	026	114	250	194	085	022	003	0+	0+	0+	0+	0+	3
	4	0+	004	035	172	229	155	061	014	001	0+	0+	0+	0+	4
	5	0+	0+	008	086	196	207	122	041	007	0+	0+	0+	0+	5
	6	0+	0+	001	032	126	207	183	092	023	002	0+	0+	0+	6
	7	0+	0+	0+	009	062	157	209	157	062	009	0+	0+	0+	7
	8	0+	0+	0+	002	023	092	183	207	126	032	001	0+	0+	8
	9	0+	0+	0+	0+	007	041	122	207	196	086	008	0+	0+	9
	10	0+	0+	0+	0+	001	014	061	155	229	172	035	004	0+	10
	11	0+	0+	0+	0+	0+	003	022	085	194	250	114	026	0+	11
	12	0+	0+	0+	0+	0+	001	006	032	113	250	257	123	008	12
	13	0+	0+	0+	0+	0+	0+	001	007	041	154	356	359	123	13
	14	0+	0+	0+	0+	0+	0+	0+	001	007	044	229	488	869	14
15	0	860	463	206	035	005	0+	0+	0+	0+	0+	0+	0+	0+	0
	1	130	366	343	132	031	005	0+	0+	0+	0+	0+	0+	0+	1
	2	009	135	267	231	092	022	003	0+	0+	0+	0+	0+	0+	2
	3	0+	031	129	250	170	063	014	002	0+	0+	0+	0+	0+	3
	4	0+	005	043	188	219	127	042	007	001	0+	0+	0+	0+	4
	5	0+	001	010	103	206	186	092	024	003	0+	0+	0+	0+	5
	6	0+	0+	002	043	147	207	153	061	012	001	0+	0+	0+	6
	7	0+	0+	0+	014	081	177	196	118	035	003	0+	0+	0+	7
	8	0+	0+	0+	003	035	118	196	177	081	014	0+	0+	0+	8
	9	0+	0+	0+	001	012	061	153	207	147	043	002	0+	0+	9

From Frederick Mosteller, Robert E. K. Rourke, and George B. Thomas, Jr., *Probability with Statistical Applications*, 2nd ed., © 1970, Addison-Wesley Publishing Company, Reading, Mass., pp. 475–477. Reprinted with permission.

TABLE 5
Poisson Probabilities $P(x) = \dfrac{\mu^x \cdot e^{-\mu}}{x!}$

						μ					
x	0.1	0.5	1.0	1.5	2.0	2.5	3.0	3.5	4.0	5.0
0	0.9048	0.6065	0.3679	0.2231	0.1353	0.0821	0.0498	0.0302	0.0183	0.0067
1	0.0905	0.3033	0.3679	0.3347	0.2707	0.2052	0.1494	0.1057	0.0733	0.0337
2	0.0045	0.0758	0.1839	0.2510	0.2707	0.2565	0.2240	0.1850	0.1465	0.0842
3	0.0002	0.0126	0.0613	0.1255	0.1804	0.2138	0.2240	0.2158	0.1954	0.1404
4		0.0016	0.0153	0.0471	0.0902	0.1336	0.1680	0.1888	0.1954	0.1755
5		0.0002	0.0031	0.0141	0.0361	0.0668	0.1008	0.1322	0.1563	0.1755
6			0.0005	0.0035	0.0120	0.0278	0.0504	0.0771	0.1042	0.1462
7			0.0001	0.0008	0.0034	0.0099	0.0216	0.0385	0.0595	0.1044
8				0.0001	0.0009	0.0031	0.0081	0.0169	0.0298	0.0653
9					0.0002	0.0009	0.0027	0.0066	0.0132	0.0363
10						0.0002	0.0008	0.0023	0.0053	0.0181
11							0.0002	0.0007	0.0019	0.0082
12							0.0001	0.0002	0.0006	0.0034
13								0.0001	0.0002	0.0013
14									0.0001	0.0005
15										0.0002

For specific details on the use of this table, see page 204.

TABLE 6
Areas of the Standard Normal Distribution

The entries in this table are the probabilities that a random variable having the standard normal distribution assumes a value between 0 and z; the probability is represented by the area under the curve shaded in the figure. Areas for negative values of z are obtained by symmetry.

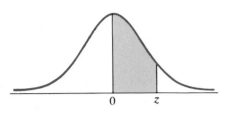

	Second decimal place in z									
z	0.00	0.01	0.02	0.03	0.04	0.05	0.06	0.07	0.08	0.09
0.0	0.0000	0.0040	0.0080	0.0120	0.0160	0.0199	0.0239	0.0279	0.0319	0.0359
0.1	0.0398	0.0438	0.0478	0.0517	0.0557	0.0596	0.0636	0.0675	0.0714	0.0753
0.2	0.0793	0.0832	0.0871	0.0910	0.0948	0.0987	0.1026	0.1064	0.1103	0.1141
0.3	0.1179	0.1217	0.1255	0.1293	0.1331	0.1368	0.1406	0.1443	0.1480	0.1517
0.4	0.1554	0.1591	0.1628	0.1664	0.1700	0.1736	0.1772	0.1808	0.1844	0.1879
0.5	0.1915	0.1950	0.1985	0.2019	0.2054	0.2088	0.2123	0.2157	0.2190	0.2224
0.6	0.2257	0.2291	0.2324	0.2357	0.2389	0.2422	0.2454	0.2486	0.2517	0.2549
0.7	0.2580	0.2611	0.2642	0.2673	0.2704	0.2734	0.2764	0.2794	0.2823	0.2852
0.8	0.2881	0.2910	0.2939	0.2967	0.2995	0.3023	0.3051	0.3078	0.3106	0.3133
0.9	0.3159	0.3186	0.3212	0.3238	0.3264	0.3289	0.3315	0.3340	0.3365	0.3389
1.0	0.3413	0.3438	0.3461	0.3485	0.3508	0.3531	0.3554	0.3577	0.3599	0.3621
1.1	0.3643	0.3665	0.3686	0.3708	0.3729	0.3749	0.3770	0.3790	0.3810	0.3830
1.2	0.3849	0.3869	0.3888	0.3907	0.3925	0.3944	0.3962	0.3980	0.3997	0.4015
1.3	0.4032	0.4049	0.4066	0.4082	0.4099	0.4115	0.4131	0.4147	0.4162	0.4177
1.4	0.4192	0.4207	0.4222	0.4236	0.4251	0.4265	0.4279	0.4292	0.4306	0.4319
1.5	0.4332	0.4345	0.4357	0.4370	0.4382	0.4394	0.4406	0.4418	0.4429	0.4441
1.6	0.4452	0.4463	0.4474	0.4484	0.4495	0.4505	0.4515	0.4525	0.4535	0.4545
1.7	0.4554	0.4564	0.4573	0.4582	0.4591	0.4599	0.4608	0.4616	0.4625	0.4633
1.8	0.4641	0.4649	0.4656	0.4664	0.4671	0.4678	0.4686	0.4693	0.4699	0.4706
1.9	0.4713	0.4719	0.4726	0.4732	0.4738	0.4744	0.4750	0.4756	0.4761	0.4767
2.0	0.4772	0.4778	0.4783	0.4788	0.4793	0.4798	0.4803	0.4808	0.4812	0.4817
2.1	0.4821	0.4826	0.4830	0.4834	0.4838	0.4842	0.4846	0.4850	0.4854	0.4857
2.2	0.4861	0.4864	0.4868	0.4871	0.4875	0.4878	0.4881	0.4884	0.4887	0.4890
2.3	0.4893	0.4896	0.4898	0.4901	0.4904	0.4906	0.4909	0.4911	0.4913	0.4916
2.4	0.4918	0.4920	0.4922	0.4925	0.4927	0.4929	0.4931	0.4932	0.4934	0.4936
2.5	0.4938	0.4940	0.4941	0.4943	0.4945	0.4946	0.4948	0.4949	0.4951	0.4952
2.6	0.4953	0.4955	0.4956	0.4957	0.4959	0.4960	0.4961	0.4962	0.4963	0.4964
2.7	0.4965	0.4966	0.4967	0.4968	0.4969	0.4970	0.4971	0.4972	0.4973	0.4974
2.8	0.4974	0.4975	0.4976	0.4977	0.4977	0.4978	0.4979	0.4979	0.4980	0.4981
2.9	0.4981	0.4982	0.4982	0.4983	0.4984	0.4984	0.4985	0.4985	0.4986	0.4986
3.0	0.4987	0.4987	0.4987	0.4988	0.4988	0.4989	0.4989	0.4989	0.4990	0.4990
3.1	0.4990	0.4991	0.4991	0.4991	0.4992	0.4992	0.4992	0.4992	0.4993	0.4993
3.2	0.4993	0.4993	0.4994	0.4994	0.4994	0.4994	0.4994	0.4995	0.4995	0.4995
3.3	0.4995	0.4995	0.4995	0.4996	0.4996	0.4996	0.4996	0.4996	0.4996	0.4997
3.4	0.4997	0.4997	0.4997	0.4997	0.4997	0.4997	0.4997	0.4997	0.4997	0.4998
3.5	0.4998									
4.0	0.49997									
4.5	0.499997									
5.0	0.4999997									

Reprinted with permission from *Standard Mathematical Tables*, 15th ed. Copyright The Chemical Rubber, Co., CRC Press, Inc.

For specific details on the use of this table, see page 218.

TABLE 7
Critical Values of Student's t Distribution

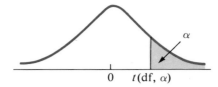

The entries in this table are the critical values for Student's t for an area of α in the right-hand tail. Critical values for the left-hand tail are found by symmetry.

df	Amount of α in one-tail					
	0.25	0.10	0.05	0.025	0.01	0.005
1	1.000	3.08	6.31	12.7	31.8	63.7
2	0.816	1.89	2.92	4.30	6.97	9.92
3	0.765	1.64	2.35	3.18	4.54	5.84
4	0.741	1.53	2.13	2.78	3.75	4.60
5	0.727	1.48	2.02	2.57	3.37	4.03
6	0.718	1.44	1.94	2.45	3.14	3.71
7	0.711	1.42	1.89	2.36	3.00	3.50
8	0.706	1.40	1.86	2.31	2.90	3.36
9	0.703	1.38	1.83	2.26	2.82	3.25
10	0.700	1.37	1.81	2.23	2.76	3.17
11	0.697	1.36	1.80	2.20	2.72	3.11
12	0.695	1.36	1.78	2.18	2.68	3.05
13	0.694	1.35	1.77	2.16	2.65	3.01
14	0.692	1.35	1.76	2.14	2.62	2.98
15	0.691	1.34	1.75	2.13	2.60	2.95
16	0.690	1.34	1.75	2.12	2.58	2.92
17	0.689	1.33	1.74	2.11	2.57	2.90
18	0.688	1.33	1.73	2.10	2.55	2.88
19	0.688	1.33	1.73	2.09	2.54	2.86
20	0.687	1.33	1.72	2.09	2.53	2.85
21	0.686	1.32	1.72	2.08	2.52	2.83
22	0.686	1.32	1.72	2.07	2.51	2.82
23	0.685	1.32	1.71	2.07	2.50	2.81
24	0.685	1.32	1.71	2.06	2.49	2.80
25	0.684	1.32	1.71	2.06	2.49	2.79
26	0.684	1.32	1.71	2.06	2.48	2.78
27	0.684	1.31	1.70	2.05	2.47	2.77
28	0.683	1.31	1.70	2.05	2.47	2.76
29	0.683	1.31	1.70	2.05	2.46	2.76
z	0.674	1.28	1.65	1.96	2.33	2.58

NOTE: For df $\geq$ 30, the critical value $t(df, \alpha)$ is approximated by $z(\alpha)$, given in the bottom row of table.

Adapted from E. S. Pearson and H. O. Hartley, *Biometrika Tables for Statisticians*, vol. I (1966), p. 146. Reprinted by permission of the Biometrika Trustees. The two columns headed "0.10" and "0.01" are taken from Table III (adapted) on p. 46 of Fisher and Yates, *Statistical Tables for Biological, Agricultural and Medical Research*, 6th ed., published by Longman Group Ltd., London, 1974 (previously published by Oliver and Boyd, Edinburgh), and by permission of the authors and publishers.

For specific details on the use of this table, see page 309.

TABLE 8
Critical Values of the χ^2 Distribution

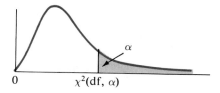

$\chi^2(df, \alpha)$

The entries in this table are the critical values for chi square for which the area to the right under the curve is equal to α.

df	\multicolumn{9}{c}{Amount of α in Right-hand Tail}
	0.995	0.990	0.975	0.950	0.900	0.100	0.050	0.025	0.010	0.005
1	0.0000393	0.000157	0.000982	0.00393	0.0158	2.71	3.84	5.02	6.64	7.88
2	0.0100	0.0201	0.0506	0.103	0.211	4.61	6.00	7.38	9.21	10.6
3	0.0717	0.115	0.216	0.352	0.584	6.25	7.82	9.35	11.4	12.9
4	0.207	0.297	0.484	0.711	1.0636	7.78	9.50	11.1	13.3	14.9
5	0.412	0.554	0.831	1.15	1.61	9.24	11.1	12.8	15.1	16.8
6	0.676	0.872	1.24	1.64	2.20	10.6	12.6	14.5	16.8	18.6
7	0.990	1.24	1.69	2.17	2.83	12.0	14.1	16.0	18.5	20.3
8	1.34	1.65	2.18	2.73	3.49	13.4	15.5	17.5	20.1	22.0
9	1.73	2.09	2.70	3.33	4.17	14.7	17.0	19.0	21.7	23.6
10	2.16	2.56	3.25	3.94	4.87	16.0	18.3	20.5	23.2	25.2
11	2.60	3.05	3.82	4.58	5.58	17.2	19.7	21.9	24.7	26.8
12	3.07	3.57	4.40	5.23	6.30	18.6	21.0	23.3	26.2	28.3
13	3.57	4.11	5.01	5.90	7.04	19.8	22.4	24.7	27.7	29.8
14	4.07	4.66	5.63	6.57	7.79	21.1	23.7	26.1	29.1	31.3
15	4.60	5.23	6.26	7.26	8.55	22.3	25.0	27.5	30.6	32.8
16	5.14	5.81	6.91	7.96	9.31	23.5	26.3	28.9	32.0	34.3
17	5.70	6.41	7.56	8.67	10.1	24.8	27.6	30.2	33.4	35.7
18	6.26	7.01	8.23	9.39	10.9	26.0	28.9	31.5	34.8	37.2
19	6.84	7.63	8.91	10.1	11.7	27.2	30.1	32.9	36.2	38.6
20	7.43	8.26	9.59	10.9	12.4	28.4	31.4	34.2	37.6	40.0
21	8.03	8.90	10.3	11.6	13.2	29.6	32.7	35.5	39.0	41.4
22	8.64	9.54	11.0	12.3	14.0	30.8	33.9	36.8	40.3	42.8
23	9.26	10.2	11.0	13.1	14.9	32.0	35.2	38.1	41.6	44.2
24	9.89	10.9	12.4	13.9	15.7	33.2	36.4	39.4	43.0	45.6
25	10.5	11.5	13.1	14.6	16.5	34.4	37.7	40.7	44.3	46.9
26	11.2	12.2	13.8	15.4	17.3	35.6	38.9	41.9	45.6	48.3
27	11.8	12.9	14.6	16.2	18.1	36.7	40.1	43.2	47.0	49.7
28	12.5	13.6	15.3	16.9	18.9	37.9	41.3	44.5	48.3	51.0
29	13.1	14.3	16.1	17.7	19.8	39.1	42.6	45.7	49.6	52.3
30	13.8	15.0	16.8	18.5	20.6	40.3	43.8	47.0	50.9	53.7
40	20.7	22.2	24.4	26.5	29.1	51.8	55.8	59.3	63.7	66.8
50	28.0	29.7	32.4	34.8	37.7	63.2	67.5	71.4	76.2	79.5
60	35.5	37.5	40.5	43.2	46.5	74.4	79.1	83.3	88.4	92.0
70	43.3	45.4	48.8	51.8	55.3	85.5	90.5	95.0	100.0	104.0
80	51.2	53.5	57.2	60.4	64.3	96.6	102.0	107.0	112.0	116.0
90	59.2	61.8	65.7	69.1	73.3	108.0	113.0	118.0	124.0	128.0
100	67.3	70.1	74.2	77.9	82.4	114.0	124.0	130.0	136.0	140.0

Adapted from E. S. Pearson and H. O. Hartley, *Biometrika Tables for Statisticians*, vol. I (1962), pp. 130–131. Reprinted by permission of the Biometrika Trustees.

For specific details on the use of this table, see page 324.

TABLE 9a
Critical Values of the F Distribution ($\alpha = 0.05$)

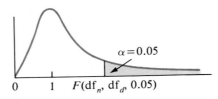

The entries in this table are critical values of F for which the area under the curve to the right is equal to 0.05.

Degrees of Freedom for Denominator	Degrees of Freedom for Numerator									
	1	2	3	4	5	6	7	8	9	10
1	161	200	216	225	230	234	237	239	241	242
2	18.5	19.0	19.2	19.2	19.3	19.3	19.4	19.4	19.4	19.4
3	10.1	9.55	9.28	9.12	9.01	8.94	8.89	8.85	8.81	8.79
4	7.71	6.94	6.59	6.39	6.26	6.16	6.09	6.04	6.00	5.96
5	6.61	5.79	5.41	5.19	5.05	4.95	4.88	4.82	4.77	4.74
6	5.99	5.14	4.76	4.53	4.39	4.28	4.21	4.15	4.10	4.06
7	5.59	4.74	4.35	4.12	3.97	3.87	3.79	3.73	3.68	3.64
8	5.32	4.46	4.07	3.84	3.69	3.58	3.50	3.44	3.39	3.35
9	5.12	4.26	3.86	3.63	3.48	3.37	3.29	3.23	3.18	3.14
10	4.96	4.10	3.71	3.48	3.33	3.22	3.14	3.07	3.02	2.98
11	4.84	3.98	3.59	3.36	3.20	3.09	3.01	2.95	2.90	2.85
12	4.75	3.89	3.49	3.26	3.11	3.00	2.91	2.85	2.80	2.75
13	4.67	3.81	3.41	3.18	3.03	2.92	2.83	2.77	2.71	2.67
14	4.60	3.74	3.34	3.11	2.96	2.85	2.76	2.70	2.65	2.60
15	4.54	3.68	3.29	3.06	2.90	2.79	2.71	2.64	2.59	2.54
16	4.49	3.63	3.24	3.01	2.85	2.74	2.66	2.59	2.54	2.49
17	4.45	3.59	3.20	2.96	2.81	2.70	2.61	2.55	2.49	2.45
18	4.41	3.55	3.16	2.93	2.77	2.66	2.58	2.51	2.46	2.41
19	4.38	3.52	3.13	2.90	2.74	2.63	2.54	2.48	2.42	2.38
20	4.35	3.49	3.10	2.87	2.71	2.60	2.51	2.45	2.39	2.35
21	4.32	3.47	3.07	2.84	2.68	2.57	2.49	2.42	2.37	2.32
22	4.30	3.44	3.05	2.82	2.66	2.55	2.46	2.40	2.34	2.30
23	4.28	3.42	3.03	2.80	2.64	2.53	2.44	2.37	2.32	2.27
24	4.26	3.40	3.01	2.78	2.62	2.51	2.42	2.36	2.30	2.25
25	4.24	3.39	2.99	2.76	2.60	2.49	2.40	2.34	2.28	2.24
30	4.17	3.32	2.92	2.69	2.53	2.42	2.33	2.27	2.21	2.16
40	4.08	3.23	2.84	2.61	2.45	2.34	2.25	2.18	2.12	2.08
60	4.00	3.15	2.76	2.53	2.37	2.25	2.17	2.10	2.04	1.99
120	3.92	3.07	2.68	2.45	2.29	2.18	2.09	2.02	1.96	1.91
∞	3.84	3.00	2.60	2.37	2.21	2.10	2.01	1.94	1.88	1.83

For specific details on the use of this table, see page 352.

TABLE 9a (Continued)

		\multicolumn{9}{c}{Degrees of Freedom for Numerator}
		12	15	20	24	30	40	60	120	∞
	1	244	246	248	249	250	251	252	253	254
	2	19.4	19.4	19.4	19.5	19.5	19.5	19.5	19.5	19.5
	3	8.74	8.70	8.66	8.64	8.62	8.59	8.57	8.55	8.53
	4	5.91	5.86	5.80	5.77	5.75	5.72	5.69	5.66	5.63
	5	4.68	4.62	4.56	4.53	4.50	4.46	4.43	4.40	4.37
	6	4.00	3.94	3.87	3.84	3.81	3.77	3.74	3.70	3.67
	7	3.57	3.51	3.44	3.41	3.38	3.34	3.30	3.27	3.23
	8	3.28	3.22	3.15	3.12	3.08	3.04	3.01	2.97	2.93
	9	3.07	3.01	2.94	2.90	2.86	2.83	2.79	2.75	2.71
	10	2.91	2.85	2.77	2.74	2.70	2.66	2.62	2.58	2.54
Degrees of Freedom for Denominator	11	2.79	2.72	2.65	2.61	2.57	2.53	2.49	2.45	2.40
	12	2.69	2.62	2.54	2.51	2.47	2.43	2.38	2.34	2.30
	13	2.60	2.53	2.46	2.42	2.38	2.34	2.30	2.25	2.21
	14	2.53	2.46	2.39	2.35	2.31	2.27	2.22	2.18	2.13
	15	2.48	2.40	2.33	2.29	2.25	2.20	2.16	2.11	2.07
	16	2.42	2.35	2.28	2.24	2.19	2.15	2.11	2.06	2.01
	17	2.38	2.31	2.23	2.19	2.15	2.10	2.06	2.01	1.96
	18	2.34	2.27	2.19	2.15	2.11	2.06	2.02	1.97	1.92
	19	2.31	2.23	2.16	2.11	2.07	2.03	1.98	1.93	1.88
	20	2.28	2.20	2.12	2.08	2.04	1.99	1.95	1.90	1.84
	21	2.25	2.18	2.10	2.05	2.01	1.96	1.92	1.87	1.81
	22	2.23	2.15	2.07	2.03	1.98	1.94	1.89	1.84	1.78
	23	2.20	2.13	2.05	2.01	1.96	1.91	1.86	1.81	1.76
	24	2.18	2.11	2.03	1.98	1.94	1.89	1.84	1.79	1.73
	25	2.16	2.09	2.01	1.96	1.92	1.87	1.82	1.77	1.71
	30	2.09	2.01	1.93	1.89	1.84	1.79	1.74	1.68	1.62
	40	2.00	1.92	1.84	1.79	1.74	1.69	1.64	1.58	1.51
	60	1.92	1.84	1.75	1.70	1.65	1.59	1.53	1.47	1.39
	120	1.83	1.75	1.66	1.61	1.55	1.50	1.43	1.35	1.25
	∞	1.75	1.67	1.57	1.52	1.46	1.39	1.32	1.22	1.00

From E. S. Pearson and H. O. Hartley, *Biometrika Tables for Statisticians*, vol. I (1958), pp. 159–163. Reprinted by permission of the Biometrika Trustees.

TABLE 9b
Critical Values of the F Distribution ($\alpha = 0.025$)

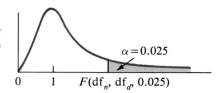

The entries in this table are critical values of F for which the area under the curve to the right is equal to 0.025.

		\multicolumn{10}{c}{Degrees of Freedom for Numerator}
		1	2	3	4	5	6	7	8	9	10
Degrees of Freedom for Denominator	1	648	800	864	900	922	937	948	957	963	969
	2	38.5	39.0	39.2	39.2	39.3	39.3	39.4	39.4	39.4	39.4
	3	17.4	16.0	15.4	15.1	14.9	14.7	14.6	14.5	14.5	14.4
	4	12.2	10.6	9.98	9.60	9.36	9.20	9.07	8.98	8.90	8.84
	5	10.0	8.43	7.76	7.39	7.15	6.98	6.85	6.76	6.68	6.62
	6	8.81	7.26	6.60	6.23	5.99	5.82	5.70	5.60	5.52	5.46
	7	8.07	6.54	5.89	5.52	5.29	5.12	4.99	4.90	4.82	4.76
	8	7.57	6.06	5.42	5.05	4.82	4.65	4.53	4.43	4.36	4.30
	9	7.21	5.71	5.08	4.72	4.48	4.32	4.20	4.10	4.03	3.96
	10	6.94	5.46	4.83	4.47	4.24	4.07	3.95	3.85	3.78	3.72
	11	6.72	5.26	4.63	4.28	4.04	3.88	3.76	3.66	3.59	3.53
	12	6.55	5.10	4.47	4.12	3.89	3.73	3.61	3.51	3.44	3.37
	13	6.41	4.97	4.35	4.00	3.77	3.60	3.48	3.39	3.31	3.25
	14	6.30	4.86	4.24	3.89	3.66	3.50	3.38	3.28	3.21	3.15
	15	6.20	4.77	4.15	3.80	3.58	3.41	3.29	3.20	3.12	3.06
	16	6.12	4.69	4.08	3.73	3.50	3.34	3.22	3.12	3.05	2.99
	17	6.04	4.62	4.01	3.66	3.44	3.28	3.16	3.06	2.98	2.92
	18	5.98	4.56	3.95	3.61	3.38	3.22	3.10	3.01	2.93	2.87
	19	5.92	4.51	3.90	3.56	3.33	3.17	3.05	2.96	2.88	2.82
	20	5.87	4.46	3.86	3.51	3.29	3.13	3.01	2.91	2.84	2.77
	21	5.83	4.42	3.82	3.48	3.25	3.09	2.97	2.87	2.80	2.73
	22	5.79	4.38	3.78	3.44	3.22	3.05	2.93	2.84	2.76	2.70
	23	5.75	4.35	3.75	3.41	3.18	3.02	2.90	2.81	2.73	2.67
	24	5.72	4.32	3.72	3.38	3.15	2.99	2.87	2.78	2.70	2.64
	25	5.69	4.29	3.69	3.35	3.13	2.97	2.85	2.75	2.68	2.61
	30	5.57	4.18	3.59	3.25	3.03	2.87	2.75	2.65	2.57	2.51
	40	5.42	4.05	3.46	3.13	2.90	2.74	2.62	2.53	2.45	2.39
	60	5.29	3.93	3.34	3.01	2.79	2.63	2.51	2.41	2.33	2.27
	120	5.15	3.80	3.23	2.89	2.67	2.52	2.39	2.30	2.22	2.16
	∞	5.02	3.69	3.12	2.79	2.57	2.41	2.29	2.19	2.11	2.05

For specific details on the use of this table, see page 352.

TABLE 9b (Continued)

		\multicolumn{9}{c}{Degrees of Freedom for Numerator}								
		12	15	20	24	30	40	60	120	∞
Degrees of Freedom for Denominator	1	977	985	993	997	1,001	1,006	1,010	1,014	1,018
	2	39.4	39.4	39.4	39.5	39.5	39.5	39.5	39.5	39.5
	3	14.3	14.3	14.2	14.1	14.1	14.0	14.0	13.9	13.9
	4	8.75	8.66	8.56	8.51	8.46	8.41	8.36	8.31	8.26
	5	6.52	6.43	6.33	6.28	6.23	6.18	6.12	6.07	6.02
	6	5.37	5.27	5.17	5.12	5.07	5.01	4.96	4.90	4.85
	7	4.67	4.57	4.47	4.42	4.36	4.31	4.25	4.20	4.14
	8	4.20	4.10	4.00	3.95	3.89	3.84	3.78	3.73	3.67
	9	3.87	3.77	3.67	3.61	3.56	3.51	3.45	3.39	3.33
	10	3.62	3.52	3.42	3.37	3.31	3.26	3.20	3.14	3.08
	11	3.43	3.33	3.23	3.17	3.12	3.06	3.00	2.94	2.88
	12	3.28	3.18	3.07	3.02	2.96	2.91	2.85	2.79	2.72
	13	3.15	3.05	2.95	2.89	2.84	2.78	2.72	2.66	2.60
	14	3.05	2.95	2.84	2.79	2.73	2.67	2.61	2.55	2.49
	15	2.96	2.86	2.76	2.70	2.64	2.59	2.52	2.46	2.40
	16	2.89	2.79	2.68	2.63	2.57	2.51	2.45	2.38	2.32
	17	2.82	2.72	2.62	2.56	2.50	2.44	2.38	2.32	2.25
	18	2.77	2.67	2.56	2.50	2.44	2.38	2.32	2.26	2.19
	19	2.72	2.62	2.51	2.45	2.39	2.33	2.27	2.20	2.13
	20	2.68	2.57	2.46	2.41	2.35	2.29	2.22	2.16	2.09
	21	2.64	2.53	2.42	2.37	2.31	2.25	2.18	2.11	2.04
	22	2.60	2.50	2.39	2.33	2.27	2.21	2.14	2.08	2.00
	23	2.57	2.47	2.36	2.30	2.24	2.18	2.11	2.04	1.97
	24	2.54	2.44	2.33	2.27	2.21	2.15	2.08	2.01	1.94
	25	2.51	2.41	2.30	2.24	2.18	2.12	2.05	1.98	1.91
	30	2.41	2.31	2.20	2.14	2.07	2.01	1.94	1.87	1.79
	40	2.29	2.18	2.07	2.01	1.94	1.88	1.80	1.72	1.64
	60	2.17	2.06	1.94	1.88	1.82	1.74	1.67	1.58	1.48
	120	2.05	1.95	1.82	1.76	1.69	1.61	1.53	1.43	1.31
	∞	1.94	1.83	1.71	1.64	1.57	1.48	1.39	1.27	1.00

From E. S. Pearson and H. O. Hartley, *Biometrika Tables for Statisticians*, vol. I (1958), pp. 159–163. Reprinted by permission of the Biometrika Trustees.

TABLE 9c
Critical Values of the F Distribution ($\alpha = 0.01$)

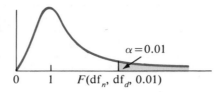

The entries in the table are critical values of F for which the area under the curve to the right is equal to 0.01.

	Degrees of Freedom for Numerator									
	1	2	3	4	5	6	7	8	9	10
1	4,052	5,000	5,403	5,625	5,764	5,859	5,928	5,982	6,023	6,056
2	98.5	99.0	99.2	99.2	99.3	99.3	99.4	99.4	99.4	99.4
3	34.1	30.8	29.5	28.7	28.2	27.9	27.7	27.5	27.3	27.2
4	21.2	18.0	16.7	16.0	15.5	15.2	15.0	14.8	14.7	14.5
5	16.3	13.3	12.1	11.4	11.0	10.7	10.5	10.3	10.2	10.1
6	13.7	10.9	9.78	9.15	8.75	8.47	8.26	8.10	7.98	7.87
7	12.2	9.55	8.45	7.85	7.46	7.19	6.99	6.84	6.72	6.62
8	11.3	8.65	7.59	7.01	6.63	6.37	6.18	6.03	5.91	5.81
9	10.6	8.02	6.99	6.42	6.06	5.80	5.61	5.47	5.35	5.26
10	10.0	7.56	6.55	5.99	5.64	5.39	5.20	5.06	4.94	4.85
11	9.65	7.21	6.22	5.67	5.32	5.07	4.89	4.74	4.63	4.54
12	9.33	6.93	5.95	5.41	5.06	4.82	4.64	4.50	4.39	4.30
13	9.07	6.70	5.74	5.21	4.86	4.62	4.44	4.30	4.19	4.10
14	8.86	6.51	5.56	5.04	4.70	4.46	4.28	4.14	4.03	3.94
15	8.68	6.36	5.42	4.89	4.56	4.32	4.14	4.00	3.89	3.80
16	8.53	6.23	5.29	4.77	4.44	4.20	4.03	3.89	3.78	3.69
17	8.40	6.11	5.19	4.67	4.34	4.10	3.93	3.79	3.68	3.59
18	8.29	6.01	5.09	4.58	4.25	4.01	3.84	3.71	3.60	3.51
19	8.19	5.93	5.01	4.50	4.17	3.94	3.77	3.63	3.52	3.43
20	8.10	5.85	4.94	4.43	4.10	3.87	3.70	3.56	3.46	3.37
21	8.02	5.78	4.87	4.37	4.04	3.81	3.64	3.51	3.40	3.31
22	7.95	5.72	4.82	4.31	3.99	3.76	3.59	3.45	3.35	3.26
23	7.88	5.66	4.76	4.26	3.94	3.71	3.54	3.41	3.30	3.21
24	7.82	5.61	4.72	4.22	3.90	3.67	3.50	3.36	3.26	3.17
25	7.77	5.57	4.68	4.18	3.86	3.63	3.46	3.32	3.22	3.13
30	7.56	5.39	4.51	4.02	3.70	3.47	3.30	3.17	3.07	2.98
40	7.31	5.18	4.31	3.83	3.51	3.29	3.12	2.99	2.89	2.80
60	7.08	4.98	4.13	3.65	3.34	3.12	2.95	2.82	2.72	2.63
120	6.85	4.79	3.95	3.48	3.17	2.96	2.79	2.66	2.56	2.47
∞	6.63	4.61	3.78	3.32	3.02	2.80	2.64	2.51	2.41	2.32

Degrees of Freedom for Denominator

TABLE 9c (Continued)

		\multicolumn{9}{c}{Degrees of Freedom for Numerator}								
		12	15	20	24	30	40	60	120	∞
Degrees of Freedom for Denominator	1	6,106	6,157	6,209	6,235	6,261	6,287	6,313	6,339	6,366
	2	99.4	99.4	99.4	99.5	99.5	99.5	99.5	99.5	99.5
	3	27.1	26.9	26.7	26.6	26.5	26.4	26.3	26.2	26.1
	4	14.4	14.2	14.0	13.9	13.8	13.7	13.7	13.6	13.5
	5	9.89	9.72	9.55	9.47	9.38	9.29	9.20	9.11	9.02
	6	7.72	7.56	7.40	7.31	7.23	7.14	7.06	6.97	6.88
	7	6.47	6.31	6.16	6.07	5.99	5.91	5.82	5.74	5.65
	8	5.67	5.52	5.36	5.28	5.20	5.12	5.03	4.95	4.86
	9	5.11	4.96	4.81	4.73	4.65	4.57	4.48	4.40	4.31
	10	4.71	4.56	4.41	4.33	4.25	4.17	4.08	4.00	3.91
	11	4.40	4.25	4.10	4.02	3.94	3.86	3.78	3.69	3.60
	12	4.16	4.01	3.86	3.78	3.70	3.62	3.54	3.45	3.36
	13	3.96	3.82	3.66	3.59	3.51	3.43	3.34	3.25	3.17
	14	3.80	3.66	3.51	3.43	3.35	3.27	3.18	3.09	3.00
	15	3.67	3.52	3.37	3.29	3.21	3.13	3.05	2.96	2.87
	16	3.55	3.41	3.26	3.18	3.10	3.02	2.93	2.84	2.75
	17	3.46	3.31	3.16	3.08	3.00	2.92	2.83	2.75	2.65
	18	3.37	3.23	3.08	3.00	2.92	2.84	2.75	2.66	2.57
	19	3.30	3.15	3.00	2.92	2.84	2.76	2.67	2.58	2.49
	20	3.23	3.09	2.94	2.86	2.78	2.69	2.61	2.52	2.42
	21	3.17	3.03	2.88	2.80	2.72	2.64	2.55	2.46	2.36
	22	3.12	2.98	2.83	2.75	2.67	2.58	2.50	2.40	2.31
	23	3.07	2.93	2.78	2.70	2.62	2.54	2.45	2.35	2.26
	24	3.03	2.89	2.74	2.66	2.58	2.49	2.40	2.31	2.21
	25	2.99	2.85	2.70	2.62	2.53	2.45	2.36	2.27	2.17
	30	2.84	2.70	2.55	2.47	2.39	2.30	2.21	2.11	2.01
	40	2.66	2.52	2.37	2.29	2.20	2.11	2.02	1.92	1.80
	60	2.50	2.35	2.20	2.12	2.03	1.94	1.84	1.73	1.60
	120	2.34	2.19	2.03	1.95	1.86	1.76	1.66	1.53	1.38
	∞	2.18	2.04	1.88	1.79	1.70	1.59	1.47	1.32	1.00

From E. S. Pearson and H. O. Hartley, *Biometrika Tables for Statisticians*, vol. I (1958), pp. 159–163. Reprinted by permission of the Biometrika Trustees.

TABLE 10
Critical Values for the Sign Test

The entries in this table are the critical values for the number of the least frequent sign for a two-tailed test at α for the binomial $p = 0.5$. For a one-tailed test, the value of α shown at the top of the table is double the value of α being used in the hypothesis test.

For specific details on the use of this table, see pages 624–628.

From Wilfred J. Dixon and Frank J. Massey, Jr., *Introduction to Statistical Analysis*, 3d ed., (New York: McGraw-Hill, 1969), p. 509. Reprinted by permission.

n	α 0.01	0.05	0.10	0.25	n	α 0.01	0.05	0.10	0.25
1					51	15	18	19	20
2					52	16	18	19	21
3				0	53	16	18	20	21
4				0	54	17	19	20	22
5			0	0	55	17	19	20	22
6		0	0	1	56	17	20	21	23
7		0	0	1	57	18	20	21	23
8	0	0	1	1	58	18	21	22	24
9	0	1	1	2	59	19	21	22	24
10	0	1	1	2	60	19	21	23	25
11	0	1	2	3	61	20	22	23	25
12	1	2	2	3	62	20	22	24	25
13	1	2	3	3	63	20	23	24	26
14	1	2	3	4	64	21	23	24	26
15	2	3	3	4	65	21	24	25	27
16	2	3	4	5	66	22	24	25	27
17	2	4	4	5	67	22	25	26	28
18	3	4	5	6	68	22	25	26	28
19	3	4	5	6	69	23	25	27	29
20	3	5	5	6	70	23	26	27	29
21	4	5	6	7	71	24	26	28	30
22	4	5	6	7	72	24	27	28	30
23	4	6	7	8	73	25	27	28	31
24	5	6	7	8	74	25	28	29	31
25	5	7	7	9	75	25	28	29	32
26	6	7	8	9	76	26	28	30	32
27	6	7	8	10	77	26	29	30	32
28	6	8	9	10	78	27	29	31	33
29	7	8	9	10	79	27	30	31	33
30	7	9	10	11	80	28	30	32	34
31	7	9	10	11	81	28	31	32	34
32	8	9	10	12	82	28	31	33	35
33	8	10	11	12	83	29	32	33	35
34	9	10	11	13	84	29	32	33	36
35	9	11	12	13	85	30	32	34	36
36	9	11	12	14	86	30	33	34	37
37	10	12	13	14	87	31	33	35	37
38	10	12	13	14	88	31	34	35	38
39	11	12	13	15	89	31	34	36	38
40	11	13	14	15	90	32	35	36	39
41	11	13	14	16	91	32	35	37	39
42	12	14	15	16	92	33	36	37	39
43	12	14	15	17	93	33	36	38	40
44	13	15	16	17	94	34	37	38	40
45	13	15	16	18	95	34	37	38	41
46	13	15	16	18	96	34	37	39	41
47	14	16	17	19	97	35	38	39	42
48	14	16	17	19	98	35	38	40	42
49	15	17	18	19	99	36	39	40	43
50	15	17	18	20	100	36	39	41	43

TABLE II
Critical Values of U in the Mann-Whitney Test

A. The entries are the critical values of U for a one-tailed test at 0.025 or for a two-tailed test at 0.05.

n_2 \ n_1	1	2	3	4	5	6	7	8	9	10	11	12	13	14	15	16	17	18	19	20
1																				
2								0	0	0	0	1	1	1	1	1	2	2	2	2
3				0	1	1	2	2	3	3	4	4	5	5	6	6	7	7	8	
4			0	1	2	3	4	4	5	6	7	8	9	10	11	11	12	13	13	
5		0	1	2	3	5	6	7	8	9	11	12	13	14	15	17	18	19	20	
6			1	2	3	5	6	8	10	11	13	14	16	17	19	21	22	24	25	27
7			1	3	5	6	8	10	12	14	16	18	20	22	24	26	28	30	32	34
8		0	2	4	6	8	10	13	15	17	19	22	24	26	29	31	34	36	38	41
9		0	2	4	7	10	12	15	17	20	23	26	28	31	34	37	39	42	45	48
10		0	3	5	8	11	14	17	20	23	26	29	33	36	39	42	45	48	52	55
11		0	3	6	9	13	16	19	23	26	30	33	37	40	44	47	51	55	58	62
12		1	4	7	11	14	18	22	26	29	33	37	41	45	49	53	57	61	65	69
13		1	4	8	12	16	20	24	28	33	37	41	45	50	54	59	63	67	72	76
14		1	5	9	13	17	22	26	31	36	40	45	50	55	59	64	67	74	78	83
15		1	5	10	14	19	24	29	34	39	44	49	54	59	64	70	75	80	85	90
16		1	6	11	15	21	26	31	37	42	47	53	59	64	70	75	81	86	92	98
17		2	6	11	17	22	28	34	39	45	51	57	63	67	75	81	87	93	99	105
18		2	7	12	18	24	30	36	42	48	55	61	67	74	80	86	93	99	106	112
19		2	7	13	19	25	32	38	45	52	58	65	72	78	85	92	99	106	113	119
20		2	8	13	20	27	34	41	48	55	62	69	76	83	90	98	105	112	119	127

B. The entries are the critical values of U for a one-tailed test at 0.05 or for a two-tailed test at 0.10.

n_2 \ n_1	1	2	3	4	5	6	7	8	9	10	11	12	13	14	15	16	17	18	19	20
1																			0	0
2					0	0	0	1	1	1	1	2	2	2	3	3	3	4	4	4
3			0	0	1	2	2	3	3	4	5	5	6	7	7	8	9	9	10	11
4			0	1	2	3	4	5	6	7	8	9	10	11	12	14	15	16	17	18
5		0	1	2	4	5	6	8	9	11	12	13	15	16	18	19	20	22	23	25
6		0	2	3	5	7	8	10	12	14	16	17	19	21	23	25	26	28	30	32
7		0	2	4	6	8	11	13	15	17	19	21	24	26	28	30	33	35	37	39
8		1	3	5	8	10	13	15	18	20	23	26	28	31	33	36	39	41	44	47
9		1	3	6	9	12	15	18	21	24	27	30	33	36	39	42	45	48	51	54
10		1	4	7	11	14	17	20	24	27	31	34	37	41	44	48	51	55	58	62
11		1	5	8	12	16	19	23	27	31	34	38	42	46	50	54	57	61	65	69
12		2	5	9	13	17	21	26	30	34	38	42	47	51	55	60	64	68	72	77
13		2	6	10	15	19	24	28	33	37	42	47	51	56	61	65	70	75	80	84
14		2	7	11	16	21	26	31	36	41	46	51	56	61	66	71	77	82	87	92
15		3	7	12	18	23	28	33	39	44	50	55	61	66	72	77	83	88	94	100
16		3	8	14	19	25	30	36	42	48	54	60	65	71	77	83	89	95	101	107
17		3	9	15	20	26	33	39	45	51	57	64	70	77	83	89	96	102	109	115
18		4	9	16	22	28	35	41	48	55	61	68	75	82	88	95	102	109	116	123
19	0	4	10	17	23	30	37	44	51	58	65	72	80	87	94	101	109	116	123	130
20	0	4	11	18	25	32	39	47	54	62	69	77	84	92	100	107	115	123	130	138

Reproduced from the *Bulletin of the Institute of Educational Research at Indiana University*, vol. 1, no 2; with the permission of the author and the publisher.

For specific details on the use of this table, see page 633.

TABLE 12
Critical Values for Total Number of Runs (V)

The entries in this table are the critical values* for a two-tailed test using $\alpha = 0.05$. For a one-tailed test, $\alpha = 0.025$ and use only one of the critical values: the smaller critical value for a left-hand critical region, the larger for a right-hand critical region.

The smaller of n_1 and n_2	\	The larger of n_1 and n_2															
		5	6	7	8	9	10	11	12	13	14	15	16	17	18	19	20
2									2/6	2/6	2/6	2/6	2/6	2/6	2/6	2/6	2/6
3			2/8	2/8	2/8	2/8	2/8	2/8	2/8	2/8	2/8	3/8	3/8	3/8	3/8	3/8	3/8
4		2/9	2/9	2/10	3/10	3/10	3/10	3/10	3/10	3/10	3/10	3/10	4/10	4/10	4/10	4/10	4/10
5		2/10	3/10	3/11	3/11	3/12	3/12	4/12	4/12	4/12	4/12	4/12	4/12	4/12	5/12	5/12	5/12
6			3/11	3/12	3/12	4/13	4/13	4/13	4/13	5/14	5/14	5/14	5/14	5/14	5/14	6/14	6/14
7				3/13	4/13	4/14	5/14	5/14	5/14	5/15	5/15	6/15	6/16	6/16	6/16	6/16	6/16
8					4/14	5/14	5/15	5/15	6/16	6/16	6/16	6/16	7/17	7/17	7/17	7/17	7/17
9						5/15	5/16	6/16	6/16	6/17	7/17	7/18	7/18	8/18	8/18	8/18	8/18
10							6/16	6/17	7/17	7/18	7/18	7/18	8/19	8/19	8/19	8/20	9/20
11								7/17	7/18	7/19	8/19	8/19	8/20	9/20	9/20	9/21	9/21
12									7/19	8/19	8/20	8/20	9/21	9/21	9/21	10/22	10/22
13										8/20	9/20	9/21	9/21	10/22	10/22	10/23	10/23
14											9/21	9/22	10/22	10/23	10/23	11/23	11/24
15												10/22	10/23	11/23	11/24	11/24	12/25
16													11/23	11/24	11/25	12/25	12/25
17														11/25	12/25	12/26	13/26
18															12/26	13/26	13/27
19																13/27	13/27
20																	14/28

From C. Eisenhart and F. Swed, "Tables for testing randomness of grouping in a sequence of alternatives," *The Annals of Statistics*, vol. 14 (1943): 66–87. Reprinted by permission.

See page 638 in regard to critical values.

For $n_1 > 20$ or $n_2 > 20$, treat V as a normal variable with a mean and a standard deviation of

$$\mu_V = \frac{2n_1 n_2}{n_1 + n_2} + 1$$

$$\sigma_V = \sqrt{\frac{2n_1 n_2 (2n_1 n_2 - n_1 - n_2)}{(n_1 + n_2)^2 (n_1 + n_2 - 1)}}$$

TABLE 13
Critical Values of Spearman's Rank Correlation Coefficient

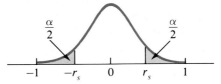

The entries in this table are the critical values of r_s for a two-tailed test at α. For a one-tailed test, the value of α shown at the top of the table is double the value of α being used in the hypothesis test.

n	$\alpha = 0.10$	$\alpha = 0.05$	$\alpha = 0.02$	$\alpha = 0.01$
5	0.900	—	—	—
6	0.829	0.886	0.943	—
7	0.714	0.786	0.893	—
8	0.643	0.738	0.833	0.881
9	0.600	0.683	0.783	0.833
10	0.564	0.648	0.745	0.794
11	0.523	0.623	0.736	0.818
12	0.497	0.591	0.703	0.780
13	0.475	0.566	0.673	0.745
14	0.457	0.545	0.646	0.716
15	0.441	0.525	0.623	0.689
16	0.425	0.507	0.601	0.666
17	0.412	0.490	0.582	0.645
18	0.399	0.476	0.564	0.625
19	0.388	0.462	0.549	0.608
20	0.377	0.450	0.534	0.591
21	0.368	0.438	0.521	0.576
22	0.359	0.428	0.508	0.562
23	0.351	0.418	0.496	0.549
24	0.343	0.409	0.485	0.537
25	0.336	0.400	0.475	0.526
26	0.329	0.392	0.465	0.515
27	0.323	0.385	0.456	0.505
28	0.317	0.377	0.448	0.496
29	0.311	0.370	0.440	0.487
30	0.305	0.364	0.432	0.478

From E. G. Olds, "Distribution of sums of squares of rank differences for small numbers of individuals," *Annals of Statistics*, vol. 9 (1938), pp. 138–148, and amended, vol. 20 (1949), pp. 117–118. Reprinted by permission.

For specific details on the use of this table, see page 642.

TABLE 14
Critical Values of r When $\rho = 0$

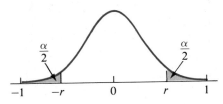

The entries in this table are the critical values of r for a two-tailed test at α.

For simple correlation. $df = n - 2$, where n is the number of pairs of data in the sample. For a one-tailed test, the value of α is shown at the top of the table is double the value of α being used in the hypothesis test.

df	0.10	0.05	0.02	0.01
1	0.988	0.997	1.000	1.000
2	0.900	0.950	0.980	0.990
3	0.805	0.878	0.934	0.959
4	0.729	0.811	0.882	0.917
5	0.669	0.754	0.833	0.874
6	0.662	0.707	0.789	0.834
7	0.582	0.666	0.750	0.798
8	0.549	0.632	0.716	0.765
9	0.521	0.602	0.685	0.735
10	0.497	0.576	0.658	0.708
11	0.476	0.553	0.634	0.684
12	0.458	0.532	0.612	0.661
13	0.441	0.514	0.592	0.641
14	0.426	0.497	0.574	0.623
15	0.412	0.482	0.558	0.606
16	0.400	0.468	0.542	0.590
17	0.389	0.456	0.528	0.575
18	0.378	0.444	0.516	0.561
19	0.369	0.433	0.503	0.549
20	0.360	0.423	0.492	0.537
25	0.323	0.381	0.445	0.487
30	0.296	0.349	0.409	0.449
35	0.275	0.325	0.381	0.418
40	0.257	0.304	0.358	0.393
45	0.243	0.288	0.338	0.372
50	0.231	0.273	0.322	0.354
60	0.211	0.250	0.295	0.325
70	0.195	0.232	0.274	0.302
80	0.183	0.217	0.256	0.283
90	0.173	0.205	0.242	0.267
100	0.164	0.195	0.230	0.254

From E. S. Pearson and H. O. Hartley, *Biometrika Tables for Statisticians*, vol. I (1962), p. 138. Reprinted by permission of the Biometrika Trustees.

For specific details on the use of this table, see page 462.

TABLE 15
Confidence Belts for the Correlation Coefficient ($1 - \alpha = 0.95$)

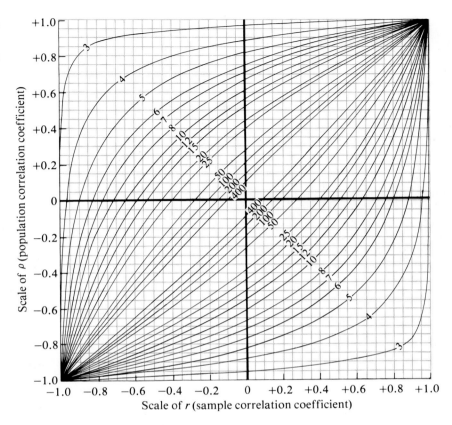

The numbers on the curves are sample sizes.

From E. S. Pearson and H. O. Hartley, *Biometrika Tables for Statisticians*, vol. I (1962), p. 140. Reprinted by permission of the Biometrika Trustees.

For specific details on the use of this table, see page 463.

Answers to Selected Exercises

Chapter 1

Section 1-3, page 12

1-2 (1) Attribute; color is an attribute.
(2) Continuous; distance is a measure.
(3) Discrete; it is a count of the number of cards.

1-3 (a) Population: all stockholders of Extro Corporation; sample: stockholders who attended the annual meeting.
(b) (1) Attribute; question requires only a yes or no.
(2) Continuous; percentages include all possible fraction values.
(3) Discrete; how many requires a count.

1-4 (a) Population: the set of all classes at our college this semester; sample: the set of classes to which you belong.
(b) Variable: the number of students enrolled in each class; discrete (a count of the number of students in a class).
(c) Experiment: ask the professor for each of the classes to which you belong for the number of students enrolled; each student might have different data.
(d) Parameter: the average value of class size for all classes in the college; statistic: the average value of class size for the set of classes to which you belong. To find this arithmetic average, you will need to add the values together and divide by the number of classes. This arithmetic average is the average value that we are all familiar with.

Section 1-5, page 15

1-5 (a) Population: all the bank's checking accounts; sampling frame: a list of all checking account numbers (*note*: there are others).
(b) Population: all the oil companies' credit card customers; sample frame: a list of all the card numbers.
(c) Population: all the persons in the labor market; sample frame: telephone directory or persons listed with employment services or U.S. census information on individuals (*note*: finding a frame for this type of problem is difficult).

1-8 (a) Yes.
(b) No, because it would exclude all noncharge customers.

Section 1-6, page 18

1-9 (a) Inappropriate sampling frame
(b) Natural bias in reporting data
(c) Bias in the collection device
(d) Nonrespondents

1-10 (a) Sampling variability (central limit theorem)
(b) Systematic bias
(c) Systematic bias
(d) Systematic bias
(e) Sampling variability

Section 1-11, page 24

1-14 (a) Simple random; checks are numbered and these numbers could be used for identifying the check randomly selected.
(b) Stratified; use each lines production as one stratum.
(c) Systematic; applications are filed in a specific order, but a list of them is unavailable. To select every 17th file does not require a listing of the sampling frame.

(d) Cluster; interviewing many doctors in each of several geographic areas would be tedious, but the other techniques would be virtually impossible.
(e) Cluster; use the cartons as the sampled cluster.

1-15 (a) Simple random; since the sampling frame is already in existence and numbered, random is easy. Just use the random number table.
(b) Systematic; the elements in the sampling frame only need to be counted as you go from one piece of data to the next.
(c) Simple random or proportional.
(d) Stratified; this allows half of the total sample to come from each part of the population.

Chapter Exercises, page 26

1-17 (a) Continuous; dollar amounts occur in fractional parts of dollars and cents (usually rounded for convenience).
(b) Discrete; a count of the number of customers.
(c) Continuous; length of time is a measurement.
(d) Attribute; color is an attribute.
(e) Attribute; fair, good, and excellent are attribute ratings.
(f) Not a variable; check number is an identification number.

1-22 (a) Cluster (b) Systematic

Chapter 2

2-2

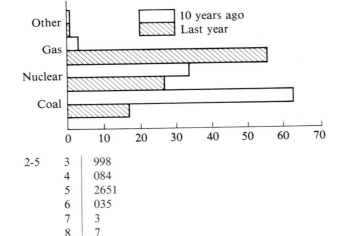

2-5
```
3 | 998
4 | 084
5 | 2651
6 | 035
7 | 3
8 | 7
```

2-9 (a)
x	f
1	5
2	6
3	19
4	7
5	7
6	2
7	3
8	0
9	1

(b) **Number of Ledger Errors**

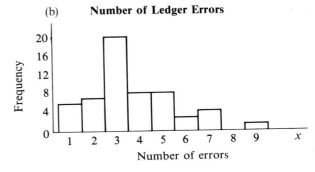

2-10 (a)
x	f	Tallies
3	1	I
4	2	II
5	2	II
6	4	IIII
7	3	III
8	4	IIII
9	5	THL
10	5	THL
11	7	THL II
12	8	THL III
13	6	THL I
14	2	II
	49	

(b) (i) 8
(ii) 1
(iii) 7.5

(c) **Number of Defective Bottles**

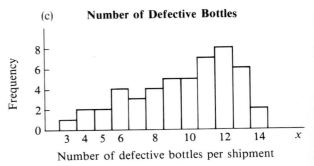

(d)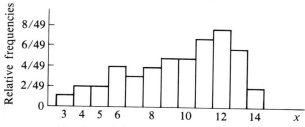
Number of Defective Bottles

2-12 (a)

Class Limits	Tallies	f
15–19	III	3
20–24	THL THL II	12
25–29	THL THL THL IIII	19
30–34	THL I	6
35–39	THL	5
40–44	II	2
45–49	II	2
50–54	I	1

(b) Class width = 5
(c) 14.5, 19.5, 24.5, 29.5, 34.5, 39.5, 44.5, 49.5, 54.5
(d) Class mark = $\frac{20 + 24}{2} = 22$; lower class limit = 20

(e)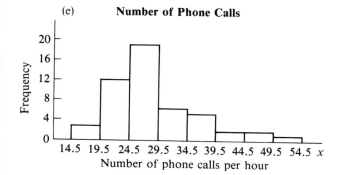
Number of Phone Calls

2-14 (a) 6.0
(b) 3
(c) No mode; no piece of data appears more than once.
(d) 6.0

2-16 (a) 3.42
(b) 3.7
(c) 3.4

2-18 (a) $\bar{x} = 6.5$
(b) $\tilde{x} = 5.5$
(c) Mode = 6
(d) Midrange = 6

2-20 (a) $\bar{x} = 16.2$
(b) $\tilde{x} = 17$
(c) Mode = 13
(d) Midrange = 16.0

2-21 $\bar{x} = 2.3$

2-24 $\bar{x} = 12.0$

2-29 (a) 8
(b) 11.5
(c) 3.4
(d) 0.57

2-30 (a) 5   (b) 3.2   (c) 1.8
(d) 0.36

2-32 (a) 30.0   (b) 22.0   (c) 4.7

2-35 (a) 14.65   (b) 1.19

2-37 Ratco: $\frac{s}{\bar{x}} = 0.09$; Warren: $\frac{s}{\bar{x}} = 0.33$; therefore Warren has a more variable price.

2-39

Number of Loans	f	x, Class Mark	xf	x²f
3– 5	2	4	8	32
6– 8	6	7	42	294
9–11	9	10	90	900
12–14	13	13	169	2,197
15–17	8	16	128	2,048
	38		437	5,441

$\bar{x} = 11.53$; $s^2 = 11.23$

2-44 $Q_1 = 5$
$Q_3 = 15$

2-46 $P_{95} = 123$

2-48 $z = -2.1$

2-51 A: $z = 1.5$; B: $z = 2.0$. B has the higher relative value.

2-53 $\bar{x} = 3.6$; $s = 1.7$
(a) $\bar{x} - s = 1.9$; $\bar{x} + s = 5.3$
(b) 39; 78%
(c) $\bar{x} - 2s = 0.2$; $\bar{x} + 2s = 7.0$
(d) 49; 98%

(e) $\bar{x} - 3s = -1.5$; $\bar{x} + 3s = 8.7$
(f) 98%
(g) 98% exceeds the 75% and 98% exceeds the 89% guaranteed by Chebyshev's theorem.
(h) 78%, 98%, and 98% do not compare very closely with the 68%, 95%, and 99.7% of the empirical rule. The distribution is mounded, but skewed.

2-56 The interval needed is $(\bar{x} - 2s)$ to $(\bar{x} + 2s)$ and is four standard deviations wide.

2-63 (a) 1.7 (b) 1 (c) 1
(d) 3 (e) Mode (f) Mean
(g) 6 (h) 2.8 (i) 1.7

2-65 Ranked data: 12, 15, 16, 17, 19, 22, 22, 22, 25, 26, 31, 32

Data summary: $n = 12$; $\sum x = 259$; $\sum x^2 = 6{,}013$
(a) 21.6
(b) 22
(c) 38.447
(d) 6.2

2-66 Summary of data: $n = 20$; $\sum x = 409{,}780$; $\sum x^2 = 8{,}533{,}282{,}164$
(a) $\bar{x} = \$20{,}489$; $s = 2{,}688.2$
(b) No. The mean calculated in (a) is the mean of 20 medians. It does not account for the different number of readers each publication has nor does it account for those people who read more than one of the publications listed.
(c) Wall Street Journal: $z = 0.46$; Forbes: $z = 1.48$; Fortune: $z = 2.02$; Barrons: $z = 1.99$; Money: $z = 0.68$; Business Week: $z = 0.89$

Chapter 3

3-1 (a) No (b) Yes (c) Yes (d) Yes

3-2 Bond Spread by Rating for 40 Bonds

Spread (× $100,000)	Rating		Total
	Aaa	Aa	
1.00–1.99	10	1	11
2.00–2.99	14	5	19
3.00–3.99	0	10	10
Total	24	16	40

3-5

Opinion	Type of Employee		Total
	Management	Labor	
Yes	3	7	10
No	7	3	10
Total	10	10	20

3-6

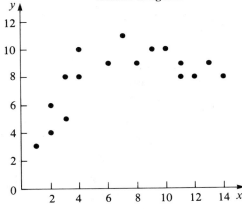

Scatter Diagram

3-11 (b) $r = 0.92$ (c) Yes

3-13 $r \simeq -0.95$; $r = -0.995$

3-15 (a)

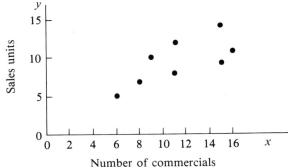

Effectiveness of TV Commercials

(b) $r \simeq +0.6$
(c) $r = +0.711$
(d) Yes. The 0.711, the calculated value, is larger than the 0.707 from the table, thereby indicating that the data does show correlation.

3-17 Quadratic: $a + bx + cx^2$

3-18 Exponential: ab^x

3-19 Linear: $a + bx$

3-20 (b) S-shaped, or logistic, curve

3-21 (a) Correlation (b) Regression
(c) Correlation (d) Correlation
(e) Regression (f) Regression

3-24 (a) Regression (b) Positive
(c) $P = a + 10D$; total sales revenue = (demand) × (price) = $(D)(a + 10D) = aD + 10D^2$; quadratic

3-27 (a) $\sum x = 170$; $\sum y = 579$; $\sum xy = 11{,}114$; $\sum x^2 = 7{,}015.125$; $\sum y^2 = 32{,}423$; $r = 0.64$
(b) $\sum x = 15.26$; $\sum y = 579$; $\sum xy = 574.87$; $\sum x^2 = 15.73$; $\sum y^2 = 32{,}423$; $r = -0.31$
(c) Difference is a better predictor.

3-28 $r = 0.70$

3-30 (a) $\sum x = 6{,}653$; $\sum x^2 = 5{,}769{,}285$; $\sum y = 320.2$; $\sum y^2 = 11{,}895.54$; $\sum xy = 219{,}087$; $n = 9$

$$r = \frac{(9)(219{,}087) - (6{,}653)(320.2)}{\sqrt{(9)(5{,}769{,}285) - (6{,}653)(6{,}653)} \times \sqrt{(9)(11{,}895.54) - (320.2)(320.2)}}$$

$$= \frac{-158{,}507.6}{\sqrt{7{,}661{,}156}\sqrt{4{,}531.82}}$$

$$= -0.8507 = -0.85$$

(b) Yes. The negative correlation coefficient of -0.85 is between the negative table value of r and -1.0, thus indicating that there is a negative correlation between the two variables.

3-32 (a)

Station Preference	Income ($1,000)				
	5–14	15–24	25–34	35–44	Sum
Music	4	14	9	3	30
News	1	5	11	3	20
Sum	5	19	20	6	50

(b)

Station Preference	Number of Flights			
	0	1–2	3 or More	Sum
Music	11	11	8	30
News	6	8	6	20
Sum	17	19	14	50

Chapter 4

4-5 The results of my 50 rolls are as follows (each set of 50 will be different):

12	65	15	32	54	12	52	63	64	62
66	44	42	45	42	35	54	66	54	32
31	12	23	33	26	33	23	32	46	64
63	63	35	54	52	55	56	26	11	44
11	61	46	11	45	55	15	33	43	11

(a) P' (black die is odd) = $\frac{24}{50} = 0.48$
(b) P' (sum is 8) = $\frac{7}{50} = 0.14$
(c) P' (same number on both) = $\frac{13}{50} = 0.26$
(d) P' (number on black exceeds number on white) = $\frac{16}{50} = 0.32$

4-7 {good, fair, poor}

4-9 {(H, 1), (H, 2), (H, 3), (H, 4), (H, 5), (H, 6), (T, 1), (T, 2), (T, 3), (T, 4), (T, 5), (T, 6)}

4-11 G = good; D = defective {(G, G), (G, D), (D, G), (D, D)}

Tree diagram: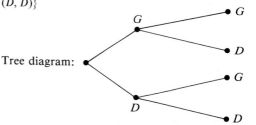

4-13 (a) On the average, approximately $\frac{1}{5}$ of the candidates interviewed get jobs.
(b) 1 to 4 in favor

4-14 (a) On the average, approximately one sale will result from each 100 advertisements mailed.
(b) 1 to 99 in favor

4-17 (a) $\frac{7}{10}$ (b) $\frac{1}{4}$ (c) $\frac{2}{5}$ (d) $\frac{5}{8}$

4-20 (a) Not mutually exclusive
(b) Mutually exclusive
(c) Not mutually exclusive
(d) Not mutually exclusive
(e) Mutually exclusive
(f) Not mutually exclusive

4-21 (a) $P(A) = 0.4$
(b) $P(A \text{ or } B) = 0.4 + 0.2 = 0.6$
(c) $P(A \text{ and } B) = 0$

4-22 No, if they were mutually exclusive, the probability that she would buy stocks or bonds would be greater than 1.

4-24 $P(A \text{ or } B) = 0.6$

4-25 $P(A \text{ or } B) = 0.4$

4-27 $P(A \text{ and } B) = 0.1$
$P(B|A) = 0.2$

4-29 (a) Independent (b) Not independent
      (c) Independent (d) Independent
      (e) Not independent

4-30 (a) 0.0025 (b) 0.000125

4-32 (a) 0.414 (b) 0.470

4-38 (a) $P = 0.625$ (b) $P = 0.25$ (c) Yes

4-42 (a) $P = \frac{2}{3}$ (b) $P = \frac{1}{2}$ (c) $P = \frac{2}{3}$ (d) $P = \frac{1}{3}$

4-43 (a) $P = 0.03$ (b) $P = 0.0199$
      (c) $P = 0.019701$

4-48 (a) $P = 0.80$ (b) $P = 0.96$ (c) $P = 0.64$

4-52 (a) $P = 0.00$ (b) $P = 0.5$ (c) $P = 0.2$
      (d) $P = 0.8$ (e) $P = 0.7$ (f) $P = 0.286$

4-55 (a) 0.84 (b) No

4-57 (a) 0.0075 (b) 0.095 (c) 0.3875 (d) 0.51

4-63 (a) 0.078, 0.389 (b) 0.25, 0.111
      (c) 0.25, 0.444 (d) No

4-65 (a) $P = 0.216$ (b) $P = 0.064$ (c) $P = 0.648$

4-66 (a-i) $P = 0.6$ (a-ii) $P = 0.4$ (a-iii) $P = 0.6$
      (b-i) $P = 0.36$ (b-ii) $P = 0.16$ (b-iii) $P = 0.60$
      (c) The more the investment is spread, the smaller the risk of a large loss and, conversely, the smaller the chance of a large gain. In general in investments, the greater the diversification (the more stocks and bonds the investment is spread over), the less speculative the investment (the smaller the chance of a large loss or profit).

4-67

	$P(A_i)$	$P(\text{predict}|A_i)$	$P(A_i \text{ and predict})$	$P(A_i|PE)$
Less than expectations, A_1	0.2	0.3	0.06	$\frac{0.06}{0.39} = 0.154$
Equal to expectations, A_2	0.5	0.6	0.30	$\frac{0.30}{0.39} = 0.769$
Exceeds expectations, A_3	0.3	0.1	0.03	$\frac{0.03}{0.39} = 0.077$
Total	1.0		0.39	1.000

Chapter 5

5-1 $D =$ stock in D-J index;
     $\bar{D} =$ stock not in D-J index
     (a) $(D, D, D), (D, D, \bar{D}), (D, \bar{D}, D), (\bar{D}, D, D),$
         $(D, \bar{D}, \bar{D}), (\bar{D}, D, \bar{D}), (\bar{D}, \bar{D}, D), (\bar{D}, \bar{D}, \bar{D})$
     (b) $(D, D, D) = 3; (D, D, \bar{D}) = 2;$
         $(D, \bar{D}, D) = 2; (\bar{D}, D, D) = 2;$
         $(D, \bar{D}, \bar{D}) = 1; (\bar{D}, \bar{D}, D) = 1;$
         $(\bar{D}, \bar{D}, \bar{D}) = 0; (\bar{D}, D, \bar{D}) = 1$
     (c) Discrete

5-3 $x = n(\text{defective sets}); x = 0, 1, 2, 3, 4, 5$

5-5 $P(x) = \dfrac{4 - x}{6}$

x	$P(x)$
1	3/6
2	2/6
3	1/6
	6/6 = 1

$x = 1, 2, 3$. Each $P(x)$ is between 0 and 1. Total of $P(x)$ is exactly 1. Therefore $P(x)$ does form a probability distribution.

5-9 $\mu = \frac{10}{6}$; $\sigma = \sqrt{20/36}$

5-15 (a) 24 (b) 5,040 (c) 1
      (d) 360 (e) 10 (f) 15
      (g) 0.0081 (h) 35 (i) 10
      (j) 1 (k) 0.4096 (l) 0.16807

5-19 (a)

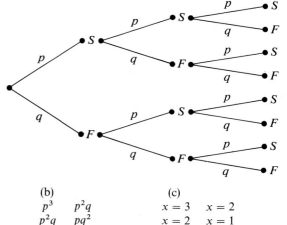

(b)
p^3 p^2q
p^2q pq^2
p^2q pq^2
pq^2 q^3

(c)
$x = 3$ $x = 2$
$x = 2$ $x = 1$
$x = 2$ $x = 1$
$x = 1$ $x = 0$

5-20 (d) $P(x) = \binom{3}{x}p^x q^{3-x}$; $x = 0, 1, 2, 3$
(c) $P(x) = \binom{6}{x}(\frac{1}{3})^x(\frac{2}{3})^{6-x}$; $x = 0, 1, 2, \ldots, 6$

5-23 (a) Yes; $n = 4$; $p = 0.05$; each stock is a trial; two outcomes: included in Dow Jones index or not included in Dow Jones index. x = number of stocks within the four selected that are included in Dow Jones index.
(b) $P(x) = \binom{4}{x}(0.05)^x(0.95)^{4-x}$; $x = 0, 1, 2, 3, 4$
(c)

x	P(x)
0	0.815
1	0.171
2	0.014
3	0^+
4	0^+

From table 4: $n = 4$; $p = 0.05$

5-25 $P(x = 5, 6, 7, 8, 9, 10)$, where $n = 12$, $p = 0.2$. It is binomial; therefore from table 4:

$P(x = 5) = 0.053$
$P(x = 6) = 0.016$
$P(x = 7) = 0.003$
$P(x = 8) = 0.001$
$P(x = 9) = 0^+$
$(+)P(x = 10) = 0^+$
          ————
          0.073

5-30 (a) $\mu = 33.3$; $\sigma = 5.3$ (b) $\mu = 30.0$; $\sigma = 5.3$
(c) $\mu = 60.0$; $\sigma = 7.3$ (d) $\mu = 180.6$; $\sigma = 8.5$

5-31 (a) 0.2240 (b) 0.4232 (c) 0.5768

5-32 (a) Continuous (b) Poisson
(c) Binomial (d) Poisson

5-33 (a) 0.2707 (b) 0.5940 (c) 0.3233

5-34 (a) 0.3679 (b) 0.0067 (c) 0.2707

5-38 $\mu = 4.0$; $\sigma = 2.2$

5-39 (a)

x	P(x)	xP(x)	x²P(x)
2	1/16	2/16	4/16
3	2/16	6/16	18/16
4	3/16	12/16	48/16
5	4/16	20/16	100/16
6	3/16	18/16	108/16
7	2/16	14/16	98/16
8	1/16	8/16	64/16
	16/16	80/16	440/16

(b) $\mu = 5.0$
(c) $\sigma = 1.6$

5-46 0.0879

5-48 0.942

5-54 0.952

5-57 (a) (1) $n = 10$ independent trials (10 physicals); (2) two outcomes, pass or fail physical; (3) $p = 0.30$ (fail) and $q = 0.70$ (pass); (4) $x = n(\text{fail}) = 0, 1, 2, \ldots, 10$
(b) $P = 0.150$

5-58 (a)

x	P(x)
0	0.058
1	0.198
2	0.296
3	0.254
4	0.136
5	0.047
6	0.010
7	0.001
8	0^+

(b) $P = 0.058$ (c) $\mu = 2.4$ (d) $\sigma = 1.296$

5-61 (a) 12 23 34 45 56
         13 24 35 46
         14 25 36
         15 26
         16

(b)

x	P(x)	xP(x)	x²P(x)
3	1/15	3/15	9/15
4	1/15	4/15	16/15
5	2/15	10/15	50/15
6	2/15	12/15	72/15
7	3/15	21/15	147/15
8	2/15	16/15	128/15
9	2/15	18/15	162/15
10	1/15	10/15	100/15
11	1/15	11/15	121/15
		105/15	805/15

(c) $\mu = \frac{105}{15} = 7.0$

$\sigma = \sqrt{\frac{805}{15} - (\frac{105}{15})^2} = \sqrt{4.666} = 2.16$

5-62 (a) Yes
(b) $P = 0.97865$

Chapter 6

6-1   A bell-shaped symmetrical distribution with a mean of 0 and a standard deviation of 1

6-2   (a) 0.4893   (b) 0.3729
      (c) 0.4997   (d) 0.4986

6-3   (a) 0.4641   (b) 0.4997
      (c) 0.4861   (d) 0.4972

6-4   (a) $0.4032 + 0.4868 = 0.8900$
      (b) $0.4940 + 0.3508 = 0.8448$
      (c) $0.4997 - 0.3849 = 0.1148$
      (d) $0.49997 - 0.20190 = 0.29807$

6-5   (a) 0.7382   (b) 0.4511   (c) 0.0387

6-7   (a) 0.9772   (b) 0.9938   (c) 0.1056
      (d) 0.7734   (e) 0.9573

6-9   (a) 1.17   (b) 0.36   (c) 1.63
      (d) 0.74   (e) 1.71   (f) 2.33

6-11 $+0.84$ or -0.84

6-13   (a) -1.15 to $+1.15$ (b) -1.96 to $+1.96$

6-15 $P = 0.2417$

6-19   (a) $P = 0.6772$ (b) 0.0228

6-23   (a) $z(0.04)$ (c) $z(0.70)$ (e) $z(0.90)$

6-24   (a) 2.33   (c) 1.96   (e) -1.28

6-28 $P = 0.1314$

6-30 $P = 0.0025$

6-32   (a) $P = 0.2389$
       (b) $P = 0.4052$

6-35   No, it would be highly skewed because there are some extremely high incomes.

6-37   (a) 1.34   (b) 0.34   (c) 0.89

6-38   (a) -1.05 (b) -0.07 (c) -2.49

6-40   (a) 1.04   (c) -0.67

6-42   (a) $\mu = 5.6$
       (b) $\mu = 3.75$, therefore reduces the average wait by 1.85 minutes

6-44   501.3 hours

6-49   (a) 0.008   (b) 0.999   (c) 0.945   (d) 0.150

6-52 $P = 0.0778$

6-54   (a) 0.0026   (b) (i) $P = 0.7482$; (ii) $P = 0.2345$

Chapter 7

7-1   (a) A sampling distribution of sample means is a collection of the means of all the possible samples of a given size.
      (b) It is one member of the entire sampling distribution.
      (c) There are exactly 10 equally likely samples; therefore each is assigned 0.1.

7-2   (a) (4, 5), (4, 6), (4, 7), (4, 8), (5, 6), (5, 7), (5, 8), (6, 7), (6, 8), (7, 8)

(b)

$\bar{x}$	$P(\bar{x})$
4.5	0.1
5.0	0.1
5.5	0.2
6.0	0.2
6.5	0.2
7.0	0.1
7.5	0.1

(c)

R	P(R)
1	0.4
2	0.3
3	0.2
4	0.1

7-3   (b)

$\bar{x}$	$P(\bar{x})$
5.00	0.1
5.33	0.1
5.67	0.2
6.00	0.2
6.33	0.2
6.67	0.1
7.00	0.1

(c)

R	P(R)
2	0.3
3	0.4
4	0.3

7-6   (a) One
      (b) $\sigma_{\bar{x}} = \sigma/\sqrt{n}$ becomes smaller as n increases

7-7   (a) Approximately 50
      (b) Approximately 2; $10/\sqrt{25}$

7-9   7.5

7-11   (a) $P = 0.9876$
       (b) $P = 0.8413$
       (c) $P = 0.8664$

7-15   (a) 6.9   (b) 0.133   (c) Normal
       (d) 0.2266   (e) 0.0668
       (f) x is continuous. The continuity correction is only used when using a continuous variable in place of a discrete variable.

7-18 (a) 0.0643 (b) 0.6103 (c) 0.9974
(e) (a) and (b) are distributions of individual x's, whereas (c) is a sampling distribution of $\bar{x}$'s.

7-21 4.90

7-23 (a) Approximately normal (b) 0.5000
(c) 0.9082 (d) 0.9962 (e) 0.00003

7-25 0.0548

7-26 $P = 0.2743$

Chapter 8

8-4 (a) Commercial is not effective.
(b) Commercial is effective.

8-5 (a) α (b) β (c) Type I (d) Type II

8-7 (a)

x	0	1	2	3	4
$P(x)$	0.0$^+$	0.001	0.006	0.022	0.061

x	5	6	7	8	9
$P(x)$	0.122	0.183	0.209	0.183	0.122

x	10	11	12	13	14
$P(x)$	0.061	0.022	0.006	0.001	0.0$^+$

(b) $x = 4, 5, 6, 7, 8, 9, 10$
(c) $x = 0, 1, 2, 3, 11, 12, 13, 14$
(d) $\alpha = P(x = 0, 1, 2, 3, 11, 12, 13, 14)$
$= 2(0.001 + 0.006 + 0.022) = 0.058$
(e) Reject H_0; fail to reject H_0; reject H_0

8-8 (a) $H_0: \mu = 55(\geq); H_a: \mu < 55$
(c) $H_0: \mu = 6; H_a: \mu \neq 6$
(e) $H_0: \mu = 2(\leq); H_a: \mu > 2$
(g) $H_0: \mu = 65; H_a: \mu \neq 65$

8-10 (a) $H_a: r > a$. Failure to reject the null hypothesis will result in the drug being marketed. Because of the current high mortality rate, the burden of proof should be on the old, ineffective treatment.
(b) $H_a: r < a$. Failure to reject the null hypothesis will result in the new drug not being marketed. Because of the current low mortality rate, the burden of proof should be on the new drug.

8-11 (a)

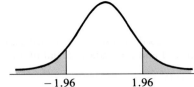

(d)

(e)

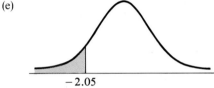

8-12 $H_a: \mu > 490$; $z^* = 0.67$; fail to reject H_0

8-13 $H_a: \mu < 1.5$; $z^* = -6.67$; reject H_0; the sample does support the claim that the training has lowered the mean below 1.5.

8-14 $H_0: \mu = 5.5; H_a: \mu > 5.5; z = 1.0$; fail to reject

8-19 (a) 0.0694 (b) 0.1977 (c) 0.242
(d) 0.0174 (e) 0.3524 (f) 0.3228

8-20 (a) Fail to reject H_0
(b) Reject H_0

8-21 $H_0: \mu = 21.5(\leq); H_a: \mu > 21.5; z^* = 1.60$;
$P = 0.0548$; fail to reject H_0

8-23 (a) 25.3
(b) $25.3 \pm (1.96)(16/\sqrt{64})$; 21.38 to 29.22

8-24 (a) $\mu = \$5{,}650$ (point estimate)
(b) $5{,}650 \pm 1.96(800/\sqrt{36})$

$5{,}650 \pm 261.3$

(5,388.7 to 5,911.3), the 0.95 confidence interval for μ

(c) $5{,}650 \pm 2.58(800/\sqrt{36})$

$5{,}650 \pm 344.0$

(5,306 to 5,994), the 0.99 confidence interval for μ

8-26 (a) $322.0 \pm (1.65)(23/\sqrt{200})$
322.0 ± 2.68
(319.32 to 324.68), the 0.90 confidence interval for μ

(b) $322.0 \pm (2.33)(23/\sqrt{200})$
322.0 ± 3.79
(318.21 to 325.79), the 0.98 confidence interval for μ
(a) 322.0 ± 2.68
(b) 322.0 ± 3.79

8-27 8.665 to 9.335

8-28 $n = \dfrac{(1.96 \times 10)^2}{4} = 96.04 = 97$

8-29 $n = 348$

8-30 $0.1 = 1.96(0.78/\sqrt{n})$
$\sqrt{n} = (1.96)(0.78)/0.1 = 1.59$
$n = 233.7 = 234$

8-33 (b) $H_a: \mu \neq 100$
(d) $z(\alpha/2) = \pm 2.58$
(e) $\mu = 100$
(f) $\bar{x} = 96$
(h) $\sigma_{\bar{x}} = \sigma/\sqrt{n} = 12/\sqrt{50} = 1.7$
(i) $z^* = (96 - 100)/1.7 = -2.35$
(j) Fail to reject H_0
(l) 0.0188

8-34 $H_a: \mu > 15$; $z^* = 7.23$; reject H_0; the evidence is sufficient to support the slow-worker claim.

8-37 $H_a: \mu > 60$; $z^* = 2.34$; reject H_0; the data support the employees' claim that the mean is now greater than 60.

8-38 Fail to reject H_0: The evidence is not sufficient to conclude that the program improved their sales effectiveness.

8-39 Reject H_0: The data supports the claim that the average length of time for payment is less than 60 days.

8-40 H_0: Gas mileage not improved
H_a: Improved gas mileage

8-41 (a) 32.0   (b) 2.4   (c) 64
(d) 0.90   (e) 1.65   (f) 0.3
(g) 0.495   (h) 32.50

8-43 (a) 487 ± 19.6 (b) 487 ± 13.07

8-47 $n = 35$

Chapter 9

9-1 (a) 1.75   (e) -1.80
(b) 1.33   (f) -2.12
(c) 3.37   (g) -2.46
(d) 2.20   (h) 1.96

9-2 (a) $t(15, 0.05) = 1.75$
(b) $\pm t(19, 0.025) = \pm 2.09$
(c) $t(8, 0.99) = -t(8, 0.01) = -2.90$
(d) $t(7, 0.10) = 1.42$
(e) $\pm t(13, 0.05) = \pm 1.77$

9-3 df = 7

9-8 (a) $t^* = 3.19$; reject H_0
(b) Prob-value $< 2(0.005) = 0.01$; reject H_0

9-9 $t^* = 1.643$; fail to reject H_0: The evidence does not show a significant increase.

9-10 (a) $\bar{x} = 11.49$; $s = 0.5175$
(b) 11.49 ± 0.29

9-11 (a) $\bar{x} = 12.69$ (b) $s = 7.26$ (c) 12.69 ± 1.99

914 $H_a: P < 0.50$; $z^* = -2.19$; reject H_0

9-15 (a) $H_a: P > 0.15$; $z^* = 1.455$; reject H_0

9-16 $p' = 0.16$; $z^* = 0.56$; fail to reject H_0

9-17 $p' - 0.60$; $z^* = 0.28$; fail to reject H_0

9-21 0.26 to 0.54

9-22 $n = 534$

9-23 $n = 369$

9-25 (a) $\chi^2(9, 0.05) = 17.0$
(b) $\chi^2(14, 0.01) = 29.1$
(c) $\chi^2(13, 0.975) = 5.01$
(d) $\chi^2(50, 0.95) = 34.8$
(e) $\chi^2(11, 0.95) = 4.58$;
$\chi^2(11, 0.05) = 19.7$
(f) $\chi^2(16, 0.975) = 6.91$;
$\chi^2(16, 0.025) = 28.9$
(g) $\chi^2(13, 0.99) = 4.11$
(h) $\chi^2(7, 0.99) = 1.24$;
$\chi^2(7, 0.01) = 18.5$

9-26 (a) 11.1 (c) 9.24

9-29 $H_a: \sigma < 2.1$ (improved); $\chi^{2*} = 17.9$; fail to reject H_0

9-31 (a) $s^2 = 52.732$ (b) 36.93 to 81.83

9-32 (c) 0.163 to 0.935

9-33 (a) $s^2 = 9.486$
(b) 6.092 to 17.173
(c) 2.468 to 4.144

9-34 (a) Taste: $\bar{x} = 2.7$; $s = 0.95$
Price: $\bar{x} = 3.5$; $s = 0.7$
Appearance: $\bar{x} = 4.6$; $s = 1.7$
(b) $H_0: \mu = 3.0(\geq)$(not great); $H_a: \mu < 3.0$(great); $\alpha = 0.05$; $t^* = -0.998$; fail to reject H_0
(c) $H_0: \mu = 4(\geq)$(not high); $H_a: \mu < 4$(high): $\alpha = 0.05$; $t^* = -2.26$; reject H_0
(d) $H_0: \mu = 4(\leq)$(pleasant); $H_a: \mu > 4$(not pleasant; $\alpha = 0.05$; $t^* = 1.12$; fail to reject H_0
(e) 2.18 to 3.22
(f) 2.78 to 4.22
(g) 3.39 to 5.81

9-36 132.40 to 147.60

9-38 7.852 to 12.548

9-39 $H_a: p > 0.40$; $z^* = 1.633$; reject H_0

9-41 0.167 to 0.233

9-43 (a) $H_a: \sigma^2 > 950$ (more than); $\chi^{2*} = 15.30$; fail to reject H_0
(b) $n = 144$

9-50 (a) $\bar{x} = 3.145$; $s = 0.805$
(b) $H_a: \mu > 2.80$; $t^* = 1.92$; reject H_0
(c) $\bar{x} = 35.05$; $s = 9.59975$
(d) $H_0: \mu = 31.0$; $H_a: \mu > 31.0$ (greater pollution); $\alpha = 0.05$; $t^* = 1.887$; reject H_0; the sample evidence supports the PILCOP.

9-51 (a) $p' = 0.8125$; 0.621 to 1.00
(b) No

Chapter 10

10-1 Independent samples

10-2 Dependent samples

10-3 Independent samples

10-6 (a) $H_a: \mu_A - \mu_B \neq 0$; $z^* = -0.36$; fail to reject H_0
(b) $z^* = -0.36$; $p = 0.640$; fail to reject H_0

10-8 -25.92 to 20.52

10-10 $H_0: \mu_A - \mu_B = 0$; $H_a: \mu_A - \mu_B \neq 0$; $\alpha = 0.10$; $z^* = 1.56$; fail to reject H_0

10-12 $H_a: \mu_B - \mu_A > 2.5$ (average output of B exceeds A by more than 2.5); $z^* = 6.0$; reject H_0

10-13 54.3 ± 7.62

10-16 (a) $F(11, 17, 0.99)$ and $F(11, 17, 0.01)$
(c) $F(7, 14, 0.05)$

10-17 (a) 2.54 (b) 2.29
(c) 2.28 (d) 3.29

10-19 (a) $H_a: \sigma_G^2/\sigma_I^2 \neq 1$; $F^* = 2.37$; fail to reject H_0
(b) $F = 2.37$; $p = 0.043$; fail to reject H_0

10-21 $H_0: \sigma_D^2/\sigma_0^2 = 1(\leq)$; $H_a: \sigma_D^2/\sigma_0^2 > 1$; $\alpha = 0.05$; $F^* = 1.92$; reject H_0

10-27 $H_a: \mu_I - \mu_{II} \neq 0$; $t^* = -3.28$; reject H_0

10-29 First, $H_0: \sigma_s^2/\sigma_R^2 = 1$; $F^* = 1.60$; assume that $\sigma_s^2 = \sigma_R^2$; then $H_a: \mu_R - \mu_s > 0$; $t^* = 2.29$; reject H_0

10-31 (a) We must use a different formula for calculating t^*.
(b) df = 9
(c) $H_a: \mu_1 \neq \mu_2$; $t^* = -1.07$; fail to reject H_0

10-33 (a) $H_a: \mu_d \neq 0$; $t^* = 1.336$; fail to reject H_0
(b) Prob-value $= 0.2 > 0.05$; fail to reject H_0

10-36 $H_a: \mu_d \neq 0$; $t^* = 1.336$; fail to reject H_0

10-38 Yes

10-39 1.917 ± 1.679

10-40 (a) $H_a: p_A - p_B \neq 0$; $z^* = -0.90$; fail to reject H_0
(b) Prob-value $= 0.3682 > 0.05$; fail to reject H_0

10-44 $H_a: p_Y - p_O \neq 0$; $z^* = -0.805$; fail to reject H_0

10-45 0.033 to 0.141

10-46 (a) -0.147 to 0.187
(b) No, the confidence interval contains the value 0, which represents no difference.

10-47 (a) $z^* = -1.06$; fail to reject H_0
(b) Prob-value $= 0.2892 > 0.05$; fail to reject H_0

10-48 0.60 to 4.80

10-52 (a) 0.41 to 0.92

10-56 -12.7 to 17.9

10-58 $t^* = 3.68$; reject H_0

10-60 $H_a: p_E - p_A > 0$ (elementary appearance more important); $t^* = 1.46$; fail to reject H_0

10-63 First, F test shows $\sigma_{ND} = \sigma_D$; second, $H_a: \mu_D - \mu_{ND} > 0$ (faster); $t^* = 2.01$; reject H_0

10-65 First, $F^* = 1.08$; therefore assume $\sigma_A = \sigma_B$; $H_a: \mu_B - \mu_A > 0$; $t^* = 0.82$; fail to reject H_0

10-67 $H_a: \mu_A - \mu_B > 0$; $t^* = 4.48$; reject H_0

Chapter 11

11-1 (a) $\chi^{2*} = 6.0$; $\chi^2(7, 0.05) = 14.1$; fail to reject H_0
(b) Prob-value $= P(x_7^2 > 6.0)$, > 0.10; therefore fail to reject H_0

11-4 $\chi^{2*} = 7.660$; $\chi^2(5, 0.05) = 11.1$; fail to reject H_0

11-6

	Employed	Not Employed	Total
In default	8 12.1	15 10.9	23
No default	75 70.9	60 64.1	135
Total	83	75	158

$x^{2*} = 3.431$; $x^2 = 3.84$; fail to reject H_0: The claim that the wife's employment status does not affect repayment cannot be rejected.

11-9

Expected Values

Age	Common	Fixed	Mixed
Over 45	203.215	119.795	161.99
Under 45	215.785	127.205	172.01

$\chi^{2*} = 2.732$; $\chi^2(2, 0.01) = 9.21$; fail to reject H_0

11-10 $H_0: \mu_A - \mu_B = 0$; $H_a: \mu_A - \mu_B \neq 0$; $\alpha = 0.10$; $z^* = 1.56$; fail to reject H_0

11-12 Expected value of option: 70.8; expected value of do not option: 129.2; $\chi^{2*} = 80.957$; $\chi^2(4, 0.01) = 13.3$; reject H_0

11-14 (a) $x^{2*} = 14.337$; $x^2(4, 0.001) = 13.3$; reject H_0
(b) Prob-value $= P(x_4^2 > 14.337) < 0.01$; reject H_0

11-18 (a) $x^{2*} = 12.115$; $x^2(6, 0.05) = 7.821$; reject H_0
(b) Prob-value $= P(x_3^2 > 12.115) < 0.01$; reject H_0

11-22 $\chi^{2*} = 20.88$; $\chi^2(2, 0.01) = 9.21$; reject H_0

11-23 $\chi^{2*} = 2.865$; $\chi^2(2, 0.10) = 4.61$; fail to reject H_0

11-24 $\chi^{2*} = 21.425$; $\chi^2(5, 0.05) = 11.1$; fail to reject H_0

11-26 $\chi^{2*} = 4.719$; $\chi^2(2, 0.10) = 4.61$; reject H_0

11-28 $\chi^{2*} = 140.34$; $\chi^2(4, 0.01) = 13.3$; reject H_0

11-29 $\chi^{2*} = 1.5623$; fail to reject H_0

Chapter 12

12-1 (a) The mean levels of the factor being tested are not all the same.
(c) The mean levels of the factor being tested are all the same.

12-3 (a) 0   (b) 3
(c) 16   (d) 60
(e) 1,232

12-4 (a) H_0: The means of each of the levels of the factor being tested are the same.
H_a: At least one of the means is different from the others.
(b) We would conclude that H_a was correct.
(c) Fail to reject H_0 would indicate that we did not find evidence that contradicted the null hypothesis; that is, no evidence to indicate that the means are different.
(d) The decision is made using an F test.

12-6 $F^* = 2.24$; $F(6, 21, 0.05) = 2.57$; fail to reject H_0

12-9

Source	SS	df	MS
Vending machine	56.82	3	18.94
Ice/no ice	0.06	1	0.06
Interaction	0.41	3	0.14
Error	0.27	16	0.02
Total	57.56	23	

(1) H_0: No differences in machines; $F^* = 947$; $F(3, 16, 0.05) = 3.24$; reject H_0; there are significant differences between machines.
(2) H_0: No difference between ice and no ice; $F^* = 3.0$; $F(1, 16, 0.05) = 4.47$; fail to reject H_0; there is no evidence that there is any difference between ice and no ice.

(3) H_0: No interaction; $F^* = 7.0$; $F(3, 16, 0.05) = 3.24$; reject H_0; there is a significant interaction between the factors.

12-12 $F^* = 0.84$; $F(3, 16, 0.01) = 5.29$; fail to reject H_0; there is no evidence of a difference between methods.

12-15

Source	SS	df	MS
Price	84.09375	3	28.03125
Error	164.875	28	5.88839
Total	248.96875	31	

$F^* = 28.03125/5.88839 = 4.76$; $F(3, 28, 0.01) = 4.58$ (by interpolation); reject H_0; there is a significant difference in the mean number of participants based on price.

12-17

Source	SS	df	MS
Entrance	401.467	5	80.2934
Error	5,692.0	24	237.167
Total	6,093.467	29	

$F^* = 0.34$; fail to reject H_0; the data do not contradict the null hypothesis that the mean number of persons entering the store is the same at each entrance.

12-23 (a)

Source	SS	df	MS
FI(sex)	328.05	1	328.05
FII(method)	510.05	1	510.05
FI × FII	11.25	1	11.25
Error	391.6	16	24.475
Total	1,240.95	19	

(1) H_0: No difference between sexes; $F^* = 13.4$; $F(1, 16, 0.10) = 9.31$; reject H_0; there is a significant difference between the sexes.
(2) H_0: No difference between methods; $F^* = 20.8$; reject H_0; there is a significant difference between methods.
(3) H_0: No interaction; $F^* = 0.5$; fail to reject H_0; there is no evidence of interaction.
(c) The interaction graph supports the results of part (a).

Chapter 13

13-1 Refer to the definition of the mean and the development of measures of dispersion in section 2-3.

13-2 $\text{covar}(x, y) = \frac{20}{7} = 2.86$

13-4 (b) $\text{covar}(x, y) = -38.0/7 = -5.43$

13-5 (a) $s_x = 2.39$; $s_y = 1.31$
 (b) 0.91   (c) 0.91

13-7 H_a: $\rho \neq 0.0$; critical value of r: ± 0.514; reject H_0; there is sufficient evidence to conclude that there is evidence of linear correlation.

13-10 For $r = +0.30$, reject H_0 at $\alpha = 0.01$.

13-11 (a) -0.47 to $+0.71$
 (c) -0.10 to $+0.91$

13-13 No. If there is no linear correlation, then the equation for the line of best fit would be of no value in predicting y based on x.

13-16 (a) $\hat{y} = 829.8 - 91.6x$ (b) 757.4

13-19 (a) $140/210 = 0.67$
 (b) $1 - (10/100) = 0.90$
 (c) $75/(25 + 75) = 0.75$

13-20 Price. A correlation coefficient of -0.7 indicates that price explains 49% of the variation in sales whereas advertisement expenditures explain only 25% ($r = 0.5$).

13-21 From formula (13-10), $R^2 = 0.77$.

13-22 From formula (13-11), $R^2 = 0.435$.

13-26 (b) $r = 0.99$ (c) Yes   (d) 0.85 to 1.0

13-27 (a) $r = -0.989$ (b) Yes   (c) -1.0 to -0.6

13-28 (a) This exercise will result in various answers; however, you should observe that most of the time the data will indicate no relationship (independence).
 (b) 0
 (c, d) The calculated value of r should be larger than the table value 10% of the time. (This is a good place to reemphasize the meaning of α.)

13-31 (b) $R^2 = 0.862$ is interpreted to mean that 86% of the variability in y is explained by the line of best fit.
 (c) $39.10

13-32 (a) $\hat{y} = 41.9 + 1.4x$
       (b) $R^2 = 0.73$
       (c) $R^2 = 0.724$
       (d) The value of $\hat{y} = 0$ is meaningless for this problem.
       (e) $b_1 = 1.4$ represents an additional cost of 1.4 million dollars for every extra 1.0 thousand man-hours used in construction.

13-33 $n = 8$; $\sum x = 54$; $\sum y = 287$; $\sum x^2 = 372$; $\sum y^2 = 10{,}723$; $\sum xy = 1{,}973.5$
       (a) $\hat{y} = 3.25 + 4.833x$ (c) 0.64
       (d) $H_a: \rho > 0$; critical value of r: 0.622; fail to reject H_0

13-34 $n = 15$; $\sum x = 202.72$; $\sum y = 300.5$; $\sum x^2 = 2{,}835.7$; $\sum y^2 = 6{,}224.27$; $\sum xy = 3{,}982.64$
       (a) $\hat{y} = 31.086 - 0.8178x$ (b) 0.31

13-35 (b) $n = 25$; $\sum x = 1{,}877$; $\sum y = 1{,}055$; $\sum x^2 = 144{,}451$; $\sum y^2 = 45{,}643$; $\sum xy = 80{,}875$; $\hat{y} = 6.73 + 0.472x$
       (c) $r^2 = 0.71$ (d) 49
       (e) x is outside range of data; cannot predict y.

Chapter 14

14-1 $n = 8$; $\sum x = 40$; $\sum y = 32$; $\sum x^2 = 240$; $\sum xy = 180$; $\sum y^2 = 140$
      (a) $\hat{y} = 1.5 + 0.5x$
      (b) $x = 2$; $\hat{y} = 2.5$
          $x = 4$; $\hat{y} = 3.5$
          $x = 6$; $\hat{y} = 4.5$
          $x = 8$; $\hat{y} = 5.5$
      (c)

Point	A	B	C	D
$\hat{y}$	2.5	2.5	3.5	3.5
e	-0.5	0.5	-0.5	0.5

Point	E	F	G	H
$\hat{y}$	4.5	4.5	5.5	5.5
e	-0.5	0.5	-0.5	0.5

      (d) $S^2 = 8(0.5)^2/6 = 2/6 = 0.33$
      (e) $S^2 = [140 - (1.5)(32) - (0.5)(180)]/6 = 0.33$

14-2 $n = 8$; $\sum x = 28$; $\sum y = 28$; $\sum x^2 = 140$; $\sum xy = 60$; $\sum y^2 = 140$
      (a) $\hat{y} = 6.67 - 0.90x$

(b)

Point	A	B	C	D
x	0	1	2	3
$\hat{y}$	6.67	5.77	4.67	3.97
(c) e	-0.67	1.23	-0.87	1.03

Point	E	F	G	H
x	4	5	6	7
$\hat{y}$	3.07	2.17	1.27	0.37
e	-1.07	0.83	-1.27	0.63

      (d) $7.6232/6 = 1.27$
      (e) $\dfrac{140 - (6.66667)(28) - (-0.9047)(60)}{6} = \dfrac{7.615}{6} = 1.269$

14-3 (b) $n = 10$; $\sum x = 48$; $\sum y = 85$; $\sum xy = 423$; $\sum y^2 = 733$; $\hat{y} = 5.9 + 0.54x$

      (d)

x	3	3	6	6
e	-0.52	0.48	-0.14	0.86

      (e) 0.3

14-4 $\sum (x - \bar{x})^2 = 40$
      (a) $0.333/40 = 0.0083$
      (b) $(8)(0.333)/320 = 0.0083$

14-5 (a) $s_{b_1} = 0.118$
      (b) $H_a: \beta_1 > 0$; $t(8, 0.05) = 1.86$; $t^* = 4.58$; reject H_0; yes, the slope b_1 is significant.

14-7 $0.54 \pm (2.31)(0.118)$; 0.565 to 1.175

14-8 1.65 to 3.73

14-9 (a) Point estimate for $\mu_{y|x=6} = 9.14$
      (b) $9.14 \pm (2.31)(0.547)\sqrt{0.1 + 0.05217} = 9.14 \pm 0.493$
      (d) $9.14 \pm (2.31)(0.547)\sqrt{1 + 0.15217} = 9.14 \pm 1.355$

14-11 The standard deviation for $\bar{x}$'s is much smaller than for individual x's (central limit theorem). Thus the confidence interval will be narrower in accordance with this.

14-13 $\tfrac{5}{9} = 0.56$

14-14 (a) $b_0 = 15.085$; $n = 8$; $\sum y = 98$; $\sum x^2 = 263$; $b_1 = -0.553$; $\sum x = 41$; $\sum xy = 473$; $\sum y^2 = 1{,}374$; $\hat{y} = 15.085 - 0.553x$
       (b) $t^* = 0.786 < 1.94$; therefore fail to reject H_0
       (c) No

14-16 (d) $\hat{y} = 3.73 - 0.39x$
(e) $s_e = 0.1827$
(f) -0.469 to -0.311
(g) 2.44 to 2.68
(h) 2.15 to 2.97

14-17 $n = 8$; $\sum x = 120$; $\sum y = 228$; $\sum x^2 = 2{,}100$;
$\sum xy = 4{,}260$; $\sum y^2 = 8{,}884$; $s_e^2 = \frac{34}{6} = 5.67$
(a) $\hat{y} = -13.5 + 2.8x$
(c) $2.8 \pm (2.45)\sqrt{5.67/\{2{,}100 - [(120)^2/8]\}}$
$2.8 \pm (2.45)\sqrt{0.0189}$
2.8 ± 0.34
(e) $56.5 \pm (2.45)(2.38)$
$\sqrt{\frac{1}{8} + \frac{8(25-15)^2}{8(2{,}100) - (120)^2}}$
$56.5 \pm (2.45)(2.38)\sqrt{0.458}$
56.5 ± 3.95

14-19 (a) Outside the domain; there are no 10-bed hospitals in New York.
(c) No, this is outside the scope of the regression. If all hospitals added 10 beds, the relationship between x and y would probably change.

14-21 1.35 ± 1.16

14-22 $n = 10$; $\sum x = 55$; $\sum x^2 = 413$; $\sum y = 485$;
$\sum y^2 = 28{,}603$; $\sum xy = 3{,}363$; $s_e^2 = 87.9385$;
$s_{b_1}^2 = 0.7958$; $b_0 = 13.9$; $b_1 = 6.29$
(a) $H_a: b_1 \neq 0$; $t(8, 0.05) = 1.86$; $t^* = 7.06$; reject H_0
(b) 6.29 ± 2.06

Chapter 15

15-1 (a) $\dfrac{0.5 - (0.7)(0.4)}{\sqrt{[1 - (0.7)^2][1 - (0.4)^2]}} = 0.34$

(b) $\dfrac{0.7 - (0.5)(0.4)}{\sqrt{[11 - (0.5)^2][1 - (0.4)^2]}} = 0.63$

(c) $\dfrac{0.4 - (0.5)(0.7)}{\sqrt{[1 - (0.5)^2][1 - (0.7)^2]}} = 0.08$

15-2 (a) 0.17 (b) -0.51 (c) 0.76

15-3 (a) 0.70 (b) 0 (c) 0

15-5 (a) 2.83 (b) 2.45 (c) 19.8

15-6 (a) The coefficient 4 suggests that each retail outlet sells 4,000 microwave ovens.
(b) The coefficient -250 suggests that for every dollar that the price of this company's microwave oven is above its competitor's price, they will lose 250,000 sales.
(c) 675,000
(d) Sales would drop by 2,500,000 units provided the number of retail outlets remained the same.

15-8 $H_a: \rho^2 \neq 0$; $F(3, 16, 0.01)$; $F^* = (0.45/3)/[(1 - 0.45)/16] = 4.36$; fail to reject H_0

15-10 (a) $\sum (y - \hat{y})^2 = 126$; $\bar{y} = 77$;
$\sum (y - \bar{y})^2 = 6{,}210$; $R^2 = 0.98$
(b) $H_a: \rho^2 \neq 0$; $F(2, 7, 0.05) = 4.74$;
$F^* = 168.9$; reject H_0

15-12 No, if we fail to reject $\rho^2 = 0$, we must conclude that all $\beta = 0$.

15-15 $s_e^2 = 150/15 = 10$;
$s_{b_1}^2 = (18)(10)/[(18)(2{,}250) - (150)^2]$
$= 0.01$
$s_{b_2}^2 = (18)(10)/[(18)(6{,}255) - (200)^2]$
$= 0.0359 = 0.04$
(a) $H_a: \beta > 0$; $t(15, 0.01) = 2.60$; $t^* = 29.0$; reject H_0

15-16 No, because of multicollinearity

15-19 (a) x_1, since it has the highest simple correlation
(b) $r_{x_2 y \cdot x_1} = 0.59$, $r_{x_3 y \cdot x_2} = 0.65$; thus x_3, since it has the highest partial correlation, holding x_1 constant

15-21 (a) $H_a: \rho^2 \neq 0$; $F(2, 25, 0.05) = 3.39$; $F^* = 59.2$; reject H_0
(b) $H_0: \beta_1 = 0$; $t^* = \dfrac{8.0 - 0}{3.0} = 2.67$; reject H_0
(c) $H_a: \beta_2 \neq 0$; $t(25, 0.025) = 2.06$; $t^* = 4.00$; reject H_0
(d) $\hat{y} = 44$

15-22 (a) 0.9896
(b) $H_0: \rho^2 = 0$; $F^* = 570.7$; reject H_0
(c) $s_e^2 = 560.8/12 = 46.73$
(d) $s_{b_1} = 0.019$
(e) $H_a: \beta_1 \neq 0$; $t(12, 0.05) = 1.78$; $t^* = 26.05$; reject H_0
(f) $H_a: \beta_2 \neq 0$; $t(12, 0.05) = 1.78$; $t^* = 0.958$; fail to reject H_0

15-24 (a) $R^2 = 0.971$
(b) $H_0: \rho^2 = 0$; $F^* = 33.5$; reject H_0

(c) $s_e^2 = 0.19490/2 = 0.09745$

(d) $s_{b_1} = \sqrt{\dfrac{5(0.09745)}{5(57,425) - (535)^2}} = 0.0233$

$s_{b_2} = \sqrt{\dfrac{5(0.09745)}{5(8.1376) - (6.34)^2}} = 0.995$

(e) $H_0: \beta_1 = 0;\ t^* = \dfrac{0.0029 - 0}{0.0233} = 0.124$

Chapter 16

16-2 (a) Time series: only sales is used
(b) Causal: relates sales to advertising
(c) Time series: uses only sales

16-3 Because different methods work for different patterns of data

16-4 Time, uncertainty, and reliance on historical data

16-8 Because positive and negative errors will incorrectly cancel out

16-9 (a)

Forecast Error	Absolute Error, |e|	Squared Error, e^2
15	15	225
12	12	144
−2	2	4
0	0	0
−12	12	144
−13	13	169
	54	686

(b) $\text{MAD} = \dfrac{\sum |e|}{n} = \dfrac{54}{6} = 9$

(c) $\text{MSE} = \dfrac{\sum e^2}{n} = \dfrac{686}{6} = 114.33$

16-11 No, ignores the question of cost and the trade-off between cost and accuracy that must be made in light of the use of the forecast

16-13 (a) MAD = 0.98/14 = 0.07
(b) MAD = 0.94/14 = 0.067
(c) The naive model does only slightly worse than the given model. Unless extreme accuracy is needed or the given model is very inexpensive and easy to use, the given model does not appear worthwhile.

16-14 (a) Long (b) Medium (c) Short
(d) Immediate (e) Short

16-17 A is positive; B is negative.

16-21 (a) The cash receipts and expenditures for the next 90 days.
(b) Past cash flow (receipts and expenditures) present bills, checks, and receipts outstanding.
(c) Time series would use only past cash balance patterns and project them forward. A causal model would look at possible relationships, such as billing, accounts payable, and company activity.
(d) MSE, since a large error could cause serious financial problems or lost opportunity.
(e) To overestimate cash on hand means you would have to borrow money to cover expenditures. Underestimating cash on hand means that money will be in an interestless checking account rather than being invested and earning interest.

16-22

Year	(a) Error	(b) Absolute Error	Squared Error
1976	−199	199	39,601
1977	−34	34	1,156
1978	+5	5	25
1979	+40	40	1,600
1980	+93	93	8,649
1981	+125	125	15,625
1982	+174	174	30,276
1983	+218	218	47,524
	+422	888	144,456

(c) MAD = 888/8 = 111
(d) MSE = 144,456/8 = 18,057
(e) That a clearly increasing trend or cyclical pattern is being missed.
(f) Yes. Since the error shows a pattern, there is a pattern being ignored.

16-23 (a) MAD(A) = 11/5 = 2.2;
MSE(A) = 82.50/5 = 16.5;
MAD(B) = 13/5 = 2.6;
MSE(B) = 42.5/5 = 8.5

16-24 (f) 0

Chapter 17

17-2 (a) Trend (b) Irregular (c) Seasonal

17-3 $(86.3)(1.02)(0.83) = 73.06$

17-4 Trend = 70,000;
cyclical = 71,000/70,000 = 1.014% = 101.4;
seasonal = 70,000/71,000 = 0.986% = 98.6;
irregular = 71,000/70,000 = 1.014% = 101.4

17-5

	Quarter	Four-Quarter Moving Total	Four-Quarter Moving Average	Four-Quarter Centered Moving Average
1980	1	110	27.5	
	2	113	28.25	27.875
	3	115	28.75	28.5
	4	117	29.25	29.0
1981	1			

17-7

	Quarter		Four-Quarter Moving Total	Four-Quarter Moving Average	Four-Quarter Centered Moving Average
1981	1	9.7			
	2	12.5			
	3	8.4	40.5	10.125	10.15
	4	9.9	40.7	10.175	10.39
1982	1	9.9	42.4	10.6	11.26
	2	14.2	47.7	11.925	12.25
	3	13.7	50.3	12.575	
	4	12.5			

17-8 First quarter, 1982: $t = 13$

	Quarter	t	T
1982	1	13	114
	2	14	117
	3	15	120
	4	16	123
1983	1	17	126
	2	18	129
	3	19	132
	4	20	135

17-11 The value of c is that part of the moving average that is not explained by the trend line. That is, c for each time period is found by dividing the centered moving average by the corresponding value of T. (T results from the trend line.)

17-14 The seasonal effect is removed.

17-16 Adjustment factor = 0.994

17-17

Quarter	Quarterly Average	Seasonal Index (adj. quarterly avg.)
1	109.6	107.28
2	95.4	93.38
3	88.5	86.62
4	115.13	112.70
	408.63	

Adjustment factor = $\dfrac{400}{408.63} = 0.9788$

17-18

Quarter	Quarterly Sales	Quarterly Seasonally Adjusted Sales
1	1,900	1,771.07
2	1,645	1,761.61
3	1,816	2,096.50
4	2,010	1,783.50

17-20 One reason for decomposition is to help understand the patterns that are in existence. A second reason is that the decomposition is the preliminary step in forecasting. Even though the actual forecasting procedures are beyond the intended level of this textbook, the decomposition procedures are included. Forecasting is likely to be part of a future course.

17-23 (b) $T = 8,330.9 + 173.5t$

Quarter		(a)	(b)	(c)	(d)	(e)
1	1					
	2					
	3	9,116.8	8,504.4	107.2	103.9	98.8
	4	9,117.6	8,677.9	105.1	100.9	99.6
2	1	9,107.0	8,851.4	102.9	97.9	102.8
	2	9,230.1	9,024.9	102.3	97.3	95.9
	3	9,403.8	9,198.4	102.2	103.9	98.3
	4	9,660.8	9,371.9	103.1	100.9	101.6
3	1	9,878.4	9,545.4	103.5	97.9	101.5
	2	9,877.8	9,718.9	101.6	97.3	104.3
	3	9,878.8	9,892.4	99.9	103.9	96.9
	4	9,930.3	10,065.9	98.7	100.9	95.4
4	1	10,182.6	10,239.4	99.4	97.9	102.0
	2	10,663.4	10,412.9	102.4	97.3	97.2
	3	10,915.1	10,586.4	103.1	103.9	104.9
	4	11,095.6	10,759.9	103.1	100.9	102.3
5	1	11,355.1	10,933.4	103.9	97.9	92.6
	2	11,542.3	11,106.9	103.9	97.3	101.5
	3					
	4					

17-27 (a)

Jan.	Feb.	Mar.	Apr.	May	June
98.0	90.0	95.0	100.0	103.0	130.0

July	Aug.	Sept.	Oct.	Nov.	Dec.
125.0	125.0	89.0	100.0	75.0	70.0

(b) Christmas jobs increase employment in November and December, whereas in the summer, students looking for jobs increase the ranks of the unemployed.

Chapter 18

18-1 To measure economic activity and to predict change in economic activity

18-2 (i) Coincidental
(ii) Lagging
(iii) Leading

18-5

Year	Indicator	Base 1978 (indicator/1,322) × 100	Base 1980 (indicator/2,334) × 100
1976	837	63.3	35.9
1977	1,183	89.5	50.7
1978	1,322	100.0	56.6
1979	1,825	138.0	78.2
1980	2,334	176.6	100.0

18-8 The term *real wages* is the value of wages expressed in terms of the dollar value in the base year.

18-11 (a) $3.00/2.50 = \$1.20$
(b) $3.20/2.80 = \$1.14$
(c) $(1.14/1.20) - 1 = -5\%$
(d) 1982 real wage $= \$1.20 \times 2.8 = \$3.36$

18-14 $\frac{131}{126} - 1 = 0.04$ or 4%. Porter Brewery increased more (5%).

18-17 $1975 = 138.3$; $1980 = 171.8$. The typical college costs were 38.3% higher in 1975 and 71.8% higher in 1980 than in 1970.

18-19 $7,565.22

18-22

Year	Real Inventory Cost
1978	$243,478
1979	253,219
1980	262,712
1981	276,151
1982	305,785

Yes, since real inventory costs are going up.

18-23 (a) No. There are no absolute prices given; only relative prices for each area are given.
(b) New York

18-25 (a) $2.20 (b) $3.91

18-28 $1979 = 100$; $1980 = 121.3$; $1981 = 120.6$

18-29 (a) $1979 = 86.5$; $1980 = 100.0$; $1981 = 100.8$
(b) $1979 = 85.8$; $1980 = 99.2$; $1981 = 100.0$

18-30 (a) 21.3%
(b) $14.30
(c) $16.07
(d) 12.4%
(e) The actual percentage change in the market basket is less because of the switch in items. There was a switch to lower-priced items.

Chapter 19

19-2 (a) $H_0: M = 24.8(\leq)$; $H_a: M > 24.8$
(1) $n(+) = 9$; $n(-) = 8$; $n = n(+) + n(-) = 17$; $x = 8$. Critical value from table is 4; since x is >4, fail to reject H_0.
(2) $x' = 8.5$; $z = 0$; fail to reject H_0
(b) $H_0: \mu = 24.8$; $H_a: \mu > 24.8$; $\bar{x} = 25.15$; $s = 0.78$; $n = 17$; $t = 1.85$; reject H_0
(c) The parametric test yields a different decision than the sign test.

19-3 $(A - B)$; $n(+) = 1$; H_0: No difference; critical value $= 0$; fail to reject H_0

19-4 $(A - B)$; $n(+) = 3$; H_0: No difference; critical value $= 2$; fail to reject H_0

19-5 45 to 52

19-7 -4 to 0

19-8 H_0: The average investment return is the same. H_a: The average investment return is not the same. Critical value is 17; $U = 17$. This implies that the two fund types return different amounts.

19-10 (a) H_0: There is no difference.
(b) H_a: There is a difference. Critical value is 127; $U = 185.5$; therefore fail to reject H_0.

19-12 H_0: The results are randomly ordered. H_a: The results are not randomly ordered. The critical values of V are 8 and 19; observed $V = 14$; therefore fail to reject H_0.

19-14 H_a: Lack of randomness (a trend, increasing absenteeism); critical value of V is 5; observed $V = 10$; therefore fail to reject H_0.

19-16 H_0: Independence between two sets of test scores; H_a: Correlation between two sets of test scores; $\sum d^2 = 114$; $r_s = 0.601$; reject H_0

19-19

Corporations									
	1	2	3	4	5	6	7	8	9
F	3.5	9.5	5.5	2	3.5	7	8	15	17
M	3	7	1.5	8	4	5	14	15	16

Corporations									
	10	11	12	13	14	15	16	17	18
F	16	18	12	14	11	5.5	9.5	1	13
M	17	18	13	11	9	6	10	1.5	12

$n = 18$; $\sum d^2 = 2{,}351$; $r_s = -1.476$, 0.476; critical values are -0.476, 0.476; reject H_0.

19-21 (b) The Mann-Whitney U test is appropriate. H_a: Average rebound height of A is higher than B. $U = 47$; the critical value is 27; fail to reject H_0.

19-22 Use sign test:
(a) $H_0: M = 50$;
$H_a: M \neq 50$;
$n(+) = 10$; $n(-) = 20$; $n(0) = 2$. Critical value from table 9 is 9; fail to reject H_0: The median age is not significantly different from 50.
(b) $H_0: M = 50$ ($\geq$); $H_a: M < 50$; $n(+) = 12$; $n(-) = 20$; $n = 32$; critical value is 10; fail to reject H_0.

19-23 (a) Rank correlation: $r_s = 0.988$; reject H_0
(b) Pearson correlation; $r = 0.823$; reject H_0

19-25 The sign test is appropriate. H_a: Average difference from 110 is positive. Critical value is 3; $n(-) = 3$; reject H_0.

19-27 Use the sign test; critical value is 5; $n(-) = 4$; reject H_0.

Chapter 20

20-1 (a) Certainty (b) Certainty
(c) Uncertainty (d) Uncertainty

20-4

	Horses	
Action	Not Injured	Injured
Write policy	$100,000	−$900,000
Do not write policy	0	0

20-5

	Project	
Action	Success	Failure
Fund project	542.5	−7.5
Do not fund project	0	0

20-6

Cakes Baked	Demand for Cakes			
	8	9	10	11
8	0	3	6	9
9	3	0	3	6
10	6	3	0	3
11	9	6	3	0

20-8 Payoff Table
States of Nature

	Novel Is			
Action	Poor(A)	Average(B)	Good(C)	Best Seller(D)
Publish (a)	−40,000	20,000	80,000	500,000
Do not publish (b)	0	0	0	0

Opportunity Losses
States of Nature

		Novel Is		
Action	Poor(A)	Average(B)	Good(C)	Best Seller(D)
Publish (a)	−40,000	0	0	0
Do not publish (b)	0	20,000	80,000	500,000

20-11 (a) Do not do project (b) Do project
(c) Do not do project

20-13 (a) Do not publish (b) Publish (c) Publish

20-14 E(write policy) $= \$50,000$; E(do not write policy) $= 0$; therefore write policy

20-17 E(do project) $= 3.5$; E(do not do project) $= 0$; therefore do project

20-18 (b) E(try) $= -480,000$; E(settle) $= -500,000$; therefore try
(c) E(try) $= E$(settle) $= -500,000$; therefore do either

20-20 E(best actions) $= (0)(0.05) + (100,000)(0.95)$
$= 95,000$; E(insure) $= \$50,000$;
EVPI $= 95,000 - 50,000 = 45,000$

20-22 7.35 million

20-25

	A	B	C
a	100	140	180
b	90	150	180
c	70	130	200

20-28 (a)

Payoff Table
State of Nature

Action	Tax Deduct. (A)	No Tax Deduct. (B)
Purchase (a)	−9,000	−9,000
Lease (b)	−7,000	−10,000

(b)
Opportunity Losses
State of Nature

Action	(A)	(B)
(a)	2,000	0
(b)	0	1,000

(c) Lease
(d) Purchase
(e) Lease

20-29 (b) Invest   (d) Do not invest

20-31 (b) $10.079   (d) 10.153   (g) 103

20-33 E(lease) $= -7,600$; E(buy) $= -9,000$; therefore lease

20-35 EOL(not invest) $= 12,000$;
EOL(invest) $= 8,000$;
therefore invest

20-37 $4,000

20-39 $200

20-42 (c) Raise prices (e) Raise prices
(g) $300,000

20-43 (c) Tax cut (d) Tax cut
(g) 10 billion

20-44 EOL(lease) $= (1 - P)(1,000)$;
EOL(buy) $= P(2,000)$;
$(1 - P)(1,000) < P(2,000)$; therefore $P > \frac{1}{3}$

Appendix A

A-1 (a) $x_1 + x_2 + x_3 + x_4$
(b) $x_1^2 + x_2^2 + x_3^2$
(c) $(x_1 + y_1) + (x_2 + y_2) + (x_3 + y_3)$
$+ (x_4 + y_4) + (x_5 + y_5)$
(d) $(x_1 + 4) + (x_2 + 4) + (x_3 + 4)$
$+ (x_4 + 4) + (x_5 + 4)$
(e) $x_1 y_1 + x_2 y_2 + x_3 y_3 + x_4 y_4 + x_5 y_5$
$+ x_6 y_6 + x_7 y_7 + x_8 y_8$
(f) $(x_1^2 f_1) + (x_2^2 f_2) + (x_3^2 f_3) + (x_4^2 f_4)$

A-2 (a) $\sum_{j=1}^{6} x_i$ (b) $\sum_{i=1}^{7} x_i y_i$
(c) $\sum_{i=1}^{9} (x_i)^2$ (d) $\sum_{i=1}^{n} (x_i - 3)$

A-4 (a) 8 (b) 68 (c) 64

A-5 (a) 8 (b) 4 (c) 12
 (d) 4 (e) 42 (f) 64
 (g) −7 (h) 32 (i) 20
 (j) 120 (k) 400

A-6 $\sum_{i=1}^{120}[12{,}000 - (100)(i-1)] \times (0.005)$

Appendix B

B-2 (1) Use a two-digit number to represent the results obtained. Let the first digit represent one of the coins and the second digit represent the other coin. Let an even digit indicate heads and an odd digit, tails. Observe 10 two-digit numbers from the table. If a 16 is observed, it represents tails and heads on the two coins. One head was therefore observed. The probabilities have been preserved.

(2) A second way to simulate this experiment is to find the probabilities associated with the various possible results. The number of heads that can be seen on two coins is 0, 1, or 2. (HH, HT, TH, TT is the sample space.)

$$P(\text{no heads}) = \tfrac{1}{4},$$
$$P(\text{one head}) = \tfrac{1}{2},$$
$$P(\text{two heads}) = \tfrac{1}{4}.$$

Using two-digit numbers, let the numbers 00–24 stand for no heads appeared, 25–74 stand for one head appeared, and 75–99 stand for two heads appeared. The probabilities have again been preserved. Observe 10 two-digit numbers.

Appendix C

C-1 (a) 13 (b) 9 (c) 9 (d) 156
 (e) 46 (f) 42 (g) 103 (h) 17

C-2 (a) 8.7 (b) 42.3 (c) 49.7
 (d) 10.2 (e) 10.4 (f) 8.5
 (g) 27.4 (h) 5.6 (i) 3.0 (j) 0.2

C-3 (a) 17.67 (b) 4.44 (c) 54.55
 (d) 102.06 (e) 93.22 (f) 18.00
 (g) 18.02 (h) 5.56 (i) 44.74
 (j) 0.67

Index

Abscissa, 216
Absolute deviation, 549
Absolute dispersion, 66
Absolute value, 59
Acceptance region, 273
Acceptance sampling, 268
Action, in decision theory, 654
Addition rule:
   general, 149
   special, 149
Adjustment factor for seasonal index, 586
All-inclusive events, 136
Alpha (α), 270, 272, 278, 287, 292, 296
Alternative hypothesis, 269, 275
Analysis of variance, 418
   assumptions for test of, 425
   degrees of freedom for, 422
   hypothesis test for, 418, 423
   mean square, 422
   single-factor, 418, 426
   sum of squares, 420
   two-factor, 432, 437
ANOVA (*see* analysis of variance)
ANOVA table, 421
Area representation of probability, 184, 218
Area sampling, 24
Assumption of consistency, 544
Asterisk notation, 287
Attribute data, 9, 10, 103
Average, 50 (*see* measures of central tendency)
Average absolute error of forecast, 554

Bar graph, 33
Base period, of index, 602
Base year, 602
Bayes's rule, 164
   tabular form for, 165
Bell-shaped curve, 216
Bell-shaped distribution, 75
Best fit, 118, 464
Beta (β):
   in line of best fit, 486
   as probability of type II error, 270, 645
Between-sample variation, 418
Bias:
   in measuring device, 17
   natural, 17
   systematic, 16
Bimodal distribution, 45
Binomial coefficient, 194, A-15
Binomial distribution, 189, 191
   mean for, 201
   normal approximation of, 232
   standard deviation for, 201
Binomial experiment, 192
Binomial parameter, 217, 316
   inferences about, 316
Binomial probability, 189, 316, 373
   by calculation, 193–196
   by normal approximation, 232
   by use of table, 197–198, A-16
Binomial probability function, 195
Binomial random variable, 192, 316, 626
Bivariate data, 103, 456
Box-Jenkins model, 541

Business indicators, 598
   coincident, 599
   lagging, 599
   leading, 599

Calculated value (*), 279
   χ^{2*}, 323, 392
   F^*, 352, 423, 435, 441, 523
   t^*, 308, 360, 362, 369, 492, 526
   z^*, 287, 317, 345, 347, 373, 626, 633, 638
Causal model, 542
Cause-and-effect relationship, 115, 501, 514
Cell, 104, 392
Census, 15
Centered moving average, 572
Central limit theorem, 249, 276, 292
   application of, 255
Central tendency, measures of, 50
Centroid, of data, 456, 470
Chart representation, of sample space, 139
Chebyshev's theorem, 74
Chi-square (χ^2), 323
   calculation of, 323, 392, 394
   critical value of, 324, 393, 397, A-23
   degrees of freedom for, 324, 396, 400
   distribution for, 323, 392
   properties of, 324

tests concerning:
  for enumerative data, 392, 400
  for population variance or standard deviation, 323
Circle graph, 33
Class, 39, 104, 392
Class boundary, 41
Classical time series, 566
  and forecasting, 540
Classification of data, procedure, 40–42
Class limits, 40
Class mark, 42, 53
Class width, 41
Cluster sampling, 23
Coefficient of determination, 472, 475, 476
Coefficient of linear correlation, 112, 459, 640
  calculation of, 112, 460
  critical values of, 462, A-34
  decision points for, 114
  estimating, 115
  negative, 112
  for population, 461
  positive, 112
  test for, 462
Coefficient of variation, 66
Coincident indicators, 599
Collecting data, 13
Comparing statistical tests, 645
Complementary event, 145
Compound event, 145
Conclusion, of hypothesis test, 279
Conditional probability, 154, 164
  and Bayes's rule, 164
Confidence belts, 462, 497
Confidence interval estimate, 292
  for correlation, A-35
  and level of confidence, 292
  lower limit for, 294
  and maximum error, 293
  for μ, 291, 313
  for $\mu_1-\mu_2$, 347, 364
  for one proportion, 319
  for paired difference, 370
  for regression, 493, 494
  for p_1-p_2, 376
  for p, 318
  for σ_1/σ_2, 356
  for sign test, 626
  and standard error, 249, 293
  upper limit for, 294
  using sign test technique, for median, 626
  for variance, standard deviation, 327
Confidence level, 292
Confidence limits, 294
Constant dollars, 603
Constant probability function, 183
Constant seasonal effect, 586
Consumer price index, 605, 614
Contingency table, 103, 399
  degrees of freedom for, 400
  expected value of, 401, 403
  test for, 400, 404
Continuity correction factor, 233
Continuous data, 9, 10
Continuous probability distribution, 216
Continuous random variable, 216
Correct decision, 270
Correlation:
  coefficient of, 112, 459, 640
  critical values, A-34
  linear, 111
  partial, 512
  rank, 640
Correlation analysis, 111
Correlation coefficient (see coefficient of linear correlation; partial correlation coefficient; rank correlation coefficient)
Cost of uncertainty, 672
Count, 9
Covariance, 457
Critical region, 271
Critical value, 272
Crossed experiment, 437
Cross-tabulation table, 103
Cumulative frequency, 46
Cumulative frequency distribution, 46
  ogive for, 46
Cumulative relative frequency distribution, 46, 131
  ogive for, 46, 77
Curvilinear regression, 119, 518
Cyclical component, of time series, 567, 584
Cyclical patterns, 545

Data, 8, 9
  attribute, 9, 10
  bivariate, 103, 456
  centroid of, 456, 470
  classification of, 42
  collection of, 13
  continuous, 9, 10
  decomposition of, 566
  deseasonalized, 589
  discrete, 9, 10
  enumerative, 392
  paired, 103, 367, 456
  patterns of, 543
  qualitative, 9, 10
  quantitative, 9, 10
  seasonally adjusted, 589
  single-variable, 32
  stationary, 545
  variable, 9, 10
Decision, in hypothesis test, 270
Decision point, 114
Decision rule, 273, 289
Decision strategies:
  maximax, 663
  maximin, 663
  minimax regret, 663
Decision theory, 654
  action in, 654
  under certainty, 654
  state of nature, 656
  under uncertainty, 655
Decomposition:
  of data, 566
  of time series, 567
Definite integral, 217
Deflating a time series, 603
Degrees of freedom:
  for χ^2, 324, 396, 400
  for contingency table, 400
  for F, 352
  for t, 309
Dependent event, 153, 157
  multiplication rule for, 154
Dependent means, 367
Dependent sample, 342
Dependent sampling, 342
Dependent variable, 103, 486
Descriptive statistics, 4
Deseasonalized data, 589
Deterministic model, 515
Deviation:

absolute, 549
from mean, 58
mean absolute, 59, 549
squared, 474
Difference between two means, 344
Difference between two proportions, test for, 372
Discrete data, 9
Discrete random variable, 180, 203
Dispersion, measure of, 58
Distribution:
  bell-shaped, 46, 216
  bimodal, 45
  binomial, 189, A-16
  chi-square, 323
  cumulative frequency, 46
  cumulative relative frequency, 46
  F, 352
  frequency, 38
  J-shaped, 46
  normal, 46, 216
  Poisson, 202, A-20
  probability, 182
  rectangular, 46
  relative frequency, 43
  sampling, 245
  skewed, 46
  standard normal, 218
  Student's t, 308
  symmetrical, 46
  uniform, 46
Distribution-free methods, 622
Dominated action, 662

Economic indicators, 598
Efficiency of statistical test, 646
Empirical probability, 130
Empirical rule, 75
Enumerative data, 392
Equally likely event, 141
Equation of straight line, 465, 486
Error:
  experimental, 427
  of forecast, 547
  mean squared, 547
  observed, 487
  in prediction, 473
  in sampling, 16

squared, 474
systematic bias in, 16, 26
type I, 270
type II, 270
Estimation, 291
  confidence interval (*see* confidence interval estimate)
  for correlation coefficient, 462
  for means:
    one, σ known, 291
    one, σ unknown, 313
    two (difference between) dependent, 370
    two (difference between) independent, 347, 364
  for median, 626
  point, 291
  for proportion:
    one, 319
    two, 376
  for regression line:
    $\mu_{y|x_0}$, 493
    slope of, 492
    y_x, 494
  for variance:
    one, 327
    two, 352
Event:
  all-inclusive, 136
  complementary, 145
  compound, 145
  dependent, 153
  equally likely, 141
  independent, 153, 154
  intersection of, 147
  mutually exclusive, 136, 145
Expected frequency, 392
Expected opportunity loss, 670
Expected value:
  for contingency table, 392
  for multinomial experiment, 396
  of payoff, 666
  of perfect information, 672
Experiment, 8, 136
  binomial, 192
  crossed, 437
  multinomial, 394
  simulation of, A-8
Experimental error, 427
Experimental probability, 130
Exponential smoothing, 557

Factorial notation, 194, A-14
Factor level, 419
Failure, 192
F distribution, 352
  calculated value for, 352
  critical value of, 352, A-24
  degrees of freedom for, 352
  hypothesis test using:
    for ANOVA, 423
    for two variances, 354
  properties of, 352
First-ace experiment, 12
First quartile, 70
Fitting an equation, 464
  for trend, 578
Food price index, 611
Forecasting, 540
  error of, 547
  models for, 541
  naive model for, 553
  patterns of data in, 543
  time frame for, 543
Forecasting techniques, 541
  best, 542, 552
Frequency, 39
  cumulative, 46
  expected, 392
  observed, 392
  relative, 43, 129
  table, 41
Frequency distribution, 38, 41
  calculations using, 52, 64
  cumulative, 46
  cumulative relative, 46
  grouped, 42
  mean of, 52
  relative, 46
  standard deviation of, 61
  table, 41
  ungrouped, 39
  variance of, 60
Frequency histogram, 43
F statistic, 352
  optional shortcut for, 355
Fulcrum, 51

General addition rule, 149
General multiplication rule, 156
Glossary of symbols, endpapers

Gosset, W.S., 308
Graphs:
　histogram, 43
　ogive, 46
　scatter diagram, 106
Greek alphabet, front endpaper
　use of, 186
Grouped frequency distribution, 42

Haphazard, 20
Histogram, 43
　bimodal, 45
　frequency, 43
　J-shaped, 46
　line, 184
　normal, 46
　for probability, 148, 181, 184
　rectangular, 46
　relative frequency, 43, 129
　shapes of, 46
　skewed, 46
　symmetrical, 46
　triangular, 46
　uniform, 46
Homogeneity, 404
Hypothesis, 269
　alternative, 269, 276, 281
　null, 269, 276
Hypothesis test, 268
　for ANOVA, 423
　classical approach, 275
　conclusion in, 273, 279
　for contingency table, 400
　critical region for, 271
　decision in, 273
　for difference between two
　　means, 344, 359
　for difference between two
　　medians, 624
　for difference between two
　　proportions, 372
　for enumerative data, 392, 400,
　　404
　for individual β's in regression, 525
　level of significance for, 272
　for linear correlation coefficient,
　　462
　for multinomial experiment, 394
　for multiple R, 523

　in nonparametric methods:
　　Mann-Whitney U, 630
　　rank correlation, 640
　　runs, 636
　　sign, 623
　for one mean, σ known, 275, 286
　for one mean, σ unknown, 308
　for one median, 623
　for one variance, 323
　for paired differences, 367
　prob-value approach, 287
　for proportion, 316
　for randomness, 636
　for rank correlation, 640
　for regression, 492
　test criteria for, 271
　for two variances, 351
　and type I and type II errors, 270
　for variance, 323

i, position number, 54
Immediate time frame, 543
Independence, 152, 153, 192, 400
Independent events, 153, 154, 157
　multiplication rule for, 156
Independent means, 344
Independent samples, 342
Independent sampling, 342
Independent trials, 192, 395
Independent variable, 103, 464, 486
Indeterminancy principle, 17
Index, 602
　base period of, 602
　seasonal, 586
Index numbers, 601
　consumer price index, 605, 614
　for Σ notation, A-2
Indicators, 598
Inferences (see estimation;
　hypothesis test)
Inferential statistics, 4
Input variable, 464, 486
Interaction, 437
Interaction graph, 441
Intercept, 465, 486
Intersection, of events, 147
Interval estimate (see confidence
　interval estimate)

Irregular component, of time series,
　568, 586
Irregular data patterns, 546

J-shaped distribution, 46
Judgment sample, 18, 19

Lagging indicators, 599
Large sample, 345
Laspeyres method, 611
Lattice grid representation, 137
Law of large numbers, 132
Laws of probability:
　addition rule, 149
　multiplication rule, 156
Leading indicators, 599
Leaf, 34
Least squares, criteria, 464
Least squares, method of, 464
Less frequent sign, 623
Level, in ANOVA, 419
Level of confidence, 292, 297
Level of significance, 272, 283
Limits:
　class, 40
　confidence, 294
Linear correlation, 111
Linear correlation coefficient (see
　coefficient of linear correlation)
Linear equation, 465, 486
Linear regression, 117, 464, 486
Line of best fit, 465
　calculation for, 466
　equation of, 466
　graphing, 469
　and method of least squares, 464
　and predictions, 118, 470, 494
　slope for, 466
　standard deviation of y about, 488
　y-intercept for, 466
Line histogram, 184
Listing, for sample space, 137
Long-term average, in probability,
　132
Long-term time frame, 543
Lower class boundaries, 41
Lower class limit, 40
Lower confidence limit, 294

Mann-Whitney U test, 630
Marginal propensity to save, 623
Marginal totals, 104
Market basket, 611
Mathematical model, 427, 432, 437, 486, 515, 516, 518
Maximax strategy, 663
Maximin strategy, 663
Maximizing expected value of payoff, 668
Maximum error of estimate, 293
   for proportions, 319
Mean:
   of binomial random variable, 201
   calculation of, 51
   of frequency distribution, 52
   inferences about (*see* estimation; hypothesis test)
   physical representation of, 51
   of population, 186
   of probability distribution, 185
   of sample, 50
   of y values, 487
Mean absolute deviation, 59; (MAD), 549
Mean difference, 369
Mean square (MS), 422
   for error, 422
   for factor, 422
   for residual, 434
Mean squared error (MSE), 550
Measure of central tendency, 50
   mean, 50
   median, 54
   midrange, 55
   mode, 54
Measure of dispersion, 58
   coefficient of variation, 66
   mean absolute deviation, 59
   range, 58
   standard deviation, 61
   variance, 60
Measurement, 10
Measure of position, 69
   percentile, 70
   quartile, 70
   z score, 72
Measure of spread, 58
Median, 54, 70
Median difference, test for, 624
Medium-term time frame, 543

Method of least squares, 464
Midrange, 55
Minimax regret strategy, 663
Minimizing expected opportunity loss, 670
Mistakes made in regression analysis, 115, 501, 514
Misuses of statistics, 6
Modal class, 45
Mode, 45, 54
Model, 515
   causal, 542
   curvilinear, 518
   deterministic, 515
   linear, 516
   probabilistic, 515
   quadratic, 518
   regression, 515, 516, 518
   time series, 542
Moving average, 571
Mu (μ), 186
Multicollinearity:
   near, 529
   perfect, 529
Multinomial experiment, 394
Multiple regression, 515
Multiple R^2, 521
Multiplication rule:
   general, 156
   special, 156
Multiplicative model, of time series, 567
Mutually exclusive events, 136, 145, 157
   addition rule for, 149

Naive model, 553
Near multicollinearity, 529
Nested notation, 436
Net regression coefficient, variance of, 518
Noncritical region, 273, 311
Nonmutual exclusiveness, 149, 157
   addition rule for, 149
Nonparametric statistics, 622
   characteristics of, 622
   comparison with parametric statistics, 622
   Mann-Whitney U test, 630

   rank correlation, 640
   runs test, 636
   sign test, 623
Nonrespondent, 17
Normal approximation to binomial, 232
   in nonparametric methods, 626, 633, 638
   rule for, 236
Normal curve, 216
Normal distribution, 46, 216
   applications of, 224
   area under, 216
   and central limit theorem, 249
   formulas for, 216
   notation for, 228
   standard, 218
Normality, 76
Normal probabilities, by table, 218, A-21
Null hypothesis, 269, 276

Observed error, 487
Observed frequency, 392
Observed probability, 130, 316
Observed value of y, 103, 486
Odds, 143
Ogive, 46, 77
One-tailed test, 281
Opportunity loss, 660
Opportunity loss table, 660
Optional formula:
   R^2, 476
   variance, 62
Ordered data (*see* ranked data)
Ordered pair, 103, 456
Ordinate, 216
Outcome, 136
Output variable (*see* predictor variable)

Paired data, 103, 367, 456
Paired difference, 367
   confidence interval for, 370
   hypothesis test for, 367
   mean for, 368
   standard deviation for, 368

Parameter, 8, 186, 268
Parametric methods, 622
Partial correlation, 513
Partial correlation coefficient, 513
Partition, 421
Patterns of data, 543
Payoff, 656
  expected value of, 666
Payoff table, 656
Pearson's product moment r, 112, 459
Percentage, 217
Percentile, 70
  procedure for finding, 70
Perfect information, expected value of, 671
Perfect multicollinearity, 529
Pie diagram, 33
Piece of data (see data)
Point estimate, 291
Poisson probability:
  distribution, 202, A-20
  experiment, 203
  function, 203, 204
Pooled estimate:
  for p, 374
  for standard deviation, 360
Population, 7
  mean of, 186
  parameter of, 186, 268
  standard deviation of, 186
  variance of, 186
Position, measures of, 69
Position number, 54
Posterior probability, 166
Power of statistical test, 645
Predicted value, 103, 118, 464, 486
Prediction equation, 118
Prediction interval
  for mean value of y given x, 494, 495
  for particular value of y given x, 494, 497
Prediction for y given x, 468, 494
  with knowledge of x, 473
  without knowledge of x, 473
Predictor variable, 103, 464, 486
Primary sampling units, 23
Prior probability, 166
Probabilistic model, 515
Probability, 13, 128
  addition rule for, 149

  area representation of, 184, 218
  basic properties of, 142
  binomial, 189, 316, 373
  of compound events, 145
  conditional, 154
  in decision strategies, 666
  empirical, 130
  of event, 130
  experimental, 130
  long-term average for, 132
  multiplication rule for, 156
  normal, 217, A-21
  observed, 130
  and odds, 143
  posterior, 166
  prior, 166
  of random sample, 19, 248
  revised, 166
  subjective, 142
  theoretical, 141
  of type I and II errors, 270
Probability distribution, 182
  binomial, 189
  continuous, 216
  discrete, 182
    mean of, 185
    variance of, 186
  histogram for, 181, 184
  line histogram for, 184
  normal, 216
  Poisson, 203
  properties of, 183
  standard deviation for, 187
Probability function, 183
  binomial, 195
  constant, 183
Probability paper, 76, 77
Probability sample, 18, 19
Probability-value, 286
Prob-value, 286, 287
  with t, 312
  with z, 286, 316
Proportion, 217
  inferences about, 316, 372
  pooled estimate for, 374
P-value, 286

Qualitative data, 9, 10
Quantitative data, 9, 10
Quartile, 70

Random, 19
Random error, 487
Randomness of data, 636
Random number, A-8
Random number table, 20, A-12
Random sample, 19, 248
Random sampling, 19, 248
Random variable, 180
  binomial, 192
  continuous, 216
  discrete, 180, 203
Range, 58
Rank correlation, 640
Rank correlation coefficient, 640
Ranked data, 54, 630
Rank number, 630 (see also position number)
Ratio:
  of standard deviations, 356
  of variances, 356
$r \times c$ contingency table, 103, 399
  degrees of freedom for, 400
Real dollars, 603
Rectangular distribution, 46
Reduced sample space, 154
Regression:
  coefficient, 516
  curvilinear, 518
  linear, 117, 486
  multiple, 515
  stepwise, 529
Regression analysis, 118
Regression line (see linear regression)
Rejection region (see critical region)
Relative frequency, 43
  in contingency table, 106
Relative frequency contingency table, 106
Relative frequency distribution, 46
Replicate, 419, 432
Representative sample, 15
Residual, 434
Response variable, 8, 103
Rho (ρ), 461
Risk, 270
Round-off rule, 61, A-10
R-square (R^2), 472, 475, 476
Run, 636
Runs test, 636
  critical values, A-32

Sample, 8, 15
　cluster, 23
　dependent, 342
　independent, 342
　judgment, 18, 19
　large, 345
　mean for, 51
　probability, 18, 19
　random, 19, 248
　representative, 15
　simple random, 19
　small, 359
　standard deviation for, 61
　statistic for, 186
　stratified, 22
　systematic, 21, 26
　variance for, 60 (see also variance)
Sample design, 18
Sample point, 136
Sample size, determination of, 296, 320
Sample space, 135, 136
　chart representation for, 139
　lattice grid representation for, 137
　listing for, 137
　reduced, 154
　tree diagram for, 137
Sample statistics, 186, 245 (see also sample)
Sample variability, 16
Sampling, 26
　acceptance, 268
　area, 24
　bias, 16
　cluster, 23
　dependent, 342
　errors in, 16–17
　independent, 342
　repeated, 244
　simple random, 19
　stratified, 22
　systematic, 21
Sampling distribution, 245, 249, 344, 368, 373
Sampling error, 16, 26
Sampling frame, 15
Sampling plan, 18
Sampling techniques (see sample design)
Sampling variability, 16, 244
Scatter diagram, 106

Seasonal component, of time series, 567
Seasonal index, 586
　adjustment factor for, 587
Seasonally adjusted data, 589
Seasonal pattern, 544
Second quartile, 70
Shortcut formula:
　for R^2, 476
　for variance, 62
Short-term time frame, 543
Sigma:
　σ, 186
　Σ, 51, 52, 64, A-2
Significance level, 272
Sign test, 623
　confidence interval procedure, 626
　critical values for, 624, A-30
　normal approximations for, 626
　zeros in, 623
Simple event (see event)
Simple random sampling, 19
Simulation, of experiments, A-8
Single-factor ANOVA, 426
Single-variable data, 32
Skewed distribution, 46
Slope (b_1), 465, 486
Slope-intercept form of line, 465, 486
Small sample, 359
Small-sample inference, 308, 351, 359, 367
Source of data, 342
Spearman, C., 640
Spearman rank correlation coefficient, 640
　critical values, A-33
Special addition rule, 149
Special multiplication rule, 156
Spread of data (see measures of dispersion)
Squared deviation, 474
Squared error, 474, 550
Square root, of a number, 61
Stable seasonals, 586
Standard chart, 41
Standard deviation, 61
　of binomial probability function, 201
　of binomial random variable, 201
　calculation of, 61
　of frequency distribution, 61

　inferences about, 323, 351
　pooled estimate for, 360
　of population, 186
　of probability distribution, 187
　ratio of, 351
　of sample, 61
　shortcut formula for, 62
Standard error, 249
　for difference between means, 344, 360, 368
　for difference between proportions, 373
　for mean, 249
　for proportion, 316
　for slope, 491
Standardized variable, 218
Standard normal distribution, 218, A-21
　application of, 224
Standard score, 72, 218
State of nature, 656
Stationary data, 545
Statistic:
　sample, 9
　test, 271, 279
Statistical deception, 78
Statistical tests (see hypothesis test)
　comparing, 645
　efficiency of, 646
　power of, 645
Statistics, 4, 13
　descriptive, 4
　inferential, 4
　misuses of, 6
　nonparametric, 622
　parametric, 622
　uses of, 6
Stem, 33
Stem-and-leaf display, 33
Straight line:
　equation for, 465
　intercept of, 465, 486
　slope of, 465, 486
Strata, 22
Stratified sampling, 22
Student's t distribution, 308
　calculated value for, 308
　critical value for, 310, A-22
　degrees of freedom for, 309
　notation for, 310
　properties of, 308

Subjective probability, 142
Subset, 8
Success, 192
Summation notation, 51, 52, 64, A-2
Sum of ranks, 631
Sum of squares, 62, 420
   for error, 420
   for factor, 420
   partitioning, 421
   residual, 434
   total, 420
Symmetrical distribution, 46
Systematic bias, 16, 26
Systematic sampling, 21, 26

Tally, 41
t distribution (*see* Student's t
   distribution)
Test criteria, 271
Test for homogeneity, 404
Test of hypothesis (*see* hypothesis
   test)
Test for independence, 400
Test for normality, 76
Test statistic, 271
   asterisk notation for, 279, 287
   calculated value of, 279
   χ^2, 323, 392, 394
   critical value of, 278
   F, 352, 423, 435, 441, 523
   t, 308, 360, 362, 369, 492, 526
   z, 287, 317, 345, 347, 373, 626,
      633, 638
Theoretical probability, 141
Third quartile, 70
Time frame, 543
Time horizon, 543
Time series, 566
   components of, 567
   decomposition of, 566
   deflating, 603
   and forecasting, 542
   multiplicative model for, 567

Tree diagram, for sample space, 137
Trend component, of time series, 578
Trend and cyclical components, of
   time series, 571
Trend line, 578
   shortcut method for, 579
Trend patterns, 544
Trial, 192
t statistic (*see* Student's t
   distribution)
Two-factor ANOVA, 432, 437
Two-tailed test, 281
Type I error, 270
Type II error, 270

Uncertainty, 543, 547, 655
Ungrouped frequency distribution,
   39
Uniform distribution, 46
Upper class boundaries, 41
Upper class limit, 40
Upper confidence limit, 294
U score, 631
U-shaped distribution, 112

Variability, 58
   sampling, 244
Variable:
   continuous, 216
   data, 9, 10
   dependent, 103, 464, 486
   discrete, 180
   dummy, 530
   independent, 103, 464, 486
   input, 464, 486
   predicted, 103, 464, 486
   predictor, 103, 464, 486
   quantitative, 530
   random, 180, 192
   response, 8, 103
Variable data, 106

Variance, 60
   calculation of, 60, 65
   of experimental error, 488
   of frequency distribution, 60
   inferences about, 323, 351
   of population, 186
   of probability distribution, 187
   ratio of, 351
   for sample, 60
   shortcut formula for, 62
   of slope, 491
   unit of measure, 62
Variation, 474
   between rows (levels), 418, 421
   coefficient of, 66
   random, 16
   within rows (levels), 418, 421
Venn diagram, 148

Weighted aggregate price formula,
   611
Weighted average of relative price
   formula, 611
Weighted price index, 610
Weights, in index numbers, 610
Within-sample variation, 418

x bar ($\bar{x}$), 50
x tilde ($\tilde{x}$), 54

y hat ($\hat{y}$), 464, 486
y-intercept (b_0), 465, 486

z score, 72
z statistic, 72, 218, 287, 317, 345,
   347, 373, 626, 633, 638
   critical value of, A-21
   notation for, 228
   as standard score, 218

$\mu_{\bar{x}}$	Mean of the distribution of all possible $\bar{x}$'s
$\mu_{y\|x_0}$	Mean of all y values at the fixed value of x, x_0
μ_V	Mean number of runs for the sampling distribution of number of runs
M	Population median
n	(Sample size) number of pieces of data in one sample
$n(\ )$	Cardinal number of
$\binom{n}{r}$	Binomial coefficient or number of combinations
O	Observed value
$P(A\|B)$	Conditional probability, the probability of A given B
$P(a < x < b)$	Probability that x has a value between a and b
P_k	kth percentile
P_{75}	Price index for year 1975; also, 75th percentile
p or $P(\ )$	Theoretical probability of an event or proportion of time that a particular event occurs
p' or $P'(\ )$	Empirical (experimental) probability or a probability estimate from observed data
p^*	Pooled estimate for the proportion
Q_1	First quartile
Q_3	Third quartile
q	($q = 1 - p$) probability that an event does not occur
q'	($q' = 1 - p'$) observed proportion of the time that an event did not occur
R	Range of the data
R^2	Coefficient of determination
ρ (rho)	Population linear correlation coefficient
r	Linear correlation coefficient for the sample data or row number
r_s	Spearman's rank correlation coefficient
r_{xy}	Partial correlation
Σ (capital sigma)	Summation notation
$SS(\ )$	Sum of squares
s^2	Sample variation
σ (lowercase sigma)	Population standard deviation
$\sigma_{\bar{x}}$	Standard deviation of the distribution of all possible $\bar{x}$'s
$\sigma_{p'}$	Standard error for proportions
$\sigma^2_{\mu_r}$	Variance among the means of the r rows